Animal Science Reviews 2012

Animal Science Reviews

Animal Science Reviews provides scientists and students in the field with timely analysis on key topics in current research. These chapters were originally published online in CAB Reviews. This volume makes available in printed form the reviews in animal science published during 2012.

Animal Science Reviews 2012

Edited by

David Hemming
CAB International, UK

www.cabi.org

CABI is a trading name of CAB International

CABI Head Office
Nosworthy Way
Wallingford
Oxfordshire OX10 8DE
UK

Tel: +44 (0)1491 832111
Fax: +44 (0)1491 833508
E-mail: cabi@cabi.org
Website: www.cabi.org

CABI North American Office
38 Chauncey St
Suite 1002
Boston, MA 02111
USA

Tel: +1 617 395 4056
Fax: +1 617 354 6875
E-mail: cabi-nao@cabi.org

ISBN: 978 1 78064 302 1

Commissioning editor: David Hemming
Production editor: Mike Bodinham

Typeset by S R Nova Pvt Ltd, Bangalore, India
Printed and bound by CPI Group (UK) Ltd, Croydon, CR0 4YY.

Contents

Breeding for disease resistance in livestock and fish

M.J. Stear[1]*, G. Nikbakht[2], Louise Matthews[1] and N.N. Jonsson[1]

Address: [1] University of Glasgow, Bearsden Road, Glasgow G61 1QH, UK. [2] Department of Microbiology and Immunology, Faculty of Veterinary Medicine, University of Tehran, Iran.

***Corresponding author:** M. J. Stear. Fax. +44 141 942 7215. Email: michael.stear@glasgow.ac.uk

Received: 18 August 2011
Accepted: 16 January 2012

Abstract

Breeding for disease resistance offers improved animal welfare and increased returns for breeders. It is now practised in cattle, sheep, pig, chicken and fish farming. This review provides an overview for the non-specialist of the strategies to breed for disease resistance, with equal weighting on traditional phenotypic methods and the changes being introduced by modern genomic tests. It starts by explaining four key concepts in selective breeding: heritability, estimated breeding value (EBV), repeatability and genetic correlation. It then explains the decisions required to create selection objectives, selection criteria and selection indices. The response to selection is then considered with emphasis on the factors that constrain the response to selection: generation interval, variation in breeding values, selection intensity, effective population size and the accuracy of selection. Quantitative genetic theory can accurately predict the response to selective breeding of production traits, but is less effective at predicting the response to selection for resistance to microbial and parasitic diseases because selecting resistant animals and culling susceptible animals can alter the rate of disease transmission. Mathematical models are therefore required to predict the effect of genetic and epidemiological changes on disease incidence.

Keywords: Disease resistance, Selective breeding, Mathematical modelling, Livestock, Quantitative genetics

Review Methodology: We searched Web of Science and used the following terms: model breeding disease; model breeding disease selection; selection disease model; selection disease model fish; selection disease fish; genetics disease fish; genetics disease livestock; breeding disease livestock; genetics disease poultry; evolution disease; mathematic* modelling disease; epidemiol* model disease; epidemiol* model disease livestock; epidemiol* gen* disease; epidemiol@* gen* livestock.

Introduction

Individuals vary in their susceptibility to disease and much of this susceptibility is genetic in origin. Farmers and breeders can exploit this genetic variation to identify and use animals that are relatively resistant to disease. There are a number of advantages in using resistant stock including increased production, improved animal welfare, reduced environmental contamination by drugs, delayed development of drug-resistant strains of pathogens or parasites and improved return on investment of time and money. These are considerable advantages and breeding for disease resistance is widely practised in the livestock industries.

Dairy breeders use somatic cell counts in milk and the incidence of clinical mastitis to select cattle with improved resistance to mastitis, especially in Scandinavia [1]; sheep breeders use faecal egg counts to select sheep with enhanced resistance to nematode infection, especially in Australia and New Zealand [2–5]; while selection of salmon for resistance against sea louse infestation and against viral diseases including infectious pancreatic necrosis (IPN) has been started by fish breeding companies in Scotland and Norway [6]. Pig and poultry breeding are dominated by a relatively small number of breeding companies and the exact details are commercial secrets, but breeding companies routinely incorporate disease resistance into their selection goals. Breeding for disease resistance is feasible, desirable and sustainable [7].

The initial stages of breeding for disease resistance generated a considerable amount of scepticism, but these concerns have been shown to be largely unfounded. One concern was that disease resistance would be poorly heritable; in contrast, the heritability of many diseases is similar to the heritability of production traits [8, 9]. Another concern was that selection for resistance to one disease would increase susceptibility to other diseases; while this is plausible under some immunological theories, there is no indication that this is the case. A further concern was that as parasites and pathogens are capable of more rapid evolution than their hosts, selection of hosts might be overwhelmed by the response of the pathogen or parasite. Again, there is no evidence that this happens [10, 11]. Another exaggerated concern was the possibility of trade-offs between disease resistance and production traits. In some circumstances, disease-resistant animals are less productive [12]; in other circumstances, disease-resistant animals are more productive [13, 14]. In themselves, trade-offs are not fatal to selection for disease resistance. Many traits in selection schemes have negative genetic correlations and selection indices maximize the responses to traits that are unfavourably correlated.

Of course, breeding for disease resistance is not a universal panacea; there is little point in selective breeding for resistance to a disease where effective and sustainable alternative control measures such as vaccination are effective. Similarly, there seems little point in selective breeding for resistance to exotic diseases that require infected animals to be identified and culled when the disease is imported into the country. Nonetheless, selection for disease resistance is a useful tool for the control of many diseases and is most effective as part of an integrated disease control strategy.

The problem is no longer whether to breed for disease resistance but how to do so. Selective breeding does pose specific problems not found in selection for production traits. For example, the traits used to determine resistance to many diseases are not normally distributed and require different statistical procedures [15–17]. The problem is less the statistical procedures than the fact that genetic theory assumes that the underlying liability to disease is normally distributed and it can be difficult to reconcile clearly non-normal distributions with quantitative genetic theory.

Selective breeding to reduce the incidence of disease is likely to have profound consequences for animal health, welfare and farming profitability. However, most reviews are written by animal breeders for animal breeders and are not readily accessible for those with little background in quantitative genetics. Yet many people outside the animal breeding community have an interest in animal health and welfare including veterinarians, agricultural consultants, farmers, those responsible for agricultural policy and scientific researchers in related fields such as parasitology, microbiology, immunology, nutrition and ecology. The inadvertent exclusion of many stakeholders has had the unfortunate consequence that much of the debate over selective breeding for disease resistance has been conducted at two levels. Quantitative geneticists and professional animal breeders have embraced the idea of selective breeding for disease resistance and the debate is about methods to achieve this goal and encourage uptake by breeders. In contrast, many outsiders are still concerned about the wisdom of selective breeding.

This review will introduce some of the key concepts in animal breeding, and then describe the process of breeding for disease resistance. Most of the published work is in sheep and cattle and the emphasis will be on the lessons those species have for other systems. The review is designed for those who lack a background in quantitative genetics and aims to provide a gentle introduction to the relevant breeding concepts.

The four key concepts of quantitative genetics are the heritability, estimated breeding value (EBV), repeatability and genetic correlation. Continuously distributed traits are assumed to be multifactorial, i.e. influenced by many different factors. Some of these factors will be different genes; some will be non-genetic factors including chance, climate and nutrition. All these non-genetic factors are referred to as the environment. For geneticists, the environment includes everything that contributes to variation in a trait that is not genetic.

Heritability

The performance of an individual for a particular trait is called the phenotypic value (P) and this is made up of a genetic component G (genotypic value) and an environmental component E (environmental deviation):

$$P=G+E.$$

The environmental deviation is defined to have a mean of zero and consequently, the average performance of a particular genotype is the genotypic value.

The breeding value of an animal is not necessarily equal to the genotypic value. The genotypic value is equal to the breeding value (A, the additive effect of the genes) plus a deviation due to dominance (D) and a deviation due to interaction between genes at different loci (I, the epistatic deviation)

$$G=A+D+I.$$

The additive genetic component is the most important for selective breeding because it is the only component of the phenotype that is inherited. The dominance and epistatic components are treated as noise though the dominance and epistatic components are utilized in crossbreeding schemes. The proportion of the variance that is due to genotypic values (V_g/V_p) is defined as heritability in the broad sense. In contrast, heritability in the narrow sense is the proportion of the variance that is attributable to variance in breeding values (V_a/V_p). It is the narrow-sense heritability (h^2) which predicts the response to selection and it is the term of most value to animal breeders. The heritability of a trait and the breeding value of an animal can vary between populations, depending on gene frequencies, type of gene action and environmental effects.

EBV

The EBV is the second key concept. If there is only a single record on one animal and no information on any of his relatives, then the EBV is the heritability multiplied by the difference between the individual observation and the population mean. For example, selection for nematode resistance in sheep uses faecal egg counts. Faecal egg counts are typically log transformed to normalize the distribution. If a lamb has a faecal egg count 2 log units (100 eggs per gram) above the flock mean and the heritability is 0.3, then the EBV=0.3×2. This corresponds to an EBV of +0.6 log-transformed units.

A single record is not particularly useful because the accuracy and the reliability are relatively low. In practice, additional information is used. This might take the form of repeated records and information on relatives, especially siblings, offspring or parents. In practice, EBV are usually calculated by specialized computer programmes, such as ASReml or VCE. These programmes can take all available information into account. The more information that is available, then the more reliable and accurate will be the results. These programmes also adjust the phenotypic observations to allow for known effects such as sex, herd, year and season. Herd, year and season are often grouped together to form a single source of environmental variation that affects herd means.

The accuracy of the EBV is the correlation between the true and the predicted breeding value. For a single measurement, the accuracy is the square root of the heritability. In the above example, the accuracy is the square root of 0.3 which is 0.55. The reliability of the EBV is the square of the accuracy. In this example, the reliability is equal to the heritability and is 0.3.

Repeatability

The third key concept is the repeatability. Sometimes, it is possible to take repeated records from the same individual, e.g. milk yield in successive lactations. The usefulness of repeated records depends upon the repeatability. The repeatability is the variance between individuals (V_b) divided by (V_b+V_w) where V_w is the variance within individuals.

$$\text{Repeatability}=V_b/(V_b+V_w).$$

When most of the variation is among individuals, then the repeatability is high. It decreases as the amount of variation within individuals increases (i.e. as the repeated records become less similar). When there are only two records with equal variances, then the repeatability is equal to the correlation.

The repeatability can be thought of as V_g (genotypic variance)+V_{pe} (permanent environmental variance) divided by the total (phenotypic) variance. In other words, the similarity between observations depends upon the genotypic component plus the permanent environmental component. The repeatability should be equal to or greater than the heritability. As the repeatability can be quickly estimated even without pedigree records, it provides a convenient way to estimate the upper limit of the heritability.

Genetic Correlation

The fourth key concept is the genetic correlation. As this is the correlation between the additive genetic effects, it is also known as the additive genetic correlation and is the correlation between the breeding values. If two traits are genetically correlated, this means that their breeding values are correlated and the value for one trait can be predicted from the other. Selection for one trait would also produce a change in the genetically correlated trait. If both traits are measured, then selection can increase one trait, while leaving the other unchanged or achieve more rapid progress in both traits. If the genetic correlation is negative, as with milk production and milk protein percentage, it is possible to increase both traits but progress is slower. For example, milk production is positively correlated with the incidence of mastitis [18, 19]. Selecting for increased milk production alone obviously increases milk production. This selection regime also increases the incidence of mastitis.

Indicator traits are those traits that are relatively unimportant in themselves but are genetically correlated with economically important traits. Indicator traits can be used to select for traits that are difficult to measure or have low heritability. Somatic cell scores are an indicator of mastitis incidence and they are now being measured in many dairy cows. Future selection indices will incorporate somatic cell scores. Selection will then either reduce the incidence of mastitis or prevent it from rising.

Selection Objective

The first decision to be made by a breeder setting up a selection scheme is the selection objective. The selection objective is the character (or characters) that we wish to improve by selection. It is also known as the breeding objective, breeding aim, breeding goal and selection goal. It is chosen irrespective of whether it is easy, difficult or even impossible to measure. Examples include reduced incidence of clinical mastitis in dairy cattle, reduced faecal nematode egg production in sheep, reduced numbers of ticks on beef cattle, reduced numbers of sea lice on

Atlantic salmon or dogs without clinical hip dysplasia. It is important to be clear about the objective. Disease resistance could mean resistance to infection or the absence of clinical signs despite infection.

As breeders usually wish to improve their return on investment, most breeding objectives aim to improve the profitability of animals. In theory, the selection objective for profitability should include every heritable trait that influences the economic return, including disease and reproductive performance. In practice, we do not have sufficient information to do this and only the most important characters are included. The different components of an objective will vary in their economic value and they are given greater or lesser weight according to their importance. In statistical terms

$$\text{Selection objective} = a_1 Y_1 + a_2 Y_2 + \ldots + a_n Y_n,$$

where $Y_1, Y_2, \ldots, Y_n$ are the traits in the selection objective and $a_1, a_2, \ldots, a_n$ are the relative economic values of the traits.

The economic values are often defined as the marginal profit resulting from a change of one unit in that trait, while all other traits remain unchanged. It is the economic values relative to each other that are important. Estimating the economic values for disease traits is not straightforward. For example, with parasitic infections such as sea louse in salmon and nematodes in sheep, the parasites are evolving resistance to the drugs used to treat them. Consequently, the economic impact will vary among sites; the impact of parasites is much greater on those sites that lack effective treatment.

Ideally, we would like to eliminate all disease but this is not feasible. Some diseases are very rare and individually unimportant enough to warrant inclusion in a selection index. There is also an opportunity cost of breeding for disease resistance. Including more traits in the selection objective can reduce the selection pressure on each trait and slow the response in other traits such as production. For rare diseases, this cost outweighs the potential benefits. For other diseases, we lack adequate information to define resistant hosts.

Mastitis is economically the most important disease of dairy cattle, while nematode infection is the most important disease of small ruminants. Unsurprisingly, selection against these diseases dominates breeding for disease resistance in these industries. In aquaculture, selective breeding has started in a variety of species for resistance to a variety of diseases [15]. In salmon breeding, sea louse infestation and viral infections including IPN are important diseases in the UK and Norway, and selection for disease resistance is focused on these diseases [6, 15]. In pigs and poultry, there are many important diseases and no disease is overwhelmingly important. Consequently, selection for resistance against a single disease is not appropriate. In these industries, selection is focused against the major causes of reduced productivity and those diseases for which useful genetic markers exist.

Selection for resistance against a specific disease usually involves testing directly for resistance to that disease. Two alternative selection strategies to reduce the impact of disease have been suggested. One is to select for enhanced immune responses [20, 21]. This approach has been applied to both pigs and cattle [22–24]. Animals with enhanced immune responses are more resistant to some diseases but more susceptible to others [23]; however, the enhanced resistance is more important than the increased susceptibility. Yorkshire pigs selected for increased immune responses matured earlier and this early maturity was valuable in itself [22]. Dairy cows classified as high responders with greater antibody-mediate immune response (AMIR) and cell-mediated immune response (CMIR) had lower odds ratio disease scores for clinical mastitis (~4×), metritis (~8×), ketosis (~3×) and retained placenta (~2.5×) (Mallard, personal communication). These individuals with both higher and more optimally balanced AMIR and CMIR, two key components of the adaptive immune system, are referred to as high immune responders. A patented test system has been developed to quickly identify these animals within dairy herds and this method is referred to as the high immune response (HIR) technology. Interestingly, HIR dairy cows showed lower occurrence of economically important infectious diseases, as well as ketosis, emphasizing the relationships between infectious disease, immune function and metabolic disorders. Dairy cows with the highest AMIR responses tended to have lower milk production, while those with the highest CMIR had higher milk production ([25], Mallard, personal communication). These results emphasize the need to select animals simultaneously for AMIR and CMIR if both health and production traits are to be maintained or improved. This is a promising approach that deserves and is receiving more attention.

The second alternative approach is to select for increased resilience [26]. Resilience is defined as the ability to be productive despite infection. Some animals develop subclinical infections and these animals could be used for breeding. To a large extent, breeders are unlikely to select solely for disease resistance and a selection objective that includes both resistance to disease and enhanced production will select for resilience as well as resistance. However, this approach may have disadvantages in situations where susceptibility to sub-clinical infection may result in economic losses for other reasons, such as increased somatic cell count in a dairy herd. Another potential disadvantage is that resilient animals may act as sources of infection for other individuals.

Selection Criteria

The second decision is the selection criteria. The selection criteria are the traits that will actually be measured. In pigs, the selection criteria might include growth rate to

90 kg liveweight, litter size and mid-back fat depth. In dairy cattle, selection criteria might be milk yield adjusted to a 305-day lactation, percentage protein and percentage butterfat. Selection criteria might also include information on relatives such as full-sibs, half-sibs or offspring. The most efficient selection will incorporate all known pedigree information in a statistically sophisticated mixed model analysis.

For disease resistance, the selection criterion could be clinical incidence. In countries such as Norway, where clinical disease is routinely recorded, the incidence of clinical mastitis in daughters is used to select bulls with superior breeding values for resistance to mastitis. For other diseases, such as infection with endoparasites or ectoparasites, clinical disease is less informative and indicator traits are used. Faecal egg counts are used to indicate susceptibility to nematode infection, while tick or louse counts can be used to identify genetically resistant cattle [27] and salmon [28, 29].

In some situations, disease resistance cannot be easily measured on potential parents. For example, in cattle breeding bulls do not produce milk nor develop mastitis. In fish breeding, counting lice on large numbers of live fish is technically challenging, especially if deliberate infection has produced large numbers of lice on each fish. In these circumstances, family-based selection is appropriate. For example, the genetic merit of bulls can be predicted from the incidence of mastitis in their daughters, while the number of sea lice on fish can be used to predict the resistance of their siblings. Modern statistical analysis can use the entire pedigree to predict the genetic merit of recorded and unrecorded related individuals. Family-based selection or progeny testing instead of individual selection has consequences for the accuracy of selection, intensity of selection and the risk of inbreeding.

There is considerable variation associated with the method of assessment in many measurements of disease. One example is faecal egg count following nematode infection [30]. A consequence of measurement variation is that the importance of genetic variation is often underestimated. Taking multiple measurements can reduce measurement variation and provide a more accurate and heritable assessment of individual breeding value [30].

When investigating resistance to tick infestation researchers have either counted the ticks on cattle or used a relatively subjective scoring approach, which has resulted in widely divergent heritability estimates for the trait [31].

The time of year at which measurements are made can influence the estimates of genetic parameters especially for seasonal infections. As lambs mature and are exposed to nematode infection, the development of the immune response leads to increased heritabilities in older animals [13, 32]. The time of year at which measurements of tick infestation are made in cattle has been shown to strongly influence heritability estimates [33]. The genetic correlations between parasitism and production traits also change over time [34].

More recently, a number of genetic markers for resistance to specific diseases have been identified and breeding companies have included these in their selection criteria. The most important of these genetic markers is the major histocompatibility complex (MHC) [27, 35]. The effect of the MHC is often among the strongest of all genetic systems or loci but it seldom accounts for more than 10% of the total phenotypic variance for any specific disease. Most loci have much smaller effects and for most diseases, the distribution of gene effects is L-shaped with most loci having very small effects [36]. The MHC is associated with resistance to mastitis [37] and nematode infection [38] but, because of its complexity, it has seldom been included in commercial breeding schemes.

Single-nucleotide polymorphisms (SNP) are the most common source of genetic variation; a single nucleotide site can have two, three or even four alleles. SNP occur approximately every 1000 bp in humans and probably at a similar frequency in most livestock species. Large SNP chips now exist or are being generated for commercially important species; these can detect variants at tens or even hundreds of thousands of sites in a single animal. These SNP chips not only assist the process of discovering genes associated with disease, they can also transform breeding. Rather than rely on indicator traits and a small number of markers, it is now possible to utilize genomic selection.

In genomic selection, each genetic region is assigned a value based on testing a large, well-characterized population. These values can then be summed to create a genomic breeding value for each individual. These genomic breeding values are usually highly correlated with the EBVs derived by traditional methods. In a separate population, these values for each polymorphism can be used to predict the breeding value of each individual [39–42]. Among other advantages, genomic selection offers the ability to identify disease-resistant individuals even in the absence of disease challenge.

Selection Index

Selection for more than one character can be carried out in three ways: tandem selection, independent culling and index selection. Tandem selection involves selecting breeding stock on only one criterion for several generations, then selecting on the next criterion for several generations. Tandem selection is inefficient and is rarely used. Independent culling levels select only those animals that exceed a particular level of performance for each criterion. In other words, an animal will not be selected if it falls below the cut-off for any one criterion, irrespective of its performance on the other criteria. The method of setting the culling levels to maximize selective improvement can be complicated, but once set, the culling levels can be applied without additional computing. The use of independent culling levels is only possible if all the traits in the selection objective can be measured in each individual. The most efficient method of combining different selection criteria is to use a selection index. The construction of an index results in a single aggregate value that weights the different selection criteria according to their relative importance. This index can be used to rank animals for their genetic merit (i.e. their breeding value).

The traits that are to be measured can be denoted as $X_1, X_2, \ldots, X_m$. The coefficients that give the most efficient response to selection are $b_1, b_2, \ldots, b_m$. Then, the

$$\text{Selection Index}=b_1X_1+b_2X_2+\ldots+b_mX_m.$$

Estimating the values of $b_1, b_2, \ldots, b_m$ requires information on how much variation exists among animals (their phenotypic variances), whether differences in one trait are associated with differences in another trait in the selection index (their phenotypic covariances), the genetic covariances between the traits and the genetic variances and covariances of the traits in the selection objective.

The equation to estimate the selection index coefficients is $\boldsymbol{Pb}=\boldsymbol{Ga}$, where $\boldsymbol{P}$ is a matrix of the phenotypic variances and covariances in the selection index and $\boldsymbol{G}$ is a matrix of the genetic covariances between the traits in the selection index and traits in the selection criterion. The vector of economic values are represented by $\boldsymbol{a}$ while the vector of coefficients to be estimated are represented by $\boldsymbol{b}$.

For a single trait that is measured only once on each animal, the selection index value is simply the trait value multiplied by the heritability. Selection indices often contain both production and fitness traits and calculate the selection index value in monetary terms, which represent the expected return from using each animal.

These equations predict the genetic benefits of selection for production traits [43] but they may underestimate the benefits of selection for disease resistance. Breeding for disease resistance confers both genetic and epidemiological benefits because resistant animals may produce fewer infectious agents and slow down the rate of disease transmission [44, 45]. Consequently, mathematical models may be needed to capture the full response to selection [46].

Response to Selection

Initially, we will describe the simple situation of selection for a single trait which is measured once on each animal. In practice, information on relatives is used to improve the accuracy of selection and selection is usually based on multiple traits. The response to selection for a single character mainly depends upon five factors: generation interval, variation in breeding values, intensity of selection, effective population size and accuracy of selection.

The generation interval (L) is the average age of parents when the offspring are born. Response to selection per year is obviously quicker if potential parents are bred at a young age. The generation interval is calculated separately for males and females, then averaged. The generation interval will be influenced by the selection method. It is likely to be shorter for individual selection or sib-testing than for progeny testing.

There is a trade-off between the generation interval and the proportion of animals selected which affects the selection intensity. It is possible to decrease the generation interval by using younger animals for breeding. However, if a greater proportion of the offspring are used to replace the parents, then the selection intensity is less. If (say) all available ewe lambs are used to replace the older ewes in a flock, the generation interval will be small but the selection intensity will be zero. Alternatively, if half the ewe lambs are selected as replacements, their mean value for the selected traits will be greater than their parents (i.e. the selection differential will be positive) but the generation interval will be larger. The situation is better in species such as fish with a large reproductive potential because each adult can produce a large number of offspring, therefore faster responses to selection are predicted and observed.

The use of artificial insemination (AI), multiple ovulation and embryo transfer (MOET) or *in vitro* fertilization (IVF) can increase the rate of genetic improvement. In the case of AI, fewer males are needed to produce a given number of offspring, although the same number of female animals (or more if the AI process is less efficient than natural insemination) will be needed. In the case of MOET and IVF, fewer females will also be needed. These methods can substantially increase the selection differential, and also shorten generation intervals as the parents can leave more offspring by the time they reach a given age.

The variation in breeding values is the basic material on which selection acts. Traits with more phenotypic variation or higher heritabilities will have greater variation in breeding values.

The selection intensity (i) is the superiority of the selected parents standardized according to the amount of variation in the trait. For selection based on the phenotype of an individual:

$$i=S/\sigma_p.$$

S is the selection differential and σ_p is the standard deviation of the phenotypic observations. It can be more convenient to talk about selecting the top 5% of animals, rather than selecting some animals such that their average performance is +10 kg over the population mean. Many textbooks on quantitative genetics supply tables of *i* [47]. For example, if the top 5% of animals are chosen, then i=2.04 (approximately). Obviously if we select and breed the top 10% for any character, then there will be a greater improvement than selecting the top 90%. In practice, the selection intensity is limited by reproductive ability, and by facilities for raising offspring. Any technique that improves reproductive rate can, if properly used, increase the selection intensity.

The effective population size (N_e) takes into account the fact that not all members of a population have an equal chance of contributing gametes to the next generation. For example, populations can vary in size between generations or the variation in family sizes can be greater than expected by chance. For farm animals, differences in the relative contribution of males and females can be important and are magnified by AI, MOET and IVF. Selection also reduces the effective population size [48]. A variety of methods for estimating effective population size exist.

There are several problems with small population sizes and this is true for conservation schemes as well as selection schemes. Accidents including disease and infertility can have a disproportionate effect by eliminating a relatively large proportion of the potential parents. Demographic stochasticity can also cause serious problems, e.g. unequal sex ratios. If the effective population size is small, then inbreeding and genetic drift can retard or prevent selective improvement.

Inbreeding, defined as the mating of related animals, is easier to avoid in large populations. It leads to increased homozygosity and decreased heterozygosity over that predicted by Hardy–Weinberg. It increases the frequency of harmful homozygous recessives, which can lead to increased incidence of Mendelian diseases and disorders. The increased frequency of deleterious recessives is one of the causes of inbreeding depression, which is the reduced performance, particularly in traits related to fertility and survival, observed in partially inbred populations. A common recommendation for breeding populations is that the rate of inbreeding should be less than 0.5% per generation. This recommendation requires an effective population size of at least 100. The development of mating schemes that maximize the response to selection while simultaneously minimizing the rate of inbreeding is an active area of research in animal breeding.

Genetic drift is variation in gene frequencies due to chance and it produces random changes in frequencies of genes, especially when the frequency of a gene is low or the population consists of few individuals. Genetic drift can result in the loss of favourable alleles. Consequently, the response to selection is decreased, the selection limit is lower and it is reached more quickly. It is easier to avoid problems associated with inbreeding than with genetic drift.

As a consequence of genetic drift and inbreeding, selection is usually more effective in large populations. Many sheep farmers have joined sire reference schemes. These have the useful effect of increasing the effective population size and increasing the variation in breeding values. Consequently, the response to selection increases. Successful sire reference schemes require the use of the same sires in different flocks to create genetic links. They consist of breeders who can agree on common goals and methods and they use sophisticated statistical methods to efficiently compare different flocks. Cattle breeders usually form part of a national breeding scheme to evaluate potential bulls.

The accuracy of selection depends upon how accurately breeding values can be estimated. The accuracy of selecting from a single measurement on a candidate is the square root of the heritability. Taking careful records decreases the environmental noise and effectively increases the heritability.

The accuracy can be increased by taking multiple measurements whenever feasible or by using records on relatives, e.g. sibs, half-sibs or progeny. The preferred statistical method for estimating breeding values is best linear unbiased predictor (BLUP), because this method is the most effective at disentangling genetic from non-genetic differences such as differences between herds or between years.

Predicting the Response to Selection for a Single Trait

The breeders' equation states that the response to selection (R), the expected improvement in a trait in the offspring, can be predicted by the selection differential (S) multiplied by the heritability (h^2). This applies to a single generation of selection for a single trait, where the response is measured in generations and the population is large enough to ignore the effects of inbreeding and drift. As selection proceeds, genes change in frequency; unfavourable genes may disappear while favourable genes may become fixed. Changes in the extent of genetic variation will influence the heritability. The heritability is the variance in breeding values divided by the phenotypic variance (V_a/V_p) and the selection differential is the selection intensity multiplied by the phenotypic standard deviation $i\ \sigma_p$. Rearranging terms gives the response to selection as $\sigma_a\ hi$, where σ_a is the variation in breeding values, h is the accuracy of selection and i is the intensity of selection.

$$R = h^2 S,$$
$$R = (V_a/V_p)\, i\sigma_p,$$
$$R = \sigma_a hi.$$

Response to Selection on Multiple Traits

The response to selection on multiple traits depends upon the additive genetic variances and covariances of the traits [49]. The equation is

$$R = \boldsymbol{G}\beta,$$

where R is the change in the multivariate phenotype, $\boldsymbol{G}$ is the additive genetic variance–covariance matrix and β is the selection gradient that is estimated from the partial regression coefficients. As with selection for a single trait, this index is best for production traits and may not capture the full benefits of selective breeding for disease resistance. Mathematical modelling is required to capture the response to selection.

Mathematical Modelling

Quantitative genetic theory can accurately predict the response to selective breeding of production traits but is less effective at predicting the response to selection for resistance to infectious diseases because selecting resistant animals and culling susceptible animals can alter the rate of disease transmission. In other words, genetic theory assumes that the environment remains unchanged but culling heavily infected or diseased animals can reduce the contamination of the environment with transmission stages. This is particularly true for terrestrial livestock. For example, selective breeding for nematode resistance on the basis of egg count reduces the number of parasite eggs shed into the environment [44–46]. Selective breeding may also reduce contamination in marine environments but this may vary with the disease and the extent to which fish farms in tidal areas are swept clean by tides.

There are a large number of epidemiological models that have explored variation among individuals in their contribution to infectious disease dynamics, especially for parasite infections [50]. However, surprisingly few have explicitly incorporated genetic variation and even fewer have modelled the epidemiological and genetic response to selection for disease resistance [51].

The production of transmission stages is more persistent and predictable for endemic metazoan parasitic infections than microbial infections. Consequently, modelling the response to selection for parasite resistance is easier than for resistance to microbial infections. To our knowledge, there are only two published models that predict the response for parasite resistance. The first models the response to selection against nematode infection of lambs [46]. The lamb model predicts relatively rapid responses to selection for lower faecal egg counts following natural nematode infection with predominantly *Teladorsagia circumcincta*. The responses observed in practice were less rapid [3]. One possible explanation is that selection in practice involves multiple traits. Another is that the model did not attempt to capture immune responses. It essentially assumed that reduced parasite numbers associated with immune responses were counterbalanced by increased larval intake. However, the immune response may play a stabilizing role in nematode infection as reduced levels of infection generate reduced immune responses [52]. The second models the response to selection against infestation with sea lice (*Lepeophtheirus salmonis*) in Atlantic Salmon [53]. The sea lice model still requires validation but resistance to sea lice develops too rapidly for acquired immunity to play an effective role.

Conclusions

Breeding for disease resistance is desirable and feasible. The practical methods are similar to existing breeding schemes, but determining the optimal weighting to give disease resistance can be challenging. It is difficult to estimate the relative economic value of disease traits because this depends upon the prevalence of infection and disease as well as the effectiveness of treatment. In addition, predicting the response to selection for resistance to infectious disease requires mathematical models. Few models exist and these models require additional development before they can accurately predict the response to infection.

Acknowledgements

We thank Professor Mallard for comments on an earlier draft. We also thank the BBSRC (BB/F015313/1), Wellcome Trust (WT091717MA) and the EC (MC ITN 264639) for funding.

References

1. Heringstad B, Klemetsdal G, Ruane J. Selection for mastitis resistance in dairy cattle: a review with focus on the situation in the Nordic countries. Livestock Production Science 2000;64:95–106.
2. Karlsson LJE, Greeff JC. Selection response in fecal worm egg counts in the Rylington Merino parasite resistant flock. Australian Journal of Experimental Agriculture 2006;46(7):809–11.
3. Kemper KE, Palmer DG, Liu SM, Greef JC, Bishop SC, Karlsson LJE. Reduction of faecal worm egg count, worm numbers and worm fecundity in sheep selected for worm resistance following artificial infection with *Teladorsagia circumcincta* and *Trichostrongylus colubriformis*. Veterinary Parasitology 2010;171:238–46.
4. Bisset SA, Morris CA, McEwan JC, Vlassoff A. Breeding sheep in New Zealand that are less reliant on anthelmintics to maintain health and productivity. New Zealand Veterinary Journal 2001;49:236–46.
5. Morris CA, Wheeler M, Watson TG, Hosking BC, Leathwick DM. Direct and correlated responses to selection for high or low faecal nematode egg count in Perendale sheep. New Zealand Journal of Agricultural Research 2005;48:1–10.
6. Moen T. Breeding for resistance to viral diseases in salmonids. In: Bishop SC, Axford RFE, Nicholas FW, Owen JB, editors. Breeding for Disease Resistance in Farm Animals. CABI, Wallingford; 2010. p. 166–79.
7. Stear MJ, Bishop SC, Mallard B, Raadsma HW. The sustainability, feasibility and desirability of breeding

livestock for disease resistance. Research in Veterinary Science 2001;71:1–7.

8. Bishop SC, Morris CA. Genetics of disease resistance in sheep and goats. Small Ruminant Research 2010;70:48–59.

9. Morris CA. Responses to selection for disease resistance in sheep and cattle in New Zealand and Australia. Proceedings of the 6th World Congress on Genetics Applied to Livestock Production 1998;27:295–302.

10. Woolaston RR, Elwin RL, Barger IA. No adaptation of *Haemonchus contortus* to genetically resistant sheep. International Journal for Parasitology 1992;22:377–80.

11. Stear MJ, Singleton DR, Matthews L. An evolutionary perspective on gastrointestinal nematodes of sheep. Journal of Helminthology 2011;85:113–20.

12. Eady SJ, Woolaston RR, Lewer RP, Raadsma HW, Swan AA, Ponzoni RW. Resistance to nematode parasites in Merino sheep: correlation with production traits. Australian Journal of Agricultural Science 1998;49:1201–11.

13. Bishop SC, Bairden K, McKellar QA, Park M, Stear MJ. Genetic parameters for faecal egg count following mixed, natural, predominantly *Ostertagia circumcincta* infection and relationships with liveweight in young lambs. Animal Science 1996;63:423–8.

14. Bouix J, Krupinski J, Rzepecki R, Nowosad B, Skrzyzala I, Roborzynski M, *et al.* Genetic resistance to gastrointestinal nematode parasites in Polish long-wool sheep. International Journal for Parasitology 1998;28:1797–804.

15. Odegard J, Baranski M, Gjerde B, Gjedrem T. Methodology for genetic evaluation of disease resistance in aquaculture species: challenges and future prospects. Aquaculture Research 2011;42:103–14.

16. Gianola D, Foulley JL. Sire evaluation for ordered categorical data with a threshold model. Genetics, Selection, Evolution 1983;15:201–23.

17. Stear MJ. Breeding for resistance to nematode infections. In: Bishop SC, Axford RFE, Nicholas FW, Owen JB, editors. Breeding for Disease Resistance in Farm Animals. 3rd edn. CABI, Wallingford, UK; 2010. p. 279–94.

18. Carlén E, Strandberg E, Roth A. Genetic parameters for clinical mastitis, somatic cell score, and production in the first three lactations of Swedish Holstein cows. Journal of Dairy Science 2004;87:3062–70.

19. van Dorp TE, Dekkers JCM, Martin SW, Noordhuizen JPTM. Genetic parameters of health disorders, and relationships with 305-day milk yield and conformation traits of registered Holstein cows. Journal of Dairy Science 1988;81:2264–70.

20. Wilkie BN, Mallard B. Selection for high immune response: an alternative approach to animal health maintenance? Veterinary Immunology and Immunopathology 1999;72:231–5.

21. Hine B, Cartwright S, Mallard B. Effect of age and pregnancy status on adaptive immune responses of Canadian Holstein replacement heifers. Journal of Dairy Science 2011;94(2):981–91.

22. Mallard BA, Wilkie BN, Kennedy BW, Gibson JP, Quinton M. Immune responsiveness in swine: eight generations of selection for high and low immune response in Yorkshire pigs. Proceedings of the 6th World Congress on Genetics Applied to Livestock Production 1998;27:257–64.

23. Wilkie BN, Mallard B. Genetic aspects of health and disease in pigs. In: Axford RFE, Bishop SC, Nicholas FW, Owen JB, editors. Breeding for Disease Resistance in Farm Animals. 2nd ed. CABI, Wallingford; 2000. p. 379–96.

24. Wagter L, Mallard B, Wilkie BN, Leslie KE, Boettcher PJ, Dekkers JCN. A quantitative approach to classifying Holstein cows based on antibody responsiveness and its relationship to peripartum mastitis occurrence. Journal of Dairy Science 2000;83:488–98.

25. Wagter LC, Mallard BA, Wilkie BN, Leslie KE, Boettcher PJ, Dekkers JCM. The relationship between milk production and antibody response to ovalbumin during the peripartum period. Journal of Dairy Science 2003;86(1):169–73.

26. Morris CA, Watson TG, Bisset SA, Vlassoff A, Douch PGC. Breeding sheep in New Zealand for resistance or resilience to nematode parasites. In: Gray GD, Woolaston RR, Eaton BD, editors. Breeding for Resistance to Infectious Diseases of Small Ruminants. ACIAR Monograph No. 34, ACIAR, Canberra, Australia; 2000. p. 77–98.

27. Stear MJ, Hetzel DJS, Brown SC, Gershwin LJ, Mackinnon MJ, Nicholas FW. The relationships among ecto- and endoparasite levels, class I antigens of the bovine major histocompatibility system, immunoglobulin E levels and weight gain. Veterinary Parasitology 1990;34:303–21.

28. Gharbi K, Glover KA, Stone LC, Macdonald ES, Matthews L, Grimholt U, *et al.* Genetic dissection of MHC-associated susceptibility to *Lepeophtheirus salmonis* in Atlantic salmon. BMC Genetics 2009;10:20.

29. Gjerde B, Ødegård J, Thorland I. Estimates of genetic variation in the susceptibility of Atlantic salmon (*Salmo salar*) to the salmon louse *Lepeophtheirus salmonis*. Aquaculture 2011;314:66–72.

30. Stear MJ, Park M, Bishop SC. The key components of resistance to *Ostertagia circumcincta* in lambs. Parasitology Today 1996;12:438–41.

31. Porto Neto LR, Jonsson NN, D'Occhio MJ, Barendse W. Molecular genetic approaches for identifying the basis of variation in resistance to tick infestation in cattle. Veterinary Parasitology 2011;180(3–4):165–72.

32. Stear MJ, Bairden K, Bishop SC, Buitkamp J, Duncan JL, Gettinby G, *et al.* The genetic basis of resistance to *Ostertagia circumcincta* in lambs. Veterinary Journal 1997;154:111–9.

33. Machado MA, Azevedo AL, Teodoro RL, Pires MA, Peixoto MG, de Freitas C, *et al.* Genome wide scan for quantitative trait loci affecting tick resistance in cattle (*Bos taurus*×*Bos indicus*). BMC Genomics 2010;11:280.

34. Mackinnon MJ, Meyer K, Hetzel DJS. Genetic variation and covariation for growth, parasite resistance and heat tolerance in tropical cattle. Livestock Production Science 1991;27:105–22.

35. Stear MJ, Belch A, Donskow-Schmelter K, Fitton LA, Innocent GT, Ishikane C, *et al.* Detection of genes with moderate effects on disease resistance using ovine MHC and resistance to nematodes as an example. Veterinary Immunology and Immunopathology 2007;120:3–9.

36. Hayes B, Goddard ME. The distribution of the effects of genes affecting quantitative traits in livestock. Genetics, Selection, Evolution 2001;33:209–29.

37. Sharif S, Mallard B, Wilkie BN, Sargeant JM, Scott HM, Dekkers JCM, *et al.* Associations of the bovine major histocompatibility complex DRB3 (BoLA-DRB3) alleles with

occurrence of disease and milk somatic cell score in Canadian dairy cattle. Animal Genetics 2000;29:185–93.

38. Stear MJ, Innocent GT, Buitkamp J. The evolution and maintenance of polymorphism in the major histocompatibility complex. Veterinary Immunology and Immunopathology 2005;108:53–7.

39. Dekkers JCM. Prediction of response to marker-assisted and genomic selection using selection index theory. Journal of Animal Breeding and Genetics 2007;124:331–41.

40. Goddard M. Genomic selection: prediction of accuracy and maximisation of long term response. Genetica 2009;136:245–57.

41. Hayes BJ, Bowman PJ, Chamberlain AJ, Goddard ME. Invited review: Genomic selection in dairy cattle: progress and challenges. Journal of Dairy Science 2009;92:433–43.

42. Muir WM. Comparison of genomic and traditional BLUP-estimated breeding value accuracy and selection response under alternative trait and genomic parameters. Journal of Animal Breeding and Genetics 2007;124:342–55.

43. Hazel LN. The genetic basis for constructing selection indexes. Genetics 1943;28:476–90.

44. Barger IA. Genetic resistance of hosts and its influence on epidemiology. Veterinary Parasitology 1989;32:21–35.

45. Gruner L, Cortet J, Sauve C, Limouzin C, Brunel JC. Evolution of nematode community in grazing sheep selected for resistance and susceptibility to *Teladorsagia circumcincta* and *Trichostrongylus colubriformis*: a 4-year experiment. Veterinary Parasitology 2002;109:277–91.

46. Bishop SC, Stear MJ. Modelling responses to selection for resistance to gastro-intestinal parasites in sheep. Animal Science 1997;64:469–78.

47. Simm G. Genetic Improvement of Cattle and Sheep. Farming Press, Ipswich; 1998.

48. Santiago E, Caballero A. Effective size of populations under selection. Genetics 1995;139:1013–30.

49. McGuigan K. Studying phenotypic evolution using multivariate quantitative genetics. Molecular Ecology 2006;15(4):883–96.

50. Cornell SJ. Modelling nematode populations: 20 years of progress. Trends in Parasitology 2006;21:542–5.

51. Bishop SC. Modelling farm animal diseases. In: Bishop SC, Axford RFE, Nicholas FW, Owen JB, editors. Breeding for Disease Resistance in Farm Animals. 3rd ed. CABI, Wallingford; 2011. p. 38–54.

52. Singleton DR, Stear MJ, Matthews L. A mechanistic model of developing immunity to *Teladorsagia circumcincta* infection in lambs. Parasitology 2011;138:322–32.

53. Gharbi K, Matthews L, Stear MJ. An experimental and modelling framework for breeding Atlantic salmon for resistance to sea lice. Final Report for DEFRA Feasibility Link Project 2011;LK0691.

Cryptosporidium and *Giardia* zoonoses: minimizing health risks from food animal production

K.A. Reynolds[1], **G.D. Di Giovanni**[2] **and K.D. Mena**[2]

Address: [1] Mel & Enid Zuckerman College of Public Health, University of Arizona, 1295 N. Martin Avenue, Tucson, AZ 85724, USA. [2] University of Texas Health Science Center at Houston, School of Public Health, 1101 N. Campbell, El Paso, TX 79902, USA.

***Correspondence:** K.D. Mena. Email: kristina.d.mena@uth.tmc.edu

Received: 1 August 2011
Accepted: 1 February 2012

Abstract

More than 10 years ago, food-borne pathogens were estimated to cause 76 000 illnesses and 5000 deaths in the USA alone. Given the under-reporting of food-borne illnesses, the numbers of illnesses and deaths associated with food are actually higher. Food-borne pathogens continue to impact public health through worldwide disease outbreaks as they have many opportunities to enter the food chain – from the pre-harvest environment to the consumer. With over half of all disease-causing micro-organisms having the ability to be transmitted zoonotically, food animal production is a key source of pathogens in the farm-to-fork continuum. The protozoan parasites, *Cryptosporidium* and *Giardia*, have both been implicated in food-borne disease, and have the potential to be transmitted zoonotically in the farm environment. Farm management strategies have been developed to minimize the transmission of protozoa (from animals) that may infect farm workers, and contaminate nearby waterways and food crops. Such strategies target protecting the health of the herd, preventing (oo)cyst transmission to surface water and vegetable crops, and providing education to animal husbandry personnel and veterinarians on best management practices. To fully address the challenges associated with food safety, post-harvest control measures should also be implemented.

Keywords: *Cryptosporidium*, *Giardia*, Zoonoses, Risk management, Farm

Review Methodology: The following databases were explored using the search terms *Cryptosporidium*, *Giardia*, zoonoses, animal production, risk management and food-borne disease: PubMed, Agricola, Medline and Google Scholar. In addition, we used appropriate peer-reviewed papers cited in the articles obtained in the initial search.

Introduction

Battling infectious disease transmitted through food continues to be a public health challenge. At least 200 illnesses are food-borne, and food-borne pathogens are estimated to cause as many as 76 000 morbidity cases in the USA each year, resulting in 5000 deaths [1, 2]. Clinical symptoms are typically self-limiting ranging from gastroenteritis, vomiting and abdominal cramps to more severe outcomes such as bloody diarrhoea, Guillain–Barré syndrome or even death in the immunocompromised. Food-borne disease statistics presented here are likely to be underestimates, since food-borne illnesses are under-reported and outbreak investigations over the last decade have improved. Through recent advances in molecular epidemiology, animal production facilities can be identified as a disease outbreak source, just as surveillance studies have linked aetiological agents to specific retailers.

Well over half of all pathogens can be transmitted zoonotically (i.e. between animals and humans) [3], and farms provide the optimal scenario for animal–human interactions. Zoonoses continue to increase as globalization of livestock and food expand, while agricultural practices have become more diverse [4]. As mentioned above, epidemiological investigations of food-borne outbreaks have identified food animal establishments and crop farms as the source of food-borne outbreaks. There are many pathways pathogens take from pre-harvest to post-harvest to reach the consumer. The pre-harvest stage of food production includes both the raising of food animals and growth of plants for processing. Some processing occurs in the farm environment, whereas much of it is processed at a manufacturing plant or slaughterhouse after it leaves the farm. Post-harvest involves distributing the food product to retailers and food service establishments, with it ultimately reaching the consumer. This 'farm-to-fork' continuum of the food production chain includes numerous opportunities for food and humans to come in contact with pathogens.

Cryptosporidium and *Giardia*

Cryptosporidium and *Giardia* are protozoan parasites transmitted through the faecal–oral route with approximately 6000–10 000 cases of cryptosporidiosis and almost 20 000 giardiasis cases reported annually in the USA from 2006 to 2008 [5–8]. They are typically thought of as water-borne pathogens, with *Cryptosporidium* the responsible agent of the largest documented water-borne outbreak in the USA [9]. However, these parasites have also been associated with food-borne outbreaks, including food compromised by contaminated water or an animal source [10–16].

Determining whether an animal is the source of *Cryptosporidium* or *Giardia* during an outbreak investigation is not always easy, given the potentially complex transmission routes. Furthermore, the dynamics of how parasites move through the environment and what serves as reservoirs is not clear. To better understand potential sources and transmission patterns of these protozoa, recent research has focused on characterizing the parasites at the molecular level and investigating the host range of different genotypes. Currently, more than 20 species and 40 genotypes of *Cryptosporidium* have been described [17]. *Cryptosporidium parvum* causes approximately half of the human cryptosporidiosis cases [18, 19] and has also been found in diverse livestock animals including calves, lambs, goats and alpacas [20]. The taxonomy of *Giardia* is not as complicated as for *Cryptosporidium*, although the zoonotic role of *Giardia* is unclear. While there are several species of *Giardia*, the only human pathogenic species is *Giardia lamblia* (synonym *Giardia duodenalis* and *Giardia intestinalis*). This species has been divided into eight different assemblages (A–H), with assemblages A and B infecting a wide range of mammals, including humans [7, 21, 22].

Hunter and Thompson [23] and Xiao and Fayer [7] provide reviews of the epidemiological investigations of possible zoonotic transmission of *Cryptosporidium* and *Giardia* (oo)cysts. Most of the investigations demonstrating zoonotic transmission of *Cryptosporidium* involved cattle and the bovine genotype of *C. parvum* [24, 25]. Information from outbreaks and case-control studies strongly support the fact that *Cryptosporidium* is a zoonotic parasite, with one of the most virulent *C. parvum* isolates (TAMU isolate) originating from a veterinary student infected following a horse necropsy [26]. Although there is the potential for *Giardia* to be transmitted zoonotically, evidence is lacking to definitively support this transmission route. Some studies suggest the potential of zoonotic transmission, particularly involving beavers, and other studies have not [7, 27, 28]. Strong molecular epidemiological support is lacking to clearly establish giardiasis as a significant zoonotic disease.

Characterizing *Cryptosporidium* or *Giardia* infections as zoonotic is further complicated by the many complex transmission routes that can take place, particularly in the farm setting. Parasites may travel from host to host either directly or through different media, such as water or food crops. Both animals and humans may directly contaminate waterways through agricultural runoff or sewage, respectively, which then may serve as a vehicle of transmission [29]. In addition, manure from food animals and wildlife may impact food crops that then enter the food chain. Infected livestock and wildlife may come in contact with uninfected food animals and transmit (oo)cysts. Protozoa transmission from humans to animals directly is theoretically possible but has been difficult to demonstrate in field studies [30].

Mitigating Risks on the Farm

Strategies to prevent the interaction between humans and zoonotic organisms have been implemented throughout the food chain, particularly at the pre-harvest stage. Food animal production has the potential to contribute considerably to the problem as it introduces zoonotic organisms at the beginning of the food chain and throughout the farm environment, including on food crops. When considering the pre-harvest environment, *Cryptosporidium* and *Giardia* (oo)cysts may enter the food chain through several ways, and management strategies that minimize transmission should target these various routes. As a proactive approach, Hazard Analysis and Critical Control Points (HACCP) programmes have been undertaken to identify where pathogen transmission occurs on the farm to inform appropriate mitigation interventions. The following describes control measures that can be taken to prevent zoonotic transmission of (oo)cysts on the farm.

Maintaining optimal health within the food animal herd is of utmost importance from both a public health standpoint and public acceptance perspective. Although food animal operations are becoming larger and perhaps more efficient than ever, this has resulted in an increasing number of challenges related to zoonoses. These 'industrial' food animal production (IFAP) establishments – or 'factory farms' – have resulted in farms with a highly concentrated number of animals housed in extremely confined spaces, thus increasing the potential for animal-to-animal transmission of disease [31]. Food animal producers must take a multi-faceted approach to address this issue in ways appropriate for specific types of herds and pathogens. In the case of cryptosporidiosis, it is important to recognize that pre-weaned calves are the most significant source of zoonotic *C. parvum*, while older calves and cattle generally have a much lower prevalence and harbour non-human-pathogenic *Cryptosporidium* species, including *C. andersoni* [32]. Calves may shed as many as 10^{10} oocysts/g faeces [33], while adult cattle have been reported to shed 25–1.8×10^4 oocysts/g faeces [34].

The number of food animals worldwide is in the billions, with cattle alone at approximately 1.3 billion [35]. Farms vary greatly in the way they practice animal husbandry, which results in a range of disease exposures for herds worldwide [36]. Since introducing new stock is an inevitable occurrence in this global economy, it is critical that producers have knowledge of the agricultural practices at the originating farm. In addition, the act of transport causes stress to the animal as well as broadens environmental exposures (such as from moving trucks, unloading docks and other equipment), which increase the chance of disease. The impact of transporting herds globally on disease transmission has been observed [37].

Preventing food animals from exposure to zoonotic pathogens is a first step in addressing the problem. Collins and Wall [36] describe an HACCP programme called the longitudinal integrated safety assurance (LISA) that provides the mechanism for producers to follow farm practices essential for food animal health. Specifically, LISA is an animal welfare-based approach that systematically focuses on herd health by starting with the acknowledgement that zoonoses are a potential problem. Farm managers monitor specific health indices of the animals throughout production to ensure animal well-being. Although the overall goal is to prevent microbial contamination of food from 'farm-to-fork', the initial objective is to prevent pathogens from entering the farm environment during pre-harvest. This programme allows farmers to advertise their products as originating or grown under optimal conditions.

Wildlife may serve as reservoirs for both *Cryptosporidium* and *Giardia*, particularly for cattle. Likewise, cattle may provide the (oo)cyst source for wildlife that may constitute the important reservoir of pathogens in the environment, including area waterways [38–40]. Some livestock and wildlife can carry and transmit the same species of both of these protozoa. Transmission of (oo)cysts on carcasses is also another possible source for contamination of the environment, including contamination of transportation vehicles and water through abattoir waste.

The use of antibiotics and vaccines has been proposed as ways to maintain livestock health and control zoonoses. Many drugs have been evaluated for their effectiveness in treating protozoan infections. The drugs decoquinate and paromomycin have been shown to minimize diarrhoea in lambs and calves, and reduce *Cryptosporidium* oocyst excretion [41, 42]. For *Giardia*, the benzimidazoles have been shown to be able to decrease infection with calves, resulting in reduced shedding, lower prevalence and elimination of trophozoites in the small intestine [43, 44]. Vaccinations have been explored as a possible mitigation strategy since they reduce some clinical symptoms; however, they do not eliminate (oo)cyst shedding so transmission may still occur [45–47].

Animal wastes from farms have traditionally been viewed as a consequential by-product of raising livestock but are now seen as a valuable resource of nutrients [48]. Manure application on land for disposal and fertilization purposes can lead to the transmission of disease-causing micro-organisms, particularly since regulations to oversee this practice are lacking [31]. Some pathogens can endure for several months in faecal material but survivability is influenced by various environmental factors such as soil temperature, pH and water content, as well as the type of micro-organism. Typically, higher temperatures and an increase in liquid content impede pathogen survival [49]. The transmission of pathogens throughout the farm from manure spreading can be extensive, either through direct contact with manure or as pathogens are aerosolized. Area waterways and food crops may become contaminated, and wildlife may come in contact and become infected. Both *Cryptosporidium* and *Giardia* have been observed to contaminate water and vegetable crops directly from animal manure [48, 50]. This increases the probability of humans coming in contact with these pathogens that originated from livestock. *Cryptosporidium* and *Giardia* have also been detected on fresh produce irrigated with protozoa-laden water [51, 52]. Therefore, providing source water protection from contamination with livestock faeces is also important in controlling the spread of zoonotic diseases.

Appropriate handling of manure may reduce the risk of protozoa transmission. Typically, micro-organisms remain in the upper portion of soil after land application of manure; however, it is not clear if hardy protozoan (oo)cysts follow that same transport pattern [48]. After land application, proper storage and containment of the manure is critical. Movement of *Cryptosporidium* oocysts is greater when found in clay loam or silty soil [53]. Rainfall and irrigation can facilitate the movement of protozoa across farms, directly impacting surface water quality, and leading to contaminated crops and infected humans and animals. One management strategy to prevent runoff from reaching critical areas is the use of vegetative filter strips [48, 54, 55]. These 'strips' provide vegetated areas that trap micro-organisms, such as protozoa, as runoff passes through.

There are a variety of other management options to control the transmission of pathogens via manure. These strategies focus on either reducing the number of pathogens occurring in faecal matter (particularly bacteria through changes in diet) or reducing the number found in manure after land application. The latter options involve various mechanisms of treating manure – such as through chemical or thermal treatment technologies – to reduce pathogen load [48].

Those practicing animal husbandry and veterinary medicine are directly exposed to animal pathogens in the pre-harvest stage, especially those working at IFAPs, where high throughput is a priority. Microbial infections among this sub-group may then be passed to other workers and workers' families. Education of personnel on appropriate animal husbandry practices and personal hygiene is critical in controlling zoonoses. However, in many situations, farms are not providing facilities that encourage proper animal handling or opportunities for routine handwashing while on the farm. Another control measure is to identify and separate sick animals to minimize transmission; however, the fact that some infected animals may be asymptomatic complicates the effectiveness of this mitigation strategy [35].

Post-harvest Controls

The goal of reducing risks of food-borne cryptosporidiosis and giardiasis in consumers must involve a best practice approach to mitigation at every step in the food production process. This includes incorporating HACCP principles during production and processing, food-handler training and consumer education. The target of pre-harvest controls is to prevent initial animal contamination, exposure, infection and further on-farm transmission of protozoan pathogens. Post-harvest controls should address food transport and distribution, storage, handling, preparation and consumption, as well as the potential for recontamination or cross-contamination of food products from handlers and processing equipment.

Recommendations promoting good sanitation and hygiene are widespread in the food processing/service industries to prevent faecal contamination of foodstuffs. Clean and convenient restroom facilities and routine cleaning and sanitization of distribution and transport environments, processing sites, and food contact surfaces are essential for preventing cross-contamination. (Oo)cysts can survive more than 4 h when dried on stainless-steel surfaces so these must be routinely sanitized between use [56]. Although resistant to chlorine, and many other liquid disinfectants, mechanical removal of oo(cysts) during surface cleaning and subsequent drying can remove up to 10^4 organisms. A 6% solution of hydrogen peroxide is also able to inactivate >3 logs of *C. parvum* on stainless-steel surfaces [57]. Similar hygiene and sanitation procedures should be practiced in consumer homes and in food service operations. Consumer education on safe food practices and critical control points during food preparation and storage provides tools for individuals to mitigate risks at the point of consumption. Consuming uncooked or unpasteurized food products presents the greatest infection risk, as heating and drying are detrimental to (oo)cyst survival.

A variety of processing technologies (i.e. freezing, pasteurization, carcass washing/steam vacuuming, irradiation, high hydrostatic pressure, electroporation and others) have been developed to inactivate or eliminate a variety of microbial pathogens from animal products; few, however, have been evaluated against control of *Cryptosporidium* and *Giardia* in meats. Control efforts at slaughter include animal cleaning and hair removal with a 'zero tolerance' approach towards inspection of visible soil [58]. Other sources of contamination are from faeces and the intestinal tract. Pre-slaughter sheering/skinning and cleaning or scalding of hides and carcasses are vital control measures [59].

The effect of freezing on protozoan pathogens has been evaluated and found to reduce levels of *Cryptosporidium* attached to the surface of lean or fat beef trimmings fated for the production of beef burgers. Under conventional packing and freezing/tempering conditions, oocysts were not completely inactivated from the meat products but were substantially reduced [60].

Current pasteurization protocols (i.e. heating to >60 °C for 15 s) are effective at inactivation of oocysts [61]. Similarly, Moriarty *et al.* [62] found that slightly higher heat treatment (75 °C, 20 s) was effective at reducing *Cryptosporidium* contamination on carcass surfaces. Based on these studies, it is recommended that wash-water temperatures be 85 °C (for bacterial pathogen control in beef carcasses) and steam vacuuming temperatures be $\geq$82 °C for *Cryptosporidium* decontamination. These same studies have led to safe cooking guidelines for consumers [63]. Various irradiation methods have been shown to be effective against *Cryptosporidium* and *Giardia* pathogens in liquid suspension and shellfish, but results have not been published related to other meat products [64, 65].

Conclusions

Best practices for minimizing health risks of *Cryptosporidium* and *Giardia* from animal production involves a multifaceted approach with controls at every step of the 'farm-to-fork' continuum. Although the use of medications and vaccines directly address animal health and oocyst shedding early in the transmission cycle, many have been shown to be ineffective. It is recommended that a combination of measures be taken to reduce risks [66]. Such strategies include protecting the health of the herd, preventing (oo)cyst transmission to surface water and vegetable crops, and providing education to animal husbandry personnel and veterinarians on best management practices. While post-harvest control measures offer additional risk reductions, preventing pathogens from entering the farm environment initially is the first critical step in risk mitigation.

References

1. Bryan FL. Diseases Transmitted by Foods (A Classification and Summary). Centers for Disease Control, Atlanta, GA; 1982.
2. Mead PS, Slutsker L, Dietz V, McCaig LF, Bresee JS, Shapiro C, *et al.* Food-related illness and death in the United States. Emerging Infectious Diseases 1999;5(5):607–25.
3. Taylor LH, Latham SM, Woolhouse ME. Risk factors for human disease emergence. Philosophical Transactions of the Royal Society 2001;356(1411):983–9.
4. Slingenbergh JI, Gilbert M, de Balogh KI, Wint W. Ecological sources of zoonotic diseases. Revue Scientifique et Technique 2004;23(2):467–84.
5. Hlavsa MC, Watson JC, Beach MJ. Giardiasis surveillance-USA, 1998–2002. Morbidity and Mortality Weekly Report Surveillance Summary 2005;54:9–16.
6. Yoder JS, Beach MJ. Cryptosporidiosis surveillance – USA, 2003–2005. Morbidity and Mortality Weekly Report Surveillance Summary 2007;56:1–10.
7. Xiao L, Fayer R. Molecular characterization of species and genotypes of *Cryptosporidium* and *Giardia* and assessment of zoonotic transmission. International Journal of Parasitology 2008;38:1239–55.
8. Yoder JS, Harral C, Beach MJ. Cryptosporidiosis surveillance – United States, 2006–2008. Morbidity and Mortality Weekly Report Surveillance Summaries 2010;59:SS–6.
9. Mac Kenzie WR, Hoxie NJ, Proctor ME, Gradus MS, Blair KA, Peterson DE, *et al.* A massive outbreak in Milwaukee of *Cryptosporidium* infection transmitted through the public water supply. New England Journal of Medicine 1994;331:161–7.
10. Karabiber N, Aktas F. Foodborne giardiasis. Lancet 1991;377:376–7.
11. Quick R, Paugh K, Addiss D, Kobayashi J, Baron R. Restaurant associated outbreak of giardiasis. Journal of Infectious Diseases 1992;166:673–6.
12. Mintz ED, Hudson-Wragg M, Mshar P, Cartr ML, Hadler JL. Foodborne giardiasis in a corporate office setting. Journal of Infectious Diseases 1993;167:250–3.
13. Millard PS, Gensheimer KF, Addiss DG, Sosin DM, Beckett GA, Houck-Jankoski A, *et al.* An outbreak of cryptosporidiosis from fresh-pressed apple cider. Journal of the American Medical Association 1994;272:1592–6.
14. Quiroz ES, Bern C, MacArthur JR, Xiao LH, Fletcher M, Arrowood MJ, *et al.* An outbreak of cryptosporidiosis linked to a foodhandler. Journal of Infectious Diseases 2000;181:695–700.
15. Harper CM, Cowell NA, Adams BC, Langley AJ, Wohlsen TD. Outbreak of *Cryptosporidium* linked to drinking unpasteurized milk. Communicable Diseases Intelligence 2002;26:449–500.
16. Blackburn BG, Mazurek JM, Hlavsa M, Park J, Tillapaw M, Parrish M, *et al.* Cryptosporidiosis associated with ozonated apple cider. Emerging Infectious Diseases 2006;12:684–6.
17. Feng Y, Deearen T, Cama V, Xiao L. 90-kilodalton heat shock protein, Hsp90, as a target for genotyping *Cryptosporidium* spp. known to infect humans. Eukaryotic Cell 2009;8:478–82.
18. Chalmers RM, Elwin K, Thomas AL, Guy EC, Mason B. Long-term *Cryptosporidium* typing reveals the aetiology and species-specific epidemiology of human cryptosporidiosis in England and Wales, 2000 to 2003. Eurosurveillance 2009;14:1–9.
19. Leoni F, Amar C, Nichols G, Pedraza-Diaz S, McLauchlin J. Genetic analysis of *Cryptosporidium* from 2414 humans with diarrhoea in England between 1985 and 2000. Journal of Medical Microbiology 2006;55:703–7.
20. Xiao L. Molecular epidemiology of cryptosporidiosis: an update. Experimental Parasitology 2010;124:80–9.
21. Lasek-Nesselquist E, Welch DM, Sogin ML. The identification of a new *Giardia duodenalis* assemblage in marine vertebrates and a preliminary analysis of *G. duodenalis* population biology in marine systems. International Journal for Parasitology 2010;40:1063–74.
22. Caccio SM, Ryan U. Molecular epidemiology of giardiasis. Molecular and Biochemical Parasitology 2008;160:75–80.
23. Hunter PR, Thompson RCA. The zoonotic transmission of *Giardia* and *Cryptosporidium*. International Journal for Parasitology 2005;35:1181–90.
24. Mallon M, MacLeod A, Wastling J, Smith H, Reilly B, Tait A. Population structures and the role of genetic exchange in the zoonotic pathogen *Cryptosporidium parvum*. Journal of Molecular Evolution 2003;56:407–17.
25. Monis PT, Thompson RCA. *Cryptosporidium* and *Giardia*-zoonoses: fact or fiction? Infection, Genetics and Evolution 2003;3:233–44.
26. Okhuysen PC, Chappell CL, Crabb JH, Sterling CR, DuPont HL. Virulence of three distinct *Cryptosporidium parvum* isolates for healthy adults. Journal of Infectious Disease 1999;180:1275–81.
27. Gray SF, Gunnell DJ, Peters TJ. Risk factors for giardiasis: a case-control study in Avon and Somerset. Epidemiology and Infection 1994;113:95–102.
28. Warburton AR, Jones PH, Bruce J. Zoonotic transmission of giardiasis: a case control study. Communicable Disease Report CDR 1994;Rev. 4:R32–6.
29. Olson M. *Cryptosporidium* and *Giardia*: Emerging zoonoses. DVM News Magazine 2002;24–30. http://veterinarynews.dvm360.com/dvm/data/articlestandard//dvm/392002/32676/article.pdf
30. Di Giovanni GD, Betancourt WQ, Hernandez J, Assadian NW, Flores Margez JP, Jaramillo Lopez E. Investigation of potential zooanthroponotic transmission of cryptosporidiosis and giardiasis through agricultural use of reclaimed wastewater. International Journal of Environmental Health Research 2006;16(6):405–18.
31. Graham JP, Leibler JH, Price LB, Otte JM, Pfeiffer DU, Tiensin T, *et al.* The animal–human interface and infectious disease in industrial food animal production: rethinking biosecurity and biocontainment. Public Health Reports 2008;123:282–99.
32. Santin M, Trout JM, Fayer R. A longitudinal study of cryptosporidiosis in dairy cattle from birth to 2 years of age. Veterinary Parasitology 2008;155:15–23.
33. Nydam DV, Wade SE, Schaaf SL, Mohammed HO. Number of *Cryptosporidium parvum* oocysts or *Giardia* spp cysts shed by dairy calves after natural

infection. American Journal of Veterinary Research 2001;62(10):1612–5.

34. Scott CA, Smith HV, Gibbs HA. Excretion of *Cryptosporidium parvum* oocysts by a herd of beef suckler cows. Veterinary Record 1994;134:172.

35. Tomley FM, Shirley MW. Livestock infectious diseases and zoonoses. Philosophical Transactions of the Royal Society 2009;364:2637–42.

36. Collins JD, Wall PG. Food safety and animal production systems: controlling zoonoses at farm level. Revue Scientifique et Technique Off int Epiz 2004;23(2):685–700.

37. Fèvre EM, Bronsvoort BM, Hamilton KA, Cleaveland S. Animal movements and the spread of infectious diseases. Trends in Microbiology 2006;14(3):125–31.

38. Heitman TL, Frederick LM, Viste JR, Guselle NJ, Morgan UM, Thompson RCA, *et al.* Prevalence of *Giardia* and *Cryptosporidium* and characterization of *Cryptosporidium* spp. isolated from wildlife, human, and agricultural sources in the North Saskatchewan River Basin in Alberta, Canada. Canadian Journal of Microbiology 2002;48(6):530–41.

39. Van Keulen H, Macechko PT, Wade S, Schaaf S, Wallis PM, Erlandsen SL. Presence of human Giardia in domestic, farm and wild animals, and environmental samples suggests a zoonotic potential for giardiasis. Veterinary Parasitology 2002;108(2):97–107.

40. Olson ME, O'Handley RM, Ralston BJ, McAllister TA, Thompson RCA. Update on *Cryptosporidium* and *Giardia* infections in cattle. Trends in Parasitology 2004;20(4):185–91.

41. De Graaf DC, Vanopdenbosch E, Ortega-Mora LM, Abbassi H, Peeters JE. A review of the importance of cryptosporidiosis in farm animals. International Journal of Parasitology 1999;29:1269–87.

42. Grinberg A, Markovics A, Galindez J, Lopez-Villalobos N, Kosak A, Tranquillo VM. Controlling the onset of natural cryptosporidiosis in calves with paromomycin sulphate. Veterinary Record 2002;151:606–8.

43. O'Handley RM, Cockwill C, Jelinski M, McAllister A, Olson ME. Effects of repeat fenbendazole treatment in dairy calves with giardiasis on cyst excretion, clinical signs and production. Veterinary Parasitology 2000;89:209–18.

44. O'Handley RM, Buret AG, McAllister TA, Jelinski M, Olson ME. Giardiasis in dairy calves: effects of fenbendazole treatment on intestinal structure and function. International Journal of Parasitology 2001;31(1):73–9.

45. Olson ME, Ceri H, Morck DW. Giardia vaccination. Parasitology Today 2000;16(5):213–7.

46. Jenkins MC. Advances and prospects for subunit vaccines against protozoa of veterinary importance. Veterinary Parasitology 2001;101:291–310.

47. Riggs MW. Recent advances in cryptosporidiosis: the immune response. Microbes and Infection 2002;4:1067–80.

48. Bicudo JR, Goyal SM. Pathogens and manure management systems: a review. Environmental Technology 2003;24(1):115–30.

49. Strauch D, Ballarini G. Hygienic aspects of the production and agricultural use of animal wastes. Journal of Veterinary Medicine 1994;41:176–228.

50. Atwill ER. Pathogens Excreted by Livestock and Transmitted to Humans through Water. Veterinary Medicine Teaching and Research Center, School of Veterinary Medicine, University of California, UCD Animal Agriculture Research Center, Davis; 1997.

51. Smith HV, Cacciò SM, Tait A, McLauchlin J, Thompson RCA. Tools for investigating the environmental transmission of *Cryptosporidium* and *Giardia* infections in humans. Trends in Parasitology 2006;22(4):160–7.

52. Amorós I, Alonso JL, Cuesta G. *Cryptosporidium* oocysts and *Giardia* cysts on salad products irrigated with contaminated water. Journal of Food Protection 2010;73(6):1138–40.

53. Mawdsley JL, Brooks AE, Merry RJ. Movement of the protozoan pathogen *Cryptosporidium parvum* through three contrasting soil types. Biology and Fertility of Soil 1996;21:30–6.

54. Lim TT, Edwards DR, Workman SR, Larson BT. Vegetated filter strip removal of cattle manure constituents in runoff. Transactions of the American Society of Agricultural Engineers 1998;41:1375–81.

55. Trask JR, Kalita PK, Kuhlenschmidt MS, Smith RD, Funk TL. Overland and near-surface transport of *Cryptosporidium parvum*. In: American Society of Agricultural Engineers Annual International Meeting, Paper No. 01-2104, St. Joseph, Michigan; 2001.

56. Deng MQ, Cliver DO. *Cryptosporidium parvum* studies with dairy products. International Journal of Food Microbiology 1999;46:113–21.

57. Barbee SL, Weber DJ, Sobsey MD, Rutala WA. Inactivation of *Cryptosporidium parvum* oocyst infectivity by disinfection and sterilization process. Gastrointestinal Endoscopy 1999;49:605–11.

58. Sofos JN. Challenges to meat safety in the 21st century. Meat Science 2008;787:3–13.

59. Fernandes R. Chilled and frozen raw meat, poultry and their products. In: Microbiology Handbook: Meat Products. RSC Publishing, Cambridge; 2009;1–44.

60. McEvoy JM, Moriarty EM, Duffy G, Sheridan JJ, Blair IS, McDowell DA. Effect of commercial freeze/tempering process on the viability of *Cryptosporidium parvum* oocysts on lean and fat beef trimmings. Meat Science 2004;67:559–64.

61. Harp JA, Fayer R, Pesch BA, Jackson GJ. Effect of pasteurization on infectivity of *Cryptosporidium parvum* oocysts in water and milk. Applied and Environmental Microbiology 1996;62:2866–8.

62. Moriarty EM, Duffy G, McEvoy JM, Caccio S, Sheridan JJ, McDowell D, *et al.* The effect of thermal treatments on the viability and infectivity of *Cryptosporidium parvum* on beef surfaces. Journal of Applied Microbiology 2005;98:618–23.

63. USDA. Fact sheets: Meat preparation. Beef ... from farm to table. United States Department of Agriculture, Food Safety and Inspection Service. Available from: URL: http://www.fsis.usda.gov/Factsheets/Beef_from_Farm_to_Table/index.asp#25 (accessed 24 January 2012).

64. Sundermann CA, Estridge BH. Inactivation of *Giardia lamblia* cysts by cobalt-60 irradiation. Journal of Parasitology 2010;96:425–8.

65. Collins MV, Flick GJ, Smith SA, Fayer R, Rubendall R, Lindsay DS. The effect of E-beam irradiation and microwave energy on Eastern oysters (*Crasssostrea virginica*) experimentally infected with *Cryptosporidium parvum*. Journal of Eukaryotic Microbiology 2005;52:484–8.

66. De Waele V, Speybroeck N, Berkvens D, Mulcahy G, Murphy TM. Control of cryptosporidiosis in neonatal calves: use of halofuginone lactate in two different calf rearing systems. Preventive Veterinary Medicine 2010;96:143–51.

Livestock infectious disease and climate change: a review of selected literature

Claire Heffernan[1]*, Mo Salman[2] and Luke York[3]

Address: [1] Smith School of Enterprise and the Environment, Hayes House, 75 George Street, Oxford OX1 2BQ, UK. [2] College of Veterinary Medicine and Biomedical Sciences, Colorado State University, 1601 Campus Delivery, Fort Collins, CO 80523-1601, USA.[3] Livestock Development Group, School of Agriculture, Policy and Development, University of Reading, Reading RG6 6AL, UK.

***Correspondence:** Email: c.l.heffernan@reading.ac.uk; claire.heffernan@smith.ox.ac.uk

Received: 1 November 2011
Accepted: 7 February 2012

Abstract

It is indisputable that climate is an important factor in many livestock diseases. Nevertheless, our knowledge of the impact of climate change on livestock infectious diseases is much less certain. Therefore, the aim of the article is to conduct a systematic review of the literature on the topic utilizing available retrospective data and information. Across a corpus of 175 formal publications, limited empirical evidence was offered to underpin many of the main arguments. The literature reviewed was highly polarized and often inconsistent regarding what the future may hold. Historical explorations were rare. However, identifying past drivers to livestock disease may not fully capture the extent that new and unknown drivers will influence future change. As such, our current predictive capacity is low. We offer a number of recommendations to strengthen this capacity in the coming years. We conclude that our current approach to research on the topic is limiting and unlikely to yield sufficient, actionable evidence to inform future praxis. Therefore, we argue for the creation of a reflexive, knowledge-based system, underpinned by a collective intelligence framework to support the drawing of inferences across the literature.

Keywords: Infectious livestock disease, animal health, systematic review, climate change, global warming

Introduction

Climate change is the dominant social, political, economic and environmental issue of our age. Both the scientific and subsequently, public debate have filtered across all societies and nations. Within the scientific literature, global climate change (GCC) is strongly associated with anthropogenic greenhouse gas emissions. Nevertheless, as Slenning [1] noted, 'GCC is not a simple process of steady and regularly warming land, sea and atmosphere'. Rather, it is a complex and dynamic set of changes involving alterations in temperature, humidity, ocean chemistry and precipitation over time [1–5].

Climate, however, is composed of discreet weather events. Therefore, accurate predictions regarding the future impact of GCC require a level of certainty over future weather patterns, which, by their very nature, are dynamic [5]. As such, global climate models are often accused of being inaccurate or inconsistent [6]. Nevertheless, some trends have been widely agreed upon, i.e. in addition to general warming in the coming decades, GCC will increase the incidence of extreme weather events (EWE) [1,4,7–10].

Clearly, factors associated with the causes of human and livestock disease are interwoven with climate/ecological drivers such as temperature, humidity, geography etc. [11–15]. As Epstein [16] noted 'climate constrains the range of infectious diseases, whereas weather affects the timing and intensity of outbreaks'. Indeed, the ecological limits and boundaries of disease processes for both humans and animals have been well studied [6,14,17–22].

Nevertheless, while there has been arguably intense global interest/speculation regarding the role of livestock in GCC, much less interest has been accorded the animal health/climate change nexus. Despite decades of research on the impact of climate on particular livestock diseases, there is a prevailing notion that the knowledge gap between animal infections and climate change remains large [23–25]. A number of authors noted the paucity of empirical data was a major challenge in attributing changes in disease dynamics to GCC [1,26–29].

As van Dijk *et al.* [23] noted 'the field of climate change and its effects on the dynamics of pathogens of veterinary importance has lagged well behind the climate-change-focused research'. Indeed, it has been argued that veterinarians, epidemiologists and other animal health experts have been relatively slow to engage with, and contribute to the debate on climate change [25,30]. The human health community has ostensibly allocated much more workforce to the topic [6,24,31,32].

The impact of climate change on human health has even been calculated in disability-adjusted life years (DALY) (see [10,31,32]).[1] While this approach has been criticized [32], such an analysis is beneficial on two counts. First, it demonstrates the disproportionate impact of climate change on the health of particular populations, i.e. those living in the Global South [31]. Second, it provides at least an initial comparative framework for analysis and thereby, a means of priority setting for resource allocation.

The number and seriousness of emerging/re-emerging livestock and zoonotic diseases over the past decade amply illustrate the enormity of the global challenge [34]. However, it is clear that in order for such disease risks to be managed, it is critical for scientists, veterinarians and other actors in the livestock sector to utilize available resources effectively. As Sutherst noted in 1998 [35], without a common framework to confront health and climate change, we risk a scientific 'tower of Babel'. Over a decade later, it is unclear if the field of animal health and climate change remains constricted by a scattergun approach or if coherent systems for knowledge generation are beginning to emerge.

To explore both the current state of knowledge regarding livestock infectious disease and climate change and the wider evidence of a framework for collective action, we performed a systematic review of 175 refereed journal articles on animal health and climate change.

Methods

The study broadly ascribed to the approach to systematic reviews outlined the PRISMA statement [36]. Therefore, the stages of the review are as follows:

(a) Establish if the existing review protocols are applicable.
(b) Outline the reporting criteria.
(c) Detail the information sources.
(d) Establish the study selection process.
(e) Describe the methods for data extraction.
(f) Identify potential biases.

Given the nature of the material, from the outset, it was unlikely that review protocols for clinical studies such as meta-analysis or the Cochrane approach would be applicable. Other protocols such as Foresight, while useful for analysing the state of predictive decision-making on a given topic, often rely on expert opinion, which was also outwith the aims of the study. Nevertheless, elements of both of the above approaches informed the article. First, similar to the wider aims of a meta-analysis, a body of literature was evaluated for evidence; however, the evidence in this case was related to the core-arguments rather than clinical data *per se*. Second, while Foresight studies often utilize historical data to create future predictions, in this case the existing evidence was evaluated for predictive decision-making frameworks themselves.

While the grey literature on climate change and infectious livestock disease is growing, the authors elected to focus on the presently available, formal, scientific literature published in English. Therefore, the initial reporting criteria for the selection of studies were as follows:

(a) Peer-reviewed journal article.
(b) Focus on livestock.
(c) Global South or developing country context.
(d) English language translation.

To gain an overview of the topic, four online search engines were utilized: Science Direct, PubMed, CAB Abstracts and Google Scholar. The following terms comprised the search: climatic change and animal health; climatic change and livestock disease; global warming and animal health; global warming and animal disease. In order to gain a historical perspective, the search included articles from 1980 to 2011.Nevertheless, as the review progressed it was clear that there were only a few studies regarding climate change and livestock disease that were specifically located within the Global South. Rather, much of the work was orientated towards the potential of tropical threats to impact livestock production or human health in the northern hemisphere. Therefore, the review was extended to include relevant articles on animal health constraints and climate change in temperate climates. Equally, articles on wildlife and human health, but with clear implications for animal health were also included.

The original search yielded 260 primary documents (241 in English and 19 in other languages). From this overview, a further selection process occurred based on the following criteria:

(a) An explicit focus on climatic variables/climate change.
(b) The inclusion of arguments of relevance to the debate on livestock disease and climate change.

[1]For example, WHO estimated that the impact of climate on DALY/million in the year 2000 in Africa was over 3000 times that in developed countries. In Southeast Asia, the DALY/million was 1700 times that of the North [10]. Over 40% of the DALY lost were associated with VBD [33]. Previous analyses have focused on poverty alleviation and DALY's [20].

Table 1 Frequency of the articles under review by year

Years	Number of articles
1980–1989	3
1990–1995	7
1996–2000	8
2001–2004	17
2005–2007	29
2008–2009	64
2010–July 2011	47
Total	175

In total, 175 articles comprised the final corpus of documents.

To 'extract the data', the articles were then coded to determine the primary topic and secondary subject area. In this manner, the coding was interpretive [37] and 'fell out' of the content. A third level of coding was also undertaken to include the orientation of the article, i.e. policy, primary evidence, review, etc. An example article might be coded as Arbovirus, Bluetongue (BT) and Historical Review. Codes were unique to the category involved and did not overlap across topic/subject/orientation. Nevertheless, some diseases clearly crossed categories, for example, arboviruses can be transmitted by both ticks and mosquitoes, and therefore, could be equally included under the wider rubric of vector-borne disease (VBD). To avoid duplicate coding, the coding reflected the primary topic of the article as explicitly detailed in the title. For example, those articles coded under the 'General' category of 'Vector-borne' diseases tended to offer a wide-ranging discussion on a variety of diseases. Only those articles that focused on a single disease were accorded a code reflecting the disease name. As such, articles on pathogens, which crossed vector species such as arboviruses were organized by the primary focus of the article.

After the articles were disaggregated by topic, the documents were then explored by core arguments and the premises evaluated. For the specific disease studies reviewed, the level of cross-referencing between key articles was also explored.

Potential biases in the article selection process are as follows:

(a) The review was limited to the formal published literature.
(b) Only articles with an English language translation were included in the review.
(c) The search queries were limited to the above key words.
(d) The use of particular on-line search engines.

All of the above biases could potential impact the 'representativeness' of the corpus of documents, as a whole. Therefore, we make no claims regarding the 'exhaustive' nature of the review and further, the objectives of the analysis were both limited and time-bound. As such, we do not offer an historical exegesis of the role of climate on infectious diseases of livestock. Rather, as noted above, our explicit focus was to review a selection of articles representing the recent literature to explore gaps, key arguments and the evidence for knowledge-based frameworks.

Table 2 Frequency of the geographic area of the articles under review

Geographic focus	Number of articles
The Arctic	4
Asia	5
Australia	5
South America	8
North America	9
Africa	23
Europe	32
Global	89
Total	175

Results

The corpus of documents

Reflecting increasing concerns regarding climate change, the majority of articles were published during 2008–2011 (Table 1).

Over 80% of articles could be classified as 'narrative' reviews, which described changes in the distribution or incidence of a range of diseases associated with climate change and other drivers. Of the total number of articles, only a few offered direct evidence and or were based on actual data. Not surprisingly, most of the reviews tended to be wide-ranging and most offered a global perspective (Table 2).

The majority of articles described both human and animal diseases, with zoonoses the explicit focus was only on a few articles.

Overall, the articles could be organized into six broad categories. Vector-borne and parasitic diseases dominated the articles. Interestingly, few of the articles offered a strictly epidemiological perspective. The topic and subject breakdown was as follows (Table 3).

The arguments

VBD

VBDs are those transmitted via indirect or direct contact with a range of agents from haematophagous arthropods to vertebrates [14,38]. Across the literature, there was strong agreement over the potential impact of climate on the lifecycle of vectors, i.e. climate impacts survival and mortality rates, vector–pathogen interactions, pathogen

Table 3 Frequency of the topic breakdown of the articles under review

Subject and topic	Total articles	Topic breakdown	Article No. by section appendix I
1. General climate change and animal health	27	*27*	*1–27*
2. Vector-borne diseases (VBDs)			
General	18	*18*	*1–18*
Leishmanisasis	1	*1*	*1*
2a. Viral diseases	36		
Arbovirus		*29*	*1,2,3,5,6,7,9,10–12, 14–28, 30,34–35*
Capripox virus		*1*	*4*
Flavivirus		*1*	*32*
Influenza virus		*3*	*8,13,29*
General		*2*	*31,36*
2b. Tick-borne diseases (TBDs)	28		
East Coast fever		*9*	*5,11,12, 13, 16,18,20, 21,23*
Genetics		*1*	*1*
Lyme disease		*2*	*15,19*
Tick-borne encephalitis		*2*	*3,26*
Tick range/habitat changes		*5*	*4, 7,8,9,10*
General		*9*	*2,6,14,17,22,24, 25,27,28*
3. Parasitic diseases	43		
Ecto-parasites		*4*	*32,39,40,43*
Helminths		*6*	*11,12,17,38,41,42*
Nematodes		*4*	*9,10,15,22*
Parasites as biological tags		*1*	*35*
Parasite–host interactions		*2*	*2,14*
Protozoan		*1*	*13*
Trematodes		*9*	*1,5,6,7,16,18,19,31,36*
General		*16*	*3,4,8,20,21,23–30,33,34,37*
4. Infectious/emerging infectious disease	9	*9*	*1–9*
5. One Health/public health	9	*9*	*1–9*
6. Bacterial diseases	4	*1*	*1*
Leptospirosis		*3*	*2,3,4*
	175	*175*	

replication and vector behaviour and distribution [39,40–43]. Nevertheless, as the following sections illustrate, the attribution of direct climate change impacts on VBDs was often contested. Indeed, anthropogenic and 'other plausible changes' were often accorded a greater role in the rising incidence of VBDs [21,44,45]. Few articles dealt with control (see [46]).

The arboviruses. Globally, arboviruses are considered one of the most important causes of human and animal disease [47,48]. Early articles tended to simply describe or list arboviruses[2] and speculate about the potential of these viruses to increase their geographic range, particularly in relation to Europe [49,50].[3]

In the later literature, with the attendant global increase in the incidence and prevalence of arbovirus-related diseases, climate change became a popular frame [47,51–53]. While most authors broadly linked arboviral diseases to climate change, there was much caution in making any direct causal associations. Rather, authors tended to list the wide range of abiotic factors implicated in disease spread [43,52,54]. As Gould and Higgs [54] noted:

> [factors important to spread include] . . . levels of socio-economic development; increasing human travel; commercial transportation; urbanization; deforestation; land reclamation; irrigation projects; human, animal and arthropod population density increase; and political and military activities that lead to mass human evacuation.

Other authors were more circumspect and limited their assessments to institutional issues such as those related to the performance of veterinary services, poor governance and ineffective legal framework for disease control [43,51].

[2]Such as Crimean Congo hemorrhagic fever (CCHF), Nairobi sheep disease virus (NSDV), Ganjam virus, Israel turkey meningoencephalitis virus, BTV and RVF to name a few.

[3]Later, the increased dispersion and spread of arboviruses in Japan and Korea was also explored [51].

As time progressed, two arboviruses began to receive greater attention in the animal health and climate change literature: BT and Rift Valley fever (RVF) (see [29,40,47,53–57]).

Bluetongue virus (BTV). BTV is an Orbivirus of ruminants transmitted by *Culicoides* biting midges. The infection can cause severe clinical disease in sheep, particularly in naïve populations [47,53]. Conversely, lesions in cattle are 'frequently subtle' [59]. In total, 24 serotypes of virus have been identified (with an additional one likely), which are referred to as BTV-1, BTV-2, etc. [57]. Wind-borne dispersal of midges (40–300 km per journey) supports the transmission of the virus over large geographic distances [60].

Historically, the virus is associated with winds from the Inter-tropical Convergence Zone (ITCZ) and was largely confined to the Global South [61]. However, since the late 1990s, the following eight serotypes (1, 2, 4, 6, 8, 9, 11 and 16) have been identified in Europe [58]. While incursions of the disease into southern Europe had occurred in the 1950s and 1960s, from 1998 to 2005, outbreaks were reported in countries in areas of the Mediterranean biome with no prior history of the disease [53]. By 2006, the virus had spread to Germany, France, the Netherlands and Luxemburg [62]. During 2007–2009, more than a million animals were infected in further outbreaks in the UK, Czech Republic, Switzerland and Hungary [29]. Thus, BT is a high-impact disease, with an estimated cost of over US$3 billion per year globally [53].

While the incursion of serotype BTV-1 into Eastern Europe is related to Indian serotypes of the virus, the incursion of BTV-2 into southern Europe is related to North African serotypes [29]. Climate change (seven of the warmest European winters on record occurred 1999–2007) has been offered to explain the successful incursion and spread of BTV-2 initially from Algeria and Tunisia to the southern tip of Sardinia and beyond. Interestingly, however, it has been hypothesized that the initial outbreaks in Algeria and Tunisia in 1999 were caused by the illegal importation of infected animals from sub-Saharan Africa (Guinea/Cote d'Ivoire) [53]. Conversely, the transmission of BTV-1 was linked to trade along the 'Eurasian livestock street' or the cross-continental trade in livestock [29]. Within the literature, BTV-2 is linked to climate change and BTV-1 to livestock trade. Zell *et al.* [63] argued three forces were involved: first, a worldwide extension of stock-breeding; second, the replacement of resistant, indigenous domestic animals by improved cattle and sheep; and third, the adaptation of the principal Old World vector of the BT disease virus, *Culicoides imicola*, to the microclimate of livestock holdings.

In 2006, however, outbreaks related to the African origin serotype (BTV-8) identified in North-eastern Europe have caused clinical disease with accompanying mortality in cattle [47,54,57,62,64]. The serotype rapidly spread and is presently estimated to occupy a zone of 2.3 million hectares [65].

The cause of the incursion of BTV-8 into northern Europe is, as yet, unknown [40,47,57]. Interestingly, with the advent of BTV-8 some authors moved away from climate factors as a primary causal factor in the spread of BTV. With regard to BTV-8, speculation includes the virus being carried by 'imported infected ruminants', infected vectors transported with horses or exotic plants, 'contaminated or unstable vaccines' [66].

Other authors went even further: for example, Tabachnick [29] suggested that the relation between climate change and the expansion of all BTV serotypes across Europe was plausible, but as of yet, unproven. Rather, as exotic wildlife can be reservoir hosts for the virus, Tabachnick speculated that increases in the number of wild animal theme parks in Europe from 1965 to 1995 was a potential factor in the spread of the disease. While a wide number of exotic species (lions, buffaloes, zebras, etc.) can carry the disease, there is no legal requirement to test imported wildlife for the virus. Nevertheless, it is clear that such an explanation suffers from a number of evidential gaps. First, if such theme parks were the epicentre of the disease, it would be presumed that BT outbreaks could be traced within the radius of any given park. The parks may simply act as a confounding variable or ecological fallacy as they are related to human and animal densities. Second, given the 30-year time-period, one would expect that outbreaks would have occurred prior to 1998. As such, while it is indeed possible that BTV may have been imported into Europe via infected exotic stock, the ability of the vector to transmit the virus under previous environmental conditions appears to have been limited. As Mellor *et al.* [53] noted:

> Long-term persistence (i.e. an endemic zone) [of BTV] is currently thought to be possible only in areas where active adult vectors are present virtually throughout the year. In such situations, if vector-free periods do occur, they must be of shorter duration than the maximum period of viraemia in the local ruminant population (up to 54 days in sheep and 60–100 days in cattle), otherwise the last infected host will have died or recovered before new adult vectors arrive on the scene.

Therefore, plausible, alternate explanations to the role of climate change in BTV transmission across Europe were rare. Nevertheless, Tabachnick [29] was correct in noting that little evidence is offered to underpin the widespread assumption of the importance of climate change to BT outbreaks.

Few authors linked BT outbreaks to wider global climate systems or offered more than a description of the general warming trend of European winters [67–69]. In an exception, Purse *et al.* [40] mapped the temperature, humidity and precipitation changes across Europe from

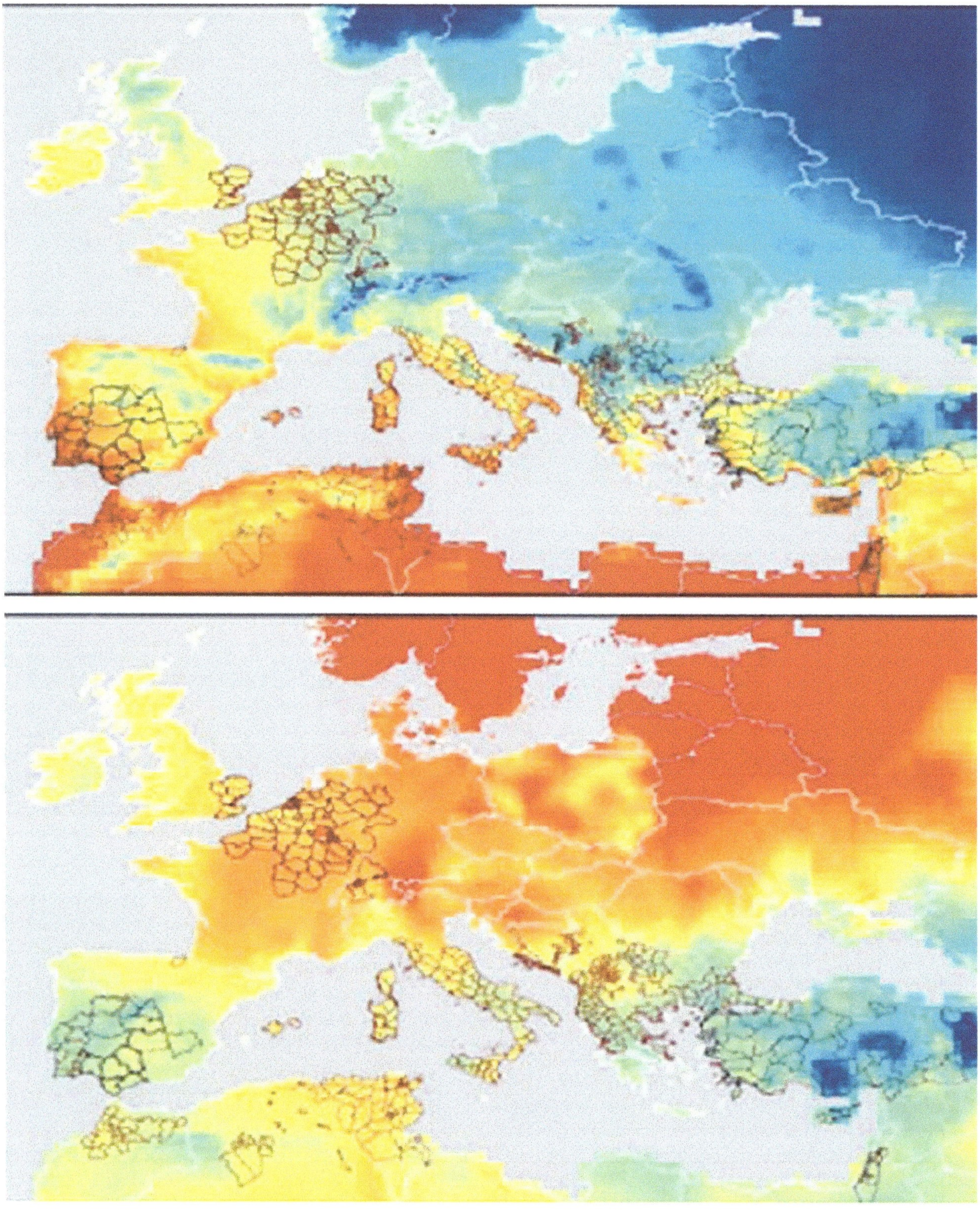

Figure 1 BT outbreaks versus temperature change in Europe (from [40])

the 1980s to 1990s to the BTV epidemic. Figure 1 maps temperature changes across the decade to historical versus 'new' BTV outbreaks.

Figure 1(a) Illustrates the average 1990s minimum temperatures (from −10°C represented by navy to 13°C represented by red). Figure 1(b) Illustrates changes to these average temperatures between the 1980s and the 1990s (a −2.0°C change is represented by dark blue while an increase of 2.0°C is represented by dark red). The maps are marked with the areas historically impacted by BT (grey) and the more recently impacted areas (red) (from [40]).

Despite earlier models, illustrating the difficulties in predicting BT outbreaks by temperature [70], Purse *et al.* [40] argued that 'the patterns of *C. imicola's* range extension during the 1990s mirrors the pattern of warming'. Further, there is evidence that *C. imicola* 'handed over' the virus for spread by local European vectors *Culicoides pulicaris* and *Culicoides obsoletus* [40]. From the maps, it is clear that the 'hand over' hypothesis is plausible for disease spread from historical incursion areas across Italy and Spain.

However, as Purse *et al.* [40] noted: 'the northern outbreaks of BTV [8] lie well outside the envelope of *C. imicola* in Europe ... the Palearctic complexes [*C. pulicaris* and *C. obsoletus*, etc.] [have a] major, if not exclusive, role as vectors.' Indeed, vector competence for BTV-8, related to African strains of the virus, appeared to be almost immediate with no lag or adaptation period required [40]. As Tabachnick [29] noted vector capacity is 'not a phylogenetic ancestral trait ... some vector capacity traits arose independently in disparate [*Culicoides*] species'. Therefore, the role of vector competence and capacity in the wake of BTV-8 requires greater explication. As MacLachlan and Guthrie [57] offered: 'determinants ... [of] vectorial capacity are only now being defined, thus many current risk analyses are based on data that are of dubious relevance and validity.'

Interestingly, BTV-8 has recently expanded southward from northern Europe into Italy and Spain (see [65]), apparently crossing the major vector barriers of the Swiss Alps and Pyrenees mountain ranges. The factors important to the expansion of BTV, has been of long interest to researchers. For example, Ward and Carpenter [71] in a simulated model of environmental conditions in Australia linked disease occurrence to temperature and rainfall i.e. 'warm, wet springs'. Conversely, in Italy, Sicily [72] found the 'best climatic indicator' for the *C. imicola* population and thereby outbreaks of disease, was the NDVI. While in Spain, a study found that the *C. imicola* population had the strongest relationship to wildlife and livestock populations followed by topo-climatic variables [18]. Finally, a model based on 2006 BTV-8 outbreak data in northern Europe, revealed local wind density to be the climatic variable with the strongest correlation to outbreaks at the municipal level [65]. According to the author: 'the spread of [BTV-8] infection was firstly affected by wind events, next by local spread by vectors and third by animal transport out of infected areas' [65]. While all of the models utilized alternate 'starting points' and approaches, overall there was little agreement as to the variables of greatest importance to the spread of the vector and/or the disease. While Baylis [70] argued that the 'climatic variables of distribution differ between *Culicoides* species', such differences do not fully explain the diversity of findings above. Further, most of the models attempted to explain specific outbreaks within a particular period of time.

Conversely, Zell *et al.* [63] explored the relationship between the Atlantic Multidecadal Oscillation (AMO) index (which details the variability of the sea surface temperature (SST) of the North Atlantic Ocean)[4] and BT outbreaks in the USA, Turkey and Europe (Figure 2).

As the figure illustrates, the 'new' BT outbreaks in Europe are linked to a positive change in the oscillation index. The results of Zell *et al.*'s work are intriguing and may offer a wider explanation for the rapid incursion of BTV in other geographic locations. For example, in the USA, eight new serotypes of BTV were discovered during 2004–2006 [73] with six in the state of Florida alone. Novel serotypes have also been identified in Australia and Israel [57].

Nevertheless, it is clear that the literature would be enhanced by a deeper and more nuanced evidence-base regarding BTV transmission and climate phenomenon. And in this manner, better predictions can be made as to the biological/behavioural capacity of both the virus and vectors under conditions of climate change. This is not to dispute the role of other factors such as livestock trade in disease transmission, rather, a deeper understanding of the inter-relationship between climate change and other abiotic factors may prove useful in underpinning future control strategies such as vaccination campaigns.

The literature on BT was dominated by 'reviews' (Table 4).

The level of cross-referencing across the articles is detailed in Table 5.

RVF. Unlike BT, a wide body of evidence links RVF directly to the global climate phenomenon termed the El Niño-Southern Oscillation (ENSO) [47,49,77]. ENSO occurs every 2–7 years and is the outcome of the weakening of the easterly trade winds, which enable the warm water of the equatorial Pacific to move into the eastern Pacific. This shift creates a constellation of opposite effects in different parts of the world, i.e. drought versus flooding. In East Africa, the El Niño effect causes heavy rainfall and flooding. The virus can lay dormant for years in dry conditions in oviposited eggs. Indeed, inter-epizootic periods can range in length from 1 to 40-plus years [54]. With the advent of heavy rainfall, the eggs hatch and mosquitoes transmit the disease. Humans often act as sentinel cases for the presence of the disease [64]. Other climatic indicators linked to RVF outbreaks, include a positive Indian Ocean dipole (IOD) or a higher than average SST in the western Indian Ocean [78].

---

[4]Zell *et al* [63] described the AMO as 'a quasi-periodic long-term fluctuation upto 50–80 years in the strength of thermohaline circulation that results in variability of sea surface temperatures in the range of 0.4–8°C. Its warm phases extended from 1860 to 1880 and 1930 to 1960, cold phases were recorded 1905–1925 and 1970–1990. In 1995, the AMO index turned positive again, indicating another warm phase.'

Table 4 Articles and evidence base: BT

Primary author	Year	Topic	Type	Evidence base
Sellers [60]	1984	Factors related to VBD spread	Model of disease spread	Literature: description of disease transmission zones related to ITCZ
Wittmann and Baylis [50]	2000	*Culicoides* and climate change	Review	Literature
Baylis *et al.* [70]	2004	Approaches to BT vector modelling	Review	Literature
Purse *et al.* [72]	2004	BT and climate variables	Predictive model	Outbreak data, NDVI and climate variables
Purse *et al.* [40]	2008	BT and climate change	Review	Literature
Mellor *et al.* [53]	2008	History of BT	Review	Literature
Purse *et al.* [40]	2008	BTV and climate change	Review	Literature
Zell *et al.* [63]	2008	Viral diseases and climate change	Review	Literature
Wilson and Mellor [69]	2008	BT and climate change	Review	Literature
Wilson *et al.* [68]	2008	BTV transmission: overwintering	Review	Literature
Gould and Higgs [54]	2009	Arboviruses and climate change	Review	Literature
MacLachlan *et al.* [76]	2009	BTV and climate change	Review	Literature
Acevedo *et al.* [18]	2010	*Culicoides* and climate change	Predictive Model	Spatial location, topography, climate, wildlife and livestock, datasets
MacLachlan [56]	2010	BT and climate change	Review	Literature
MacLachlan and Guthrie [57]	2010	Re-emergence of BT	Review	Literature
MacLachlan [58]	2011	BT epidemiology	Review	Literature
Ducheyne [65]	2011	BT epidemiology	Predictive model	Pan-European outbreak data and Wind density maps

Table 5 Article cross-referencing: BT

References	Mellor *et al.* (2008) [53]	Purse *et al.* (2008) [40]	Zell *et al.* (2008) [63]	Wilson *et al.* (2008) [68]	Gould and Higgs (2008) [54]	MacLachlan (2009) [76]	Acevedo *et al.* (2010) [18]	MacLachlan (2011) [58]	Ducheyene *et al.* (2011) [65]
Sellers (1984) [60]					x				
Wittmann and Baylis (2000) [50]	x								x
Baylis *et al.* (2004) [70]							x		
Purse *et al.* (2004) [72]							x		
Purse *et al.* (2005) [67]	x	x			x		x	x	
Mellor *et al.* (2008) [53]			x					x	
Purse *et al.* (2008) [40]						x	x	x	
Zell *et al.* (2008) [63]									
Wilson and Mellor (2008) [69]						x		x	
Wilson *et al.* (2008) [68]						x			
Gould and Higgs (2009) [54]						x			
MacLachlan *et al.* (2009) [76]								x	
Acevedo *et al.* (2010) [18]									
MacLachlan (2010) [56]								x	
MacLachlan and Guthrie, (2010) [57]								x	
MacLachlan (2011) [58]									
Ducheyne *et al.* (2011) [65]									

Historically, the virus has been confined to Africa [55,64]. In 2000, an outbreak occurred in Saudi Arabia and Yemen, owing to live animals transported from Africa to feed pilgrims on the Hajj. Authors revealed a level of anxiety over the disease reaching Europe [55,64,77,79]. An expert consultation ranked RVF as a potential risk to

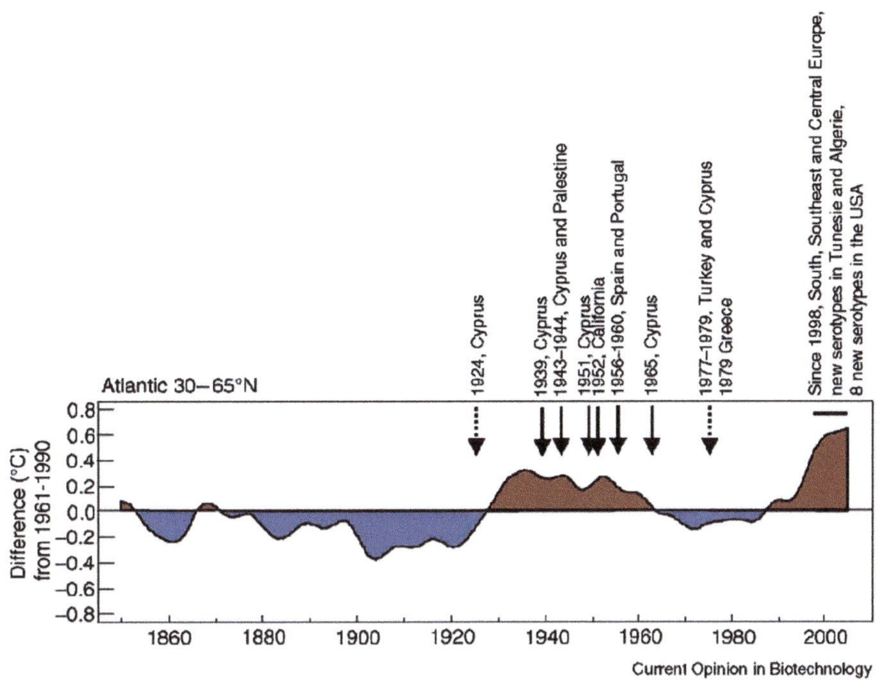

Figure 2 BT outbreaks in Europe and the USA versus the AMO index from 1961 to 1990 mean (from [63])

Europe under conditions of future climate change [80]. As Artz *et al.* [64] noted:

> The greatest threat from RVF is that animal movements and changes in virus–vector–host dynamics will facilitate extension of the disease's range into Europe and beyond, with calamitous veterinary and human public health consequences . . .Long neglected by the human and veterinary medical communities, the disease is now on the move through animal trade and in facilitated transmission mode owing to climatic changes.

Similarly to BTV above, it appears that notions regarding the disease being in 'facilitated transmission mode due to climatic change' is presumed rather than proven. As the pathogenesis of RVF differs from BT, potential spread may be more confined. Indeed, RVF is limited in its viraemic stage, host susceptibility and potential vectors to transmit the virus. Nevertheless, while authors offered opinions regarding the potential expansion of RVF under conditions of climate change, predictive models based on expected global temperature increases or with the expected variability in ENSO, were rare.

Thus, while literature on RVF amply illustrates the primary effects of climate on the disease (Table 6), little attention has been accorded to the evidence-base regarding RVF and climate change. While Clements *et al.* [81] created a 'disease atlas' for RVF based on the findings of published serological surveys. Revisiting such concepts via the lens of climate change may be beneficial to future developments in the field.

Across the articles reviewed, there was little cross-referencing (Table 7).

Tick-borne diseases

Across the articles under review, authors were in wide agreement about the importance of climate to tick populations [1,28,86–93]. As Estrada-Peña [88] noted:

> Both temperature and relative humidity (RH) are key factors governing the main aspects of the tick life-cycle. The 'growth periods' (i.e. oviposition or moulting) are temperature-mediated. Furthermore, tick mortality is the result of the complex interactions between temperature, RH, rainfall or evaporation.

Consequently, researchers within the field of TBDs have long been interested in modelling climatic effects versus tick populations. As Estrada-Peña [88] offered:

> The ability to predict the range of a given tick species is an old challenge in tick research . . . .

And indeed, much work has broadly focused on risk mapping as a means of identifying potential future 'hot spots' of tick populations and therefore, areas at risk for greater disease transmission [87,88,90,94–96]. Despite historical interest in the topic, the lack of data emerged as an issue for some authors. Eisen [28] argued that current empirical models were of 'variable data quality' and the lack of long-term empirical datasets were a crucial issue

Table 6 Articles and evidence base: RVF

Primary author	Year	Topic	Type	Evidence base
Linthicum *et al.* [82]	1991	RVF and climate variables	Predictive model	Remote sensing: NDVI, NOAA* and LANDSAT* data
Anyamba *et al.* [49]	2001	RVF and climate change	Review	Literature
Caminade *et al.* [83]	2006	RVF and malaria	Risk mapping	Primary data
Clements [84]	2006	RVF disease mapping	Spatial model	Secondary data
Martin *et al.* [41]	2008	RVF and climate change	Review of early warning systems	Remote sensing: NDVI and NOAA data
Gale *et al.* [80]	2009	Emerging viruses	Qualitative risk assessment	Expert opinion and literature
Gould and Higgs [54]	2009	Arboviruses and climate change	Review	Literature
Tourre *et al.* [85]	2009	RVF and environmental Variables	Risk mapping	Remote sensing: rainfall, hydrology, vegetation and cattle population
Artz *et al.* [64]	2010	Globalization and animal health	Review	Literature
Chevalier *et al.* [79]	2010	RVF and climate change	Review	Literature
Weaver and Reisen [47]	2010	General arbo-virus	Review	Literature

*National Oceanographic and Atmospheric Administration, US Department of Commerce.
*LANDSAT Satellites, NASA, USA.

Table 7 RVF: author cross-referencing

Primary authors and year	Martin *et al.* (2008) [41]	Gale *et al.* (2009) [80]	Gould and Higgs (2009) [54]	Tourre *et al.* (2009) [85]	Artz *et al.* (2010) [64]	Chevalier *et al.* (2010) [79]	Weaver and Reisen (2010) [47]
Linthicum *et al.* (1991) [82]							
Anyamba *et al.* (2001) [49]							
Clements *et al.* (2006) [84]							
Caminade *et al.* (2006) [83]		x					
Martin *et al.* (2008) [41]					x		
Gale *et al.* (2009) [80]							
Gould and Higgs (2009) [54]		x					
Tourre *et al.* (2009) [85]							
Artz *et al.* (2010) [64]							
Chevalier *et al.* (2010) [79]							
Weaver and Reisen (2010) [47]							

with regard to assessments of TBDs and climate change. Kivaria [97] also noted:

> We must now provide new and updated data, collected in different geographical and climatological areas . . .National surveys of ticks, hosts and TBDs should be revived and undertaken regularly so that changes are captured in scales.

However, across the articles, the relationship between TBDs and climate change was often contested. For example, in Europe, the increasing incidence and prevalence of human and animal TBDs has been the cause for concern (see [98–100]). However, some authors disputed causal links between climate change and the rise in TBDs [1,91,92]. Prospective hypotheses were offered as to the negative impact of temperature on the life cycle of ticks. For example, Slenning [1] noted within the context of warming temperatures, tick species such as *Ixodes ricinus* are likely to begin questing behaviour too early, i.e. before the birthing season of their mammalian hosts. Therefore, survival rates will decline in the face of climate change. Randolph [91] noted that while warming temperatures increase tick populations, declining moisture has the opposite effect (the drier the atmosphere, the more questing behaviour is interrupted, thereby decreasing the chance of acquiring a host). As such, warmer spring weather may not favour ticks, as this increases the probability of a tick using up energy stores too early.

Interestingly, Randolph [91] offered that the increase incidence of human TBDs in Central and Eastern Europe (CEE) were positively correlated to 'national mean figures on perceived poverty and the proportion of total expenditure spent on food'. Survey data in Latvia illustrated that low socio-economic standing and unemployment predisposed people to visit forests. Randolph concluded that climate change is not sufficient in and of itself to explain changes in the incidence of TBDs in Europe. In later work, Randolph [92], in an exploration of over 40 years of climate data in Europe, concluded that the links between climate change and TBDs were more 'coincidence than causal'.

Randolph's work raises an alternate point that few studies within the animal health literature have explored, i.e. the inter-relationship between socio-economic factors, livestock disease and climate change. While Heffernan [101] in a study in Kenya found that the severity of impact of TBDs varied depending on the socio-economic standing of the household involved, the study did not link such outcomes to climate change. Conversely, in Kenya, Gachohi et al. [102] explored farm level variables and climate to the incidence and prevalence of TBDs, but socio-economic factors were not included.

East Coast fever (ECF). ECF is a major tick-borne disease of cattle in Africa and infects over 1 million animals annually [87,103]. In naïve populations, mortality can approach 95% [87]. Caused by the protozoan parasite *Theileria parva*, and transmitted by the tick species *Rhipicephalus appendiculatus*, during 1989, it was estimated that ECF cost African farmers US$168 million [104–106]. Later estimates nearly double this figure to US$300 million per year [103]. Despite the importance of ECF to African agriculture, comparatively little attention has been accorded to the disease in the climate change and animal health literature.

Few authors explored direct links between ECF and GCC (see [65,105]). Fandamu et al. [107] found a strong association between ENSO and an elevated ECF disease hazard in southern Zambia. The authors followed herds over a 9-year period in two discreet geographic zones: a high rainfall, ECF endemic area and a lower rainfall ECF epidemic area. In the arid/semi-arid areas, during normal years the challenge levels for the disease were low, thereby decreasing the immunity of the susceptible population. Consequently, during El Niño years, higher rainfall supported an increased tick population, which increased exposure of young animals to ECF with ensuing epidemics.

Despite this overall lack of attention in terms of GCC, the impact of climate on ECF has been well studied. Indeed, it may be argued that ECF was the subject of some of the earliest work exploring and modelling the interaction between livestock disease and the environment. For example, Yeoman [108] elucidated the links between the epidemiological status of ECF versus agro-ecological zone. Branagan [109], in field trials in Kenya, found temperature was the most critical factor to the length of the life cycle of *R. appendiculatus*. King et al. [86] utilized this work in combination with 20 years of climate data to create a predictive model of tick life-cycle development.

Alternatively, during the 1990s, Norval et al. [87] explored the potential for the tick species to colonize the highlands of Ethiopia. Using vegetative and climate models, Norval et al. predicted that much of the Ethiopian highlands were eminently suitable for colonization by *R. appendiculatus* and this was a likely occurrence in the future. Reasons offered as to why the disease had not spread to Ethiopia were twofold. First, the highlands were 'naturally isolated' by the wide tsetse fly belt on the western and southern margins. And second, the wide band of desert and scrubland on the Kenyan and Somali borders also prevented incursion of the disease.

Later work by Olwoch et al. [106] also utilized a risk-mapping approach to explore changes to the range of the tick species under climate change. Under conditions of a 2-degree temperature rise, the authors predicted that the future range of *R. appendiculatus* would decline by 23% across sub-Saharan Africa [106]. However, on the national level, large variations were predicted, e.g. the range of *R. appendiculatus* would increase in southern Africa (South Africa, Botswana, Malawi and Namibia), while declining in a variety of other countries (Mozambique, Tanzania, Kenya, Uganda, DCR and Zambia) [106]. Ethiopia, however, was not included in the analysis, owing to the continued absence of the disease.

The two studies illustrate two key points in the climate and animal health debate. First, with regarding to ECF and the related tick vector, Norval et al. [87] illustrated the challenges in making correct climate change and disease predictions. Indeed, from the results of the study, it may be argued that climate and vegetative changes impacting the fortunes of the tsetse fly may be more important to the spread of *R. appendiculatus* that those impacting the tick species itself. Conversely, Olwoch et al. [106] illustrate how early work in the field is rarely built upon or further explored over time. Indeed, the lack of *R. appendiculatus* in the highlands of Ethiopia may be just as vital to elucidating the drivers to animal health and climate change as those TBDs with present changes in geographic distribution.

Authors exploring ECF and climate change based their work on a number of climate models, with a pre-dominant focus on tick ecology and predictions of changing distributions (Table 8). Overall, there was little cross-referencing across the recent literature (from 2008 onwards) (see Tables 9–11).

Parasitic diseases

Across the literature on parasitic diseases, a similar duality in 'frames' was noted, i.e. for some authors, impact of climate change on such diseases was largely viewed as a 'given' and therefore, the starting point for further analysis [11,14,26,115,120,121]. Others, viewed climate change as part of a wider aggregate of forces [122–126]. As Patz et al. [122] offered:

> Deforestation and ensuing changes in land use, human settlement, commercial development, construction of roads, water control systems (dams, canals, irrigation systems, reservoirs), and climate

Table 8 Articles and evidence base: East Coast fever

Primary author	Year	Topic	Type	Evidence base
King *et al.* [86]	1988	ECF and climate factors	Model	Primary data
Norval *et al.* [87]	1991	ECF spread, Horn of Africa	Predictive model	Climex climate model/NDVI East Africa
Randolph [110]	2000	Tick distribution	Review	Literature: remote sensing models
Randolph [111]	2004	Tick ecology	Review	Literature
Cumming and Van Vuuren[89]	2006	Ticks and climate change	Model	Image 2.2 model and tick database
Fandamu *et al.* [105]	2005	ECF and ENSO	Disease incidence	Primary and secondary data
Fandamu *et al.* [107]	2006	ECF and ENSO	Disease transmission	Primary and secondary data
Olwoch *et al.* [106]	2008	ECF and climate change	Predictive model	Darlam climate model and secondary livestock data
Gachohi *et al.* [102]	2010	Sero-prevalence and risk factors, TBDs in Kenya	Disease incidence	Primary data
Kivaria [97]	2010	ECF and climate change	Editorial	Literature

Table 9 Author cross-referencing: ECF

Primary authors and year	Olwoch *et al.* (2008) [106]	Gachohi *et al.* (2010) [102]	Kivaria (2010) [97]
King *et al.* (1988) [86]			
Norval *et al.* (1991) [87]			
Deem *et al.* (1993) [104]		x	
Randolph (2000) [110]	x		
Randolph (2004) [111]			
Cumming and Van Vuuren (2006) [89]	x		
Fandamu *et al.* (2005) [105]		x	
Fandamu *et al.* (2006) [107]			
Olwoch *et al.* (2008) [106]		x	x

singly and in combination have been accompanied by global increases in morbidity and mortality from a number of emergent parasitic diseases.

Some authors quite happily drew on both frames, thereby avoiding any contention [14,121]. For example, Genchi *et al.* [14], in a review of climate and *Dirofilaria* infection in Europe, opined the following:

> . . . the link between animal and human infections and the global environment is unquestionable . . . global warming tends to extend the risk season for infection and to maintain a high incidence level . . . other important drivers for the (re)-emergence and spread of vector-borne parasites include habitat changes, alterations in water storage and irrigation habits, pollution, development of insecticide and drug resistance, globalization and the increase in international trade, tourism and travel.

Conversely, Taraschewski [127] explored many ecological factors that supported host and therefore, parasite colonization. The author described four potential outcomes of the invasion of hosts/parasites into a new geographic area:

1. While colonizing novel areas, hosts may lose the parasite of their source area.
2. A host may establish in a colonized area together with a parasite, with or without a lag time of the latter.
3. Host and parasite co-invade but, due to the low host-specificity of the parasite, it successfully colonizes populations of novel hosts, native or introduced, in the recipient area.
4. Parasites colonize a target region without their hosts.[5]

Thus, all of the above potential outcomes are possible under conditions of climate change. Interestingly, in item four, the author argued that 'a new evolutionary factor', i.e. human transportation, was supporting the creation of new parasite ranges [127]. The author noted that 'the invasion success of many species depends more on their ability to respond to natural selection than on a broad physiological tolerance or plasticity . . .' Rather, 'high levels of additive genetic variance' was the main criteria of successfully invading species [127]. Across the articles, few authors explored the role of genetic variation among parasite species in terms of success and failure under conditions of climate change. Equally, host immunological response to parasite

[5]A key example offered by the author was the swimbladder nematode *A. crassus*, which has invaded Europe without its main host the Japanese eel (*A. japonica*). For example, transfers of eel stock around the UK by lorry, rapidly transferred the exotic nematodes around the country, with driver rest stops where the eels's water was changed, key infection points in the nematode epidemiology [127].

Table 10 Articles and evidence base: Fasciolosis

Primary author	Year	Topic	Type	Evidence base
Mallone *et al.* [19]	1998	Risk mapping *Fasciola* in East Africa	Risk model	Existing models and literature
Valero *et al.* [112]	1999	*Fasciola* morphology high versus low altitude	Morphological study	Primary data
Fuentes [113]	2001	Validation of risk maps	GIS model exploring NDVI usefulness	Primary data
Poulin [114]	2006	Impact of temperature on trematode cercarial emergence	Review of experimental data	Compilation experimental data
Mas-Coma *et al.* [115]	2008	Climate change and helminthiasis	Review	Literature
McCann *et al.* [116]	2010	Spatial distribution of *Fasciola*	Risk model	Primary data
van Dijk *et al.*[117]	2010	Climate change and helminthiasis	Review	Literature
Bennema *et al.* [75]	2011	*Fasciola* management versus environmental factors	Comparison of models	Primary data
Fairweather Fairweather [118]	2011	*Fasciola* control	Review	Literature
Fox *et al.* [119]	2011	Risk mapping for *Fasciola* in UK	Risk model	Primary data

Table 11 Article cross-referencing: fasciolosis

	Mas-Coma *et al.* (2008) [115]	Mas-Coma *et al.* (2009) [9]	McCann *et al.* (2010) [116]	van Dijk (2010) [117]	Bennema *et al.* (2011) [75]	Fairweather (2011) [118]	Fox *et al.* (2011) [119]
Mallone *et al.* (1998) [19]							
Valero *et al.* (1999) [112]							
Fuentes *et al.* (2001) [113]	x	x					
Poulin (2006) [114]	x	x		x			
Mas-Coma *et al.* (2008) [115]		x			x		
Mas-Coma *et al.* (2009) [9]							
McCann *et al.* (2010) [116]					x	x	
van Dijk *et al.* (2010) [117]					x	x	
Bennema, *et al.* (2011) [75]							
Fairweather (2011) [118]							
Fox *et al.* (2011) [119]						x	

burdens in the face of climate change was rarely mentioned [42].

Nevertheless, the direct impact of climate on the various life stages of a wide range of parasites has been well studied, over many decades [17,128]. More recent authors have explored the robustness of the exogenous life-cycle stages under temperature extremes [9,22,115,129]. For example, Gajadhar and Allen [129] in an exploration of the 'thermal death rates' of water-borne parasites found the exogenous stages can withstand extreme temperature fluctuations. The authors found the eggs of *Echinococcus granulosus* can survive in temperatures as low as −50°C and as high as 70°C [129]. The lack of humidity, however, can be problematic for parasites. For example, Mas-Coma *et al.* [115] relayed, the eggs of *Echinococcus multilocularis* can 'can survive for 478 days at 95% RH and 4°C, but for only 24 h at 27% RH and 25°C.'

Across the literature, fasciolosis was often mentioned as a disease whose changing distribution was directly linked to climate change [9,27,75,115,118,127].

Fasciolosis. Fasciolosis (also known as fascioliasis) is caused by the trematode species *Fasciola* (mainly *Fasciola hepatica* and *Fasciola gigantica)*, whose life-cycle includes certain species of snails as intermediate hosts (family Lymnaeidae). According to Bennema *et al.* [75], assessments of the changing distribution of *Fasciola* have been topic of exploration since the 1960s. Across the articles, there was general agreement that fascioliasis was on the rise globally [23,27,75,115,118].

Fasciola species have one of the 'widest latitudinal, longitudinal and altitudinal distribution known' [115]. Indeed, the parasite is endemic in areas from below sea level to the 4,000 m Altiplano in Bolivia [9,113,127]. However, Taraschewski [127] argued that parasites with 'low host specificity were better colonizers than those with more specific requirements' and that *Fasciola hepatica* fits this description. Nevertheless, it is often presumed that adaptation to such a wide geographic range (particularly in relation to the high altitude) and is an illustration of parasite/intermediate host responses to evolutionary pressure with consequent genetic divergence [127]. Interestingly, however, the genetic testing of both snails and *F. hepatica* on the Bolivian Altiplano revealed an absence of polymorphism among snail populations and low rates of morphological divergence among *Fasciola* populations [112,127,130]. While little work has been done in this area, the findings leave open the question of the role of end-host genetic and immuno-susceptibility to *Fasciola* to the increased incidence of fasciolosis in both human and livestock populations.

Many authors have developed spatial risk models for *Fasciola* species (see [116,118]). Such models link environmental and climate variables to the incidence and prevalence of the disease, the risk factors are then linked to geographic information systems (GIS) to predict future patterns of endemnicity [113,116]. For example, McCann *et al.* [116], in a recent exploration of the spatial epidemiology of *F. hepatica* in the UK, explored five variables: climate, environment, soil, livestock and pasture, which were linked to local geographies. However, other authors noted that climate and vegetative models were often incorrect. As Mas-Coma *et al.* [115] offered 'the most useful climate indices known for fascioliasis indicated surprisingly that is disease should not be present on the Altiplano [of Bolivia]'. Fuentes *et al.* [113] also noted in Bolivia that predictions, which estimated snail populations using the NDVI, were often inaccurate.

Bennema *et al.* [75] argued that management issues are a key driver of infections and are often not represented in climate models of the disease. Clearly poorly implemented management and control strategies could have influenced the spread of the disease on the Bolivian Altiplano (F. Lizon, personal communication). As Bennema *et al.* [75] offered:

> Integrating management in spatial distribution models poses additional difficulties as the collection and updating of the data is more time-consuming than for instance remote sensing of meteorological data and it causes a decrease of spatial resolution, as the management data are point measurements. On the other hand, we observed spatial patterns in the distribution of management factors . . . In the future, the use of environmental, land use or land cover data as proxy variables for management should be investigated.

Equally problematic, as Fairweather [118] noted, is an increase in liver fluke resistance to anthelmintics such as triclabendazole (TCBZ). Legal limits on TCBZ usage in some countries, is an additional consideration.

The risk of infection from fasciolosis, however, has been linked to the El Niño – the Southern Oscillation phenomenon [9]. Indeed, the increased rainfall associated with El Niño is expected to increase the time period of contamination for humans and animals in endemic areas. Further, El Niño-associated droughts and floods can increase the distribution and survival rates of limnaeids [9]. In the hyper-endemic disease areas on the Bolivian Altiplano, fasciolosis is linked to the ability of the snail species, *Galba trunculata*, to inhabit water bodies filled with run-off from the Eastern Andean mountain range [9]. As glacier melting in the Andes has been well-documented (see [131]), the impact is likely to change the epidemiology of the disease over both the short and longer term. In temperate areas, authors have also linked outbreaks to increases in rainfall [75,116].

Discussion

Undoubtedly, climate change and infectious livestock disease are an emerging field. Two-thirds of the articles under review were published from 2008 onwards. Despite the relative newness of the literature, the following core arguments divided the studies under review:

(a) Climate change is an important driver to the presumptive increase in both endemic and epidemic disease threats of livestock with equally compelling predicted increases in zoonotic diseases.
(b) The complexity of host–pathogen interactions prevents any clear-cut conclusions being made. Climate change is a largely unproven driver of infectious diseases and at most, one factor among a range of other factors.

However, across both arguments, authors tended to repeat or quote the findings of other descriptive reviews. Over time, statements tended to take on increasing surety. Noticeably absent were frameworks or approaches to resolve this polarity.

With regard to content, there was little cross-referencing within and between articles. As such, authors on arboviruses, tended to reference the literature on viral diseases and climate change, while those writing on TBDs cited the tick literature, etc. In this manner, knowledge 'silos' could be identified across the articles under review. While the literature on arboviruses and climate change was arguably the most alarmist, authors writing on TBDs tended to be the most contentious, while those writing on parasitic diseases appeared to be the most convinced regarding the impact of climate change. Interestingly, the literature on parasitic disease and climate change drew more frequently on approaches from population ecology [27,127,132] than that for the VBDs.

Across the corpus, authors rarely drew upon related studies in the medical literature. For example, while evidence of the link between ENSO and infectious disease [74,133], sparked off a wide body of research by the medical community on the topic, the veterinary community appeared slower to respond. While the relationship between ENSO and RVF were well explicated, across the review, only a few articles explored the linkage between other VBDs and parasitic diseases and global oscillation indices (see [9,49,63,105,107,134]). While the authors linked ENSO or the AMO index to a range of disease processes, there was little follow-up or building on the findings across subsequent work.

Other biases emerged in relation to topic, orientation and geographic location of the studies under review. First, relatively little attention has been accorded climate change and the potential emergence of new infectious diseases. Rather, the shifting geographic distribution of well-known VBDs and parasitic disease was the main focus of the articles. Unlike the wider climate change literature, little attention was accorded mitigation and adaptation strategies, particularly at the farm level. Second, few methods or approaches were offered to underpin the field as a whole. Historical explorations or systematic reviews were equally rare.

The more detailed exploration of articles on particular diseases, i.e. BT, RVF, ECF and fasciolosis, was revealing on a number of levels. First, across the diseases a dominance of thinking and/or approaches was evident. For example, the literature on BT and climate change was dominated by narrative reviews. While the level of cross-referencing was acceptable, the finding may simply be the result of the propensity of particular authors to co-author large numbers of works together. Conversely, the articles related to RVF tended to describe models based on remote sensing, with only one author collecting field-level data (see [83]). Despite the similarity of approaches, the level of cross-referencing among RVF authors was extremely low. Conversely, while the ECF literature ostensibly appeared more balanced, as noted above, key findings regarding ENSO and ECF were largely ignored. Finally, the fasciolosis literature had the strongest primary evidence base. Yet while authors noted the influence of socio-economic and management factors, most models were focused almost exclusively on environmental variables.

Across the articles, future predictions regarding animal infectious disease and climate change were most frequently biased towards the positive, i.e. focused on increases in disease distribution rather than decreases. The potential decline of only a small number of pathogens in the face of GCC warranted mention i.e. influenza viruses and *Yersinia pestis* (see [1,135]). Finally, on a geographic level, much of the work focused on the potential of tropical livestock diseases impacting populations of naïve animals in temperate countries, rather than any explicit impacts of such diseases in their native ranges.

Across the corpus, authors also suggested a wide and disparate variety of data gaps and researchable constraints. For example, Tabachnik [29], in relation to VBDs, noted 'the genetic systems controlling vector competence' and the impact of the environment on vector genotypes were critical gaps. Conversely, Mills *et al.* [136] identified six general areas important to furthering knowledge of VBZD ranging from pathogen identification to enhancing surveillance to the generation of empirical data. In relation to TBDs, Eisen [28] offered that 'comprehensive field data on spatial patterns of tick distribution' was lacking. For parasitic disease, Polly and Thomson [27] argued that the 'rigorous assessment' of data was needed to better explicate the 'ecosystem-new-climate-host-parasite matrix'.

However, additional research confined to the above knowledge silos is likely to simply amplify the 'tower of Babel' that presently characterizes the literature. The ultimate outcome of continuing on this 'silo-based' pathway is a globally disjointed policy response with ineffective and inefficient resource allocation for animal diseases. Thus, it is clear that a more integrated approach for exploring the issues surrounding climate change and animal health is urgently required.

Recently, two frameworks for assessing animal health/infectious disease constraints have been described [137,138]. For example, Wilkinson *et al.* [138] argued that understanding the drivers of infectious disease required a multi-disciplinary approach with the reframing of 'key issues' via the incorporation of natural and social sciences. However, how problems are 'framed' indicates a world view, i.e. a frame is defined as a cognitive structure that determines how we describe reality [139]. Therefore, reframing in this case appears to simply entail re-describing the issues related to infectious diseases from an apparently wider discursive framework. While a multi-disciplinarity perspective is valuable, it is unclear how such an approach will bridge the chasms related to a 'fragmented knowledge base'.

Conversely, in an alternate approach, Heffernan [137] described the Mosaic Model. The Mosaic Model identifies the unknowns surrounding a specific disease and coordinates the different types of knowledge produced by research to solve/resolve outstanding gaps [137]. As Heffernan [137] noted:

> . . . the Mosaic Model unites three broad pathways for change: research which produces new knowledge, that which produces information, and that which explores beliefs or tests assumptions. Within traditional approaches to research, projects are designed to produce only one of these outputs. The Mosaic Approach is unique in combining all three forms of knowledge.

Thus, a disease 'Mosaic' would evaluate both the form and content of existing knowledge, to identify areas for coordinated action. While the aim of the model was to support the eradication of particular animal health constraints, it may be useful in dissolving the knowledge silos built up around animal health and climate change. Further, the Mosaic Model is underpinned by the concept of collective intelligence (see [140]). The notion of collective intelligence has informed developments in the fields of computer science and business studies. Yet there are key elements that may help resolve current divisions in the field of animal health and climate change.

A first step towards this end is the development of a knowledge-based system (KBS) to identify synergies, inter-relationships and key gaps across the totality of literature on the topic. KBSs can underpin decision-making, particularly in relation to 'high-impact questions that involve uncertainty, risk, several possible perspectives and multiple competing objectives' [141]. A KBS on Climate Change and the Infectious Diseases of Livestock would be beneficial on two counts. First, it would enable the animal health community to identify priority areas with regard to research and funding allocation. Second, such an approach would enable the cross-fertilisation of ideas, methods and approaches under the single umbrella of animal health and climate change.

Conclusions

The review was based on a selection of articles, as such the corpus was not exhaustive and the authors make no claims to the contrary. Equally, all reviews of this nature suffer from some form of 'selection bias' i.e. decisions must be made over the criteria for the inclusion/exclusion of documents in addition to the sources of the literature itself. Despite such limitations, the corpus of documents was revealing on a number of levels.

First, the findings of the review bolstered the contention detailed in the introduction, questioning the state of knowledge regarding animal health and climate change. Indeed, the analysis revealed a largely descriptive literature with a focus on those diseases with well-known and well-studied climate components. Methodologically, the literature was weak and authors offered little in the way of new evidential frameworks or approaches to explore the topic. Given the likely pace of future change, there is an urgent need for frameworks and/or approaches that support collective learning and approaches within this emerging field.

Second, there is an urgent need to enhance our abilities in predictive decision-making. The review flagged up gaps in the following critical areas important to furthering this capacity:

(a) A greater understanding of the role of extra-climate factors on animal health, i.e. management issues and socio-economic factors and how these interplay with climate change impacts.
(b) A greater cross-fertilization across topics/disciplines and methods within and between the field of animal health and allied subjects.
(c) A stronger evidence-base via the increased collection of empirical data in order to inform both scenario planning and future predictions regarding animal health and climate change.
(d) The need for new and improved methods to both elucidate uncertainty and explicate the direct and indirect causal relationships between climate change and animal infections.

Clearly, the success of our efforts in this regard will largely be down to our ability to collect primary data on the above factors. In terms of the future success of the field, as a whole, the overall lack of primary data, in colloquial terms, may be described as the elephant in the room.

It is also important to remember that enhancing our understanding of, and better predicting changes to, the host–pathogen–ecosystem nexus within a context of GCC will only ever be one piece of a larger puzzle. Clearly the impacts of future disease threats can be greatly ameliorated by the adoption of appropriate and effective control strategies. Yet across the review, little attention was accorded either parameter: control measures or their adoption. As efforts at control will only be as good as their level of uptake, enhancing our understanding of farmer behaviour is likely to be vital to meeting the challenges posed by future disease threats. Therefore, strengthening the evidence base requires the generation of data on a wide range of topics from 'blue-sky' to applied.

Finally, advancing the field and confronting the future demands that climate change will place on livestock and human health services, demands collective action. Such collective action should be based on collaborative KBSs for solving research problems. Without an integrated approach, research on animal health and climate change will likely remain underutilized and consequently underfunded in comparison with other areas of climate research. Addressing this issue is likely to be one of the most critical challenges facing veterinarians, scientists, epidemiologists and others involved in the animal health sub-sector for decades to come.

Acknowledgements

We thank Elsevier and the OIE Scientific and Technical Review for providing permission to reprint figures 1 and 2. We are also grateful to the USAID/Colorado State Livestock-Climate Change CRSP for supporting early thinking/work in this area.

References

1. Slenning B. Global climate change and the implications for infectious disease. Veterinary Pathology 2010;47(1):28–33.
2. Soulé M. The social and public health implications of global warming and the onslaught of alien species. Journal of Wilderness Medicine 1992;3(2):118–27.
3. Epstein P. Climate change and emerging infectious diseases. Microbes and Infection 2001;3:747–54.
4. Delecluse P. The origin of climate changes. Revue Scientifique et Technique (International Office Epizootics) 2008;27:309–17.

5. Stone D. Predicted climate changes for the years to come and implications for disease impact studies. Revue Scientifique et Technique (International Office Epizootics) 2008;27:319–30.

6. Rosenthal J. Climate change and the geographic distribution of infectious diseases. Ecohealth 2009;6:489–95.

7. Patz J, Reisen W. Immunology, climate change and vector-borne diseases. Trends in Immunity, 2001;22(4):171–2.

8. IPCC. Climate Change 2007: Synthesis Report. In Pachauri R, Reisinger A, editors. Contribution of Working Groups I, II and III to the Fourth Assessment Report of the Intergovernmental Panel on Climate Change. IPCC, Geneva, Switzerland; 2007, p. 104.

9. Mas-Coma S, Valero MA, Bargues MD. Climate change effects on trematodiases, with emphasis on zoonotic fascioliasis and schistosomiasis. Veterinary Parasitology 2009;163:264–80.

10. Shuman E. Global climate change and infectious diseases. New England Journal of Medicine 2010;362(12):1061–3.

11. Brooks DR, Hoberg E. How will global climate change affect parasite–host assemblages? Trends in Parasitology 2007;23(12):571–4.

12. de La Rocque S, Rioux JA, Slingenbergh J. Climate change: effects on animal disease systems and implications for surveillance and control. Revue Scientifique et Technique (International Office Epizootics) 2008;(27):339–54.

13. Lovejoy T. Climate change and biodiversity. Revue Scientifique et Technique (International Office Epizootics) 2008;27:331–8.

14. Genchi C, Rinaldi L, Mortarino M, Genchi M, Cringoli G. Climate and *Dirofilaria* infection in Europe. Veterinary Parasitology 2009;163:286–92.

15. Lau C, Smythe LD, Craig SB, Weinstein P. Climate change, flooding, urbanisation and leptospirosis: fuelling the fire? Transactions of the Royal Society of Tropical Medicine and Hygiene 2010;104(10):631–8.

16. Epstein P. Climate change and public health: emerging infectious diseases. Encyclopedia of Energy 2004;1:381–92.

17. Marquardt W, Senger C, Seghetti L. The effect of physical and chemical agents on the oocysts of *Eimeria zuernii* (Protozoa, Coccidia). Journal of Protozoology 1960;7:186–9.

18. Acevedo P, Ruiz-Fons F, Estrada R, Marquez A, Miranda M, Gortazar C *et al.* A broad assessment of factors determining *Culicoides imicola* abundance: modelling the present and forecasting its future in climate change scenarios. PLoS ONE 2010;5(12):e14236.

19. Mallone J, Gommes R, Hansen J, Yilma J, Slingenberg J, Snijdes F *et al.* A geographic information system on the potential distribution and abundance of *Fasciola hepatica* and *F. gigantica* in east Africa based on Food and Agriculture Organization databases. Veterinary Parasitology 1998;78(2):87–101.

20. Harrus S, Baneth G. Drivers for the emergence and re-emergence of vector-borne protozoal and bacterial diseases. International Journal of Parasitology 2005;35(11–12):1309–18.

21. Rogers D, Randolph S. Climate change and vector-borne diseases. Advances in Parasitology 2006;62:345–81.

22. King B, Monis P. Critical processes affecting *Cryptosporidium* oocyst survival in the environment. Parasitology 2007;134:309–23.

23. van Dijk J, David GP, Baird G, Morgan E. Back to the future: developing hypotheses on the effects of climate change on ovine parasitic gastroenteritis from historical data. Veterinary Parasitology 2008;158(1–2):73–84.

24. Pinto J, Bonacic C, Hamilton-West C, Romero J, Lubroth J. Climate change and animal diseases in South America. Revue Scientifique et Technique (International Office Epizootics) 2008;27(2):599–613.

25. Vallat B. Climate change: impact on the epidemiology and control of animal diseases. Revue Scientifique et Technique (International Office Epizootics) 2008;27(2):297–8.

26. Mouritsen K, Tompkins D, Poulin R. Climate warming may cause a parasite-induced collapse in coastal amphipod populations. Oecologia 2005;146:476–83.

27. Polley L, Thompson R. Parasite zoonoses and climate change: molecular tools for tracking shifting boundaries. Trends in Parasitology 2009;25:285–91.

28. Eisen L. Climate change and tick-borne diseases: a research field in need of long-term empirical studies. International Journal of Medicine and Microbiology 2008;298:12–18.

29. Tabachnick W. Challenges in predicting climate and environmental effects on vector-borne disease episystems in a changing world. Journal of Experimental Biology 2010;213:946–54.

30. Reiter P. Climate change and mosquito-borne disease: knowing the horse before hitching the cart. Revue Scientifique et Technique (International Office Epizootics) 2008;27(2):383–98.

31. Haines A, Kovats R, Campbell-Lendrum D, Corvalan C. Climate change and human health: impacts, vulnerability and mitigation. Lancet 2006;367:2101–9.

32. Yang G-J, Utzinger J Lv S, Qian Y-J, Li S-Z, Wang Q, Bergquist R *et al.* The regional network for Asian Schistosomiasis and other helminth zoonoses (RNAS(+)) target diseases in face of climate change. Advances in Parasitology 2010;73:101–35.

33. Pickett J, Birkett M, Dewhirst S, Logan J, Omolo M, Torto B *et al.* Chemical ecology of animal and human pathogen vectors in a changing global climate. Journal of Chemical Ecology 2010;36(1):113–21.

34. Rock M, Buntain B, Hatfield J, Hallgrimsson B. Animal–human connections, 'one health' and the syndemic approach to prevention. Social Science and Medicine 2009;68(6):991–5.

35. Sutherst R. Implications of global change and climate variability for vector-borne diseases: generic approaches to impact assessments. International Journal of Parasitology 1998;28(6):935–45.

36. PRISMA. PRISMA 2009 Checklist. Preferred Reporting Items for Systematic Reviews and Meta-Analyses; 2009. Available from: URL: http://www.prisma-statement.org/2.1.2%20-%20PRISMA%202009%20Checklist.pdf

37. Punch K. Introduction to Social Research: Quantitative and Qualitative Approaches. Sage Publications, London, UK; 1999.

38. Beugnet F, Marie J. Emerging arthropod-borne diseases of companion animals in Europe. Veterinary Parasitology 2009;163:298–305.

39. Martens WM, Jetten TH, Rotmans J, Niessen L. Climate change and vector-borne diseases: a global modelling perspective. Global Environmental Changing 1995;5(3):195–209.

40. Purse B, Brown H, Harrup L, Mertens P, Rogers D. Invasion of bluetongue and other orbivirus infections into Europe: the role of biological and climatic processes. Revue Scientifique et Technique (International Office Epizootics) 2008;27:427–42.

41. Zhang Y, Bi P, Hiller J. Climate change and the transmission of vector-borne diseases: a review. Asian Pacific Journal of Public Health 2008;20:64–76.

42. Morgan ER, Wall R. Climate change and parasitic disease: farmer mitigation? Trends in Parasitology 2009;25(7):308–13.

43. Gummow B. Challenges posed by new and re-emerging infectious diseases in livestock production, wildlife and humans. Livestock Science 2010;130(1–3):41–6.

44. Molyneux D. Common themes in changing vector-borne disease scenarios. Transactions of the Royal Society of Tropical Medicine and Hygiene 2003;97(2):129–32.

45. Colwell DD, Dantas-Torres F, Otranto D. Vector-borne parasitic zoonoses: emerging scenarios and new perspectives. Veterinary Parasitology 2011;182(1):14–21.

46. Azhahianambi P, Ghosh S. Genetic transformation of arthropod vectors for control of vector borne diseases. Indian Journal of Biotechnology 2007;(6):305–14.

47. Weaver S, Reisen W. Present and future arboviral threats. Antiviral Research 2010;85:328–45.

48. Ramasamy R, Surendran S. Possible impact of rising sea levels on vector-borne infectious diseases. BMC Infectious Diseases 2011;11:18.

49. Anyamba A, Linthicum K, Tucker C. Climate-disease connections: Rift Valley Fever in Kenya. Cadernos de Saúde Pública 2001;17:133–40.

50. Wittmann EJ, Baylis M. Climate change: effects on *Culicoides* transmitted viruses and implications for the UK. Veterinary Journal 2000;160(2):107–17.

51. Forman S, Hungerford N, Yamakawa M, Yanase T, Tsai HJ, Joo YS *et al.* Climate change impacts and risks for animal health in Asia. Revue Scientifique et Technique (International Office Epizootics) 2008;27(2):581–97.

52. Gould E, Higgs S, Buckley A, Gritsun T. Potential arbovirus emergence and implications for the United Kingdom. Emerging Infectious Diseases 2006;12:549–55.

53. Mellor P, Carpenter S, Harrup L, Baylis M, Mertens P. Bluetongue in Europe and the Mediterranean Basin: history of occurrence prior to 2006. Preventive Veterinary Medicine 2008;87:4–20.

54. Gould E, Higgs S. Impact of climate change and other factors on emerging arbovirus diseases. Transactions of the Royal Society of Tropical Medicine and Hygiene 2009;103:109–21.

55. Ahmed J, Bouloy M, Ergonul O, Fooks A, Paweska J, Chevalier V *et al.* International network for capacity building for the control of emerging viral vector-borne zoonotic diseases: arbo-zoonet. Euro Surveillance 2009;12(14):1–4.

56. MacLachlan N. Global implications of the recent emergence of bluetongue virus in Europe. Veterinary Clinics of North America: Food Animal Practice 2010;26(1):163–71.

57. Maclachlan N, Guthrie A. Re-emergence of bluetongue, African horse sickness and other Orbivirus diseases. Veterinary Research 2010;41(35):2–8.

58. Maclachlan N. Bluetongue: history, global epidemiology, and pathogenesis. Preventive Veterinary Medicine 2011;102(2):107–111. doi:10.1016/j.prevetmed.2011.04.005

59. Mellor P Baylis M, Mertens P Bluetongue. Elsevier, London; 2009.

60. Sellers R. Bluetongue in Africa, the Mediterranean region and Near East: disease, virus and vectors. Preventive Veterinary Medicine 1984;2:371–8.

61. Sellers R. Weather, host and vector: their interplay in the spread of insect-borne animal virus diseases. Journal of Hygiene 1980;85(1):65–102.

62. Vellema P. Bluetongue in sheep: question marks on bluetongue virus serotype 8 in Europe. Small Ruminant Research 2008;76(1–2):141–8.

63. Zell R, Krumbholz A, Wutzler P. Impact of global warming on viral diseases: what is the evidence? Current Opinion in Biotechnology 2008;19(6):652–60.

64. Artz J, White R, Thomsen B, Brown C. Agricultural diseases on the move early in the third millennium. Veterinary Pathology 2010;47:15.

65. Ducheyne E, Lange M, Van der Stede Y, Meroc E, Durand B, Hendricks G. A stochastic predictive model for the spread of bluetongue. Preventive Veterinary Medicine 2011;99(1):48–59.

66. EFSA. Scientific opinion of the scientific panel on animal health and welfare on the EFSA Selfmandate on bluetongue origin and occurrence. EFSA Journal 2007;480:1–20.

67. Purse B, Mellor P, Rogers D, Samuel A, Mertens P, Baylis M. Climate change and the recent emergence of bluetongue in Europe. Nature Reviews Microbiology 2005;3:171–81.

68. Wilson A, Darpel K, Mellor P. Where does bluetongue virus sleep in the winter? PLoS Biology 2008;6(8):e210.

69. Wilson A, Mellor P. Bluetongue in Europe: vectors, epidemiology and climate change. Parasitology Research 2008;103:s69–77.

70. Baylis M, O'Connell L, Purse B. Modeling the distribution of bluetongue vectors. Veterinary Italy 2004;40:176–81.

71. Ward M, Carpenter T. Simulation modeling of the effect of climatic factors on bluetongue virus infection in Australian cattle herds: I model formulation, verification and validation. Preventive Veterinary Medicine 1996;27(1–2): 1–12.

72. Purse B, Tatem A, Caracappa S, Rogers D, Mellor P, Baylis M, Torina A. Modelling the distributions of Culicoides bluetongue virus vectors in Sicily in relation to satellite-derived climate variables. Medical and Veterinary Entomology 2004;18(2): 90–101.

73. Gibbs E, Tabachnick W, Holt T, Stallknecht D. US concerns over bluetongue. Science 2008;320:872.

74. Yang X, Scherm H. El Nino and infectious disease. Science 2000;275(5301):739.

75. Bennema S, Ducheyne E, Vercruysse J, Claerebout E, Hendrickx G, Charlier J. The use of bulk-tank milk ELISAs to

assess the spatial distribution of *Fasciola hepatica*, *Ostertagia ostertagi* and *Dictyocaulus viviparus* in dairy cattle in Flanders (Belgium). Veterinary Parasitology 2011;41(2):225–33.

76. Maclachlan N, Drew C, Darpel K, Worwa G. The pathology and pathogenesis of bluetongue. Journal of Comparative Pathology 2009;141:1–16.

77. Martin V, Chevalier V, Ceccato P, Anyamba A, De Simone L, Lubroth J, de La Rocque S, Domenech J. The impact of climate change on the epidemiology and control of Rift Valley fever. Revue Scientifique et Technique (International Office Epizootics) 2008;27(2):413–26.

78. Senior K. Could Rift Valley fever reach Europe? Lancet 2009;9(2):85.

79. Chevalier V, Pépin M, Plée L, Lancelot R. Rift Valley fever – a threat for Europe? Euro Surveillance 2010;15(10):pii=19506. Available from: URL: http://www.eurosurveillance.org/ViewArticle.aspx?ArticleId=19506

80. Gale P, Drew T, Phipps LP, David G, Wooldridge M. The effect of climate change on the occurrence and prevalence of livestock diseases in Great Britain: a review. Journal of Applied Microbiology 2009;106(5):1409–23.

81. Clements A, Pfeifer D, Martin V, Otte J. A rift valley fever atlas for Africa. Preventive Veterinary Medicine 2007;82:72–82.

82. Linthicum K, Bailey C, Tucker D, Angleberger T, Cannon T, Logan T *et al.* Towards real-time predictions of Rift Valley fever epidemics in Africa. Preventive Veterinary Medicine 1991;11:325–334.

83. Caminade C, Ndione JA, Kebe CMF, Jones AE, Danuor S, Tay S, *et al.* Mapping Rift Valley fever and malaria risk over West Africa using climatic indicators. Atmospheric Science Letters 2011;12(1): 96–103.

84. Clements AC, Pfeiffer DU, Martin V. Application of knowledge-driven spatial modelling approaches and uncertainty management to a study of Rift Valley fever in Africa. International Journal of Health Geography 2006;5:57.

85. Tourre Y, Lacaux, J-P, Vignolles C, Lafaye M. Climate impacts on environmental risks evaluated from space: a conceptual approach to the case of Rift Valley Fever in Senegal. Global Health Action 2009. doi:10.3402/gha.v2i0.2053.

86. King D, Gettinby G, Newson R. A climate-based model for the development of the ixodid tick, *Rhipicephalus appendiculatus* in East Coast Fever zones. Veterinary Parasitology 1988;29:41–51.

87. Norval R, Perry B, Gebreab F, Lessard P. East coast fever: a problem of the future for the horn of Africa. Preventive Veterinary Medicine 1991;10(3):163–72.

88. Estrada-Peña A. Forecasting habitat suitability for ticks and prevention of tick-borne diseases. Veterinary Parsitology 2001;98:111–32.

89. Cumming G, Van Vuuren D. Will climate change affect ectoparasite species ranges? Global Ecology and Biogeography 2006;15:486–97.

90. Ogden N, Maarouf A, Barker I, Bigras-Poulin M, Lindsay L, Morshed M *et al.* Climate change and the potential for range expansion of the Lyme disease vector *Ixodes scapularis* in Canada. International Journal of Parasitology 2006;36(1):63–70.

91. Randolph S. Dynamics of tick-borne disease system: minor role of recent climate change. Revue Scientifique et Technique (International Office Epizootics) 2008;(27):367–81.

92. Randolph S. To what extent has climate change contributed to the recent epidemiology of tick-borne diseases? Veterinary Parasitology 2009;167(2–4):92–4.

93. Gale P, Adkin A, Drew T, Wooldridge M. Predicting the impact of climate change on livestock disease in Great Britain. Veterinary Records 2008;162(7):214–15.

94. Perry B, Lessard P, Norval RAI, Kundert K, Kruska R. Climate, vegetation, and the distribution of *Rhipicephalus appendiculatus* in Africa. Parasitology Today 1990;6:100–94.

95. Estrada-Peña A, Horak I, Petney T. Climate changes and suitability for the ticks *Amblyomma hebraeum* and *Amblyomma variegatum* (Ixodidae) in Zimbabwe (1974–1999). Veterinary Parasitology 2008;151(2–4):256–67.

96. Jaenson T, Lindgren E. The range of *Ixodes ricinus* and the risk of contracting Lyme borreliosis will increase northwards when the vegetation period becomes longer. Ticks and Tick-borne Disease 2011;2:44–9.

97. Kivaria F. Climate change and the epidemiology of tick-borne diseases of cattle in Africa. Veterinary Journal 2010;184(1):7–8.

98. Brinkley C, Nolskog P, Golovjova I, Lundkvist A, Bergstrom T. Tick-borne encephalitis virus natural foci emerge in western Sweden. International Journal of Medicine and Microbiology 2008;298(1):73–80.

99. Walder G, Falkensammer B, Heinz F, Holzmann H, Dierich M, Wurzner R. Tick-borne encephalitis in the Tyrol (Austria): changes in incidence and endemicity 2000–2006. International Journal of Medicine and Microbiology 2008;298(1):88–93.

100. Gilbert L. Altitudinal patterns of tick and host abundance: a potential role for climate change in regulating tick-borne diseases? Oecologia 2009;162(1):217–25.

101. Heffernan C. Panzootics and the poor: devising a global livestock disease prioritisation framework. Revue Scientifique et Technique (International Office Epizootics) 2010;28(3):897–907.

102. Gachohi J, Ngumi P, Kitala P, Skilton R. Estimating seroprevalence and variation to four tick-borne infections and determination of associated risk factors in cattle under traditional mixed farming system in Mbeere District, Kenya. Preventive Veterinary Medicine 2010;95:208–223.

103. Patel E, Lubembe D, Gachanja J, Mwaura S, Spooner P, Toye P. Molecular characterization of live *Theileria parva* sporozoite vaccine stabilates reveals extensive genotypic diversity. Veterinary Parasitology 2011;179(1–3):62–8.

104. Deem SL, Perry BD, Katende JM, McDermott JJ, Mahan SM, Maloo SH *et al.* Variations in prevalence rates of tick-borne diseases in Zebu cattle by agroecological zone: implications for East Coast fever immunization. Preventive Veterinary Medicine 1993;16:171–87.

105. Fandamu P, Duchateau L, Speybroeck N, Marcotty T, Mbao V, Mtambo J *et al.* *Theileria parva* seroprevalence in

traditionally kept cattle in southern Zambia and El Niño. International Journal of Parasitology 2005;1(35):391–6.

106. Olwoch J, Reyers B, Engelbrecht F, Erasmus B. Climate change and the tick-borne disease, theileriosis (East Coast fever) in sub-Saharan Africa. Journal of Arid Environment 2008;72(2):108–20.

107. Fandamu P, Duchateau L, Speybroeck N, Mulumba M, Berkvens D. East Coast fever and multiple El Niño southern oscillation ranks. Veterinary Parasitology 2006;135:147–52.

108. Yeoman G. Field vector studies of epizootic East Coast fever. A quantitative relationship between *R. appendiculatus* and the epizooticity of East Coast fever. Bulletin of Epizootic Diseases of Africa 1966;14:5–27.

109. Branagan D. Observations on the development and survival of the Ixodid tick *Rhipicephalus appendiculatus* Neumann, 1901 under quasi-natural conditions in Kenya. Tropical Animal Health and Production 1973;5:153–65.

110. Randolph S. Ticks and tick borne disease systems in space and from space. Advances in Parsitology 2000;47:217–47.

111. Randolph S. Tick ecology: processes and patterns behind the epidemiological risk posed by ixodid ticks as vectors. Parasitology 2004;129(suppl. 1):s37–65.

112. Valero M, Marcos M, Comes A, Sendra M, Mas-Coma S. Comparison of adult liver flukes from highland and lowland populations of Bolivian and Spanish sheep. Journal of Helminthology 1999;73:341–5.

113. Fuentes M, Malone J, Mas-Coma S. Validation of a mapping and prediction model for human fasciolosis transmission in Andean very high altitude endemic areas using remote sensing data. Acta Tropica 2001;79(1):87–95.

114. Poulin R. Global warming and temperature-mediated increases in cercarial emergence in trematode parasites. Parasitology 2006;132(1):143–51.

115. Mas-Coma S, Valero MA, Bargues M. Effects of climate change on animal and zoonotic helminthiases. Revue Scientifique et Technique (International Office Epizootics) 2008;27:443–57.

116. McCann C, Baylis M, Williams D. The development of linear regression models using environmental variables to explain the spatial distribution of *Fasciola hepatica* infection in dairy herds in England and Wales. International Journal of Parasitology 2010;40(9):1021–8.

117. van Dijk J, Sargison ND, Kenyon F, Skuce PJ. Climate change and infectious disease: helminthological challenges to farmed ruminants in temperate regions. Animal 2010;4(03):377–92.

118. Fairweather I. Reducing the future threat from (liver) fluke: realistic prospect or quixotic fantasy? Veterinary Parasitology 2011;180:133–43.

119. Fox N, White P, McClean C, Marion G, Evans A, Hutchings M. Predicting impacts of climate change on *Fasciola hepatica* risk. PLoS ONE 2011;6(1):e16126.

120. Chomel B. Control and prevention of emerging parasitic zoonoses. International Journal of Parasitology 2008;38(11):1211–7.

121. Kutz S. The Arctic as a model for anticipating, preventing, and mitigating climate change impacts on host–parasite interactions. Veterinary Parasitology 2009;163(3):217–28.

122. Patz JA, Graczykb TK, Gellera N, Vittor A. Effects of environmental change on emerging parasitic diseases. International Journal of Parasitology 2000;30:1395–405.

123. Sutherst R. The vulnerability of animal and human health to parasites under global change. International Journal of Parasitology 2001;31:933–48.

124. Petney T. Environmental, cultural and social changes and their influence on parasite infections. International Journal of Parasitology 2001;31(9):919–32.

125. Patz J, Olson S, Uejio C, Gibbs H. Disease emergence from global climate and land use change. Medical Clinics of North America 2008;92(6):1473–1491.

126. Petney T, Taraschewski H. Waterborne parasitic diseases: hydrology, regional development and control. In: Wilderer P editor. Treatise on Water Science, vol. 3; 2011 pp 303–366, Elsevier, Oxford.

127. Taraschewski H. Hosts and parasites as aliens. Journal of Helminthology 2006;80:99–128.

128. Farr M, Wehr E. Survival of *Eimeria acervulina E. tenella* and *E. maxima* oocysts on soil and under various conditions. Annals of the New York Academy of Sciences 1949;52:468–72.

129. Gajadhar A, Allen J. Factors contributing to the public health and economic importance of waterborne zoonotic parasites. Veterinary Parasitology 2004;126(1–2):3–14.

130. Jabbour-Zahab R, Pointier J, Jourdane J, Jarne P, Oviedo J, Bargues M *et al.* Phylogeography and genetic divergence of some lymnaeid snails, intermediate hosts of human and animal fascioliasis with special reference to lymnaeids from the Bolivian Altiplano. Acta Tropica 1997;64:191–203.

131. Rosenzweig C, Casassa G, Karoly DJ, Imeson A, Liu C, Menzel A *et al.* Assessment of observed changes and responses in natural and managed systems. In: Parry ML, Canziani OF, Palutikof JP, van der Linden PJ, Hanson CE editors, Climate Change 2007: Impacts, Adaptation and Vulnerability. Cambridge University Press, Cambridge, UK; 2007, p. 79–131.

132. Timi J. Parasites as biological tags for stock discrimination in marine fish from South American Atlantic waters. Journal of Helminthology 2007;81:107–11.

133. Nicholls N. El Niño Southern oscillation and vector-borne diseases. Lancet 1993;342:1284–5.

134. Mouritsen K, Poulin R. Parasitism, climate oscillations and the structure of natural communities. Okois 2002;97(3):462–8.

135. Vandegrift K, Sokolow S, Daszak P, Kilpatrick A. Ecology of avian influenza viruses in a changing world. Annals of the New York Academy of Sciences 2011;1195:113–28.

136. Mills J, Gage K, Khan A. Potential influence of climate change on vector-borne and zoonotic diseases: a review and proposed research plan. Environmental Health Perspective 2010;18(11):1507–14.

137. Heffernan C. Building synergies in development research: is it time for the Mosaic approach? Development in Practice 2008;18(6):686–700.

138. Wilkinson K, Grant W, Green L, Hunter S, Jeger M, Lowe P *et al.* Infectious diseases of animals and plants: an interdisciplinary approach. Philosophical Transactions of the Royal Society 2011;366:1933–42.

139. Misturelli F. Never the twain shall meet: a comparative analysis of the discourses of development professionals and the narratives of poor livestock keepers in Kenya [PhD thesis]. Livestock Development Group, University of Reading, Reading, UK; 2007.

140. Levy P. From social computing to reflexive collective intelligence: the IEML research program. Information Sciences 2008;180:71–94.

141. ITA. What is Decision Science? Institute of Technology Assessment, University of Harvard, Cambridge, MA, USA; 2011. Available from: URL: http://www.mgh-ita.org/index.php/Frequently-Asked-Questions/decision-science.html

Appendix I: The Corpus of Documents by Topic

General Climate Change and Animal Health (27)

1. Artz J, White R, Thomsen B, Brown C. Agricultural diseases on the move early in the third millennium. Veterinary Pathology 2010;47:15.
2. Black PF, Murray JG, Nunn MJ. Managing animal disease risk in Australia: the impact of climate change. Revue Scientifique et Technique (International Office Epizootics) 2008;27(2):563–80.
3. Conraths FJ, Mettenleiter T C Globalisation and change of climate: growing risk for livestock epidemics in Germany. Zuchtungskunde 2011;83(1):21–6.
4. De La Rocque S, Rioux JA, Slingenbergh J. Climate change: effects on animal disease systems and implications for surveillance and control. Revue Scientifique et Technique (International Office Epizootics) 2008;27(2):339–54.
5. Delecluse P. The origin of climate changes. Revue Scientifique et Technique (International Office Epizootics) 2008;27(2):309–17.
6. Dufour B, Moutou F, Hattenberger AM, Rodhain F. Global change: impact, management, risk approach and health measures – the case of Europe. Revue Scientifique et Technique (International Office Epizootics) 2008;27(2):529–50.
7. Ebi KL, Burton I. Identifying practical adaptation options: an approach to address climate change-related health risks. Environmental Science and Pollution 2008;11(4):359–69.
8. Forman S, Hungerford N, Yamakawa M, Yanase T, Tsai H J, Joo Y S *et al.* Climate change impacts and risks for animal health in Asia. Revue Scientifique et Technique (International Office Epizootics) 2008;27(2):581–97.
9. Gale P, Adkin A, Drew T, Wooldridge M Predicting the impact of climate change on livestock disease in Great Britain. Veterinary Records 2008;162(7):214–5.
10. Gale P, Drew T, Phipps LP, David G, Wooldridge M. The effect of climate change on the occurrence and prevalence of livestock diseases in Great Britain: a review. Journal of Applied Microbiology 2009;106(5):1409–23.
11. Haines A, Kovats R, Campbell-Lendrum D, Corvalan C. Climate change and human health: impacts, vulnerability and mitigation. Lancet 2006;367:2101–9.
12. Hoberg EP, Polley L, Jenkins EJ, Kutz SJ. Pathogens of domestic and free-ranging ungulates: global climate change in temperate to boreal latitudes across North America. Revue Scientifique et Technique (International Office Epizootics) 2008;27(2):511–28.
13. Horeth-Bontgen D, Bohle W. "Exotic Animal Diseases" – what we should be prepared for. Prakt Tierarzt 2010;91:506.
14. Lovejoy T. Climate change and biodiversity. Revue Scientifique et Technique (International Office Epizootics), 2008;27:331–8.
15. Lips KR, Diffendorfer J, Mendelson JR, Sears MW. Riding the wave: reconciling the roles of disease and climate change in amphibian declines. PLoS Biology 2008;6(3):e72.
16. Martin V, De Simone L, Lubroth J, Ceccato P, Chevalier V. Perspectives on using remotely-sensed imagery in predictive veterinary epidemiology and global early warning systems. Geospatial Health 2007;2(1):3–14.
17. McIntyre KM, Setzkorn C, Baylis M, Waret-Szkuta A, Caminade C, Morse A P *et al.* Impact of climate change on human and animal health. Veterinary Records 2010;167(15):586.
18. Morand S, Guegan JF. How the biodiversity sciences may aid biological tools and ecological engineering to assess the impact of climatic changes. Revue Scientifique et Technique (International Office Epizootics) 2008;27(2):355–66.
19. Patz J, Olson S, Uejio C, Gibbs H. Disease emergence from global climate and land use change. Medical Clinics of North America 2008;92(6):1473–91.
20. Pinto J, Bonacic C, Hamilton-West C, Romero J, Lubroth J. Climate change and animal diseases in South America. Revue Scientifique et Technique (International Office Epizootics) 2008;27(2):599–613.
21. Singh SP, Aditya M. Impact of climatic variation on animal health and productivity: a review. Indian Dairyman 2007;59:47–54.
22. St. Louis ME, Hess JJ. Climate Change: Impacts on and Implications for Global Health. American Journal of Preventive Medicine 2008;35(5):527–38.
23. Stone D. Predicted climate changes for the years to come and implications for disease impact studies. Revue Scientifique et Technique (International Office Epizootics) 2008;27(2):319–30.
24. Vallat B. Climate change: impact on the epidemiology and control of animal diseases. Revue Scientifique et Technique (International Office Epizootics) 2008;27(2):297–8.
25. Van den Bossche P, Coetzer JA. Climate change and animal health in Africa. Revue Scientifique et Technique (International Office Epizootics) 2008;27(2):551–62.
26. Wieliczko A, Staroniewicz Z. Impact of global warming on the spread of infectious diseases. Mediterranean Weather 2010;66:363–5.
27. Zinsstag J. Animal Health Research. Science 2007;315(5816):1193.

VBDs (18)

1. Caminade C, Ndione JA, Kebe CMF, Jones AE, Danuor S, Tay S *et al.* Mapping Rift Valley fever and malaria risk over West Africa using climatic indicators. Atmospheric Science Letters 2011;12(1):96–103.
2. Colwell DD, Dantas-Torres F, Otranto D. Vector-borne parasitic zoonoses: emerging scenarios and new perspectives. Veterinary Parasitology 2011;182(1):14–21.

3. Curto de Casas SI, Carcavallo RU. Climate change and vector-borne diseases distribution. Social Science and Medicine 1995;40(11):1437–40.

4. Gubler DJ. Vector-borne diseases. Revue Scientifique et Technique (International Office Epizootics) 2009;28(2):583–8.

5. Harrus S, Baneth G. Drivers for the emergence and re-emergence of vector-borne protozoal and bacterial diseases. International Journal of Parasitology 2005;35(11–12):1309–18.

6. Martens WM, Jetten TH, Rotmans J, Niessen L. Climate change and vector-borne diseases: a global modelling perspective. Global Environmental Change 1995;5(3):195–209.

7. Mills J, Gage K, Khan A. Potential influence of climate change on vector-borne and zoonotic diseases: a review and proposed research plan. Environmental Health Perspectives 2010;18(11):1507–14.

8. Molyneux D. Common themes in changing vector-borne disease scenarios. Transactions of the Royal Society of Tropical Medicine and Hygiene 2003;97(2):129–132.

9. Nakazawa Y, Williams R, Peterson AT, Mead P, Staples E, Gage KL. Climate change effects on plague and tularemia in the United States. Vector Borne Zoonotic Disease 2007;7(4):529–40.

10. Pickett J, Birkett M, Dewhirst S, Logan J, Omolo M, Torto B *et al.* Chemical ecology of animal and human pathogen vectors in a changing global climate. *Journal of Chemical Ecology* 2010;36(1):113–21.

11. Ramasamy R, Surendran S Possible impact of rising sea levels on vector-borne infectious diseases. BMC Infectious Disease 2011;11:18.

12. Reiter P. Climate change and mosquito-borne disease: knowing the horse before hitching the cart. Revue Scientifique et Technique (International Office Epizootics) 2008;27(2):383–98.

13. Rogers D, Randolph S. Climate change and vector-borne diseases. Advances in Parasitology 2006;62:345–81.

14. Sellers R. Weather, host and vector: their interplay in the spread of insect-borne animal virus diseases. Journal of Hygiene 1980;85(1):65–102.

15. Sutherst R. Implications of global change and climate variability for vector-borne diseases: generic approaches to impact assessments. International Journal of Parasitology 1998;28(6):935–45.

16. Sutherst RW, Ingram JI, Scherm H. Global change and vector-borne diseases. Parasitology Today 1998;14(8):297–9.

17. Tabachnick W. Challenges in predicting climate and environmental effects on vector-borne disease episystems in a changing world. Journal of Experimental Biology 2010;213:946–54.

18. Zhang Y, Bi P, Hiller J. Climate change and the transmission of vector-borne diseases: a review. *Asian Pacific Journal of Public Health* 2008;20:64–76.

Leishmaniasis

1. Ready PD Leishmaniasis emergence and climate change. Revue Scientifique et Technique (International Office Epizootics) 2008;27(2):399–412.

Viral Diseases (36)

1. Acevedo P, Ruiz-Fons F, Estrada R, Marquez A, Miranda M, Gortazar C *et al.* A broad assessment of factors determining *Culicoides imicola* abundance: modelling the present and forecasting its future in climate change scenarios. PLoS ONE 2010;5(12):e14236.

2. Ahmed J, Bouloy M, Ergonul O, Fooks A, Paweska J, Chevalier V *et al.* International network for capacity building for the control of emerging viral vector-borne zoonotic diseases: arbozoonet. Euro Surveillance 2009;12(14):1–4.

3. Anyamba A, Linthicum K, Tucker C. Climate-disease connections: Rift Valley Fever in Kenya. Cad. Saúde Pública 2001;17:133–40.

4. Babiuk S, Bowden TR, Boyle DB, Wallace DB, Kitching RP. Capripoxviruses: an emerging worldwide threat to sheep, goats and cattle. Transbound Emerging Disease 2008;55(7):263–72.

5. Baylis M, O'Connell L, Purse B. Modeling the distribution of bluetongue vectors. Veterinary Italy 2004;40:176–81.

6. Chevalier V, Pépin M, Plée L, Lancelot R. Rift Valley fever – a threat for Europe? Euro Surveillance 2010;15(10):pii=19506. Available from: URL: http://www.eurosurveillance.org/ViewArticle.aspx?ArticleId = 19506

7. Clements AC, Pfeiffer DU, Martin V. Application of knowledge-driven spatial modelling approaches and uncertainty management to a study of Rift Valley fever in Africa. International Journal of Health Geography 2006;5:57.

8. Cook RA. Emerging diseases at the interface of people, domestic animals and wildlife. The role of wildlife in our understanding of highly pathogenic avian influenza. Yale Journal of Biology and Medicine 2005;78(5):343–53.

9. Ducheyne E, Lange M, Van der Stede Y, Meroc E, Durand B, Hendricks G. A stochastic predictive model for the spread of bluetongue. Preventive Veterinary Medicine 2011;99(1):48–59.

10. EFSA Scientific opinion of the scientific panel on animal health and welfare on the EFSA Selfmandate on bluetongue origin and occurrence. EFSA Journal 2007;480:1–20.

11. Gale P, Estrada-Pena A, Martinez M, Ulrich RG, Wilson A, Capelli G *et al.* The feasibility of developing a risk assessment for the impact of climate change on the emergence of Crimean-Congo haemorrhagic fever in livestock in Europe: a review. Journal of Applied Microbiology 2010;108(6):1859–70.

12. Gale P, Brouwer A, Ramnial V, Kelly L, Kosmider R, Fooks A *et al.* Assessing the impact of climate change on vector-borne viruses in the EU through the elicitation of expert opinion. Epidemiology and Infection 2009;138(2):214–25.

13. Gilbert M, Slingenbergh J, Xiao X. Climate change and avian influenza. Revue Scientifique et Technique (International Office Epizootics) 2008;27(2):459–466.

14. Gould E, Higgs S, Buckley A, Gritsun T. Potential arbovirus emergence and implications for the United Kingdom. Emerging Infectious Diseases 2006;12:549–55.

15. Gould E, Higgs S. Impact of climate change and other factors on emerging arbovirus diseases. Transactions of the Royal Society for *Tropical* Medicine and Hygiene 2009;103:109–21.

16. Koslowsky S, Staubach C, Kramer M, Wieler LH. Risk assessment of bluetongue disease incursion into Germany

using geographic information system (GIS). Berl Munch Tierarztl Wochenschr 2004;117(5–6):214–225.

17. Linthicum K, Bailey C, Tucker D, Angleberger T, Cannon T, Logan T *et al.* Towards real-time predictions of Rift Valley fever epidemics in Africa. Preventive Veterinary Medicine 1991;11:325–34.
18. Maclachlan N, Drew C, Darpel K, Worwa G. The pathology and pathogenesis of bluetongue. Journal of Comparative Pathology 2009;141:1–16.
19. MacLachlan N. Global implications of the recent emergence of bluetongue virus in Europe. Veterinary Clinics of North America: Food Animal Practice 2010;26(1):163–71.
20. Maclachlan N. Bluetongue: history, global epidemiology, and pathogenesis. Preventive Veterinary Medicine 2011;doi:10.1016/j.prevetmed.2011.04.005
21. Maclachlan N, Guthrie A. Re-emergence of bluetongue, African horse sickness and other Orbivirus diseases. Veterinary Research 2010;41(35):2–8.
22. Martin V, Chevalier V, Ceccato P, Anyamba A, De Simone L, Lubroth J *et al.* The impact of climate change on the epidemiology and control of Rift Valley fever. Revue Scientifique et Technique (International Office Epizootics) 2008;27(2):413–26.
23. Mellor P, Carpenter S, Harrup L, Baylis M, Mertens P. Bluetongue in Europe and the Mediterranean Basin: history of occurrence prior to 2006. Preventive Veterinary Medicine 2008;87(1–2):4–20.
24. Purse B, Mellor P, Rogers D, Samuel A, Mertens P, Baylis M Climate change and the recent emergence of bluetongue in Europe. Nature Reviews Microbiology 2005;3(2):171–181.
25. Purse B, Brown H, Harrup L, Mertens P, Rogers D. Invasion of bluetongue and other orbivirus infections into Europe: the role of biological and climatic processes. Revue Scientifique et Technique (International Office Epizootics) 2008;27(2):427–42.
26. Russell RC. Mosquito-borne arboviruses in Australia: the current scene and implications of climate change for human health. International Journal of Parasitology 1998;28(6):955–69.
27. Sellers R. Bluetongue in Africa, the Mediterranean region and Near East: disease, virus and vectors. Preventive Veterinary Medicine 1984;2:371–8.
28. Tourre Y, Lacaux, J-P, Vignolles C, Lafaye M. Climate impacts on environmental risks evaluated from space: a conceptual approach to the case of Rift Valley Fever in Senegal. Global Health Action 2009;2.
29. Vandegrift K, Sokolow S, Daszak P, Kilpatrick A. Ecology of avian influenza viruses in a changing world. Annals of the New York Academy of Sciences 2011;1195:113–28.
30. Walker PJ, Winton JR. Emerging viral diseases of fish and shrimp. Veterinary Research 2010;41(6):23.
31. Weaver S, Reisen W. Present and future arboviral threats. Antiviral Research 2010;85(2):328–45.
32. Weissenböck H, Hubálek Z, Bakonyi T, Nowotny N. Zoonotic mosquito-borne flaviviruses: worldwide presence of agents with proven pathogenicity and potential candidates of future emerging diseases. Veterinary Microbiology 140(3–4):271–80.
33. Wilson A, Darpel K, Mellor P. Where does bluetongue virus sleep in the winter? PLoS Biology 2008;6(8):e210.
34. Wilson A, Mellor P Bluetongue in Europe: vectors, epidemiology and climate change. Parasitology Research 2008;103:s69–77.
35. Wittmann EJ, Baylis M. Climate change: effects on *Culicoides* transmitted viruses and implications for the UK. Veterinary Journal 2000;160(2):107–17.
36. Zell R, Krumbholz A, Wutzler P. Impact of global warming on viral diseases: what is the evidence? Current Opinion in Biotechnology 2008;19(6):652–60.

Tick-Borne Diseases (28)

1. Azhahianambi P, Ghosh S. Genetic transformation of arthropod vectors for control of vector borne diseases. Indian Journal of Biotechnology 2007;(6):305–14.
2. Beugnet F, Marie J. Emerging arthropod-borne diseases of companion animals in Europe. Veterinary Parasitology 2009;163:298–305.
3. Brinkley C, Nolskog P, Golovjova I, Lundkvist A, Bergstrom T. Tick-borne encephalitis virus natural foci emerge in western Sweden. International Journal of Medicine and Microbiology 2008;298(1):73–80.
4. Cumming G, Van Vuuren D. Will climate change affect ectoparasite species ranges. Global Ecology and Biogeography 2006;15:486–97.
5. Deem SL, Perry BD, Katende JM, McDermott JJ, Mahan SM, Maloo SH *et al.* Variations in prevalence rates of tick-borne diseases in Zebu cattle by agroecological zone: implications for East Coast fever immunization. Preventive Veterinary Medicine 1993;16:171–87.
6. Eisen L. Climate change and tick-borne diseases: a research field in need of long-term empirical studies. International Journal of Medicine and Microbiology 2008;298:12–18.
7. Estrada-Peña A. Climate warming and changes in habitat suitability for *Boophilus microplus* (Acari: Ixodidae) in Central America. Journal of Parasitology 2001;87(5):978–87.
8. Estrada-Peña A. Forecasting habitat suitability for ticks and prevention of tick-borne diseases. Veterinary Parasitology 2001;98(1–3):111–32.
9. Estrada-Peña A, Corson M, Venzal JM, Mangold AJ, Guglielmone A. Changes in climate and habitat suitability for the cattle tick *Boophilus microplus* in its southern Neotropical distribution range. Journal of Vector Ecology 2006;31(1):158–67.
10. Estrada-Peña A, Horak I, Petney T. Climate changes and suitability for the ticks *Amblyomma hebraeum* and *Amblyomma variegatum* (Ixodidae) in Zimbabwe (1974–1999). Veterinary Parasitology 2008;151(2–4):256–67.
11. Fandamu P, Duchateau L, Speybroeck N, Mulumba M, Berkvens D. East Coast fever and multiple El Niño southern oscillation ranks. Veterinary Parasitology 2006;135:147–52.
12. Fandamu P, Duchateau L, Speybroeck N, Marcotty T, Mbao V, Mtambo J *et al.* *Theileria parva* seroprevalence in traditionally kept cattle in southern Zambia and El Niño. International Journal of Parasitology 2005;1(35):391–6.
13. Gachohi J, Ngumi P, Kitala P, Skilton R. Estimating seroprevalence and variation to four tick-borne infections and determination of associated risk factors in cattle under

traditional mixed farming system in Mbeere District, Kenya. Preventive Veterinary Medicine 2010;95:208–23.

14. Gilbert L. Altitudinal patterns of tick and host abundance: a potential role for climate change in regulating tick-borne diseases? Oecologia 2009;162(1):217–25.
15. Jaenson T, Lindgren E. The range of *Ixodes ricinus* and the risk of contracting Lyme borreliosis will increase northwards when the vegetation period becomes longer. Ticks and Tick-borne Disease 2011;2:44–9.
16. King D, Gettinby G, Newson R. A climate-based model for the development of the ixodid tick, *Rhipicephalus appendiculatus* in East Coast Fever zones. Veterinary Parasitology 1988;29:41–51.
17. Kivaria F. Climate change and the epidemiology of tick-borne diseases of cattle in Africa. Veterinary Journal 2010;184(1):7–8.
18. Norval R, Perry B, Gebreab F, Lessard P. East coast fever: a problem of the future for the horn of Africa. Preventive Veterinary Medicine 1991;10(3):163–72.
19. Ogden N, Maarouf A, Barker I, Bigras-Poulin M, Lindsay L, Morshed M *et al.* Climate change and the potential for range expansion of the Lyme disease vector *Ixodes scapularis* in Canada. International Journal of Parasitology 2006;36(1):63–70.
20. Olwoch J, Reyers B, Engelbrecht F, Erasmus B. Climate change and the tick-borne disease, Theileriosis (East Coast fever) in sub-Saharan Africa. Journal of Arid Environment 2008;72(2):108–20.
21. Olwoch JM, Van Jaarsveld AS, Scholtz CH, Horak IG. Climate change and the genus *Rhipicephalus* (Acari: Ixodidae) in Africa. Onderstepoort Journal of Veterinary Research 2007;74(1):45–72.
22. Otranto D, Wall R. New strategies for the control of arthropod vectors of disease in dogs and cats. Medical and Veterinary Entomology 2008;22(4):291–302.
23. Perry B, Lessard P, Norval RAI, Kundert K, Kruska R. Climate, vegetation, and the distribution of *Rhipicephalus appendiculatus* in Africa. Parasitology Today 1990;6:100–94.
24. Randolph S. Ticks and tick borne disease systems in space and from space. Advances in Parsitology 2000;47:217–47.
25. Randolph S. Tick ecology: processes and patterns behind the epidemiological risk posed by ixodid ticks as vectors. Parisitology 2004;129(s1):s37–65.
26. Randolph S. Dynamics of tick-borne disease system: minor role of recent climate change. Revue Scientifique et Technique (International Office Epizootics) 2008;(27):367–81.
27. Randolph S. To what extent has climate change contributed to the recent epidemiology of tick-borne diseases? Veterinary Parasitology 2009;167(2–4):92–4.
28. Walder G, Falkensammer B, Heinz F, Holzmann H, Dierich M, Wurzner R. Tick-borne encephalitis in the Tyrol (Austria): Changes in incidence and endemicity 2000–2006. International Journal of Medicine and Microbiology 2008;298(1):88–93.

Parasitic Diseases (43)

1. Bennema S, Ducheyne E, Vercruysse J, Claerebout E, Hendrickx G, Charlier J. Relative importance of management, meteorological and environmental factors in the spatial distribution of *Fasciola hepatica* in dairy cattle in a temperate climate zone. International Journal of Parasitology 2011;41(2):225–33.
2. Brooks DR, Hoberg E. How will global climate change affect parasite–host assemblages? Trends in Parasitology 2007;23(12):571–4.
3. Chomel B. Control and prevention of emerging parasitic zoonoses. International Journal of Parasitology 2008;38(11):1211–7.
4. Coles GC. The future of veterinary parasitology. Veterinary Parasitology 2001;98(1–3):31–9.
5. Fairweather I. Reducing the future threat from (liver) fluke: realistic prospect or quixotic fantasy? Veterinary Parasitology 2011;180:133–43.
6. Fox N, White P, McClean C, Marion G, Evans A, Hutchings M. Predicting impacts of climate change on *Fasciola hepatica* risk. PLoS ONE 2011;6(1):e16126.
7. Fuentes M, Malone J, Mas-Coma S. Validation of a mapping and prediction model for human fasciolosis transmission in Andean very high altitude endemic areas using remote sensing data. Acta Tropica 2001;79(1):87–95.
8. Gajadhar A, Allen J. Factors contributing to the public health and economic importance of waterborne zoonotic parasites. Veterinary Parasitology 2004;126(1–2):3–14.
9. Genchi C, Rinaldi L, Mortarino M, Genchi M, Cringoli G. Climate and *Dirofilaria* infection in Europe. Veterinary Parasitology 2009;163(4):286–92.
10. Jenkins EJ, Veitch AM, Kutz SJ, Hoberg EP, Polley L. Climate change and the epidemiology of protostrongylid nematodes in northern ecosystems: *Parelaphostrongylus odocoilei* and *Protostrongylus stilesi* in Dall's sheep (Ovis d. dalli). Parasitology 2006;132(Pt 3):387–401.
11. Jenkins EJ, Schurer JM, Gesy KM. Old problems on a new playing field: Helminth zoonoses transmitted among dogs, wildlife, and people in a changing northern climate. Veterinary Parasitology 2011;182(1):54–69.
12. Kenyon F, Sargison ND, Skuce PJ, Jackson F. Sheep helminth parasitic disease in south eastern Scotland arising as a possible consequence of climate change. Veterinary Parasitology 2009;163(4):293–7.
13. King B, Monis P. Critical processes affecting Cryptosporidium oocyst survival in the environment. Parasitology 2007;134:309–23.
14. Kutz S. The Arctic as a model for anticipating, preventing, and mitigating climate change impacts on host–parasite interactions. Veterinary Parasitology 2009;163(3):217–28.
15. Kutz SJ, Hoberg EP, Polley L. A new lungworm in muskoxen: an exploration in Arctic parasitology. Trends in Parasitology 2001;17(6):276–80.
16. Mallone J, Gommes R, Hansen J, Yilma J, Slingenberg J, Snijdes F *et al* A geographic information system on the potential distribution and abundance of *Fasciola hepatica* and *F. gigantica* in east Africa based on Food and Agriculture Organization databases. Veterinary Parasitology 1998;78(2):87–101.
17. Mas-Coma S, Valero MA, Bargues M. Effects of climate change on animal and zoonotic helminthiases. Revue

Scientifique et Technique (International Office Epizootics) 2008;27(2):443–57.

18. Mas-Coma S, Valero MA, Bargues MD. Climate change effects on trematodiases, with emphasis on zoonotic fascioliasis and schistosomiasis. Veterinary Parasitology 2009;163(4):264–80.
19. McCann C, Baylis M, Williams D. The development of linear regression models using environmental variables to explain the spatial distribution of *Fasciola hepatica* infection in dairy herds in England and Wales. International Journal of Parisitology 2010;40(9):1021–8.
20. Morgan ER, Milner-Gulland EJ, Torgerson PR, Medley GF. Ruminating on complexity: macroparasites of wildlife and livestock. Trends in Ecology and Evolution 2004;19(4):181–8.
21. Morgan ER, Wall R. Climate change and parasitic disease: farmer mitigation? Trends in Parasitology 2009;25(7):308–13.
22. Morley NJ. Aquatic molluscs as auxiliary hosts for terrestrial nematode parasites: implications for pathogen transmission in a changing climate. Parasitology 2010;137(7):1041–56.
23. Mouritsen K, Tompkins D, Poulin R. Climate warming may cause a parasite-induced collapse in coastal amphipod populations. Oecologia 2005;146:476–83.
24. Mouritsen K, Poulin R. Parasitism, climate oscillations and the structure of natural communities. Okois 2002;97(3):462–8.
25. Patz JA, Graczykb TK, Gellera N, Vittor A. Effects of environmental change on emerging parasitic diseases. International Journal of Parasitology 2000;30:1395–405.
26. Patz J, Olson S, Uejio C, Gibbs H. Disease emergence from global climate and land use change. Medical Clinics of North America 2008;92(6):1473–91.
27. Patz J, Reisen W. Immunology, climate change and vector-borne diseases. Trends in Immunology 2001;22(4):171–2.
28. Petney T. Environmental, cultural and social changes and their influence on parasite infections. International Journal of Parasitology 2001;31(9):919–32.
29. Petney T, Taraschewski H. Waterborne parasitic diseases: hydrology, regional development and control. In Wilderer P, editor. Treatise on Water Science, Vol 3; 2011. p. 303–66.
30. Polley L, Thompson R. Parasite zoonoses and climate change: molecular tools for tracking shifting boundaries. *Trends* in *Parasitology* 2009;25:285–91.
31. Poulin R. Global warming and temperature-mediated increases in cercarial emergence in trematode parasites. Parasitology 2006;132(1):143–51.
32. Rose H, Wall R. Modelling the impact of climate change on spatial patterns of disease risk: sheep blowfly strike by Lucilia sericata in Great Britain. International Journal of Parasitology 2011;41(7):739–46.
33. Sutherst R. The vulnerability of animal and human health to parasites under global change. International Journal of Parasitology 2001;31(9):933–48.
34. Taraschewski H. Hosts and parasites as aliens. Journal of Helminthology 2006;80(2):99–128.
35. Timi J. Parasites as biological tags for stock discrimination in marine fish from South American Atlantic waters. Journal of Helminthology 2007;81(2):107–11.
36. Valero M, Marcos M, Comes A, Sendra M, Mas-Coma S Comparison of adult liver flukes from highland and lowland populations of Bolivian and Spanish sheep. Journal of Helminthology 1999;73:341–5.
37. van Dijk J, David GP, Baird G, Morgan E. Back to the future: Developing hypotheses on the effects of climate change on ovine parasitic gastroenteritis from historical data. Veterinary Parasitology 2008;158(1–2):73–84.
38. van Dijk J, Sargison ND, Kenyon F, Skuce PJ. Climate change and infectious disease: helminthological challenges to farmed ruminants in temperate regions. Animal 2010;4(03):377–92.
39. Wall R, Ellse LS. Climate change and livestock parasites: integrated management of sheep blowfly strike in a warmer environment. Global Change in Biology 2011;17(5):1770–7.
40. Wall R, Rose H, Ellse L, Morgan E. Livestock ectoparasites: integrated management in a changing climate. Veterinary Parasitology 2011;180(1–2):82–9.
41. Weaver HJ, Hawdon JM, Hoberg EP. Soil-transmitted helminthiases: implications of climate change and human behavior. Trends in Parasitology 2010;26(12):574–81.
42. Yang GJ, Utzinger J, Lv S, Qian YJ, Li SZ, Wang Q *et al.* The Regional Network for Asian Schistosomiasis and Other Helminth Zoonoses (RNAS(+)) target diseases in face of climate change. Advances in Parasitology 2010;73:101–35.
43. Zhang B, McGraw E, Floate KD, James P, Jorgensen W, Rothwell J. Wolbachia infection in Australasian and North American populations of *Haematobia irritans* (Diptera: Muscidae). Veterinary Parasitology 2009;162(3–4):350–3.

Infectious/Emerging Infectious Disease (9)

1. Endy TP, Rochford R, Yuen K-Y, Lei H-Y. Emerging infectious diseases as a global health threat. Experimental Biology and Medicine 2011;236(8):897–8.
2. Epstein P. Climate change and emerging infectious diseases. Microbes and Infection 2001;3(9):747 – 54.
3. Gummow B. Challenges posed by new and re-emerging infectious diseases in livestock production, wildlife and humans. Livestock Science 2010;130(1–3):41–6.
4. Khasnis AA, Nettleman MD. Global warming and infectious disease. Archives of Medical Research 2005;36(6):689–96.
5. Rosenthal J. Climate change and the geographic distribution of infectious diseases. Ecohealth 2009;6:489–95.
6. Shuman E. Global climate change and infectious diseases. New England Journal of Medicine 2010;362(12):1061–3.
7. Slenning B. Global climate change and the implications for infectious disease. Veterinary Pathology 2010;47(1): 28–33.
8. Wilkinson K, Grant W, Green L, Hunter S, Jeger M, Lowe P *et al.* Infectious diseases of animals and plants: an interdisciplinary approach. Philosophical Transactions of the Royal Society 2011;366:1933–42.
9. Zell R. Global climate change and the emergence/re-emergence of infectious diseases. International Journal of Medicine and Microbiology 2004;37:16–26.

One Health/Public Health (9)

1. Ahmed JS, Sparagano O, Seitzer U. One health, one medicine: tackling the challenge of emerging diseases. Transbound Emerging Disease 2010;57(1–2):1–2.
2. Cardiff RD, Ward JM, Barthold SW. 'One medicine–one pathology': are veterinary and human pathology prepared? Laboratory Investigation 2008;88(1):18–26.
3. Coker R, Rushton J, Mounier-Jack S, Karimuribo E, Lutumba P, Kambarage D *et al*. Towards a conceptual framework to support one-health research for policy on emerging zoonoses. Lancet Infectious Disease 2011;11(4):326–31.
4. Conrad PA, Mazet JA, Clifford D, Scott C, Wilkes M. Evolution of a transdisciplinary "One Medicine–One Health" approach to global health education at the University of California, Davis. Preventive Veterinary Medicine 2009;92(4):268–74.
5. Haines A, Kovats RS, Campbell-Lendrum D, Corvalan C. Climate change and human health: impacts, vulnerability and public health. Public Health 2006;120(7):585–96.
6. Rock M, Buntain B, Hatfield, J, Hallgrimsson B. Animal-human connections, 'one health' and the syndemic approach to prevention. Social Science and Medicine 2009;68(6):991–5.
7. Singer MC. Doorways in nature: syndemics, zoonotics, and public health. A commentary on Rock, Buntain, Hatfield and Hallgrimsson. Social Science and Medicine 2009;68(6):996–9.
8. Slenning B. One Health and climate change: linking environmental and animal health to human health. North Carolina Medical Journal 2010;71(5):434–437.
9. Soulé, M. The social and public health implications of global warming and the onslaught of alien species. Journal of Wilderness Medicine 1992;3(2):118–27.

Bacterial Diseases (4)

1. Joyner TA, Lukhnova L, Pazilov Y, Temiralyeva G, Hugh-Jones ME, Aikimbayev A *et al*. Modeling the potential distribution of *Bacillus anthracis* under multiple climate change scenarios for Kazakhstan. PLoS ONE 2010;5(3).
2. Lau C, Smythe LD, Craig SB, Weinstein P. Climate change, flooding, urbanisation and leptospirosis: fuelling the fire? Transactions of the Royal Society of Tropical Medicine and Hygiene 2010;104(10):631–8.
3. Rock C, Horgan M. Leptospirosis – on the increase due to global warming. Ireland Medical Journal 2010;103(4):101–2.
4. Victoriano AF, Smythe LD, Gloriani-Barzaga N, Cavinta LL, Kasai T, Limpakarnjanarat K *et al*. Leptospirosis in the Asia Pacific region. BMC Infectious Diseases 2009;9:147.

Intensification of animal production from grassland and ecosystem services: a trade-off

Gilles Lemaire*

Address: UR0004 P3F Unité de Recherche Pluridisciplinaire Prairies et Plantes Fourragères, INRA route de Saintes, 86600 Lusignan, France.

***Corresponding author:** Email: gilles.lemaire@lusignan.inra.fr

Received: 6 September 2011
Accepted: 7 March 2012

Abstract

Grasslands have to be considered not only as a means for human food production but also as a means for providing ecosystem services. Coupling of C and N cycles through soil–vegetation interactions gives grassland ecosystems a high capacity for reduction of environmental impacts. Moreover, the high biodiversity within grasslands gives them a high-adaptive capacity to changes and perturbations. Nevertheless, the C–N decoupling by grazing animals annihilates partly these ecosystem services as the intensification and hence the production increase. Research is necessary for optimizing this trade-off through spatial and temporal interactions with other land use systems at farm and landscape level.

Keywords: Grasslands, Ecosystem services, Biogeochemical cycles, Biodiversity.

Grasslands, as defined as permanent herbaceous vegetation used by wild or domestic herbivores, cover about one-quarter of the Earth's land surface [1]. Except within eco-geographical areas where the vegetation is climatically maintained at herbaceous stage, most of the grasslands around the world are the result of the human management of domestic herbivores [2, 3]. So, even if the long-term life history of grassland species has been determined by their co-evolution with wild herbivores [4], the more recent evolution of these plant communities results from the different grassland management programmes [5]. Grasslands contribute to the livelihoods of over 800 million people [6] and provide different goods and services to support flora, fauna and human populations.

The primary goal of twentieth century agriculture has been to meet the food demands of increasing human population [7]. The exploitation of the vegetation resource and its transformation in animal products has been the main focus for 'grassland science' during the last century (see [3]). Nowadays, in facing the problem of sustainable development, it is necessary also to consider (i) the important role of grassland ecosystems in the dynamics of atmosphere, hydrosphere and continental surface interactions that drive global changes and environmental hazards, and (ii) how the functioning of the grassland ecosystems can adjust to these changes. So grasslands must be studied as an important land use and management system in conjunction with forests and croplands for the regulation of biogeochemical cycles, the control and the mitigation of environmental fluxes, the preservation of biodiversity and the production of safe food for human society.

These multi-functional roles of grasslands are encapsulated within the concept of ecosystem services [8]. These ecosystem services are of different natures:

- *provisioning* food, drinking water, timber, fibre and energy;
- *regulating* climate, floods, diseases, wastes and water quality;
- *providing cultural amenities*; and
- *supporting functions* such as soil-quality dynamics, primary production, biogeochemical cycles and dynamics of biodiversity, necessary for other services.

The concept of sustainable development implies that any single ecosystem service cannot be achieved at the expense of others. In consequence, it is no longer possible to optimize solely food production from grassland without taking into account the trade-offs with other ecosystem services.

The objective of this paper is not to review in detail all the different ecosystem services potentially provided by grasslands, but to focus on ecosystem processes and functions determining the responses of grasslands to both management and environment forcings that are at the base of the trade-offs between the necessary intensification of food production and the necessity to reduce or to mitigate the deterioration of environment by agriculture systems.

Grassland Soil–Vegetation Interactions: Coupling of Biogeochemical Cycles

The C cycle in grasslands, as in other ecosystems, is intimately process-linked with other nutrient dynamics, particularly with N [9, 10]. In grassland ecosystems, the permanency of soil–vegetation interactions leads to strong coupling between the different cycling elements, carbon, nitrogen, phosphorus, etc. These couplings operate: (i) in plants where N and P are linked to C for plant tissue synthesis through light interception, photosynthesis, N and P assimilation, and plant growth processes; and (ii) in soils through the dynamics of organic matter with the capacity of microbes to recapture and recycle mineral N and P. These C–N coupling processes contribute to the atmospheric CO_2 sink in soils, and to reduce the emissions of reactive N to the hydrosphere and atmosphere. So grasslands can be considered as a land-use system that is very favourable for the preservation of environment [11].

As shown by Parsons et al. [12], the relative distribution of C between atmosphere and biosphere is proportional to the relative mean residence time of C in each compartment. The residence time of C in soils depends of the C:N ratio of the fresh organic matter inputs. As litter deposition in grasslands is predominantly dead leaf and root material with high C:N ratio [13] the mean residence times of C within grassland soils are generally long, and grassland ecosystems are considered in general sink for CO_2 [14]. As N and P are linked to C by stoichiometric relations within the different compartments of soil organic matter and within soil micro-organism pools C flows through the ecosystem determine flows of N and P and *vice versa*. The most notable change in C:N ratio occurring in grassland biosphere is the remobilization of N during the process of leaf senescence, from C:N of approximately 12 : 1 for living green leaves to approximately 50 : 1 for leaf litter [12]. So this constant flow of high C:N organic compounds to soil via leaf litter deposition associated with the corresponding flow of root litter represents a high demand for N relative to C for microbe communities in soil having a relatively low functional C:N stoichiometric ratio [12]. As a consequence mineral N provided in soil through the gross mineralization of soil organic matter is rapidly immobilized by soil micro-organisms. Tlustos et al. [15] found that permanent grassland soils exhibited mineralization–immobilization turnover rates 30 times higher than those in arable soils, although net N mineralization rates were similar between systems. So this strong coupling between C and N allows a relatively high conservative cycle of N within the grassland ecosystem avoiding too large an accumulation of inorganic N in soils. Given this very low nitrate concentration in soil, the risk of nitrate leaching under grassland vegetation is generally low, even with relatively ample N fertilizer applications [16]. In the same way, denitrification being dependent on the concentration of available mineral N in soil, the N_2O losses under grasslands should be low in general because of the very low residence time of nitrate in grassland soil and hence of the low nitrate content in soil [17]. Flechard et al. [18] show that most of the N_2O emissions monitored from grassland soils usually occur in 'hot spots' associated with urine patches or just after fertilizer or manure spreading when there is spatial and temporal concordance between high nitrate accumulation and anoxic conditions.

So, as stated above, the permanency of a strong C flow through vegetation to soil by means of light interception, photosynthesis, plant tissue elaboration and leaf and root litter deposition with a high C:N ratio allow a strong coupling of C and N cycles. This coupling reduces then N loss flows to atmosphere and hydrosphere, and determines a long residence time of C within stabilized fractions of soil organic matter that gives a high potential capacity of sequestration of atmospheric CO_2 within soils. However, such a capacity of grassland ecosystems for promoting processes and functions favourable for providing some ecosystem services such as quality of atmosphere and quality of hydrosphere are also strongly dependent on the management procedures through which grasslands are used within animal production systems.

Intensification of Animal Production from Grasslands and 'Decoupling' C and N Cycles

Grazing animals, by consuming a variable proportion of shoot biomass, contribute to the partial 'uncoupling' of N from C. Approximately 70% of the C ingested by grazing animals is released back to the atmosphere as CO_2 in respiration and as methane (CH_4) in rumen fermentation [12]. For N, 70–80% is excreted to the soil surface as urine, while only 20–30% remains coupled with C within dungs. So as shown by Ledgard et al. [19], the risk of N leaching under pasture is more fundamentally linked to stocking density via the density of urine patches and not directly to the amount of N fertilizer inputs.

Intensification of animal output production by grasslands can be achieved by two ways: (i) an increase in herbage production through applications of N and P fertilizers, and (ii) an increase in the proportion of herbage production consumed by animals through grazing management. The application of N fertilizers increases interception of light by swards and leaf photosynthesis leading to an increased flow of C in the ecosystem, as shown by Thornley [20] through a modelling approach. But this positive effect on C sequestration should be partially offset in the long term because the application of N fertilizers tends to decrease in C:N ratio in soils and then should decrease the mean residence time of C within the system [21]. Conant et al. [22] through a meta-analysis of 115 temperate pastures showed that soil C levels tend to increase with improved management (primarily N fertilization) in 74% of situations. Soussana et al. [17] indicate that both moderate intensification of nutrient-poor permanent grasslands and reduction of N fertilizer inputs in highly intensive leys are two options for increasing the C sequestration capacity of grassland soils. These authors reported that, according to soil and climate conditions, grasslands could pass from C sink to C source and *vice-versa* in relation to changes in the degree of intensification.

The second way for improving the productivity of grassland is to increase the herbage use efficiency by increasing stocking density. For a given level of herbage production, any increase in stocking density reduces the proportion of leaf tissues being recycled in soil via litter deposition as demonstrated by Parsons et al. [12] and Lemaire et al. [23]. In low productive grasslands with low stocking densities, the efficiency of herbage production can be less than 20% and then 80% of the C used for herbage production return to soil accompanied with a high C:N ratio and then a long residence time of C within soil organic matter. In contrast, in highly intensified pasture (high N fertilizer application and high stocking density) with optimum grazing management for maximizing herbage use efficiency, up to 70% of herbage produced is consumed by animals and only 30% returned to soil as litter with a low C:N ratio and a shorter residence time of C in soil. As a result, the overall positive effect of grassland intensification on the absorption of atmospheric CO_2 by vegetation is partially off-set by (i) an increasing loss of C by CO_2 and CH_4 emissions by animals and (ii) a reduction of residence time of C in soil as a consequence of lower C:N ratio. As a consequence of this trade-off, as proposed by Soussana et al. [17] optimizing the CO_2 sequestration capacity of grassland should require an intermediary level of intensification. As stated by Soussana et al. [17] there is a range of grassland management practices for reducing C losses and for increasing C sequestration: (i) avoiding soil tillage and the conversion of grasslands to arable use; (ii) moderately intensifying nutrient-poor permanent grasslands; (iii) using light grazing instead of heavy grazing; and (iv) increasing duration of grass leys.

This trade-off between production and environmental outputs can be analysed through the concept of 'environmental carrying capacity' of grasslands, which is the maximum stocking density a grassland area can support for maintaining a positive balance between the N–C 'coupling' capacity of soil–vegetation system, and the 'uncoupling' activity of grazing animals. For each grassland situation (soil, climate and vegetation), a level of intensification (and then of stocking density) should exist beyond which the negative environmental impacts (CO_2 balance, CH_4 and N_2O emissions, nitrate leaching, etc.) associated with animal grazing activities become too important as compared with positive impacts associated with soil–vegetation N–C coupling. Modelling of responses of grassland ecosystems to manipulations of management programmes should allow the quantitative prediction of environmental consequences of any change in land use and management (see [12]).

Biodiversity within Grassland Ecosystem: From a Conservative to a Functional Approach

Grasslands are considered as hot spots of biodiversity because they host many specialized plants, animals and microbes, all nested in a great variety of microhabitats. This is particularly true for semi-natural grasslands, but agricultural intensification through reseeding monospecific swards has resulted in a loss of biodiversity [24]. At the same time, over-exploitation of natural grasslands in rangeland systems in semi-arid regions may have also diminished the species richness of these ecosystems [25]. In semi-natural grasslands, nutrient availability and disturbance regime are two factors influencing vegetation dynamics that are constrained by agricultural management [26].

Biodiversity of grassland vegetation must be analysed according to a conservative view as an important ecosystem service, but its functional role has also to be questioned as necessary for promoting other services. Hence, the loop of interactions between grassland biodiversity and soil organic matter dynamics, biogeochemical cycles and regulation of environmental fluxes to hydrosphere and atmosphere has to be studied with a long-term perspective. Moreover, understanding these interactive processes should help in forecasting the evolution of biodiversity and ecosystem functioning under climate changes scenarios. As stressed by Silvertown et al. [27] this approach requires long-term experimental platforms. It implies that one not only needs to analyse the biodiversity at plant community level but also the associated biodiversity of soil microbe and soil fauna communities within an integrated approach.

The functional approach of plant diversity by means of functional traits [28] allows the understanding of the dynamics of grassland communities under different drivers such as management systems and climate changes. According to Lavorel and Garnier [29], screening for plant traits allows two questions to be addressed: (i) which traits are associated with a given set of plant populations within a community? and (ii) how do the traits of the selected plant populations control ecosystem functioning and productivity? As shown by Louault et al. [26], plant traits can be related to management practices. A decline in herbage use efficiency and then a lower disturbance regime reduced individual leaf mass, specific leaf area and shoot digestibility, but increased leaf C and dry matter content. So management of permanent pastures promotes plant types with some common attributes. Diaz et al. [30] have demonstrated that similar plant traits, in different floras, were associated with specific response to grazing. In addition, plant response generally involved functional trade-offs, implying association between plant traits [31]. However, until now, these plant traits have been mainly determined at the leaf level and they are more related to the productive capacity of plant species (efficiency of use of light, water, N, minerals, etc.). It is necessary now to determine also root functional traits for foraging and capturing soil resources in interaction with other living soil organisms, bacteria, fungi and micro–meso-fauna.

As reported in the literature, biodiversity is known to affect productivity of ecosystems [32, 33]. This positive relationship between biodiversity and productivity of grassland is observed when biodiversity is enhanced in a given environment. But when environment is improved through an increased nutrient availability, then we observe in general a drastic reduction of the biodiversity [34]. This negative effect is the result of (i) the reduction of heterogeneity of habitats [34], (ii) the enhancement of competition for light and (iii) the recruitment limitation [35]. So the relationship between species diversity and grassland productivity is not univocal and it depends on the factors manipulated. The general tendency is to get a decrease of species diversity with grassland intensification, but with a possibility to get over-yielding when we increase biodiversity by adding new species [36]. More important is probably the effect of species diversity on grassland stability, i.e. the ability of grassland ecosystem to maintain its productivity despite fluctuations of environment. Tilman et al. [37] found that the temporal stability of grasslands increased with species diversity, despite a lower temporal stability of individual species. Such a property of grassland ecosystem is very important because changes in environment are not the only source of instability, modification of management practices are also important and then species diversity can be considered as a source of flexibility for grassland systems [24].

The major species in grasslands are out-breeding and heterozygous, and even within populations there is considerable genetic variation. Such diversity was considered as a handicap for plant breeding, and as a consequence, the genetic progress for grassland species was very limited as compared with crop species [38]. However, as stressed by Fridley and Grime [39], this intraspecific variability has a high-functional role in plant community dynamics. Grasslands have to grow all year along in continually changing conditions, and it is not the same genotype that performs well in winter, spring, summer or autumn – in cold or warm, dry or wet conditions – under low- or high-grazing intensity, within refused urine of dung or well grazed patches, or under cutting regime, etc. Moreover, the traits for adaptation to these diverse temporal and spatial conditions cannot be regrouped within the same genotype because some of these traits are physiologically incompatible. So the paradigm for genetic improvement of grasslands is not, as for arable crop, to introduce all good traits within the same genotype , but to introduce these traits within the same population or community and then to analyse their adaptation and evolution capacity according to environment and management. For arable crops, the objective is to adapt 'elite' plants to a narrow range of conditions and then to control environment for matching plant requirements through crop management procedures, while for grasslands, the objective is to improve and to adapt the intra- and interspecific diversity facing the huge and unpredictable variation of environmental conditions and the diversity of management programmes.

The plant species diversity tends to decline as intensive management are imposed to grasslands, so there is a trade-off between increasing animal outputs through grassland production intensification and maintaining a high level of biodiversity unless specific management programmes were used. As discussed above for ecosystem services related to the regulation of C and N cycles and of environmental fluxes, there probably exists for each type of grassland and each environmental condition a threshold level of intensification above which the ecosystem services related to biodiversity would deteriorate. The challenge for grassland sciences should be to determine in each situation this 'threshold level of intensification' for these two types of ecosystem services, those related to regulation of biogeochemical cycles and those related to biodiversity dynamics and to detect and analyse possibilities of synergy for a more sustainable grassland management.

Integration of Grassland Ecosystem at Landscape Level: Interactions with Other Land-use Systems

By using large areas as agriculture lands, human societies have changed the kind and the amount of ecosystem services provided by these areas [40]. In many situations, poor agriculture practices or too intensive production methods have severely affected the ability of landscapes to provide services such as maintenance of soil fertility, biotic regulation, nutrient cycling, sequestration of atmospheric CO_2 and maintenance of genetic diversity [41]. So, some of the impacts of grasslands can be analysed only at the landscape level through spatial and temporal interactions with other land-use systems. For impacts such as soil erosion, water quality and landscape biodiversity, where spatially explicit processes are important, it is impossible to develop only an aggregative approach from field plots to catchments and landscapes. The regional impact is not the sum or the average of local impacts because of the topographical interactions between the different spatial components.

Moreover, grassland management affects also the landscape biodiversity by means of the trophic network, which can be sustained and the habitat provided for vertebrate populations such as birds or mammals [42]. So the contribution of grasslands to ecosystem services results of both the direct impacts of grassland areas and the indirect effects of grasslands through spatial and temporal interactions with other land-use systems for regulating biotic and abiotic fluxes. This is the case for mixed farming systems, where the introduction of grassland areas within arable cropping systems can result in a more sustainable agricultural production system than arable cropping and livestock production separately [43]. Grassland leys included within arable rotations can contribute to mitigate some of the negative environmental impacts provided by intensive arable cropping systems: attenuation of soil erosion risk, sequestration of pesticides and heavy metals, improvement of nutrient cycling and reduction of fertilizer use, maintenance and restoration of soil fertility, weed community control and reduction of herbicide use, controlling insect and diseases dissemination and reduction of pesticide use, etc. [43]. But as showed by Björklund et al. [7], the intensification of agriculture production has been accompanied by an increase in farm and field sizes and by a huge specialization of production systems, leading to strong simplification of the land-use systems at the landscape level. So the integration between crop and livestock production system that was at the base of the agriculture production system in Europe since the sixteenth century disappeared from most of the regions in Europe, with intensive livestock systems concentrated within some regions, while intensive cereal cropping systems are concentrated within others, without any local interactions. This excessive specialization associated with an overly intensive production leads to unacceptable deterioration of environment.

The specialization of farms is strongly determined by socio-economic constraints and it is probably difficult come back to the traditional farm. But it should be possible to avoid excessive specialization at the landscape level by favouring local interactions among specialized neighbouring farms, re-creating at landscape level a meta agro-ecosystem with high land-use diversity. In the same way, interactions between forest and grassland areas could be mediated through relevant livestock production systems for optimizing both their own productivity and their overall capacity to provide ecosystem services [44].

In more fragile systems, such as rangelands or semi-arid pastoral systems, the sustainability of resources facing excessive human pressure and climate changes could also require more substantial integration at the regional level with some local cropping or forest areas. All these integrated systems have to be conceived on the basis of well-documented knowledge on the different processes driving the interactions between landscape components and how they can lead to environmental and production enhancement. Then the place and role of grasslands as a component of these systems could be well-defined according to local and regional conditions.

Conclusions

Grassland ecosystems have, in general, a high potential for coupling strongly C and N, and for regulation of biogeochemical cycles and environmental fluxes. This ecosystem property gives grasslands a high capacity to provide some of the ecosystem services related to the protection of environment, such as sequestration of atmospheric CO_2 and protection of surface and ground water quality. Moreover, grasslands can be considered as a source of biodiversity, not only directly through the diversity of plant species within grassland communities but also indirectly through the diversity of habitats and trophic networks these ecosystems can host. So some other ecosystem services related to biodiversity such as pollination, biotic regulation and stability or resilience of productivity are also potentially provided by grassland ecosystems. Nevertheless, the level at which these different services can be provided depends on the management practices. In general, a moderate intensification of grassland production through both an increase in herbage production and an increase in stocking density should be favourable for most of the ecosystem services. For each condition (grassland type, soil, climate, etc.) a 'threshold level of intensification' (herbage production and stocking density) exists for each situation, a challenge beyond which any further increase in intensification of production would lead to a deterioration of the ecosystem services. So the capacity of identification of this threshold for each situation is a challenge for grassland science and is a prerequisite tool for adapting management strategy for solving the trade-off between food security and environment objectives.

References

1. Ojima DS, Parton WJ, Schimel DS, Scurlock JMO, Kittel TGF. Modelling the effects of climatic and CO_2 changes on grassland storage of soil C. Water, Air, and Soil Pollution 1993;70:643–57.

2. Lauenroth WK. Grassland primary production: North American Grasslands in perspective. In French NR, editor. Perspectives in Grassland Ecology. Ecological Studies. Springer-Verlag, New York; 1979. p. 3–24.

3. Lemaire G, Wilkins R, Hodgson J. Challenges for grassland science: managing research priorities. Agriculture, Ecosystems and Environment 2005;108:99–108.

4. Harper JL. Population Biology of Plants. Academic Press, London; 1977. p. 892.

5. White RP, Vannaselt W. Still and silent ecosystems: declining grassland biodiversity. In: Carol Rosen, editor. Pilot Analysis of Global Ecosystems: Grassland Ecosystems. Earth Trends 2001. World Resources Institute, Washington, DC; 2001.

6. Reynolds SG, Batello C, Baas S, Mack S. Grasslands and forage to improve livehoods and reduce poverty. In: Gilloway DA, editor. 'Grassland: A Global Resource', Proceedings of the 20th International Grassland Congress, Dublin, Ireland. Wageningen Academic Publishers, Wageningen, The Netherlands; 2005. p. 323–38.

7. Björklund J, Limburg KE, Rydberg T. Impact of production intensity on the ability of the agricultural landscape to generate ecosystem services: an example from Sweden. Ecological Economics 1999;29:269–91.

8. Millennium Ecosystem Assessment. Ecosystems and Human Well-Being: Synthesis. Island Press, Washington, DC; 2005.

9. Bol R, Poirier N, Balesdent J, Gleixner G. Molecular turnover time of soil organic matter in particle-size fractions of an arable soil. Rapid Communication in Mass Spectrometry 2009;23:2551–8.

10. Bol R, Dunn RM, Pilgrim ES. Managing C and N in grassland systems: the adaptive cycle theory perspective. In: Lemaire G, Hodgson J, Chabbi A, editors. Grassland Productivity and Ecosystem Services. CAB International, Wallingford, UK; 2011. p. 73–82.

11. Lemaire G, Hodgson J, Chabbi A. Food Security and Environmental Impacts: Challenge for Grassland Science. In: Lemaire G, Hodgson J, Chabbi A, editors. Grassland Productivity and Ecosystem Services. CAB International, Wallingford, UK; 2011. pp. xiii–vii.

12. Parsons AJ, Rowarth J, Thornley J, Newton P. Primary production of grasslands, herbage accumulation and use, and impacts of climate change. In: Lemaire G, Hodgson J, Chabbi A, editors. Grassland Productivity and Ecosystem Services. CAB International, Wallingford, UK; 2011. p. 3–18.

13. Rasse DP, Rumpel C, Dignac M-F. Is soil carbon mostly root carbon? Mechanisms for a specific stabilisation. Plant and Soil 2005;269:341–56.

14. Goudriaan J. Biosphere structure, carbon sequestering potential and the 14C carbon atmospheric record. Journal of Experimental Botany 1992;43:1110–20.

15. Tlustos P, Willison TW, Baker JC, Murphy DV, Pavlikova D, Goulding KWT, *et al.* Short-term effects of nitrogen on methane oxidation in soils. Biology and Fertility Soils 1998;28:64–70.

16. Ledgard SF, Luo J, Monagham RM. Managing mineral N leaching in grassland systems. In: Lemaire G, Hodgson J, Chabbi A, editors. Grassland Productivity and Ecosystem Services. CAB International, Wallingford, UK; 2011. p. 83–91.

17. Soussana JF, Tallec T, Blanfort V. Mitigating the greenhouse gas balance of ruminant production systems through carbon sequestration in grasslands. Animal 2010;4(3): 334–50.

18. Flechard CR, Neftel A, Jocher M, Ammann C, Fuhrer J. Bi-directional soil/atmosphere N_2O exchange over two mown grassland systems with contrasting management practices. Global Change Biology 2005;11:2114–27.

19. Ledgard SF, Schils R, Eriksen J, Luo J. Environmental impacts of grazed clover/grass pastures. Irish Journal of Agricultural Research 2009;48:209–26.

20. Thornley JHM. Grassland Dynamics: an Ecosystem Simulation Model. CAB International, Wallingford, UK; 1998.

21. Sanaullah M, Chabbi A, Lemaire G, Charrier X, Rumpel C. How does plant leaf senescence of grassland species influence decomposition kinetics and litter compounds dynamics? Nutrient Cycling in Agroecosystems 2010;88:159–71.

22. Conant RT, Paustian K, Eliott ET. Impacts of periodic management and conversion into grasslands: effects on soil carbon. Ecological Applications 2001;11:343–55.

23. Lemaire G, Da Silva SC, Agnusdei M, Wade M, Hodgson J. Interactions between leaf life span and defoliation frequency in temperate and tropical pastures : a review. Grass and Forage Science 2009;64:341–53.

24. Soussana JF, Duru M. Grassland science in Europe facing new challenges: biodiversity and global environment change. CAB International Reviews: Perspectives in Agriculture, Veterinary Science, Nutrition and Natural Resources 2007;1:72.

25. Briske DD, Derner JD, Brown JR, Fuhlendorf SD, Teague WR, Havstad KM, *et al.* Rotational grazing on rangelands: reconciliation of perception and experimental evidence. Rangeland Ecology and Management 2008;61:3–17.

26. Louault F, Pillar VD, Aufrère J, Garnier E, Soussana J-F. Plant traits and functional types in response to reduced disturbance in a semi-natural grassland. Journal of Vegetation Science 2005;16:151–60.

27. Silvertown J, Tallowin J, Stevens C, Power SA, Morgan V, Emmett B, *et al.* Environmental myopia: a diagnosis and a remedy. Trends in Ecology and Evolution 2010;25:556–61.

28. Garnier E, Navas M-L. A trait-based approach to comparative functional plant ecology: concepts, methods and applications. Agronomy for Sustainable Development 2011; doi:1007/ s13593-011-0036-y.

29. Lavorel S, Garnier E. Predicting changes in community composition and ecosystem functioning from plant traits: revisiting the Holy Grail. Functional Ecology 2002;16:545–56.

30. Diaz S, Noy-Meir I, Cabido M. Can grazing response of herbaceous plants be predicted from simple vegetative traits? Journal of Applied Ecology 2001;38:497–508.

31. Poorter H, Garnier E. Ecological significance of inherent variation in relative growth rate and its components. In: Pugnaire FI, Valladares F, editors. Handbook of Functional Plant Ecology. Marcel Dekker, New York, USA; 1999. p. 81–122.

32. Tilman D, Wedin D, Knops J. Productivity and sustainability influenced by biodiversity in grassland ecosystems. Nature 1996;379:718–20.

33. Hector A, Bazeley-White E, Loreau M, Otway S, Schmid B. Overyielding in grassland communities: testing the sampling effect hypothesis with replicated biodiversity experiments. Ecology Letters 2002;5:502–11.

34. Benton TG, Vickery JA, Wilson JD. Farmland biodiversity: is habitat heterogeneity the key? Trends in Ecology and Evolution 2003;18:182–8.

35. Tilman D. Community invisibility, recruitment limitation, and grassland biodiversity. Ecology 1997;78:81–92.

36. Trenbath BR. Biomass productivity of mixtures. Advances in Agronomy 1974;26:177–210.

37. Tilman D, Reich PB, Knops J. Biodiversity and ecosystem stability in a decade-long grassland experiment. Nature 2006;441:629–31.

38. Parsons AJ, Edwards GR, Newton PDC, Chapman DF, Caradus JR, Rasmussen S, *et al.* Past lessons and future prospects: plant breeding for yield and persistence in cool-temperate pastures. Grass and Forage Sciences 2011;66:153–72.

39. Fridley JD, Grime JP. Community- and ecosystem-level consequences of intraspecific genetic diversity in grassland microcosms of varying species diversity. Ecology 2010;91:2272–83.

40. Daily CG. Nature's Services. Societal Dependence on Natural Ecosystem Services. Island press, Washington, DC; 1997. 392 p.

41. Matson PA, Parton WJ, Power AG, Swift MJ. Agricultural intensification and ecosystem properties. Science 1997;277:504–9.

42. Bretagnolle V, Inchausti P. Modelling population reinforcement at a large spatial scale as a conservation strategy for the declining little bustard (*Tetrax tetrax*) in agricultural habitats. Animal Conservation 2005;8(1):59–68.

43. Franzluebbers AJ. Integrated crop–livestock systems in the southeastern USA. Agronomy Journal 2007;99:361–72.

44. Balent G, Gibon A. Interactions of grasslands with forest at the landscape level: topical issues and challenges attached to land management in agro-silvo-pastoral landscapes. In: Lemaire G, Hodgson J, Chabbi A, editors. Grassland Productivity and Ecosystem Services. CAB International, Wallingford, UK; 2011.

A review of the structure and regulation of the laying hen's reproductive system and the interaction between the stress and reproductive axes

J.B. Golden* and D.V. Arbona

Address: Department of Poultry Science, North Carolina State University, Scott Hall, Rm. 215, Raleigh, NC, USA.

***Correspondence:** jackie_golden@ncsu.edu

Received: 8 November 2011
Accepted: 27 January 2012

Abstract

The laying hen's ovary is a unique biological model for studying the interaction of the reproductive (hypothalamic-pituitary-gonadal (HPG)) and stress (hypothalamic-pituitary-adrenal (HPA)) axes because of the highly regulated, grossly apparent hierarchical organization of ovarian follicles destined for maturation and ovulation. Additionally, the histological structure of the hen's pre-ovulatory hierarchical follicles allows for the manual separation of the theca and granulosa tissues with ease, making it an ideal model for conducting gene expression studies that are tissue-specific in avian species. Through follicular gene expression studies, the processes of vitellogenesis, follicular maturation, selection and ovulation have been thoroughly characterized in the laying hen. More recently, an interest in understanding the impact of stress on follicular development has arisen in the poultry industry given producers' desires to maximize production efficiency. Specifically, the role of glucocorticoids and the expression of glucocorticoid receptors (GRs) on avian ovarian follicles have been examined to closely assess the interaction of the HPG and the HPA axes. By understanding the interaction of these axes, a better understanding of the impact of stress on layer hens' ovarian function will be gleaned. In this review, the hen's reproductive system as well as reproductive and stress axes are covered extensively in order to provide insight into the dynamic interactions that exist between the hen's ovarian follicles and primary stress hormone, corticosterone.

Keywords: Layers, Corticosterone, Ovarian follicle, Stress, Glucocorticoid receptor

Review Methodology: In this review, the main search engines used were PubMed and Google Scholar, and databases such as Cab Abstracts and Agricola were used through resources at the NC State University Libraries system. Main key terms used in the search included, avian reproduction, glucocorticoid receptor, reproductive axis, and layer hen stress.

Part I: Structure, Development, and Regulation of the Domestic Hen's Reproductive System

Structure of the Domestic Hen's Ovary

The laying hen's ovary is a particularly useful biological model for the study of follicular maturation and differentiation, selection, and ovulation because of its unique properties. One of the most prominent differences in the development of the hen's ovary in comparison to mammalian species is the development of only the left ovary as a result of the inhibitory effects of Müllerian inhibiting substance (MIS) [1], a glycoprotein growth factor produced by male and female gonads of avian and mammalian species during early embryological development [2]. In mammalian male embryos, MIS prevents the formation of the Müllerian ducts, which later develop into the fallopian tubes. In the laying hen, however, MIS leads to the regression of the right ovary while the left ovary reaches full maturation [2]. The development of the left ovary is protected from the effects of MIS, because of oestrogen's suppressive action on MIS in conjunction with increased expression of oestrogen receptors on the left ovary [3–5].

In addition to the unique asymmetrical development of the avian ovaries, the hen's ovary also serves as an excellent biological model because of the large size of the ovarian follicles, which allows for easy visualization of a follicle's stage of maturation based upon differences in size. The avian ovary consists of several pools of follicles connected and innervated by a follicular stalk to the ovarian tissue [6]. The ovary contains five to eight large yellow-yolk-filled follicles that are arranged in a hierarchy on the basis of size, which inherently represents the sequence in which the follicles will be ovulated. Each of these large pre-ovulatory hierarchical follicles consists of yellow yolk and has a diameter ranging from approximately 12–40 mm, based upon their stage of maturation. The largest of the pre-ovulatory follicles is designated as the F1 follicle and will be ovulated within the next 24 h. After the F1 follicle, the F2 follicle is the next largest in size and the remaining pre-ovulatory hierarchical follicles are named in a subsequent fashion. After the F1 follicle is ovulated, the F2 follicle continues to mature and is now designated as the F1 follicle. Similarly, the succeeding follicles in the hierarchy each advance one place and an additional follicle is recruited into the hierarchy from a pool of small, yellow, non-hierarchical follicles each time an F1 follicle is ovulated.

The small, non-hierarchical follicles are arranged in a pool of small, yellow-yolk-filled follicles (SYF) that are approximately 5–12 mm in diameter and a pool of white follicles that are less than 5 mm in diameter. The white follicles consist of white yolk and are further classified by size into two groups: the large white follicles (LWF, 2–5 mm) and the small white follicles (SWF, <2 mm). Recruitment of follicles into the hierarchy occurs following each consecutive ovulation of the F1 follicle. One follicle is recruited from the pool of SYF and thus enters the group of pre-ovulatory hierarchical follicles that are destined for ovulation while several LWF advance to become SYF. Follicular recruitment into the hierarchy is, however, a highly selective process. In fact, it has been estimated that for every 20 ovarian follicles in the hen's ovary that grow to a size of 6–8 mm in diameter, only one will be selected into the pre-ovulatory hierarchy [7]. The remaining follicles will undergo follicular atresia [7]. Follicular atresia is defined as the breakdown or re-absorption of the ovarian follicle and is normally characterized by rupture of the follicular wall [6].

Histological organization of the ovarian follicles

Each ovarian follicle consists of several concentric layers of tissue. The theca and granulosa tissues comprise the outermost layers of the follicle with a basal lamina separating them. The theca layer, or most external layer, can be divided into the theca externa and theca interna, both of which contain much of the follicle's vasculature. Deep in the theca interna lies a basement membrane that separates the theca interna from the granulosa cell layer. Beneath the granulosa cells is the perivitelline substance followed by the plasma membrane of the ovum [8].

In the white-yolk containing non-hierarchical follicles the theca layer is less developed, whereas the granulosa cells comprise several layers. In contrast, yellow-yolk-containing hierarchical follicles have a monolayer of granulosa cells that can be easily dissected away and separated from the theca cell layer. The granulosa cells of non-hierarchical SYF and LWF are arranged in several densely packed layers and are cuboidal in shape [9]. As the follicles mature and grow, granulosa cells spread out significantly, leaving hierarchical follicles with a granulosa cell layer that is only one cell thick.

The histological structure of the hen's pre-ovulatory hierarchical follicles allows for the manual separation of the theca and granulosa tissues with ease and minimal contamination [10], making it an ideal model for the conduction of expression studies that are tissue-specific in avian species. Such studies are not as feasible in mammals where the theca and granulose cells of the follicle are more intermixed and not easy to separate manually.

The germinal disc region (GDR) of the pre-ovulatory hierarchical follicles is also easily identified, as it is white in colour and stands out on the surface of the yellow-yolk-containing follicles. The GDR consists of the germinal disc, also known as the blastodisc, where the sperm will penetrate during fertilization and the embryo will begin to form, in addition to the granulosa cells that lie over it. The GDR of pre-ovulatory hierarchical follicles contains all of the hen's genetic material needed for procreation. Destruction of the GDR induces apoptosis and atresia of the follicle, suggesting that the GDR produces a factor that is essential for ovulation and follicle viability [10]. Apoptosis is defined as programmed cell death and is the mechanism by which the granulosa cells of an ovarian follicle die and detach from the basal lamina. The oocyte of each immature follicle dies and at this point it is considered to have undergone atresia, at which point it is termed an atretic follicle. Eventually the entire structure undergoes phagocytosis and resorption.

Follicular development

Maturing follicles undergo vitellogenesis, or yolk accumulation, which occurs in three phases: (1) a slow growth phase occurs in follicles 60–100 μm in diameter over the course of several months and possibly years whereby the yolk deposited into the follicle consists mainly of white neutral lipids to comprise the LWF; (2) an intermediate growth phase where the diameter of the follicle increases from 2 to 6 mm as yolk protein begins to be deposited into the follicle over a period of 60 days and the follicle matures into a small yellow follicle; and (3) a rapid growth phase, which occurs during the final days before ovulation when the bulk of yellow yolk consisting of nutrients, proteins, and lipids is deposited into the follicle and the diameter and volume of the follicle increase significantly per day [6]. The number of follicles in the rapid growth phase remains consistent as these follicles make up the hierarchical follicles. A follicle in the intermediate growth phase only enters the hierarchy and the rapid growth phase after the largest follicle (F1) is ovulated [11].

Regulation of growth, differentiation and ovulation of avian pre-ovulatory hierarchical and non-hierarchical follicles is dependent upon the timely release of the gonadotropin hormones: follicle-stimulating hormone (FSH) and luteinizing hormone (LH) [12] from the anterior pituitary gland. LH primarily acts upon the F1 and F2 follicles in order to prepare these follicles for ovulation, whereas the F3–F5 follicles undergo significant follicular growth as a result of FSH stimulation [10].

Reproduction: ovulation of the ovarian follicles

Avian species serve as excellent biological models for reproductive studies in comparison with mammalian species because of their consecutive ovulations, which occur, depending upon the species, approximately every 24 or 48 h. Ovulation is easily predicted in domesticated poultry because it occurs approximately 20–40 min after ovipositioning, or egg laying. Similar to maturation and proliferation, ovulation is highly dependent upon the regulatory activities of gonadotropin releasing hormone (GnRH) and its stimulatory effect on FSH and LH, all of which are secreted as part of the hypothalamic-pituitary-gonadal (HPG) axis. The HPG axis is comprised of the synergistic relationship between the hypothalamus, the anterior pituitary gland and the ovaries (female gonads). The HPG axis modulates growth and ovulation through the interaction of steroids and hormones secreted by these three components. Ovulation is induced by a rise in FSH secretion from the anterior pituitary gland 12–16 h prior to ovulation [13] and a pre-ovulatory surge of LH also from the anterior pituitary gland 4–6 h prior to ovulation [14]. The LH surge is accompanied by simultaneous increases in progesterone (P4) and oestrogen secretion from the F1 follicle [9].

The HPG axis

The HPG axis plays an essential role in regulating female reproduction as it mediates follicular maturation, development and ovulation [15]. GnRH also formerly known as luteinizing hormone releasing hormone (LHRH) is synthesized and stored in the hypothalamus. Specifically, GnRH is secreted by neurons in the pre-optic and arcuate nuclei of the hypothalamus into the hypophyseal portal system, or blood vessels, which link the hypothalamus and the pituitary gland [15]. When secreted, GnRH binds to its receptors at the anterior pituitary gland stimulating the synthesis and release of FSH and LH. At the level of the ovary, FSH regulates granulosa cell differentiation and has been detected in the follicle 15 h prior to ovulation in the domestic hen [13]. LH administration has been found to increase plasma concentrations of progesterone, androgens and oestrogens and can result in follicular atresia or premature ovulation of follicles, depending upon the follicle's position within the ovulatory sequence [7].

In addition to acting on several target tissues in the reproductive and central nervous system, the gonadal steroids secreted from the follicles: progesterone (granulosa cells), androgens and oestrogens (theca cells) exert feedback at the level of the anterior pituitary gland further mediating release of FSH and LH [15]. Because of the detection of increased LH plasma concentrations in hens 4–7 h prior to ovulation in conjunction with increased levels of progesterone [16], researchers were led to investigate the relationship between increased levels of gonadal steroids and the pre-ovulatory release of LH in the laying hen [17]. Intramuscular injections of progesterone and oestrogen in ovariectomized hens were observed to exert a positive feedback action on LH secretion, leading to the conclusion that LH release is facilitated by the combined actions of oestrogen and progesterone [17]. The HPG axis and its effects on ovulation and follicular maturation are not only modulated by feedback from the gonadal steroids but also by interactions with the hypothalamic–pituitary–adrenal (HPA) axis, also known as the stress axis [18].

Part I: Summary

Overall, the histological structure and hierarchy of the hen's ovary provide an excellent paradigm for studying the modulation of follicular selection, growth, development, survival and atresia, given the highly regulated and visually apparent avian follicular hierarchy and the ease with which the histological structure can be manipulated and analysed. Understanding the growth phases of follicular development, and the influence of the hormones, which regulate maturation and ovulation of the avian ovarian follicles leads to analysis of the relationship between the hormones of the HPG axis and their collaborative roles in influencing reproduction. In order to fully understand how gonadal function is modulated, it is also important to examine the interactions of the HPG axis with the HPA axis.

Part II: Interaction and Dysfunction of the HPG (Reproductive) and HPA (Stress) Axes

The HPA axis

The HPA axis involves interactions among the hypothalamus, the anterior pituitary gland, and the adrenal glands [19]. Interaction of these endocrine glands regulates the body's neuroendocrine response to stress and influences essential bodily functions such as digestion, immunity, reproduction, and metabolism of nutrients [20].

Corticotropin-releasing hormone (CRH) is released from the hypothalamic neurons in response to stressors [20]. CRH stimulates the release of adrenocorticotropic hormone (ACTH) from the anterior pituitary gland. As a result, CRH and ACTH directly regulate daily variation in the production of glucocorticoids; cortisol (in mammalian species) or corticosterone (in avian species) at the level of the adrenal glands [21], which are located on top of the kidneys. In turn, these glucocorticoids may exert negative feedback control on the hypothalamic release of CRH and the anterior pituitary gland's release of ACTH. Increased plasma concentrations of glucocorticoids have been found to be inhibitory to reproductive function in many species [18] indicating that the stress axis modulates the normal functions of the HPG axis.

Dysfunction of the HPA axis and impacts on the HPG axis

Research suggests that the HPA axis plays an important role in regulating reproduction through a variety of endocrine regulatory points [22]. For example, stress can hinder reproductive function by inhibiting or interfering with the timing of reproductive hormone release in the HPG axis [22]. Reproductive hormones or gonadotropins such as LH and FSH, which have the ability to affect growth of the gonads, are essential in the process of steroidogenesis at the level of the ovary [6]. LH stimulates androstenedione production in theca cells and using theca cells as a substrate, androstenedione is aromatized into oestrogens and androgens.

Studies in Wistar rats have shown that hypo- and hyper-corticosteronisms can impair ovarian endocrine and exocrine function [23]. Corticosterone administered to Wistar rats resulted in a decrease in FSH, LH and oestradiol levels [23]. Such interference occurs at the level of the hypothalamus when secretion of CRH from the hypothalamus exerts an inhibitory effect on GnRH secretion, blocking secretion of LH and FSH from the anterior pituitary gland and limiting the production of progesterone and oestrogen at the level of the ovary [15, 24, 25].

As a result, limited production of progesterone and oestrogen leads to disruptions in functional follicular maturation [26, 27].

In addition to exerting an effect at the level of the hypothalamus, the HPA axis can also inhibit the HPG axis at the level of the target tissues [15]. The avian ovary is not able to produce corticosterone *de novo,* and as a result, corticosterone that impacts the ovary is synthesized in the adrenal cortex and secretion of it is mediated by the HPA axis [28]. Anatomically, glucocorticoid receptors (GRs) and ovarian CRH receptors, in addition to components of the HPA axis are located near reproductive tissues and modulate their activity [15].

Given the close anatomic location of the adrenal gland to the left ovary of the laying hen, the left ovary is highly innervated by the adrenal gland, which further suggests interactions between the two [29]. When high levels of corticosterone are induced from lack of food and water (dietary stress) or infused daily, ovarian regression is observed and a decrease in the gonadotropin, LH is also observed [30]. The total number of large yellow-yolk-filled hierarchical follicles decreases and the number of atretic, pre-hierarchical follicles increases [30].

Dysfunction of the HPG axis

As the laying hen ages, and levels of gonadotropin hormones fluctuate, dysfunction of the HPG axis may occur [31]. LH positive-feedback mechanisms normally modulate the release of progesterone, but as ovaries age and lose function, regulating steroids such as oestrogen and progesterone are less concentrated in the blood to feedback into the HPG axis, resulting in lowered levels of LH [31]. Lowered function of the LH positive feedback mechanism is thought to be responsible for the fall in the rate of lay towards the end of the hens' laying cycle [31]. Ovariectomized hens showed lowered function of the LH positive-feedback mechanism until injected with proper doses of progesterone and oestrogen [31]. This led researchers to conclude that even with the onset of age, the ovary may potentially be fully functional and that changes in the rate of lay and fluctuations of gonadotropin levels may also be because of functional changes in the central nervous system (HPG axis) [31].

Influence of glucocortcoids on avian ovarian function

Given the significant interactions between the HPA and HPG axes, it is important to understand the role of corticosterone, the avian glucocorticoid, in avian ovulation and egg production. Glucocorticoids affect ovarian function through three primary modes: (1) glucocorticoids act on the HPG axis and exert an inhibitory effect at the level of the anterior pituitary gland [22]; (2) glucocorticoids may influence levels of ovarian growth factors and metabolic hormones, which are essential to ovarian development and function [32, 33]; and (3) glucocorticoids can exert direct effects at the level of the ovary [27].

The effects of glucocorticoids at the level of the anterior pituitary gland can result in disruption of follicular maturation because of increased amounts of progesterone and oestrogen in the theca and granulosa tissues of the follicle. Additionally, receptors in the ovary for the enzyme P450 aromatase, involved in oestrogen biosynthesis, and LH are necessary prerequisites for proper follicular maturation [27]. Glucocorticoids have been shown to suppress the expression of the enzyme P450 aromatase [34] and LH receptors [35] in cultured rat granulosa cells, impacting follicular growth and development. Similar results have been found in cultured bovine [26] and swine [36] granulosa cells.

Steroidogenesis occurs primarily in the granulosa and theca cells of the ovarian follicle [6]. The granulosa cells are considered to be the primary source of secreted progesterone. In contrast, the theca cells produce oestrogens and androgens [6]. Glucocorticoids are believed to interfere in this process by suppressing the aromatization of androstenedione into oestrogen in cultured rat granulosa cells [34]. Glucocorticoids have also been found to stimulate the production of progesterone in cultured rat granulosa cells and in bovine granulosa and theca cells [26, 34, 37]. These results suggest that glucocorticoids cause both inhibitory and stimulatory effects at the level of the ovary [34]. Because increased levels of progesterone play a role in the induction of ovulation [38], these results also suggest that glucocorticoids play a more important role in ovulation than in follicular maturation [27].

Glucocorticoids can also affect the ovulatory process by inhibiting the synthesis and expression of enzymes produced by follicles, which are essential for ovulation, such as prostaglandin E2 (PGE2) [39], and enzymes such as phospholipase A2 and cyclooxygenase 2 that are responsible for PGE2 production [40]. According to these findings, excessive concentrations of glucocorticoids may thus block ovulation [27].

Additionally, glucocorticoids exert a direct effect on the ovary by playing a role in mediating the processes of follicular rupture and apoptosis. Follicular rupture can occur when proteases act on the follicular wall. Tissue-type plasminogen (tPA) is an example of a protease that glucocorticoids have been shown to increase activity and expression of in cultured rat granulosa cells [41]. Alternatively, confounding studies have found that glucocorticoids prevented apoptosis in human and rat granulosa cells [42, 43], while also inducing apoptosis in invading leukocytes, which protect follicular cells [42]. These findings have led researchers to suggest that glucocorticoids may be involved in minimizing damage caused by the inflammatory ovulatory process by promoting repair and luteinization of ovulated follicles [27, 44].

Regulation of glucocorticoid secretion via circadian rhythms

Previous studies have shown that hens maintained on a constant lighting schedule (e.g. 14 h of light and 10 h of darkness) exhibit pre-ovulatory surges of LH during the period of darkness, suggesting circadian regulation of ovulation [31, 44, 45]. Corticosterones may be part of the mechanism responsible for modulating the responsiveness of the hypothalamus to circadian rhythms because of findings that include a daily pattern of circulating corticosterone levels in chickens [6, 46–49]. In hens exposed to 14 h of light and 10 h of darkness, a surge of corticosterone was detected 2 h preceding ovulation during the dark period and was concluded to be the result of coincident ovipositioning [14]. Another minimally invasive study on circadian rhythms involved the insertion of a cannula into individual hens to take blood samples of corticosterone, which showed a clear, daily rhythm of corticosterone [50].

In studies using metyrapone, a drug that blocks corticosterone synthesis, to manipulate the effectiveness of exogenous ACTH, results suggested that environmental stimuli such as light can act via the adrenal gland to modulate the timing of the pre-ovulatory release of LH in the hen, therefore demonstrating the possible inverse relationship between plasma concentrations of LH and corticosterone [51].

Enzymatic regulation of corticosterone production and activity

The ovary modulates normal levels of active glucocorticoids locally during follicular maturation and ovulation with two isoforms of the enzyme 11β-hydroxysteroid dehydrogenase (11β-HSD1 and 11β-HSD2), which catalyse the conversion between active and inactive glucocorticoids. 11β-HSD1 is an oxidoreductase that is expressed in the liver, adipose tissue, brain, lung, vasculature, ovary, central nervous system and other glucocorticoid target tissues [52, 53]. In order for 11β-HSD1 to modulate glucocorticoid activity, it must be activated when there is a substantial pool of substrate 11-ketosteroids (cortisone, 11-dehydrocorticosterone) available which are inactive precursors to glucocorticoids. The oxo-reductase activity of 11β-HSD1 increases levels of active glucocorticoids during ovulation while in contrast the dehydrogenase activity of 11β-HSD2 suppresses active glucocorticoids during follicular maturation [27]. 11β-HSD2 is expressed in aldosterone-selective target tissues such as the distal nephron, colon, sweat glands and the placenta [54] and has been shown to be highly down-regulated in human and rat granulosa cells during the ovulatory process triggered by the HPG axis, whereas 11β-HSD1 is up-regulated [55, 56]. The presence of both isoforms has been characterized in various ovarian tissues such as humans [55, 57, 58], rats [56, 59–61], mice [62] and cattle [63].

The GR

Glucocorticoids modulate ovarian function by binding to GRs. The fact that glucocorticoids can be inhibitory and stimulatory at the same time in ovarian tissue may be explained by identifying the GR as the mediator of glucocorticoid action in the ovary [64]. Therefore, analysis and characterization of the expression of functional GRs in target tissues can be a direct measurement of the potential action of glucocorticoids on that tissue [65].

The GR is a type I nuclear receptor, member of subfamily 3, group C, member 1 that binds the mammalian glucocorticoid, cortisol, and the avian glucocorticoid, corticosterone. GR expression has been characterized in the rat [56, 61, 64, 66] ovarian follicles and *corpora lutea*.

The complete cDNA for mammalian GR in the human [67], mouse [68], rat [69] and rabbit [70] has been characterized and cloned in addition to teleost fish such as trout [71]. The entire cDNA for avian species GR had not been cloned until the cDNA for chicken GR was cloned and characterized in its entirety from the chicken kidney by [72]. GR was detected and found to be widely expressed in 12 avian tissues: brain, pituitary, lung, heart, liver, kidney, intestine, pancreas, breast muscle, spleen, testis and the ovary [72]. Relatively high expression of GR was detected in pituitary, muscle, ovary and kidney tissues [72].

While the tissue-specific characterization of GR in the chicken ovary has yet to be attempted, the GR has been characterized in the follicles and *corpus luteum* of the human [57] and rat [63, 64, 67]. In humans, the GR was characterized in the ovarian surface epithelial cells [57] whereas in the rat, GR mRNA was detected mainly in preantral follicle granulosa cells [64]. Additionally, bovine GR mRNA has been detected in granulosa and theca tissues [73]. This study also examined differences in expression levels of GR in dominant and small developing follicles, and even though GR expression in granulosa and thecal tissues did not differ between different types of bovine follicles [73], it is important to continue to examine for this possibility.

Part II: Summary and Conclusions

The HPA (stress) axis regulates normal reproduction by modulating the release of hormones within the HPG axis. Dysfunction of either of these two axes can occur for a variety of reasons and can impact the level of hormones released that may act directly on target tissues of the ovary. Glucocorticoids, the primary class of stress hormones released by way of the HPA axis, affect ovulation and egg production by modulating the HPG axis and influencing the levels of growth factors and hormones acting on the ovary. Secretion of glucocorticoids via the HPA axis is regulated via circadian rhythms and enzymatic activity.

In order to better understand the influence of glucocorticoids in modulation of gonadal function, characterization of the GR has occurred in a variety of species in several tissues including the reproduction tract. Further tissue-specific characterization is necessary in the laying hen's hierarchical and non-hierarchical follicles in order to examine the role of glucocorticoids in regulation of follicular recruitment and development. Additionally, future examination of differences in GR expression in hens reared in different environmental management methods may provide insight into how such management methods impact reproductive fitness because of potentially adverse effects of glucocorticoids on follicular development, maturation and ovulation.

References

1. Mishra A, Koene P, Schouten W, Spruijt B, van Beek P, Metz JHM. Environment, well-being, and behavior: temporal and sequential structure of behavior and facility usage of laying hens in an enriched environment. Poultry Science 2005;84:979–91.
2. Hutson J, Ikawa H, Donahoe PK. The ontogeny of Mullerian inhibiting substance in the gonads of the chicken. Journal of Pediatric Surgery 1981;16(6):822–7.
3. Hutson JM, Ikawa H, Donahoe K. Estrogen inhibition of Mullerian inhibiting substance in the chick embryo. Journal of Pediatric Surgery 1982;17(6):953–9.
4. Teng CS. Differential activation of MAPK during Mullerian duct growth and apoptosis: JNK and p38 stimulation by DES blocks tissue death. Cell and Tissue Research 2001;306(1):27–34.
5. Johnson PA, Kent TR, Urick ME, Giles JR. Expression and regulation of anti-Mullerian hormone in an oviparous species, the hen. Biology of Reproduction 2008;78(1):13–9.
6. Johnson AL. Review: Steroidogenesis and actions of steroids in the hen ovary. Poultry Biology 1990;2(4):319–46.
7. Gilbert AB, Perry MM, Waddington D, Hardie MA. Role of atresia in establishing the follicular hierarchy in the ovary of the domestic hen (*Gallus domesticus*). Journal of Reproduction and Fertility 1983;69:221–7.
8. Whittow GC, editor. Sturkie's Avian Physiology, 5th ed. Academic, San Diego; 2000.
9. Etches R. The ovulatory cycle of the hen. Critical Reviews in Poultry Biology 1990;2(4):293–318.
10. Hernandez A. Thesis: Extraovarian and Intraovarian Control of Follicular Selection in the Chicken Ovary 2001.
11. Zakaria A, Sakai H, Imai K. Time of follicular transformation to the rapid growth phase in relation to the ovulatory cycle of laying hens. Poultry Science 1984;63:1061–3.
12. Von Engelhardt N, Groothius TGG. Measuring steroid hormones in avian eggs. Annals of the New York Academy of Sciences 2005;1046:181–92.
13. Scanes CG, Godden PMM, Sharp PJ. An homologous radioimmunoassay for chicken follicle stimulating hormone: observations on the ovulatory cycle. Journal of Endocrinology 1977;73:473–81.
14. Johnson AL, van Tienhoven A. Plasma concentration of six steroids and LH during the ovulatory cycle of the hen, *Gallus domesticus*. Biology of Reproduction 1980;23:386–93.
15. Chrousos GP, Torpy DJ, Gold PW. Interactions between the hypothalamic-pituitary-adrenal axis and the female reproductive system: clinical implications. Annals of Internal Medicine 1998;129(3):229–40.

16. Lague PC, van Tienhoven A, Cunningham FJ. Concentrations of estrogens, progesterone and LH during the ovulatory cycle of the laying chicken. Biology of Reproduction 1975;12:590–8.

17. Wilson S, Sharp P. Induction of luteinizing hormone release by gonadal steroids in the ovariectomized domestic hen. Journal of Endocrinology 1976;71:87–98.

18. Gore AC, Attardi B, DeFranco DB. Glucocorticoid repression of the reproductive axis: effects on GnRH and gonadotropin subunit mRNA levels. Molecular and Cellular Endocrinology 2006;256:40–8.

19. Harvey S, Phillips JG, Rees A, Hall TR. Stress and adrenal function. The Journal of Experimental Zoology 1984;232:633–45.

20. Harvey S, Hall TR. Hormones and stress in birds: activation of the hypothalamo-pituitary-adrenal axis. Progress in Clinical Biological Research 1990;342:453–60.

21. Carsia RV, Harvey S. Adrenals. In: Whittow G, editor. Sturkie's Avian Physiology. 5th ed. Academic Press, New York; 2000. p. 489–537.

22. Dobson H, Smith RF. What is stress, and how does it affect reproduction? Animal Reproduction Science 2000;60–61:743–52.

23. Valli G, Sudha S, Sankar R, Govindarajulu P, Srinivasan N. Altered corticosterone status impairs steroidogenesis in the granulosa and thecal cells of Wistar rats. The Journal of Steroid Biochemistry and Molecular Biology 2000;73:153–8.

24. Saketos M, Sharma N, Santoro NF. Suppression of the hypothalamic-pituitary-ovarian axis in normal women by glucocorticoids. Biology of reproduction 1993;49:1270–6.

25. Matsuwaki T, Kayasuga Y, Yamanouchi K, Masugi N. Maintenence of gonadotropin secretion by glucocorticoids under stress conditions through the inhibition of prostaglandin synthesis in the brain. Endocrinology 2006;147(3):1087–93.

26. Kawate N, Inaba T, Mori J. Effects of cortisol on the amounts of estradiol-17beta and progesterone secreted and the number of luteinizing hormone receptors in cultured bovine granulosa cells. Animal Reproduction Science 1992;32:15–25.

27. Tetsuka M. Review Article: Actions of glucocorticoid and their regulatory mechanisms in the ovary. Animal Science Journal 2007,78:112–20.

28. Omura T, Morohashi K. Gene regulation of steroidogenesis. Journal of Steroid, Biochemistry, and Molecular Biology 1995;53(1–6):19–25.

29. Etches RJ, Petite JN, Anderson-Langmur CE. Interrelationships between the hypothalamus, pituitary gland, ovary, adrenal gland, and the open period for LH Release in the hen (*Gallus domesticus*). The Journal of Experimental Zoology 1984;232:501–11.

30. Etches RJ, Williams JB, Rzasa J. Effects of corticosterone and dietary changes in the hen on ovarian function, plasma LH and steroids and the response to exogenous LH-RH. Journal of Reproduction and Fertility 1984;70:121–30.

31. Williams J, Sharp P. Age-dependent changes in the hypothalamo-pituitary-ovarian axis of the laying hen. Journal of Reproduction and Fertility 1978;53:141–6.

32. Thakore JH, Dinan TG. Mini review: growth hormone secretion: the role of glucocorticoids. Life Sciences 1994;55(14):1093–9.

33. Kritsch KR, Murali S, Adamo ML, Ney DM. Dexamethasone decreases serum and liver IGF-I and maintains liver IGF-I mRNA in parenterally fed rats. American Journal of Physiology: Regulatory, Integrative and Comparative Physiology 2002;282:R528–36.

34. Hsueh AJW, Erickson GF. Glucocorticoid inhibition of FSH-induced estrogen production in cultured rat granulosa cells. Steroids. 1978;32(5):639–48

35. Schoonmaker JN, Erickson GF. Glucocorticoid modulation of follicle-stimulating hormone-mediated granulosa cell differentiation. Endocrinology 1983;113:1356–63.

36. Danisova A, Sebokova E, Kolena J. Effect of corticosteroids on estradiol and testosterone secretion by granulosa cells in culture. Experimental and Clinical Endocrinology 1987;89(2):165–73.

37. Spicer LJ, Chamberlain CS. Influence of cortisol on insulin- and insulin-like growth factor 1 (IGF-1)-induced steroid production and on IGF-1 receptors in cultured bovine granulosa cells and thecal cells. Endocrine 1998;9:153–61.

38. Richards JS, Russell DL, Ochsner S, Espey LL. OVULATION: New dimensions and new regulators of the inflammatory-like response. Annual Review of Physiology 2002;64(1):69–92.

39. Davis BJ, Lennard DE, Lee CA, Tiano HF, Morham SG, Wetsel WC, *et al.* Anovulation in cyclooxygenase-2-deficient mice is restored by prostaglandin E2 and interleukin-1beta. Endocrinology 1999;140:2685–95.

40. Ben-Schlomo I, Kol S, Ando M, Altman KR, Putowski LT, Rohan RM, *et al.* Ovarian expression, cellular localization, and hormonal regulation of rat secretory phospholipase a2: increased expression by interleukin-1 and by gonadotropins. Biology of Reproduction 1997;57:217–25.

41. Wang C, Leung A. Glucocorticoids stimulate plasminogen activator production by rat granulosa cells. Endocrinology. 1989;124(4):1595–601.

42. Amsterdam A, Sasson R. The anti-inflammatory action of glucocorticoids is mediated by cell type specific regulation of apoptosis. Molecular and Cellular Endocrinology 2002;189:1–9.

43. Sasson R. Pleiotropic anti-apoptotic activity of glucocorticoids in ovarian follicular cells. Biochemical Pharmacology 2003;66(8):1393–401.

44. Hillier SG, Tetsuka M. An anti-inflammatory role for glucocorticoids in the ovaries. Journal of Reproductive Immunology 1998;39:21–7.

45. Wilson S, Sharp P. Variations in plasma LH levels during the ovulatory cycle of the hen, *Gallus Domesticus*. Journal of Reproduction and Fertility 1973;35:561–4.

46. Seigel HS, Mitchell BW, Gould NR, Latimer JW, Wilson RL, editors. Circadian Rhythms for Corticosterone, Corticosteroid Binding Capacity, Plasma Glucose, Heart Rate, Respiration Rate and Body Temperature in White Rock Males. Proceedings of the 5th European Poultry Congress, Malta; 1976

47. Etches R. Plasma concentrations of progesterone and corticosterone during the ovulation cycle of the hen (*Gallus domesticus*). Poultry Science 1979;58:211–6.

48. Wilson S, Cunningham F. Effect of photoperiod on the concentrations of corticosterone and luteinizing hormones in the plasma of the domestic hen. Journal of Endocrinology 1981;91:135–43.

49. Wilson S, Jennings R, Cunningham F. Developmental changes in the diurnal rhythm of secretion of corticosterone and LH in the domestic hen. Journal of Endocrinology 1984;101:299–304.

50. Beuving G, Vonder GMA. Daily rhythm of corticosterone in laying hens and the influence of egg laying. Journal of Reproduction and Fertility 1977;51:169–73.

51. Wilson S, Cunningham F. Modification by metyrapone of the 'open period' for pre-ovulatory LH release in the hen. British Poultry Science 1980;21:351–61.

52. Monder C. The forms and functions of 11-betaHSD. Journal of Steroid, Biochemistry, and Molecular Biology 1993;45(1–3):161–5.

53. Stewart PM, Krozowski ZS. 11beta hydroxysteroid dehydrogenase. Vitamins and Hormones 1999;57:249–324.

54. Seckl JR. Minireview: 11 -hydroxysteroid dehydrogenase type 1– a tissue-specific amplifier of glucocorticoid action. Endocrinology 2001;142(4):1371–6.

55. Tetsuka M, Thomas FJ, Thomas MJ, Anderson RA, Mason JI, Hillier SG. Differential expression of messenger ribonucleic acids encoding 11B-hydroxysteroid dehydrogenase types 1 and 2 in human granulosa cells. Journal of Clinical Endocrinology and Metabolism 1997;82:2006–9.

56. Tetsuka M, Milne M, Simpson GE, Hillier SG. Expression of 11beta-hydroxysteroid dehydrogenase, glucocorticoid receptor, and mineralcorticoid receptor genes in rat ovary. Biology of Reproduction 1999;60:330–5.

57. Rae MT. Antiinflammatory steroid action in human ovarian surface epithelial cells. Journal of Clinical Endocrinology and Metabolism 2004;89(9):4538–44.

58. Albiston AL, Obeyesekere VR, Smith RE, Krozowski ZS. Cloning and tissue distribution of the human 11beta-HSD type 2 enzyme. Molecular and Cellular Endocrinology 1994;105:R11–R7.

59. Benediktsson R, Yau JLW, Low S, Brett LP, Cooke BE, Edwards CRW, *et al.* 11beta-Hydroxysteroid dehydrogenase in the rat ovary: high expression in the oocyte. Journal of Endocrinology 1992;135:53–8.

60. Roland BL, Funder JW. Localization of 11B-hydroxysteroid dehydrogenase type 2 in rat tissues: *in situ* studies. Endocrinology 1996;137:1123–8.

61. Waddell BJ, Benediktsson R, Seckl JR. 11B-Hydroxysteroid dehydrogenase type 2 in the rat *corpus luteum*: induction of messenger ribonucleic acid expression and bioactivity coincident with luteal regression. Endocrinology 1996;137:5386–91.

62. Condon J, Ricketts ML, Whorwood CB, Stewart PM. Ontogeny and sexual dimorphic expression of mouse type 2 11beta-HSD. Molecular and Cellular Endocrinology 1997;127:121–8.

63. Tetsuka M, Yamamoto S, Hayashida N, Hayashi KG. Expression of 11beta-HSD in bovine follicle and *corpus luteum*. Journal of Endocrinology 2003;177:445–52.

64. Schreiber JR, Nakamura K, Erickson GF. Rat ovary glucocorticoid receptor: identification and characterization. Steroids 1982;39(5):569–83.

65. Bamberger CM, Bamberger AM, Castro M, Chrousos GP. Glucocorticoid receptor beta, a potential endogenous inhibitor of glucocorticoid action in humans. The Journal of Clinical Investigation 1995;95:2435–41.

66. Towns R, Menon KMJ, Brabec RK, Silverstein AM, Cohen JM, Bown JM, *et al.* Glucocorticoids stimulate the accumulation of lipids in the rat *corpus luteum*. Biology of Reproduction 1999;61:416–21.

67. Hollenberg SM, Weinberger C, Ong ES, Cerelli G. Primary structure and expression of a functional human glucocorticoid receptor cDNA. Nature 1985;318(19):635–41.

68. Francke U, Gehring U. Chromosome assignment of a murine glucocorticoid receptor gene (Grl-1) using intraspecies somatic cell hybrids. Cell 1980;22:657–64.

69. Miesfeld R, Rusconi S, Godowski PJ, Maler BA. Genetic complementation of a glucocorticoid receptor deficiency by expression of cloned receptor cDNA. Cell 1986;46:389–99.

70. James ER. Presence of a transcriptionally active glucocorticoid receptor in lens epithelial cells. Investigative Ophthalmology and Visual Science 2003;44(12):5269–5276.

71. Gao X, Kalkhoven E, Peterson-Maduro J, van der Burg B, Destree OH. Expression of the glucocorticoid receptor gene is regulated during early embryogenesis of *Xenopus laevis*. Biochimica et Biophysica Acta 1994;1218:194–8.

72. Kwok AHY, Wang Y, Wang CY, Leung FC. Cloning of the chicken glucocorticoid receptor (GR) and characterization of is expression in pituitary and extrapituitary tissues. Poultry Science 2007;86:423–30.

73. Tetsuka M, Nishimoto H, Miyamoto A, Okuda K, Hamano S. 11HSD in the bovine follicle: gene expression of 11beta-HSD and glucocorticoid receptor in the bovine (*Bos taurus*) follicle during follicular maturation and atresia: the role of follicular stimulating hormone. Journal of Reproduction and Development 2010;Published online.

Communicable diseases and the human–animal relationship - Mini Review

Roger Webber

Abstract

Communicable diseases continue to be a major health problem especially in developing countries, with respiratory infections causing the greatest mortality. The method of transmission is the key to grouping similar infections and the focus for investigating new diseases. Several new infections have resulted in species transfer, as has occurred recently with influenza. Although restricted to a few countries and individuals coming in close contact with poultry, avian influenza A(H5N1) remains a continuing threat in that it might mutate with an existing strain of human influenza to produce a highly lethal infection.

Keywords: Communicable diseases, Infection, Influenza, Avian

The spectrum of communicable diseases is extensive and can become a major challenge to the health worker trying to master the subject. There are the old scourges such as plague and malaria, complex diseases in themselves, and then suddenly, as if out of nowhere, comes a new frightening infection such as swine 'flu' A(H1N1), or West Nile virus appearing in North America, when (as its name suggests) it is normally restricted to Africa. Alarm is raised and they become the topic of the moment, instilling unease in the population as people seek to understand the implications. However, the burden of known diseases is still considerable and many show evidence of having inflicted populations from antiquity.

Most are related to poverty so that they are commoner in the poorer parts of the world, particularly in developing countries. Globally, the five most important are illustrated in Table 1. The respiratory diseases including tuberculosis are by far the commonest cause of death and morbidity from communicable diseases in the world [1] Children, in particular, bear the brunt of these infections.

Biological processes do not follow simple rules, so it is never easy to categorize them, but some order needs to be brought to the vast number of communicable diseases (in excess of 350) if any progress is to be made in understanding them and finding a method of control. Indeed, since control is the objective then the means of transmission, which is generally the main place at which control is aimed, becomes the key. Most diarrhoeal diseases are associated with food or water or both so by providing a clean water supply and cooking food properly could be immediate ways to control the infection before the full epidemiology of the disease has been worked out. Infections transmitted by the respiratory route have a similar means of transmission, so a way now becomes apparent of grouping the number of communicable diseases into manageable units.

When a new infection arises the immediate question is always: where has it come from? It often seems to have self-generated, until one looks more closely at the history. Here again, the key is transmission, how is it spread from person to person and is it possible that it entered the human population from a non-human source. Since humans are animals then some infecting organisms are species-specific, such as malaria, while others find no problem in infecting either, such as rabies. So it is often worthwhile to see if an organism has crossed the species barrier, as was classically shown by new variant Creutzfeldt–Jakob Disease (CJD). Originally a disease of sheep, called scrapie, when cows were fed infected offal they developed bovine spongiform encephalopathy (BSE) and in their turn humans consuming infected beef developed new variant CJD. An even less obvious route than the eating of infective material can occur as seems to be the origin of new strains of influenza. Avian influenza A(H5N1) is an infection predominantly of chickens and ducks, is highly infectious and results in an almost 100% mortality of domestic fowls. The infection is maintained in wild waterfowl, predominantly ducks and geese, which can spread the infection over long distances when they migrate. The virus circulates asymptomatically with individual birds remaining infected for about a month, excreting large quantities of virus in their droppings. Congregations of large numbers of birds prior to migration provide ideal conditions for transmission and mixing of avian influenza virus. While migrating wildfowl might have been the initiating cause of the outbreak of A(H5N1) the main method of continued transmission was probably by the transport of birds, contaminated clothing and equipment. Bird excrement is highly infectious so humans coming into contact with sick birds or their droppings are at risk of infection. The first human cases occurred in Hong Kong in 1997 with a high mortality, and it has since spread to several more countries (Figure 1) [2].

Table 1 The five most important communicable diseases in the world. Data from the World Health Organization, Geneva [1]

Disease	DALYS* (000)	%	Deaths (000)	%
Lower respiratory infections	94 511	25.42	4109	32.9
Diarrhoeal disease	72 777	19.57	2127	17.03
HIV/AIDS	58 513	15.74	2040	16.33
Tuberculosis	34 217	9.20	1464	11.72
Malaria	33 976	9.14	1021	8.17

*DALY, disability-adjusted life year.
The DALY is a calculation of the morbidity and mortality of the particular disease averaged out over the expected life of a person, so it reflects the prevalence of the disease and the disability it produces.

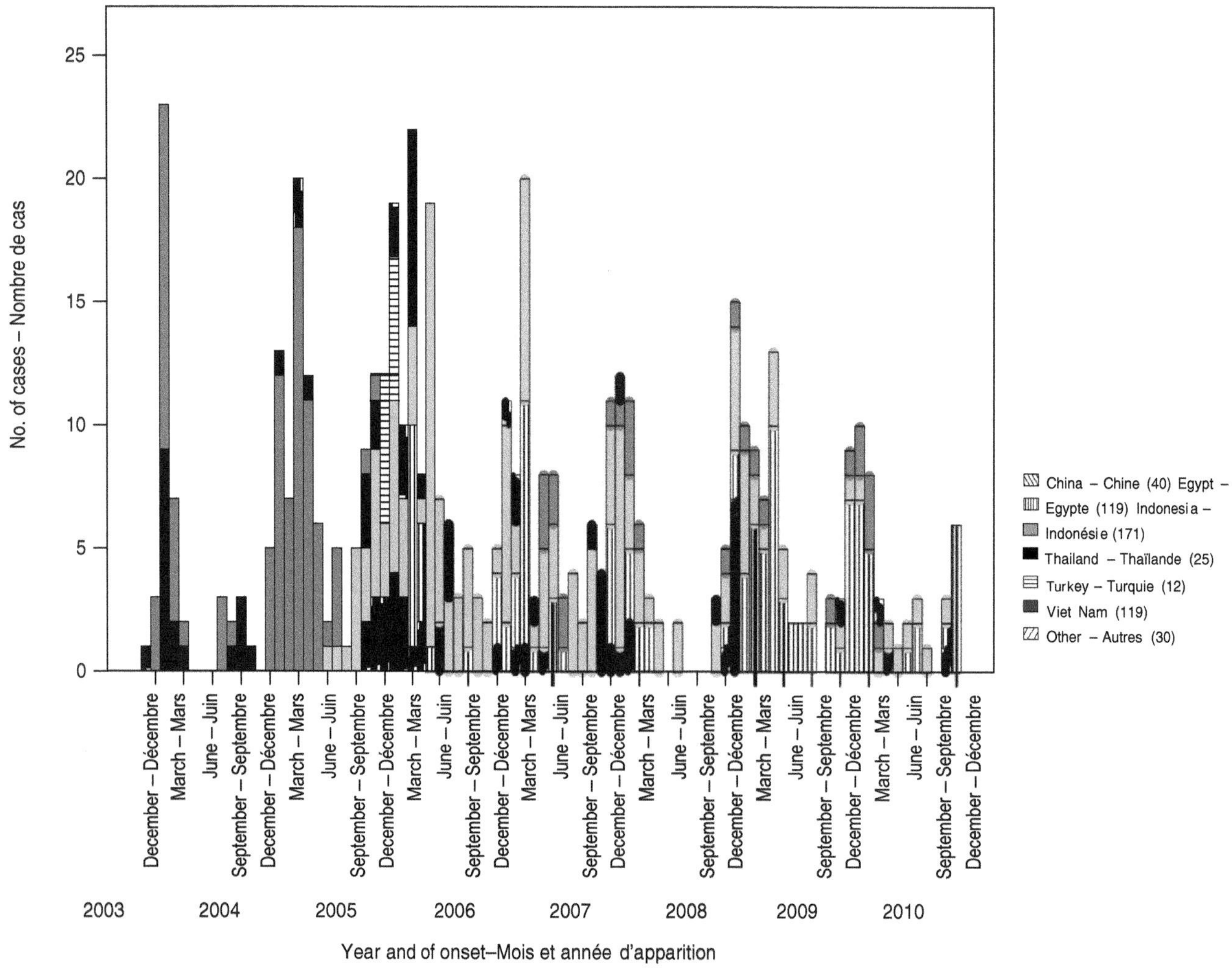

Figure 1 Number of confirmed human cases of infection with influenza A(H5N1) virus by month and country, 2003–2010 (reproduced with the permission of the World Health Organization, Geneva) [2]

It seems likely that this close association of humans with poultry might well be the origin of pandemic influenza. The so-called 'Asian flu' epidemic of 1957 was found to have originated in chickens in Guangdong province of southern China, followed in 1968 by the Hong Kong influenza epidemic, which also originated from the same area. Although the 1918 epidemic was termed 'Spanish flu', it is thought that this too might have started originally in southern China. There is a very close association between people and animals of all kinds in the southern part of China where they are held in cages, generally in unhygienic and restricted conditions, prior to being eaten. Such conditions provide the ideal environment for constant transmission of disease organisms so in time the chance of a new disease, not found before in humans (such as severe acute respiratory syndrome (SARS)) or the development of a new strain of an existing infection (such as influenza) could take place. There is considerable concern that avian influenza A(H5N1) might mutate with seasonal strains of influenza to produce a new strain of infection, with probably a high mortality.

This continues as a dangerous source of infection that could cross from animal-to-human and then continue as human-to-human transmission. Considerable effort needs to be made to control this trade with the implementation of good practice, the wearing of protective clothing, use of antiseptics to clean out cages and all other methods of prevention. Where people have come in contact with an infected bird then they can be given a neuramidase inhibitor (oseltamir, zanamivir, etc.) as a prophylactic. An A(H5N1) vaccine has been developed, which could be used to protect persons in high-risk occupations involved with poultry or its preparation, and will also be valuable should an epidemic occur.

References

1. World Health Organization. World Health Statistics 2011. WHO, Geneva, Switzerland; 2011.
2. World Health Organization. Update on human cases of highly pathogenic avian influenza A(H5N1) virus infection, 2010. Weekly Epidemiological Record 2011;86(17):161–6.

Analysing the genetic basis of milk production traits

Nicolas Lopez-Villalobos*

Address: Institute of Veterinary, Animal and Biomedical Sciences, Massey University, Private Bag 11222, Palmerston North 4442, New Zealand.

***Correspondence:** Email: N.Lopez-Villalobos@massey.ac.nz

Received: 9 February 2012
Accepted: 11 April 2012

Abstract

Genetic analyses of milk production traits have been based on a polygenic model that assumes that a trait is affected by an infinitesimal number of genes each with a very small effect. An updated review on the genetic parameters (heritabilities and genetic correlations) for milk components estimated using the polygenic model is provided in this review. There are many studies reporting estimates of genetic parameters for component yields of milk, fat and protein and concentrations of fat and protein. Few studies have considered lactose yield and percentage. A limited number of studies have reported estimates of genetic parameters for milk fat and protein composition. At the start of the new millennium, advances in DNA technology permitted the sequencing of bovine genome, the identification of large numbers of genetic markers across the genome in the form of single nucleotide polymorphisms (SNP) and the genotyping of thousands of these SNPs on individual animals at low costs. Based on these advances, genomic selection emerged as a new methodology to perform genetic analysis and estimate breeding values for milk production traits using genome-wide dense markers. Systems biology is becoming the new method to perform genetic analysis of milk production traits integrating results from genomics, transcriptomics, proteomics, metabolomics and bioinformatics studies. A conclusion from this review is that the genetic analysis of milk production traits has been changing dramatically during recent years from a polygenic infinitesimal model towards a systems biology approach.

Keywords: milk components, fat composition, protein composition, genetic parameters, quantitative trait loci, genome-wide association, genomic selection, systems biology

Introduction

Milk production is a major component of global food production and the implications of a change in milk composition are very significant. Factors to be considered include firstly the efficiency of the dairy cow in converting plant material, inedible to humans, into a human food of high nutritional value, and secondly, the strong genetic correlations between the fat content, the protein content and the yield of milk per cow [1].

Milk contains proteins, fats, lactose and various minerals and vitamins, and is produced by the mammary glands of all mature female mammals after they give birth and serves as nourishment for their offspring. For many years, milk has been considered as a food for humans of the most versatile composition, with a high nutrient density per energy unit, providing possibilities for many types of dairy products and also food products in which milk is an ingredient increasing the nutritional value of other foods.

There are many situations in which it may be advantageous to change the composition of milk. For instance, dairy farmers might be interested in changing the concentration of a particular milk component for economic reasons. Consumer demand for milk fat and milk protein changes over time and producers may wish to maximize their income by using nutritional strategies to alter the protein:fat ratio in milk. Likewise, dairy processors may be interested in strategies that increase casein for maximization of cheese yield, or in methods that increase

Table 1 Productivity and milk composition from three breeds of dairy cows in the United States (from [2]) and New Zealand (from [3])

	USA			New Zealand		
	Holstein	Jersey	Ayrshire	Holstein	Jersey	Ayrshire
Days in milk	305	305	305	219	219	226
Milk (kg)	11 630	8285	8264	3419	2466	3101
Fat (kg)	425	379	316	151	140	137
Protein (kg)	350	296	259	124	101	111
Fat (%)	3.65	4.58	3.82	4.42	5.68	4.42
Protein (%)	3.01	3.57	3.13	3.63	4.08	3.59
Protein:fat ratio	0.82	0.78	0.82	0.82	0.72	0.81

unsaturated fatty acids, resulting in a softer butter. Another example is the desire to change milk composition in ways that promote better human health.

Table 1 shows the average production of milk, fat and protein and the concentrations of fat and protein and the protein:fat ratio between different cow breeds and two contrasting countries, the USA and New Zealand. The production systems in the USA are mainly indoors where cows are fed total mixed ration, and the breeding objective has been focused on production of milk and components per cow with the majority of the milk destined for a large internal market. The production systems in New Zealand are under grazing conditions where cows receive relatively low levels of supplements, mainly in the form of maize and grass silage, palm kernel expeller, crushed grains, meals, commercial concentrates and winter brassica crops. The breeding objective for New Zealand dairy cattle has been focused on developing a dairy cow with a high capacity to harvest and transform pasture into milk solids (fat and protein), which are transformed into more than 700 dairy products sold in the international dairy market. Breed differences for milk production create opportunities for farmers to alter milk composition in a faster way than selection within breed. The Jersey breed is characterized by a low protein:fat ratio that may be best suited for butter and anhydrous milk fat.

Yields of milk and its components are quantitative traits that show a continuous variation in the population of dairy cows. The analyses of these traits assume that the variation is caused by the effects of multigene loci, by environmental influences and by genotype–environment interactions. The genetics of milk production traits in dairy cattle have been studied without identifying the genes involved. Traditionally, the total phenotypic variation has been estimated as the sum of the genetic variation plus the residual environmental variation.

In general, there are four ways of changing the genetic control of milk production in dairy cows [4, 5]. The first approach is selective breeding, in which parents of the next generation are selected based on estimated breeding values that are calculated from phenotypic records and pedigrees using a polygenic infinitesimal model; this approach has been improved during the 20th century by the introduction of advanced statistical and computational genetics procedures to better assess heritability of traits and breeding values of animals. The second approach is marker-assisted selection, in which parents are selected on a mix of breeding values and favourable polymorphisms of genetic markers associated with milk production traits; this approach has been around since the 1980s. The third approach is genetic modification that consists of single-gene addition or deletion. The fourth approach is incorporation of 'omics' technologies into selection and breeding; this 'omics' approach could fundamentally change the practice of animal breeding, moving away from the polygenic model towards an approach that considers the regulatory networks and pathways underlying the expression of milk production traits. The objective of this paper is to provide an updated review on the genetic basis of milk production traits focused in dairy cattle.

Partitioning of Hereditary Variance

After the discovery of Mendel laws [6], Fisher in 1918 [7] established a polygenic regression model to link the phenotype of an individual with its genotype, defined as the sum of many genes acting according to Mendelian inheritance. Fisher's polygenic model established the mathematical link between the phenotype and the genotype as a function of how much an allele substitution in a locus adds to the phenotype (additive effects), how two alleles of a gene interact with each other when they occur simultaneously (dominance effects) and how alleles of a gene interact with alleles of another gene (epistatic effects).

By the 1940s, researchers concluded that the synthesis of milk is under the genetic control of thousands of genes [8] and Fisher's [7] polygenic linear model was used to obtain estimates of heritability [9] and genetic correlations between milk components [10] using estimates of variance components based on dam-daughter records.

The polygenic model proposed by Fisher in 1918 [7] was subsequently extended by Fisher in 1925 [11] to study the partitioning of the total phenotypic variance into genetic and residual variances, which has provided the framework to study the genetics of milk production traits in dairy cattle. Cockerham [12] and Kempthorne [13] extended the concept of partitioning the total genetic variance into additive effects, dominance deviations and epistatic effects. The contribution of Fisher [11] for the analysis of variance has been one of the most useful statistical developments for animal breeding. Apart from its use to test mean differences, its application for the estimation of genetic and environmental variances has provided the foundation for genetic analyses used for parameter estimation in animal breeding.

A large number of papers have been dedicated to the estimation of variance and covariance components for milk production traits [14, 15, 16]. Unequal and unbalanced subclass number required special adaptations of the analysis of variance procedures [17] according to the expectation of mean squares of the fixed and random factors included in the linear model to represent the variance of the trait. Henderson [18] proposed an approach to account for adjustment of fixed effects, plus the adaptation of maximum likelihood in variance component estimation to obtain best linear unbiased predictors (BLUP) of random effects.

The importance of genetic parameters in estimating the true genetic value of an animal for various production traits of economic importance led to the development of several statistical tools such as restricted maximum likelihood (REML) [19, 20].

General mixed models are now routinely used by animal breeders for variance component estimation using REML and subsequent estimation of breeding values using BLUP [21]. These procedures rely on accurate measurements of phenotypic records and adequate pedigree structures of the animals under evaluation. Many computer packages use REML for the estimation of variance components, e.g. ASReml [22], MTDFREML [23], WOMBAT [24], DMU [25], REMLF90 [26] and VCE-5 [27] using large datasets because computer capabilities have increased significantly in the last decade.

In matrix notation, the general mixed model can be written as:

$$\mathbf{y}=\mathbf{Xb}+\mathbf{Zu}+\mathbf{e},$$

where **y** is the vector of phenotypes for the trait under study, **X** and **Z** are design matrices relating the fixed and random effects to the phenotypes, **b** is the vector of fixed effects such as the contemporary group defined as cows calving in the same herd, year and month, **u** is the vector of random effects; commonly sire, animal or cow permanent effects, and **e** is the vector of residual errors not accounted for the fixed and random effects. The simultaneous solution of fixed and random effects and estimates of variances can be obtained solving the following mixed model equations:

$$\begin{bmatrix} \mathbf{X'R^{-1}X} & \mathbf{X'R^{-1}Z} \\ \mathbf{Z'R^{-1}X} & \mathbf{Z'R^{-1}Z+G^{-1}} \end{bmatrix} \begin{bmatrix} \mathbf{b} \\ \mathbf{u} \end{bmatrix} = \begin{bmatrix} \mathbf{X'R^{-1}y} \\ \mathbf{Z'R^{-1}y} \end{bmatrix},$$

where $\mathbf{R}^{-1}$ is the inverse of the residual variance matrix and $\mathbf{G}^{-1}$ is the inverse of the variance matrix for **u**. For analysis of a single trait, it is generally assumed that the residual variances for each record are equal (homogeneous) and uncorrelated (independent). In this case, the residual variance is $\mathbf{R}^{-1}=\mathbf{I}\sigma_e^2$ and $\mathbf{R}^{-1}=\mathbf{I}/\sigma_e^2$. Similarly **G** can be defined as $\mathbf{G}=\mathbf{A}\sigma_a^2$ and $\mathbf{G}^{-1}=\mathbf{A}^{-1}/\sigma_a^2$ where **A** is the numerator relationship matrix, σ_a^2 is the additive genetic variance and σ_e^2 is the residual variance. The mixed model equations can be simplified by multiplying both sides by σ_e^2 as follows:

$$\begin{bmatrix} \mathbf{X'X} & \mathbf{X'Z} \\ \mathbf{Z'X} & \mathbf{Z'Z+A^{-1}}\sigma_e^2/\sigma_a^2 \end{bmatrix} \begin{bmatrix} \mathbf{b} \\ \mathbf{u} \end{bmatrix} = \begin{bmatrix} \mathbf{X'y} \\ \mathbf{Z'X} \end{bmatrix}.$$

These mixed model equations allow the simultaneous estimation of variances, σ_a^2 and σ_e^2, and solutions for fixed (**b**) and random (**u**) effects by iterative methods. Breeding values for large number of animals can be estimated using known estimates of σ_a^2 and σ_e^2, but solving the mixed model equations only to obtain best linear unbiased estimates of **b** and BLUP of **u**.

Currently, breeding values for lactation yields of milk, fat and protein of cows and bulls in countries with advanced systems of genetic evaluation are estimated using random regression models. A random regression model is intended for use on longitudinal data where observations for a trait are collected several times during the course of an animal's life period such as herd-test records for milk, fat and protein of the dairy cow during the lactation. A random regression model is a statistical procedure that considers all genetic and environmental effects directly on a test-day basis. The first estimates of variance components for test-day milk yields obtained by random regression models were published by Jamrozik and Schaeffer [28] in 1997. Random regression models allow the estimation of heritabilities for productions traits at each day of the lactation.

Estimation of Genetic Parameters for Milk Production Traits

Accurate estimates of genetic parameters including phenotypic and genetic standard deviations, heritabilities, and genetic and phenotypic correlations for milk production traits are necessary to predict the response to selection for these traits or combinations of traits. Reviews on estimates of heritabilities appear elsewhere including Tyler [29], Maijala and Hanna [30] and Wilcox [16].

The heritability of a trait is a measure of the fraction of the total variation attributable to genetic differences between animals. The response to selection of a trait is largely a function of its heritability. The estimation of heritability values for milk production traits has been the subject of considerable research since breeding programmes for the genetic improvement of dairy cattle were implemented in several countries. Heritabilities of yields, percentages and ratios of milk constituents are

Table 2 Estimates of heritability for milk production traits[1] reported in dairy cattle

	Robertson *et al.* [34]	Gacula *et al.* [14]	Maijala and Hanna [30]	Wilcox [16]	Welper and Freeman [35]	Schopen *et al.* [36]
Milk yield	0.25	0.38	0.26	0.25	0.29	0.41
Fat yield		0.31	0.23	0.25	0.28	0.39
Protein yield		0.34	0.27	0.25	0.25	0.24
Lactose yield					0.25	0.43
Casein yield						0.26
LM yield				0.25		
SNF yield		0.36		0.25		
TS yield		0.33		0.25		
Fat%	0.32	0.72	0.47	0.60	0.51	0.50
Protein%	0.48	0.57	0.47	0.50	0.45	0.66
Casein%	0.58					
Lactose%	0.36				0.48	0.62
Minerals%	0.50					
SNF%	0.53	0.59		0.55		
TS%		0.58		0.60		
Casein number						0.70
Protein:fat ratio			0.47	0.55		
SNF:fat ratio				0.60		

[1]LM, lactose plus minerals; SNF, solids-not-fat; TS, total solids.

shown in Table 2. Heritabilities of yield traits, including milk yield, are about 0.25. Heritabilities of percentage traits are at least twice as high as those for yield and range from 0.50 to 0.60. Heritabilities of ratios of solids-not-fat (SNF):Fat and Protein:Fat range from 0.55 to 0.60 and are similar to those of the percentage traits. These relatively high heritabilities suggest that changes in milk composition are possible through selective breeding. The rate of genetic gains in the USA dairy cow population from 1980 to 2008 have been significant, 108 kg milk, 3.5 kg fat and 3.3 kg protein per cow per year. However, these positive trends have been accompanied with detrimental changes in fertility [31]. The decline in the breeding values for daughter pregnancy rate since 1980 until 2000 was 2.5% [32], but since 2002, breeding values for daughter pregnancy rate have remained constant and started to improve from 2006 [31]. In New Zealand dairy cattle, annual genetic gains from 1996 to 2004 were 36 kg milk, 1.88 kg fat, 1.71 kg protein, 0.85 kg cow live weight, 0.06% fertility, a small change in somatic cell count and 9.5 days in longevity.

Phenotypic and genetic correlation between yields and concentrations of milk components are shown in Table 3. Literature regarding genetic parameters for lactose, and its relationship with other production traits in dairy cattle, is scarce. The Canadian national breeding programme started collecting data for lactose in 2001 in approximately one-third of milk-recorded farms [33]. Analysis of lactose was introduced to include this trait in the calculation of the energy value of milk for the purpose of establishing energy requirements for feeding recommendations. Many breeding programmes are using Fourier transform infrared spectroscopy to determine the concentration of fat, protein, lactose and somatic cell count in milk samples used for herd-testing. However, genetic evaluation for lactose is not conducted because lactose has neither been included in the payment system nor in the breeding objectives.

Genetic Variation for Fatty Acids

The quality of cow's milk for human health is influenced by its lipid content and fatty acid composition. Diets rich in saturated fatty acids (SFA), such as the lauric (C12:0), myristic (C14:0) and palmitic acids (C16:0), are correlated with an increased risk of atherosclerosis, obesity and coronary heart diseases [40]. Cow milk fat typically contains 70% SFA, 25% monounsaturated fatty acids and 5% polyunsaturated fatty acids [41]. A fatty acid composition more favourable to human health would be approximately 30% SFA, 60% monounsaturated fatty acids and 10% polyunsaturated fatty acids [42]. The fatty acid profile of cow milk is therefore far from optimal.

The fatty acid composition also influences the technological properties of butter [43]. Increasing the concentrations of C18:1 and C18:2 and reducing the concentrations of C8:0 to C14:0 fatty acids improve the solid fat content at 10°C and therefore the spreadability of butter. The concentration of fatty acids in milk fat is affected by diet [44, 45] and the natural genetic variation between the individual cows [46, 47, 48, 49] and breeds [44, 46, 50, 51, 52, 53].

Table 4 shows estimates of heritabilities for individual fatty acids. The estimates, while not consistent among the studies, tend to be of moderate value, suggesting that genetic variability of fatty acids should be sufficient to implement a selection programme for the improvement of fatty acid composition. A limitation of this implementation is that the determination of individual

Table 3 Genetic parameters[1] for milk production traits[2] in dairy cattle

	Row	MY	FY	PY	LY	F%	P%	L%	SCS	P:F	LWT
MY	1	**0.27**	0.88	0.95		−0.20	−0.19				
	2	**0.47**	0.79	0.81	0.99	−0.36	−0.52	−0.08	−0.04		
	3	**0.29**	0.73	0.91					−0.13		0.25
	4	**0.36**	0.79	0.94	0.99	−0.36	−0.52	0.25	−0.08		
	5	**0.26**									
FY	1	0.81	**0.24**	0.93		0.24	−0.04				
	2	0.71	**0.28**	0.85	0.79	0.27	−0.06	−0.02	−0.07		
	3	0.54	**0.29**	0.82					−0.09		0.24
	4	0.57	**0.36**	0.81	0.77	0.26	−0.25	0.16	−0.06		
	5	0.76	**0.31**								
PY	1	0.85	0.85	**0.27**		−0.01	0.06				
	2	0.93	0.79	**0.25**	0.94	−0.20	−0.09	0.01	−0.04		
	3	0.86	0.69	**0.30**					−0.96		0.30
	4	0.89	0.68	**0.40**	0.92	−0.24	−0.21	0.21	−0.06		
	5	0.90	0.87	**0.17**							
LY	1										
	2	0.92	0.68	0.89	**0.25**	−0.30	−0.27	0.20	−0.08		
	3										
	4	0.98			**0.50**	−0.37	−0.53	0.38	−0.11		
	5	0.99	0.77	0.86	**0.37**						
F%	1	−0.31	0.21	0.08		**0.47**	0.49				
	2	−0.40	0.35	−0.21	−0.35	**0.51**	0.51	0.11	−0.06		
	3										
	4	−0.43				**0.54**	0.47	−0.17	0.05		
	5	−0.21	0.49	0.19	−0.25	**0.53**					
P%	1	−0.28	0.14	0.23		0.58	**0.47**				
	2	−0.47	−0.04	−0.12	−0.37	0.59	**0.45**	0.29	−0.01		
	3										
	4	−0.54				0.66	**0.57**	−0.25	0.09		
	5	−0.56	−0.18	−0.11	−0.62	0.63	**0.53**				
L%	1										
	2	−0.30	−0.16	−0.21	0.10	0.16	0.29	**0.48**	−0.15		
	3							**0.49**			
	4	0.10				0.06	0.02	**0.50**	−0.23		
	5	0.68	0.56	0.37	0.75	−0.2	−0.74	**0.33**			
SCS	1										
	2	0.15	0.12	0.18	0.13	−0.06	0.01	−0.11	**0.13**		
	3	0.04	0.12	0.06					**0.18**		0.04
	4	0.00	−0.05	0.01	−0.02	−0.11	0.09	−0.20	**0.27**		
	5	−0.22	−0.64	−0.88	−0.46	−0.49	−0.33	−0.18	**0.29**		
P:F Ratio	1									**0.47**	
	2										
	3										
	4										
	5	−0.21	−0.66	−0.25	−0.23	−0.70	0.09	−0.47	0.32	**0.42**	
LWT	1										
	2										
	3	0.28	0.33	−0.36					−0.01		**0.39**
	4										
	5										

[1]Genetic correlations shown below the diagonal, phenotypic above, and heritabilities in boldface along the diagonal. First row from Maijala and Hanna [30], second row from Welper and Freeman [35], third row from Pryce and Harris [37], fourth row from Miglor *et al.* [33] and fifth row from Roman *et al.* [38] (heritabilties) and Roman and Wilcox [39] (genetic correlations).
[2]MY, milk yield; FY, fat yield; PY, protein yield; LY, lactose yield; F%, fat percentage; P%, protein percentage; L%, lactose percentage; SCS, somatic cell score; P:F ratio. protein to fat ratio; LWT, live weight.

concentration of each fatty acid in individual cows by gas chromatographic analyses is expensive and time consuming. Several studies [54, 55, 56] have shown the possibility to estimate milk fatty acids concentrations by Fourier transform infrared spectrometry, which is faster and cheaper than the reference chemical analysis.

Breed differences in the milk fatty acid profile have been reported by several studies summarized by Arnould and Soyeurt [42]. An updated review considering only Holstein (or Holstein–Friesian) and Jersey cows is presented in Table 5. In general, concentrations of C6:0, C8:0, C10:0, C10:1, C12:0, C12:1, C14:0, C16:0

Table 4 Estimates of heritability for fatty acids in milk fat[1] of dairy cattle

	Stoop *et al.* [47]	Bobe *et al.* [48]	Garnsworthy *et al.* [49]
C4:0	0.35	0.00 ± 0.06	0.10 ± 0.074
C6:0	0.39	0.00 ± 0.08	0.27 ± 0.103
C8:0	0.48	0.18 ± 0.12	0.27 ± 0.099
C10:0	0.54	0.22 ± 0.13	0.20 ± 0.088
C12:0	0.35	0.18 ± 0.12	0.13 ± 0.073
C14:0	0.49	0.00 ± 0.08	0.09 ± 0.068
C14:1 *cis*-9			0.28 ± 0.098
C16:0	0.31	0.09 ± 0.12	0.06 ± 0.064
C16:1		0.49 ± 0.13	0.08 ± 0.065
C18:0	0.19	0.24 ± 0.11	0.04 ± 0.053
C18:1		0.06 ± 0.10	
C18:1 *cis*-9	0.18		0.12 ± 0.073
C18:1 *cis*-11	0.12		
C18:1 *trans*-4–8	0.18		
C18:1 *trans*-9	0.11		
C18:1 *trans*-10			0.13 ± 0.106
C18:1 *trans*-11	0.12		0.00 ± 0.05
C18:2		0.00 ± 0.11	
C18:2 *cis*-9,12	0.13		0.10 ± 0.074
C18:2 *cis*-9, *trans*-11 (CLA)	0.21		0.02 ± 0.059
C18:3 *cis*-9,12,15	0.09		0.05 ± 0.059
C20:0			0.00 ± 0.005
Groups of fatty acids			
C6–12	0.49		
C6:0–C14:0		0.04 ± 0.09	0.15 ± 0.079
C14–16	0.11		
C18u	0.18		
Saturated fatty acids		0.05 ± 0.10	0.14 ± 0.081
Monounsaturated fatty acids		0.08 ± 0.10	0.09 ± 0.069
Polyunsaturated fatty acids		0.00 ± 0.11	0.03 ± 0.069
Ratio saturated fatty acids:unsaturated fatty acids	0.20		

[1]Fatty acids were expressed as percentage of the total milk fat fraction.

and C18:0 are significantly lower ($P < 0.05$) in Holstein (or Holstein–Friesian) cows than in Jersey cows, but concentrations of unsaturated fatty acids are significantly higher ($p < 0.05$) in Holstein (or Holstein–Friesian) than in Jersey cows.

Genetic Variation for Milk Proteins

The protein fraction of milk from New Zealand Holstein–Friesian cows typically contains 77.1% of casein proteins (α_{S1}-casein, α_{S2}-casein, β-casein and κ-casein), 19.8% of whey proteins (α-lactalbumin, β-lactoglobulin, bovine serum albumin and immunoglobulin G) [52], and more than 30 other proteins with partly known and unknown functions have been reported [57].

Several authors have investigated protein composition in relation to cheese-making properties of milk and have shown significant associations of absolute concentrations of κ-casein as well as κ-casein as a fraction of the total casein with rennet clotting time, curd firmness and cheese production efficiency [58, 59, 60]. In addition to its relevance for cheese production, protein composition might be of relevance for other aspects of dairy products and dairy processing; for example, production of infant formula.

Table 6 shows estimates of heritabilities for individual milk proteins reported by several studies. The estimates are not consistent among the studies but tend to be of moderate or high value, suggesting that there are opportunities to alter the composition in the cow's milk using selective breeding to enhance the suitability of milk for cheese making or for the production of specific milk products. A limitation of the implementation of such programmes is the requirement of available measurements of individual milk proteins on a large scale. The determination of concentration of milk proteins has been typically performed using electrophoretic techniques and high-performance liquid chromatography (HPLC) [52], and more recently reverse-phase HPLC (RP-HPLC) [61] and capillary zone electrophoresis (CZE) [62]. However, large-scale application of RP-HPLC or CZE is not feasible, because these techniques are time consuming and require expensive analytical equipment and skilled operators. Similar to fatty acids, some studies have shown the possibility to estimate the concentrations of individual proteins in individual cows by Fourier transform infrared spectrometry [63, 64], but with much lower accuracy.

Differences between Holstein–Friesian and Jersey cows in milk protein reported in studies with New Zealand grazing cattle [52, 65] and Australian dairy cattle [66] are presented in Table 7. Compared with Holstein–Friesian cows, Jersey cows produced milk with higher concentrations of nearly all types of proteins measured in

Table 5 Differences in the fatty acid profile of milk fat from Holstein (or Holstein–Friesian) and Jersey cows obtained in various studies

	DePeters *et al.* [50] (total mixed ration)		Beaulieu and Palmquist [51] (total mixed ration)		Palladino *et al.* [53] (grazing ryegrass)		Auldist *et al.* [52] (grazing ryegrass)	
	(*N*=16)	Jersey (*N*=23)	Holstein (*N*=8)	Jersey (*N*=8)	Holstein–Friesian (*N*=27)	Jersey (*N*=27)	Holstein–Friesian (*N*=29)	Jersey (*N*=29)
Milk yield (kg)	30.3	17.6**	36.1	24.0**	21.10	14.50**		
Composition (%)								
Fat	3.74	4.81**	2.97	3.99**	3.79	5.09**	4.47	5.82**
Protein	3.16	3.95**	2.97	3.61**	3.43	4.01**	3.55	3.98**
Fatty acid (g/100 g of fat)								
C4 : 0 (butyric)	4.08	3.88	4.10	4.00	6.45	6.17	4.92	5.07*
C6 : 0 (caproic)	2.71	2.80*	2.40	2.80*	3.88	3.92	3.00	3.14**
C8 : 0 (caprylic)	1.59	1.71**	1.30	1.80	2.20	2.21	1.73	1.81**
C10 : 0 (capric)	3.68	4.18**	3.00	4.30**	4.20	4.46	3.62	3.86**
C10 : 1							0.32	0.28
C12 : 0 (lauric)	4.26	4.98**	3.50	5.00**	4.29	4.44	3.79	3.99
C12 : 1							0.08	0.05**
C14 : 0 (myrsitic)	13.56	13.88	11.60	12.60*	11.67	11.63	11.03	11.42*
C14 : 1					1.18	0.97**	0.86	0.73*
C15 : 0					1.27	1.17*	1.97	1.58**
C16 : 0 (palmitic)	35.33	34.89	32.40	30.20*	23.75	25.53**	25.50	26.69
C16 : 1					1.38	1.34	1.17	0.99**
C16 : 1 *trans*	0.44	0.41						
C16 : 1 *cis*	1.99	1.80						
C17 : 0					0.45	0.42	1.44	1.32**
C17 : 1							0.22	0.17
C18 : 0 (stearic)	8.93	9.52**	7.20	8.10	7.97	8.30	10.96	12.62**
C18 : 1							23.83	21.63**
C18 : 1 *cis*-9 (oleic)	17.88	16.18**	17.30	15.10**	15.81	13.93*		
C18 : 1 *trans*-9	0.21	0.23						
C18 : 1 *trans*-11 (vaccenic)	1.51	1.54			4.34	4.50		
C18 : 2 *cis*-9,*cis*-12 (linoleic)	2.53	2.57	2.20	2.20	0.59	0.47**	1.69	1.54**
C18 : 2 *cis*-9, *trans*-11 (conjugated linoleic acid)					1.72	1.56	1.53	1.08**
C18 : 3 *n*-3 (linolenic)	1.29	1.49	0.48	0.40	0.66	0.60	0.77	0.69**
C18 : 1/C18 : 0	2.06	1.72**	2.54	2.02*				
C20 : 0 (arachidic)							0.09	0.10**
UFA[1]	25.86	24.18**						
UFA[2]					27.36	24.60*		
SFA[3]					65.88	68.90*		
UFA[2]:SFA[3]					0.05	0.04**		
SCFA[4]	8.38	8.38			12.46	12.41		
MCFA[5]	21.50	23.04*						
MCFA[6]					48.56	50.43		
LCFA[7]	32.35	31.48						
LCFA[8]					32.18	30.71		

[1]Unsaturated fatty acids; sum of C16 : 1 *trans*, C16 : 1 *cis*, C18 : 1 *cis* 9 + 10, Cl8 : 1 *trans*-9, C18 : 1 *trans*-11, C18 : 2 and C18 : 3.
[2]Unsaturated fatty acids; sum of C10 : 1, C14 : 1, C16 : 1, C17 : 1, C18 : 1, vaccenic acid, linoleic acid, conjugated linoleic acid and linolenic acid.
[3]Saturated fatty acids; sum of C4 : 0, C6 : 0, C8 : 0, C10 : 0, C11 : 0, C12 : 0, C13 : 0, C14 : 0, C15 : 0, C16 : 0, C18 : 0, C20 : 0, C21 : 0, C22 : 0 and C24 : 0.
[4]Short-chain fatty acids; sum from C4 : 0 to C8 : 0.
[5]Medium-chain fatty acids; sum of C10 : 0, C12 : 0 and C14 : 0
[6]Medium-chain fatty acids; sum of C10 : 0, C12 : 0, C14 : 0, C15 : 0 C16 : 0 and C16 : 1.
[7]Long-chain fatty acids; sum of C18 : 0, C18 : 1 *trans*-11, Cl8 : 1 *cis*-9 + 10, C18 : 2 and C18 : 3.
[8]Long-chain fatty acids; sum from C17 : 0, C18 : 0, C18 : 1 *cis*-9, C18 : 1 *trans*-11, vaccenic acid, linoleic acid, conjugated linoleic acid, linolenic acid, C20 : 0, C20 : 3*n*-6, C20 : 4*n*-6, and C22 : 6*n*-3.
* $P<0.05$, ** $P<0.01$.

Table 6 Estimates of heritability for milk proteins[1]

	Kroeker *et al.* [67]	Bobe *et al.* [68]	Graml and Pirchner [69]	Schopen *et al.* [36]	Bonfatti *et al.* [70]	Huang *et al.* [71][5]
α_{S1}-casein	0.02	0.18	0.27–0.37	0.47	0.68	0.33
α_{S2}-casein		0.01	0.17	0.73	0.28	
β-casein	0.03	0.01	0.32–0.34	0.25	0.69	0.33
γ-casein					0.18	
κ-casein	0.01	0.28	0.22–0.28	0.64	0.63	0.66
α-lactoalbumin	0.14	0.00		0.55		0.33
β-lactoglobulin	0.24	0.36	0.26–0.35	0.80	0.34	0.69
Bovine serum albumin	0.15					
Immunoglobulin	0.02					
Σcasein[2]				0.41		
Σwhey[3]				0.71		
Casein index[4]				0.70		

[1]Proteins were expressed as percentage of the total protein fraction.
[2]Σcasein=α_{S1}-casein + α_{S2}-casein + β-casein + κ-casein.
[3]Σwhey=α-lactoalbumin + β-lactoglobulin.
[4]Casein index=Σcasein/(Σcasein + Σwhey) × 100.
[5]Proteins expressed as mg/100 μL of milk.

Table 7 Differences in concentration of proteins in milk from Holstein–Friesian and Jersey cows obtained in various studies

	Auldist *et al.* [52]		Back and Lopez-Villalobos [65]		McLean *et al.* [66]	
	Holstein–Friesian (*N*=29)	Jersey (*N*=29)	Holstein–Friesian (*N*=20)	Jersey (*N*=20)	Holstein–Friesian (*N*=238)	Jersey (*N*=262)
Milk yield (kg)			21.71	15.22**	3177	3168
Protein (g/kg)	35.5	39.8**	35.7	40.6**	31.6	39.9**
Casein (g/kg)	27.4	31.2**	27.5	32.2**	24.5	31.2**
Casein:Protein	0.771	0.785**	0.770	0.793**	0.775	0.782
α_{s1}-casein (g/kg)					8.50	10.00*
α_{s2}-casein (g/kg)					2.80	4.00
α-casein (g/kg)	11.50	11.90	16.67	19.97**		
β-casein (g/kg)	11.00	13.50**	12.87	14.32**	8.90	10.80*
κ-casein (g/kg)	3.80	4.10*	4.17	5.55**	2.60	3.90**
γ-casein (g/kg)					1.90	2.50
A-lactalbumin (g/kg)	1.30	1.50	0.90	1.02*	0.95	1.13
B-lactoglobulin (g/kg)	4.90	5.30*	4.64	5.03**	2.96	3.67
BSA[1] (mg/kg)	183	151**	209	176		
IgG[2] (mg/kg)	634	590	477	558		

[1]Bovine serum albumin.
[2]Immunoglobulin G.
$^*P<0.05$, $^{**}P<0.01$.

lower milk yields. Therefore it would be expected that the higher concentrations of milk proteins would contribute to Jersey milk, generating higher cheese yields than Holstein–Friesian milk. However, Auldist *et al.* [52] demonstrated that the differences in cheese-making properties between Jersey and Holstein–Friesian milk were significantly related to the concentrations of solids in milk rather than concentrations of individual proteins.

Association between Blood and Protein Polymorphisms and Milk Components

During the mid-1950s information on biochemical polymorphisms in blood became available in the literature and animal breeders started studying the association of these polymorphisms and milk production traits [72]. However, the proportion of genetic variance attributable to significant blood group associations was very small. The discovery of abundant polymorphisms in electrophoretic mobility of proteins was the leading edge of a wave of studies uncovering genetic variation at the molecular level [73, 74] providing evidence of linkage between milk production and milk protein polymorphisms.

An extensive research on genetic polymorphism of milk proteins has been associated with concentration of α_{S1}-casein, α_{S2}-casein, β-casein, κ-casein, α-lactalbumin and β-lactoglobulin, and yield of milk and its components [75, 76, 77, 78] and technological properties of milk [79, 80]. The general conclusion by Heck *et al.* [78] was that selection for both the β-lactoglobulin genotype B and the β-κ-casein haplotype A^2B will result in cows that produce milk that is more suitable for cheese production.

The genes coding for milk proteins have been thoroughly investigated in cattle and goats, and a noticeable genetic variation has been identified and characterized [81]. The major milk proteins of the bovine milk (α_{S1}-casein, α_{S2}-casein, β-casein, κ-casein, α-lactalbumin and β-lactoglubulin) are coded by six structural genes [82]. The four casein genes have been mapped on chromosome 6 in cattle [83], whereas the genes encoding α-lactalbumin and β-lactoglobulin have been mapped on chromosomes 5 [83] and 11 [84].

Finding quantitative trait loci (QTLs) for milk production traits using microsatellites

By the 1930s, scientists [8, 85] realized that yield of milk components was under the genetic control of many genes with additive, dominance and epistatic effects as proposed by Fisher [7]. However, it was not possible to identify the number and the location of the genes regulating the synthesis of milk. The genetic architecture of milk production was confined to estimates of the heritability, which was defined as the fraction of the total phenotypic variance attributable to additive genetic variance. A more effective selection for improved milk production and composition can be achieved if the number, location and properties of the genes regulating the synthesis of milk are known.

The principle of mapping QTLs was noted about 90 years ago [86]. If a QTL is linked to a marker locus, there will be a difference in mean values of the quantitative trait among individuals with different genotypes at the marker locus. The theory was known, but no suitable markers were available in dairy cattle [87].

Advances in DNA technology by the end of the 1980s revealed an abundant number of molecular polymorphisms segregating in most species, at the level of variation at single nucleotides (single-nucleotide polymorphisms or SNPs), short di-, tri- or tetra-nucleotide tandem repeats (microsatellites), longer tandem repeats (minisatellites), small insertions/deletions and insertion sites of transposable elements.

By the 1990s, QTL experiments with dairy cattle were carefully planned using cross-breeding between two purebreds [88, 89], selective genotype [90], and daughter [72] and granddaughter designs [91]. Many dairy cattle breeding programmes invested enormous amount of resources into genotyping and phenotyping to identify QTL for milk production traits, particularly milk yield, protein yield, fat yields, protein percentage and fat percentage but with few commercial applications of this work possible. Publication of the results of the first genome-wide scan in the US Holstein population by Georges *et al.* [92] in 1995 was followed by Spelman *et al.* [93] in 1996 in the Dutch Holstein population and Vilkki *et al.* [94] in 1997 in the Finnish Ayrshire population. After these studies, there was a flood of QTLs being reported for dairy cattle traits.

The type of DNA markers in 1990s were microsatellites and about 1500 were available in dairy cattle with an average spacing between them along the bovine chromosome of 2 centiMorgans (cM, equivalent to about 2% recombination). The distribution of the markers was, however, not uniform, so some areas of the genome were more densely marked, and some were less densely marked [95]. This limited number of markers resulted in a poor mapping resolution with confidence intervals for the location of the QTL in the order of 20–30 cM, i.e. as much as 20–30 million base pairs and probably as many as 500–1000 genes [95].

By the start of 2000, commercial application of QTL in dairy cattle breeding was limited. The main reasons for this poor application have been discussed elsewhere [96, 97, 98] and included (i) the QTL effect explained a limited amount of genetic variation for the trait; (ii) a limited number of genetic markers were available and the cost of genotyping was relatively high; (iii) several results were false positives or were discovered in experimental populations and did not replicate in populations of interest; (iv) marker or QTL effects were estimated on a within-family basis, which made it more difficult to incorporate them into breeding programmes. However, some QTLs affecting milk production have been identified consistently across many studies and the underlying causative mutation for the QTL has been identified; diacylglycerol-acyltransferase 1 (DGAT1) [99] on chromosome 14, ATP binding, cassette super-family G, member 2 (ABCG2) [100, 101] on chromosome 6 and growth hormone receptor (GHR) [102] on chromosome 20. These QTLs have significant effects on fat and protein concentrations, which have the highest heritabilities of all traits routinely analysed in dairy cattle.

A considerable number of studies reporting QTL for production, reproduction and health traits in dairy cattle were accumulated between 1995 and 2003. Khatkar *et al.* [103] conducted a meta-analysis in 2004 using 55 publications to obtain consensus on QTL location and allelic substitution effect of these QTL. The meta-analysis indicated a number of regions, the most striking being two distinct regions affecting milk yield on chromosome 6 at 49 and 87 cM explaining 4.2 and 3.6% of the genetic variance of milk yield, respectively. The first of these regions affects five separate milk production traits (protein yield, protein percentage, fat yield, fat percentage, in addition to milk yield). By 2009, Lemay *et al.* [104] summarized that 238 milk trait QTLs were identified for these five traits in 59 references. Of the 238 QTLs, 63 were reported with flanking markers having a median interval size of approximately 17 million base pairs. Some QTLs were reported for more than one milk trait; thus, these QTLs span only 168 unique genome locations. These milk trait

QTLs span all 29 autosomes, with the highest densities of QTLs occurring on chromosomes 6, 14, 20 and 27. Possible differences in genetic architecture were most obvious between fat and protein percentage traits, where fat percentage QTLs were present on fewer chromosomes with lower QTL density and protein percentage QTL were present on all but two chromosomes, most with higher QTL density. Lemay *et al.* [104] suggested that fat percentage may be controlled by relatively fewer genes each with larger effects, whereas protein percentage may be controlled by far more genes, each with smaller effects. This is credible as fat products are derived from metabolites, whereas protein products are all gene derivate products and therefore protein production requires more direct genetic control.

Finding QTL for Milk Production Traits using SNPs

Meuwissen *et al.* [105] in 2001 proposed that fine QTL maps can be produced using genome-wide associations (GWA) between phenotypes and SNPs if these markers are evenly distributed within the genome in the order of thousands. The basic design for QTL identification using GWA was summarized by Goddard and Hayes [106]. A sample of animals is recorded for a trait of interest and assayed for a genome-wide panel of markers to detect statistical associations between the trait and any of the markers. Design parameters include the choice and number of animals and markers. Most commonly, one SNP at a time is included in a simple linear model that considers the effect of the SNP and other fixed effects, such as the contemporary group and the polygenic breeding value of each animal, which is the result of all the other genes affecting the trait. The first GWA studies using a panel of 10 000 SNPs found significant SNPs associated with yield of milk components: 31 for milk, seven for fat and 22 for protein in the US [107] and Canadian [108] Holstein populations.

In January 2008, Illumina Inc. (San Diego, CA) released the Infinium BovineSNP50 Bead-Chip, which included 54 001 SNPs approximately evenly spaced across the entire bovine genome. A flood of studies emerged using this bead-chip to identify QTL for milk production, e.g., in the USA [109], Australia [110], the Netherlands [111, 112], Ireland [113] and China [114]. High-density chips with 777 000 SNPs are commercially available (http://www.illumina.com) but a chip with two million SNPs can be available for GWA studies [115].

The focus of many research projects on gene mapping at the present is on identification of genes contributing to the genetic variation of milk components aiming to increase our understanding of the genetic control of the synthesis of milk components. Extensive reviews on the use of DNA analysis in dairy cattle breeding include the mapping of genes for complex traits [106, 116], genomic selection [117, 118, 119] and genetical genomics [5, 120, 121].

The traditional definition of milk composition based on concentration of fat, protein, lactose and minerals is now changing to a more detailed description in terms of concentration of individual fatty acids in milk fat and concentration of individual proteins in the protein fraction. These phenotypes are being used in GWA studies to identify the location of the genes controlling the production of these individual milk components.

Schopen *et al.* [111] identified genomic regions on BTA 5, 6, 11 and 14 with large effects on milk protein composition or protein percentage. In some cases, these genomic regions explained more than 80% of the genetic variation in the milk proteins (e.g. BTA 6 for β-casein and BTA 11 for β-lactoglobulin). The largest effects for six milk proteins were found in the region where the gene coding for the corresponding milk protein is located: for α_{S1}-casein, α_{S2}-casein, β-casein and κ-casein on BTA 6; for α-lactalbumin on BTA 5; and for β-lactoglobulin on BTA 11. The number of chromosomal regions affecting a specific milk protein was quite different, varying from three regions for β-casein and β-lactoglobulin to 12 regions for α_{S2}-casein. A conclusion from this study was that milk protein composition is regulated by few genes with large effects on all milk proteins and several genes with smaller effects on specific milk proteins.

Results from the Dutch study [111] were confirmed by another GWA study in the USA with Holstein and (Holstein×Jersey)×Holstein backcross cows [71]. In that study, the association between a comprehensive list of 53 SNPs within and close to four milk protein casein genes (CSN1S1, CSN1S2, CSN2 and CSN3) and the two major whey protein genes (LALBA and LGB) and the milk protein profile measured by RP-HPLC was evaluated. Several SNPs were significantly associated with concentrations of α_S-casein, β-casein, κ-casein, α-lactalbumin and β-lactoglobulin in milk. Importantly, the significant SNPs explained a large proportion of the phenotypic variation of milk protein composition.

QTLs for short-, medium- and long-chain fatty acids have been reported in Holstein Dutch cattle [122, 123]. Significant QTLs explaining 3–19% of the phenotypic variance were found for concentration of C6 : 0 and C8 : 0 on BTA6, fat percentage, all uneven chain fatty acids, C14 : 0, C16 : 0 and C16 : 1 on BTA14, C14 : 0 on BTA19 and the monounsaturated fatty acids and their unsaturation indices on BTA26. Additional analyses revealed that the QTL on BTA14 was most likely caused by a mutation in DGAT1, whereas the QTL on BTA26 was most likely caused by a mutation in the stearoyl-CoA desaturase 1 (SCD1) gene. In a following study with the same data set, Bouwman *et al.* [112] performed another GWA analysis and found a total of 54 regions on 29 chromosomes that were significantly associated with one or more fatty acids. BTAs 14, 19 and 26 showed highly significant associations with 7–10 traits, explaining a relatively large percentage of the total additive genetic variation. Many additional regions were significantly associated with the fatty acids. Some of the genes already identified to be involved in fat synthesis were confirmed in this study: ABCG2 and PPARGC1A (peroxysome proliferator activated receptor-gamma coactivator-1alpha) on BTA 6; ACSS2 (acyl-CoA synthetase short-chain family member 2) on BTA 13; DGAT1 on BTA 14; ACLY (ATP citrate lyase), SREBF1 (sterol regulatory element binding transcription factor 1), STAT5A (signal transducer and activator of transcription 5A), GH, and FASN (fatty acid synthase) on BTA 19; SCD1 on BTA26; and AGPAT6 (1-acylglycerol-3-phosphate *O*-acyltransferase 6) on BTA 27. The number of studies reporting QTL for milk components will increase substantially in the near future because the aim of many genomic studies will be the actual identification of the genes controlling the synthesis of milk components in greater definition, e.g. fatty acids, proteins and oligosaccharides. Some QTL databases are accumulating information on genetic maps displaying genetic markers, genes and QTLs and this information can be accessed at the following web sites:

Bovine QTL viewer
The University of Adelaide, SA, Australia (http://genomes.sapac.edu.au/bovineqtl/)
Polineni *et al.* [124]

Cattle QTLdb
National Animal Genome Research Program, United States Department of Agriculture, Washington, DC (http://www.animalgenome.org/QTLdb/cattle)
Hu *et al.* [125]

QTL map
The University of Sidney, NSW, Australia (http://firefly.vetsci.usyd.edu.au/reprogen/QTL_Map)
Khatkar *et al.* [103]

Bovine Assembly 4.0 on Chromosomes http://genomes.arc.georgetown.edu/cgi-bin/gbrowse/bovine40_chromosome/

Bovine Genome Database
USDA National Institute of Food and Agriculture, Cattle Genome Mapping Project http://www.marc.usda.gov/genome/cattle/cattle.html
Agricultural Research Service, United States Department of Agriculture

Bos taurus (cattle) genome view
http://www.ncbi.nlm.nih.gov/mapview/map_search.cgi?taxid=9913
National Center for Biotechnology Information, National Library of Medicine, Bethesda, MD 20894.

Genomic Selection

Selection of cows and bulls to produce progeny with superior genetic potential for milk, fat and protein has been based on their estimated breeding values, which are obtained by BLUP procedures using phenotypic records. The estimated breeding values are predictors of the true breeding values and the objective of the animal breeder is to predict, as accurately as possible and early in life, the true breeding values of the animals available for selection.

High-density chips with thousands of SNPs have allowed the calculation of genomic breeding values as new predictors of true breeding values for traits of economic importance in dairy cattle. Genomic selection as proposed by Meuwissen *et al.* [105] has been implemented commercially in breeding programmes in countries including the USA [126], Australia [117], Ireland [127], New Zealand [128], France [129] and the Nordic countries [130]. Reliabilities of genomic breeding values are 0.50–0.60 for production traits [126, 128], but bulls can have their genomic breeding values just after birth and can be widely used in the population when they are 1-year old (once they start producing semen). In contrast, sires that are progeny tested with 85 daughters have reliabilities of 0.75–0.85 for production traits but they are at least 5 years old when the progeny test is completed. Genomic selection in dairy cattle may lead to a much greater rate of genetic gain, especially when it is applied in combination with a reduction of the generation interval [131].

The model proposed by Meuwissen *et al.* [105] is first applied to a set of individual phenotypes that make up the training population using the following equation [118],

$$y_i=\mu+\Sigma_j X_{ij} b_j+e_i,$$

where y_i is the phenotype (or progeny mean phenotype) of individual *i*, μ represents fixed effects, Σ_j is the sum of all genotyped SNPs, X_{ij} is the number (0, 1 or 2) of copies of allele '1' (versus '0') that individual *i* carries at SNP*j*, b_j is the allele substitution effect for SNP *j*, and e_i is a random residual. The main statistical challenge for this model is that the number of markers (>30 000 in 2009) is much greater than the number of animals in the training population with phenotypic records that are available to estimate their effects (< 2000). Several methods to obtain the estimated b_js have been compared, including the G-BLUP and Bayesian methods [105] and other versions and implementations of these methods [132, 133]. The second step is to apply the prediction model to calculate the genomic breeding values for individuals that make up the validation population and estimate the correlation between the resulting genomic breeding values with the phenotypes of these individuals. The square of this correlation is a measure of the reliability of the estimation of genomic breeding values.

Genomic selection allows the estimation of breeding values for traits that are sex-limited (e.g. milk production and reproduction), traits that can only be obtained late in life (e.g. longevity and reproductive performance) and traits that require killing the animal (e.g. meat quality), or that are expensive to record (e.g. disease resistance and feed conversion efficiency). However, Goddard and Hayes [106] indicated that the widespread implementation of genomic selection is still limited by two factors. Firstly, genotyping of individual animals is still expensive (about US$200 per animal) to be used on a large number of animals, and secondly, estimation of genomic breeding values for new traits requires the estimation and identification of new SNPs. Therefore, each new panel of SNPs will have to be genotyped across a reference population of tens of thousands of animals with phenotypic records for the new trait.

Genetic Analysis of Milk Production Traits using Systems Biology

The human genome was completed in 2003 and it took 13 years to determine the sequence of three billion nucleotide base-pairs of DNA, coding for 20 000–25 000 genes. The sequence of the bovine genome was completed in 2009 and took 6 years to determine the sequence of three billion base pairs [134]. A list of web sites providing information on genome sequence of the bovine and other livestock species is provided by the National Animal Genome Research Program (http://www.animalgenome.org).

In May 2011, Illumina announced that it had reduced its full genome sequencing service to US$5000 per human genome and US$4000 per genome for bulk orders of 50 or more. In March 2010, Pacific Biosciences stated that their second-generation machine, which is scheduled for release in 2015, will be capable of providing a full genome sequence for a person in just 15 min, for less than US$100.

The bovine genome assemblage, so far, has served as the backbone for GWA studies with milk production traits but still there is lack of a comprehensive understanding of what the 20 000–25 000 genes actually do. Several recent studies have suggested that an approach integrating expression profiling with genotyping, referred to as expression of QTL (eQTL) analysis, may provide knowledge on the functional architecture of the genes controlling the synthesis of milk. Illumina is releasingthis technology along with software packages for gathering, formatting, processing and analysing eQTL data.

The introduction of microarray technology enabled expression profiling of thousands of genes simultaneously. Functional genomics approaches can be used to generate information about gene function, as well as data on genetic interactions, not only among and between gene complexes but also in response to environmental stimuli. The outcome of bioinformatics analyses of microarray experiments are functional networks of genes reflecting interactions at the level of DNA, RNA or proteins. The results enable the derivation of new hypotheses on the nature of a trait. In livestock, expression microarrays have been used with regard to development and reproduction traits, growth, immune response and host–pathogen interaction [5, 121, 135].

In dairy cattle, the large gap between extensive knowledge of the physiological pathways controlling milk synthesis and genetic architecture of these processes is being reduced through the rapid development of high-throughput RNA-transcriptome technologies [57, 136]. An insight into the genes involved in bovine milk oligosaccharide metabolism was obtained by analysing the oligosaccharide profiles of Holstein and Jersey cows at four different stages of lactation [136]. The high abundance of sialylated and galactosylated oligosaccharides in bovine milk highlight the importance of genes coding for enzymes involved in metabolic processes related to sialylation and galactosylation of oligosaccharides in milk somatic cells. These candidate genes will guide the design of a targeted breeding strategy to optimize the content of beneficial oligosaccharides in bovine milk. Another example of gene networks was given by Bionaz and Loor [137] to illustrate how the biosynthesis of bovine milk fat is regulated by many genes belonging to several pathways.

The biggest challenges are to assemble the high-throughput 'omic' data, integrate them with phenotypic data and interpret this information in a biological and practical sense. A holistic and systemic approach using systems biology will be very useful to overcome this challenge. The new methods to perform genetic analysis of complex traits are embedded into a new discipline called systems biology [5, 121, 138] that is defined as an inter-disciplinary study of genomics, transcriptomics, proteomics, metabolomics and bioinformatics.

An illustrative example on the genetic analysis of milk components using a system biological approach was given by Flori *et al.* [139]. A whole genome scan for differentiation using 42 486 SNPs in the three major French dairy cattle breeds (Holstein, Normande and Montbéliarde) was carried out to identify the main physiological pathways and regions of the genome that were affected by artificial selection for increased milk yield. Thirteen highly significant regions subjected to strong recent positive selection for milk yield were identified and some of them contained genes with strong effects on milk production traits or coat colouration. A comprehensive network analysis considering genes that play a central role in the somatotropic and gonadotropic axes provided an insight into the antagonism, at the genome level, between milk production and reproduction traits in high producing dairy cows, as found in traditional quantitative analysis estimating genetic correlations.

Genetic Modification

The application of genetic modification to produce transgenic animals has been advanced to practical application in biomedicine, in particular, for the production of pharmaceutical proteins in milk (gene pharming: GTC Biotherapeutics, Pharming and Biosidus) and xenotransplantation, i.e. the production of porcine organs and tissue in human organ transplantation to cope with the shortage of human organs for use in terminally-ill patients [135].

Potential uses and economic benefits of transgenic dairy cows producing milks with altered composition have been reviewed by Karatzas and Turner [4] and Karatzas [140]. The main reasons for genetic modification of milk include the removal of allergenic proteins (β-lactoglobulin in infants), reduction of lactose concentration, increasing the copy number of β-casein and κ-casein proteins to increase cheese yield, inserting anti-pathogen molecules and inserting high-quality lipids.

The concept of modifying milk composition by augmenting the protein content of milk through increased casein gene dosage in the cow genome was illustrated experimentally by Brophy *et al.* [141] generating eight transgenic cows through a combination of genetic engineering and nuclear transfer techniques. The cows produced milk with elevated levels of β-casein and κ-casein proteins. The innovative results reported by Brophy *et al.* [141] clearly demonstrated that it is possible to generate transgenic cattle producing 'novel' milks but considerable effort and time would be required to propagate the new genetics into commercial dairy herds, in addition to the high costs of successfully producing a transgenic cow. Only 3–25% of offspring born after pronuclear injection are transgenics and with the target transgene in the specific location of the chromosome [142]. The costs of production of a transgenic cow were calculated to be in the area of US$250 000 per cow [143].

With advances in reproductive and DNA technologies, the potential to introduce or remove genes with known functions are emerging. Gene targeting is a process in which a DNA molecule introduced into a cell replaces the corresponding chromosomal segment by homologous recombination, and thus presents a precise way to manipulate the genome. In the past, the application of gene targeting mammalian cells was limited by its low efficiency. Zinc finger nucleases show promise in improving the efficiency of gene targeting by introducing DNA double-strand breaks in target genes, which then stimulate the cell's endogenous homologous recombination machinery. This technique has been applied in the modification of the β-lactoglobulin protein of dairy cattle [144].

A very important factor limiting the application of gene modification in dairy cattle is that products from genetically modified animals are perceived as a risk by the consumer. It is critical that a clear benefit from genetic modification can be demonstrated for this technology to have any chance of having a commercial application. Unfortunately, there are few examples that demonstrate a clear benefit for the consumer. Increased casein content for example, is not a benefit to the consumer unless accompanied by increased availability or reduced pricing of cheese products.

Future genetic studies will likely investigate (among others) the effects of casein overexpression in mammary gland biology, casein micelle structure and composition, cheese-making processes and nutritional value of milk in addition to animals' physiology and nutritional requirements [135, 141].

Precision Technology to Measure Fat and Protein Composition

The quality of milk is difficult to define because it includes different aspects depending on the perspective of the evaluation. For example, the producer usually defines milk quality based on the basis of payment, which is usually dependent on its contents of protein, fat and somatic cells. The processing plant defines milk quality as composition, taste, durability and manufacturing abilities. The consumer further considers sight, smell, taste, freshness, amount of fat and protein, composition of fat and protein and safety (i.e. in relation to hormones and drugs).

Measuring individual fatty acids and proteins for each cow is expensive and the economic values for the most relevant fatty acids and proteins have not yet been estimated. At the same time that genomic selection is progressing, new interpretation of Fourier-transformed and mid-infrared spectra are offering fast and low-cost predictions of fatty acids [54, 55], proteins [63, 64] and milk coagulation properties [145]. This large number of phenotypes can be used as training data for the prediction of genomic breeding values. Future breeding objectives of dairy cattle may consider the inclusion of individual fatty acids and proteins with their relative economic values obtained from more complex models that consider farm costs, processing costs and market prices of the new enhanced dairy products. Research will be required to define the most desirable milk profile of fatty acids and proteins that results in an enhanced quality of milk from the processing and consumer point of view.

Summary

Genetic analysis of milk components started in the 1930s using a polygenic model proposed by Fisher in 1918. Estimates of genetic and environmental variances and covariances for milk components were obtained using the analysis of variance proposed by Fisher in 1925. Lush [146], in his 1945 book, established the principles of breeding plans based on breeding values estimated from adjusted phenotypic records. The focus of animal breeding plans was to manipulate genetics by selective breeding, to obtain desired phenotypes. Up to the present time, there are vast number of studies reporting estimates of genetic parameters (heritabilities and genetic correlations) for yields of milk, fat and protein and concentration of fat and protein. Estimates of genetic parameters for lactose yield and percentage are scarce because the economic value of lactose has been neglected by the dairy industries. The number of studies reporting genetic analysis for individual fatty acids and milk proteins is limited. The main reason of this is because the world-wide payment systems to reward producers for their milk supply are based on total fat and protein. Moreover, the selective breeding for fatty acids and individual milk proteins is limited because of the high cost of laboratory analysis for the determination of the concentration of fatty acids and proteins profiles. Fourier-transform infrared technology is offering fast and low-cost predictions with high reliability (high coefficient of determination) for fatty acids, but with low reliability for milk proteins.

Mapping of QTLs for milk traits started in the 1990s using a few hundred genetic markers distributed unevenly across the genome with experimental designs mainly utilizing cross-breeding between two purebreds and the daughter and granddaughter designs.

At the starting of the new millennium, Meuwissen *et al.* [105] published a landmark paper proposing a new methodology to estimate breeding values for complex traits using thousands of SNPs evenly distributed in the genome. Based on this paper, GWA studies started flooding the literature on genetic analysis for milk production traits and the actual genes controlling the synthesis of milk are being identified and reported in the literature.

Genomic selection is already being used successfully to select dairy cattle carrying desirable alleles for milk yield and other traits, and the theoretical estimation indicates that the rate of genetic gain for milk production traits will be increased mainly through shorter generation intervals. Dairy cattle breeding programmes are dramatically modifying the selection schemes to integrate genomic selection into the breeding scheme, to increase the rate of genetic gain for milk production traits and also to focus on farm profitability as part of the overall breeding objective. Cows of high economic performance producing milk with desired milk composition will be obtained faster and at low cost, creating opportunities to increase farm profitability.

Acknowledgements

The author acknowledges his PhD students Nicholas Sneddon and Gonzalo Tuñon and his colleagues Steve Davis and Penny Back for reading and correcting the manuscript, which has contributed to many aspects of this review.

References

1. Maijala K. Cow milk and human development and well-being. Livestock Production Science 2000;65:1–18.
2. Norman HD, Ross Jr FA. State and national standardized lactation averages by breed for cows calving in 2007. AIPL Research Report K2-07 (2-09) 2007. Available from: URL: http://aipl.arsusda.gov/publish/dhi/dhi09/laall.shtml
3. Dairy NZ. New Zealand Dairy Statistics 2009–10. DairyNZ, Hamilton, New Zealand; 2010.
4. Karatzas CN, Turner JD. Toward altering milk composition by genetic manipulation: current status and challenges. Journal of Dairy Science 1997;80:2225–32.
5. Kadarmideen HN, van Rohr P, Janss LLG. From genetical genomics to systems genetics: potential applications in quantitative genomics and animal breeding. Mammalian Genome 2006;17:548–64.
6. Bateson W. Mendel's Principles of Heredity – A Defense. Cambridge University Press, Cambridge, UK; 1902.
7. Fisher RA. The correlation between relatives on the supposition of Mendelian inheritance. Transactions of the Royal Society of Edinburgh 1918;52:399–433.
8. Lush JL. The bull index problem in the light of modern genetics. Journal of Dairy Science 1933;16:501–22.
9. Lush JL, Straus FS. The heritability of butterfat production in dairy cattle. Journal of Dairy Science 1942;25:975–82.
10. Touchberry BW. Genetic correlations between five body measurements, weight, type and production in the same individual among Holstein cows. Journal of Dairy Science 1951;34:242–55.
11. Fisher RA. Statistical Methods for Research Workers. Oliver and Boyd, Edinburgh, UK; 1925.
12. Cockerham CC. An extension of the concept of partitioning hereditary variance for analysis of covariances among relatives when epistasis is present. Genetics 1954;39:859–82.
13. Kempthorne O. An Introduction to Genetic Statistics. Wiley, New York, 1957.
14. Gacula MCJ, Gaunt SN, Damon RAJ. Genetic and environmental parameters of milk constituents for five breeds. II. Some genetic parameters. Journal of Dairy Science 1968;51:438–44.
15. Legates JE, Warwick EJ. Breeding and Improvement of Farm Animals, 8th edn. McGraw-Hill, New York; 1990.
16. Wilcox CJ. Genetics: basic concepts. In: Van Horn HH, Wilcox CJ, editors. Large Dairy Herd Management. ADSA, Champaign, IL, USA; 1992. p. 1–7.
17. Harvey WR. Estimation of variance and covariance components in the mixed model. Biometrics 1970;26:485–504.
18. Henderson CR. Estimation of variance and covariance components. Biometrics 1953;9:226–52.
19. Hartley HO, Rao JNK. Maximum likelihood estimation for the mixed analysis of variance model. Biometrics 1967;54:93–108.
20. Patterson HD, Thompson R. Recovery of interblock information when block sizes are unequal. Biometrika 1971;58:545–54.
21. Henderson CR. Best linear unbiased estimation and prediction under a selection model. Biometrics 1975;31:423–47.

22. Gilmour AR, Gogel BJ, Cullis BR, Thompson R. ASReml User Guide Release 3.0. VSN International Ltd, Hemel Hempstead, UK; 2009.

23. Boldman KG, Kriese LA, Van Vleck LD, Kachman SD. A Manual for use of MTDFREML. USDA-ARS, Clay Center, Nebraska; 1993.

24. Meyer K. WOMBAT-A tool for mixed model analyses in quantitative genetics by restricted maximum likelihood (REML). Journal of Zhejiang University Science B 2007;8:815–21.

25. Jensen J, Madsen P. DMU: A Package for the Analysis of Multivariate Mixed Models. Version 6, Release 4.7. Faculty Agricultural Sciences, University of Aarhus, Denmark; 2008.

26. Misztal I. Reliable computing in estimation of variance components. Journal of Animal Breeding and Genetics 2008;125:363–70.

27. Kovacs MA, Groeneveld E. VCE-5 User's Guide and Reference Manual, Version 5.1. Institute of Animal Husbandry and Animal Behaviour, Federal Research Center of Agriculture, Mariensee, Germany; 2003.

28. Jamrozik J, Schaeffer LR. Estimates of genetic parameters for a test day model with random regressions for yield traits of first lactation Holsteins. Journal of Dairy Science 1997;80:762–70.

29. Tyler WJ. Genetic aspects of solids-not-fat and its components. Journal of Dairy Science 1958;41:447–9.

30. Maijala K, Hanna M. Reliable phenotypic and genetic parameters in dairy cattle. In Proceedings of the 1st World Congress on Genetics Applied to Livestock Production, Volume 1, Madrid, Spain; 1974. p. 541–63.

31. Norman HD, Wright JR, Hubbard SM, Miller RH, Hutchison JL. Reproductive status of Holstein and Jersey cows in the United States. Journal of Dairy Science 2009;92:3517–28.

32. Shook GE. Major advances in determining appropriate selection goals. Journal of Dairy Science 2006;89:1349–61.

33. Miglior F, Sewalem A, Jamrozik J, Bohmanova J, Lefebvre DM, Moore RK. Genetic analysis of milk urea nitrogen and lactose and their relationships with other production traits in Canadian Holstein cattle. Journal of Dairy Science 2007;90:2468–79.

34. Robertson A, Waite R, White JCD. Variations in the chemical composition of milk with particular reference to the sollds-not-fat. II. The effect of heredity.Journal of Dairy Research 1956;23:82.

35. Welper RD, Freeman AE. Genetic parameters for yield traits of Holsteins, including lactose and somatic cell score. Journal of Dairy Science 1992;75:1342–8.

36. Schopen GCB, Heck JML, Bovenhuis H, Visker MHPW, van Valenberg HJF, Van Arendonk JAM. Genetic parameters for major milk proteins in Dutch Holstein-Friesians. Journal of Dairy Science 2009;92:1182–91.

37. Pryce JE, Harris BL. Genetics of body condition score in New Zealand dairy cows. Journal of Dairy Science 2006;89:4424–32.

38. Roman RM, Wilcox CJ, Martin FG. Estimates of repeatability and heritability of productive and reproductive traits in a herd of Jersey cattle. Genetics and Molecular Biology 2000;23:113–9.

39. Roman RM, Wilcox CJ. Bivariate animal model estimates of genetic, phenotypic, and environmental correlations for production, reproduction, and somatic cell in Jerseys. Journal of Dairy Science 2000;83:829–35.

40. Haug A, Hostmark AT, Harstad OM. Bovine milk in human nutrition – a review. Lipids in Health and Disease 2007;6:25.

41. Grummer RR. Effect of feed on the composition of milk fat. Journal of Dairy Science 1991;74:3244–57.

42. Arnould VM-R, Soyeurt H. Genetic variability of milk fatty acids. Journal of Applied Genetics 2009;209:29–39.

43. MacGibbon AKH, McLennan WD. Hardness of New Zealand patted butter: seasonal and regional variations. New Zealand Journal of Science and Technology 1987;22:143–56.

44. Palmquist DL, Beaulieu AD, Barbano DM. Feed and animal factors influencing milk fat composition. Journal of Dairy Science 1993;76:1753–71.

45. Dewhurst RJ, Shingfield KJ, Lee MRF, Scollan ND. Increasing the concentrations of beneficial polyunsaturated fatty acids in milk produced by dairy cows in high-forage systems. Animal Feed Science and Technology 2006;131:168–206.

46. Soyeurt H, Dardenne P, Gillon A, Croquet C, Vanderick S, Mayeres P, *et al.* variation in fatty acids contents of milk and milk fat within and across breeds. Journal of Dairy Science 2006;89:4858–65.

47. Stoop WM, Van Arendonk JAM, Heck JML, Van Valenberg HJF, H. B. Genetic parameters for major milk fatty acids and milk production traits of Dutch Holstein-Friesians.Journal of Dairy Science 2008;91:385–94.

48. Bobe G, Bormann JAM, Lindberg GL, Freeman AE, Beitz DC. Short communication: Estimates of genetic variation of milk fatty acids in US Holstein cows. Journal of Dairy Science 2008;91:1209–13.

49. Garnsworthy PC, Feng S, Lock AL, Royal MD. Short communication: Heritability of milk fatty acid composition and stearoyl-CoA desaturase indices in dairy cows. Journal of Dairy Science 2010;93:1743–8.

50. DePeters EJ, Medrano JF, Reed BA. Fatty acid composition of milk fat from three breeds of dairy cattle. Canadian Journal of Animal Science 1995;75:267–9.

51. Beaulieu AD, Palmquist DL. Differential effects of high fat diets on fatty acid composition in milk of Jersey and Holstein cows. Journal of Dairy Science 1995;78:1336–44.

52. Auldist MJ, Johnston KA, White NJ, Fitzsimons WP, Boland MJ. A comparison of the composition, coagulation characteristics and cheesemaking capacity of milk from Friesian and Jersey dairy cows. Journal of Dairy Research 2004;71:51–7.

53. Palladino RA, Buckley F, Prendiville R, Murphy JJ, Callan J, Kenny DA. A comparison between Holstein-Friesian and Jersey dairy cows and their F_1 hybrid on milk fatty acid composition under grazing conditions. Journal of Dairy Science 2010;93:2176–84.

54. Soyeurt H, Dardenne P, Dehareng F, Lognay G, Veselko D, Marlier M, *et al.* Estimating fatty acid content in cow milk using mid-infrared spectrometry. Journal of Dairy Science 2006;89:3690–5.

55. Rutten MJM, Bovenhuis H, Hettinga KA, van Valenberg HJF, Van Arendonk JAM. Predicting bovine milk fat composition using infrared spectroscopy based on milk samples collected

in winter and summer. Journal of Dairy Science 2009;92:6202–9.

56. Coppa M, Ferlay A, Leroux C, Jestin M, Chilliard Y, Martin B, *et al.* Prediction of milk fatty acid composition by near infrared reflectance spectroscopy. International Dairy Journal 2010;20:182–9.

57. Casado B, Affolter M, Kussmann M. OMICS-rooted studies of milk proteins, oligosaccharides and lipids. Journal of Proteomics 2009;73:196–208.

58. Wedholm A, Larsen LB, Lindmark-Mansson H, Karlsson AH, Andren A. Effect of protein composition on the cheese-making properties of milk from individual dairy cows. Journal of Dairy Science 2006;89:3296–305.

59. Bonfatti V, Di Martino G, Cecchinato A, Degano L, Carnier P. Effects of β-κ-casein (CSN2–CSN3) haplotypes, β-lactoglobulin (BLG) genotypes, and detailed protein composition on coagulation properties of individual milk of Simmental cows. Journal of Dairy Science 2010;93:3809–17.

60. Penasa M, Cassandro M, Pretto D, De Marchi M, Comin A, Chessa S, *et al.* Short communication: influence of composite casein genotypes on additive genetic variation of milk production traits and coagulation properties in Holstein–Friesian cows. Journal of Dairy Science 2010;93:3346–9.

61. Bonfatti V, Grigoletto L, Cecchinato A, Gallo L, Carnier P. Validation of a new reversed-phase high-performance liquid chromatography method for separation and identification of bovine milk protein genetic variants. Journal of Chromatography A 2008;1195:101–6.

62. Heck JML, Olieman C, Schennink A, van Valenberg HJF, Visker MHPW, Meuldijk RCR, *et al.* Estimation of variation in concentration, phosphorylation and genetic polymorphism of milk proteins using capillary zone electrophoresis. International Dairy Journal 2008;18:548–55.

63. Bonfatti V, Di Martino G, Carnier P. Effectiveness of mid-infrared spectroscopy for the prediction of detailed protein composition and contents of protein genetic variants of individual milk of Simmental cows. Journal of Dairy Science 2011;94:5776–85.

64. Rutten MJM, Bovenhuis H, Heck JML, Van Arendonk JAM. Predicting bovine milk protein composition based on Fourier transform infrared spectra. Journal of Dairy Science 2011;94:5683–90.

65. Back PJ, Lopez-Villalobos N. Breed and heterosis effects for milk protein composition estimated in two stages of lactation in New Zealand dairy cows. Proceedings of the New Zealand Society of Animal Production 2007;67:399–402.

66. McLean DM, Graham ERB, Ponzoni RW. Effects of milk protein genetic variants on milk yield and composition. Journal of Dairy Research 1984;51:531–46.

67. Kroeker EM, Ng-Kwai-Hang KF, Hayes JF, Moxley JE. Heritability of relative percentages of major bovine casein and serum proteins in test-day milk samples. Journal of Dairy Science 1985;68:1346–8.

68. Bobe G, Beitz DC, Freeman AE, Lindberg GL. Effect of milk protein genotypes on milk protein composition and its genetic parameter estimates. Journal of Dairy Science 1999;82:2797–804.

69. Graml R, Pirchner F. Effects of milk protein loci on content of their proteins. Archiv Tierzucht, Dummerstorf 2003;46:331–40.

70. Bonfatti V, Cecchinato A, Gallo L, Blasco A, Carnier P. Genetic analysis of detailed milk protein composition and coagulation properties in Simmental cattle. Journal of Dairy Science 2011;94:5183–93.

71. Huang W, Peñagaricano F, Ahmad KR, Lucey JA, Weigel KA, Khatib H. Association between milk protein gene variants and protein composition traits in dairy cattle. Journal of Dairy Science 2012;95:440–9.

72. Neimann-Sørensen A, Robertson A. The association between blood groups and several production characters in three Danish cattle breeds. Acta Agriculturae Scandinavica 1961;11:163–96.

73. Geldermann H, Pieper U, Roth B. Effects of marked chromosome sections on milk performance in cattle. Theoretical and Applied Genetics 1985;70:138–46.

74. Gonyon DS, Mather RE, Hines HC, Haenlein GFW, Arave CW, Gaunt SN. Associations of bovine blood and milk polymorphisms with lactation traits: Holsteins. Journal of Dairy Science 1987;70:2585–98.

75. Hill JP. The relationship between β-lactoglobulin phenotypes and milk composition in New Zealand dairy cattle. Journal of Dairy Science 1993;76:281–6.

76. Ng-Kwai-Hang KF, Hayes JF, Moxley JE, Monardes HG. Variation in milk protein concentrations associated with genetic polymorphism and environmental factors. Journal of Dairy Science 1987;70:563–70.

77. Ikonen T, Ojala M, Ruottinen O. Associations between milk protein polymorphism and first lactation milk production traits in Finnish Ayrshire cows. Journal of Dairy Science 1999;82:1026–33.

78. Heck JML, Schennink A, van Valenberg HJF, Bovenhuis H, Visker MHPW, van Arendonk JAM, *et al.* Effects of milk protein variants on the protein composition of bovine milk. Journal of Dairy Science 2009;92:1192–202.

79. Ng-Kwai-Hang KF. A review of the relationship between milk protein polymorphism and milk composition/milk production. In: Federation ID, editor. Milk Protein Polymorphism. International Dairy Federation, Brussels, Belgium; 1997. p. 22–37.

80. Hill JP, Boland MJ, Harris P, Paterson G. Impact of genetic polymorphism on milk powder manufacture and processing. In: Agnew RE, Agnew KW, Fearon AM, editors. Milk Composition. British Society of Animal Science, Belfast, Northern Ireland; 2000. p. 87–92.

81. Caroli AM, Chessa S, Erhardt GJ. Invited review: Milk protein polymorphisms in cattle: Effect on animal breeding and human nutrition. Journal of Dairy Science 2009;92:5335–52.

82. Martin P, Szymanowska M, Zwierzchowski L, Leroux C. The impact of genetic polymorphisms on the protein composition of ruminant milks. Reproduction Nutrition Development 2002;42:433–59.

83. Hayes HC, Petit EJ. Mapping of the beta-lactoglobulin gene and of an immunoglobulin-M heavy chain-like sequence to homoeologous cattle, sheep, and goat chromosomes. Mammalian Genome 1993;4:207–10.

84. Hayes HC, Petit EJ, Bouniol C, Popescu P. Localisation of the alpha-S2-casein gene (CASAS2) to the homologous cattle, sheep and goat chromosomes 4 by *in situ* hybridization. Cytogenetics and Cell Genetics 1993;64:282–5.

85. Castle WE. An improved method of estimating the number of genetic factors concerned in cases of blending inheritance. Science 1921;54:223.

86. Sax K. The association of size differences with seed-coat pattern and pigmentation in *Phaseolus vulgaris*. Genetics 1923;8:552–60.

87. Soller M. Genetic mapping of the bovine genome using deoxyribonucleic acid-level markers to identify loci affecting quantitative traits of economic importance. Journal of Dairy Science 1990;73:2628–46.

88. Soller M, Brody T, Genizi A. On the power of experimental designs for the detection of linkage between marker loci and quantitative loci in crosses between inbred lines. Theoretical and Applied Genetics 1976;47:35–9.

89. Lebowitz RJ, Soller M, Beckmann JS. Trait-based analyses for the detection of linkage between marker loci and quantitative trait loci in crosses between inbred lines. Theoretical Applied Genetics 1987;73:556–62.

90. Darvasi A, Soller M. Selective genotyping for determination of linkage between a marker locus and a quantitative trait locus. Theoretical and Applied Genetics 1992;85:353–9.

91. Weller JI, Kashi Y, Soller M. Power of daughter and granddaughter designs for determining linkage between marker loci and quantitative trait loci in dairy cattle. Journal of Dairy Science 1990;73:2525–37.

92. Georges M, Nielsen D, Mackinnon M, Mishra A, Okimoto R, Pasquino AT, *et al*. Mapping quantitative trait loci controlling milk production in dairy cattle by exploiting progeny testing. Genetics 1995;139:907–20.

93. Spelman RJ, Coppieters W, Karim L, Van Arendonk JAM, Bovenhuis H. Quantitative trait loci analysis for five milk production traits on chromosome six in the Dutch Holstein–Friesian population. Genetics 1996;144:1799–808.

94. Vilkki HJ, de Koning D-J, Elo K, Velmala R, Mäki-Tanila A. Multiple marker mapping of quantitative trait loci of Finnish dairy cattle by regression. Journal of Dairy Science 1997;80:198–204.

95. Haley CS, Visscher PM. Strategies to utilize marker-quantitative trait loci associations. Journal of Dairy Science 1998;81:85–97.

96. Dekkers JCM. Commercial application of marker- and gene-assisted selection in livestock: strategies and lessons. Journal of Animal Science 2004;82(E- Suppl.):E313–28.

97. Goddard ME, Hayes BJ. Genomic selection. Journal of Animal Breeding and Genetics 2007;124:323–30.

98. Boichard D, Fritz S, Rossignol MN, Guillaume F, Colleau JJ, Druet T. Implementation of marker assisted selection: practical lessons from dairy cattle. In 8th World Congress on Genetic Applied to Livestock Production, 13–18 August 2006, Belo Horizonte, MG, Brazil; 2006.

99. Grisart B, Coppieters W, Farnir F, Karim L, Ford C, Berzi P, *et al*. Positional candidate cloning of a QTL in dairy cattle: Identification of a missense mutation in the bovine DGAT1 gene with major effect on milk yield and composition. Genome Research 2002;12:222–31.

100. Cohen-Zinder M, Seroussi E, Larkin DM, Loor JJ, Wind AE-vd, Lee J-H, *et al*. Identification of a missense mutation in the bovine ABCG2 gene with a major effect on the QTL on chromosome 6 affecting milk yield and composition in Holstein cattle. Genome Research 2005;15:936–44.

101. Olsen HG, Lien S, Gautier M, Nilsen H, Roseth A, Berg PR, *et al*. Mapping of a milk production quantitative trait locus to a 420-kb region on bovine chromosome 6. Genetics 2005;169:275–83.

102. Blott S, Kim J-J, Moisio S, Schmidt-Küntzel A, Cornet A, Berzi P, *et al*. Molecular dissection of a quantitative trait locus: a phenylalanine-to-tyrosine substitution in the transmembrane domain of the bovine growth hormone receptor is associated with a major effect on milk yield and composition. Genetics 2003;163:253–66.

103. Khatkar MS, Thomson PC, Tammen I, Raadsma HW. Quantitative trait loci mapping in dairy cattle: review and meta-analysis. Genetics, Selection, Evolution 2004;36:163–90.

104. Lemay DG, Lynn DJ, Martin WF, Neville MC, Casey TM, Rincon G, *et al*. The bovine lactation genome: insights into the evolution of mammalian milk. Genome Biology 2009;10:R43.

105. Meuwissen THE, Hayes BJ, Goddard ME. Prediction of total genetic value using genome-wide dense marker maps. Genetics. 2001;157:1819–29.

106. Goddard ME, Hayes BJ. Mapping genes for complex traits in domestic animals and their use in breeding programmes. Nature Reviews 2009;10:381–91.

107. Daetwyler HD, Schenkel FS, Sargolzaei M, Robinson JAB. A genome scan to detect quantitative trait loci for economically important traits in Holstein cattle using two methods and a dense single nucleotide polymorphism map. Journal of Dairy Science 2008;91:3225–36.

108. Kolbehdari D, Wang Z, Grant JR, Murdoch B, Prasad A, Xiu Z, *et al*. A whole genome scan to map QTL for milk production traits and somatic cell score in Canadian Holstein bulls. Journal of Animal Breeding and Genetics 2009;126:216–27.

109. Cole JB, VanRaden PM, O'Connell JR, van Tassell CP, Sonstegard TS, Schnabel RD, *et al*. Distribution and location of genetic effects for dairy traits. Journal of Dairy Science 2009;92:2931–46.

110. Bolormaa S, Pryce JE, Hayes BJ, Goddard ME. Multivariate analysis of a genome-wide association study in dairy cattle. Journal of Dairy Science 2010;93:3818–33.

111. Schopen GCB, Visker MHPW, Koks PD, Mullaart E, van Arendonk JAM, Bovenhuis H. Whole-genome association study for milk protein composition in dairy cattle. Journal of Dairy Science 2011;94:3148–58.

112. Bouwman AC, Bovenhuis H, Visker MHPW, van Arendonk JAM. Genome-wide association of milk fatty acids in Dutch dairy cattle. BMC Genetics 2011;12:43–54.

113. Meredith BK, Kearney JF, Berry DP. 2010. Genome-wide associations for milk production and somatic cell count in Irish Holstein-Friesian cattle. In Proceedings of the 9th World Congress of Genetics Applied to Livestock Production, 1–2 August 2010, Communication 737, Leipzig, Germany; 2010.

114. Jiang L, Liu J, Sun D, Ma P, Ding X, Yu Y, *et al*. Genome wide association studies for milk production traits in Chinese Holstein population. PLoS ONE 2010;5:e13661.

115. Eck SH, Benet-Pages A, Flisikowski K, Meitinger T, Fries R, Strom TM. Whole genome sequencing of a single *Bos taurus* animal for single nucleotide polymorphism discovery. Genome Biology 2009;10:R82.

116. Weller JI, Ron M. Invited review: Quantitative trait nucleotide determination in era of genomic selection. Journal of Dairy Science 2011;94:1082–90.

117. Hayes BJ, Bowman PJ, Chamberlain AJ, Goddard ME. Invited review: Genomic selection in dairy cattle: progress and challenges. Journal of Dairy Science 2009;92:433–43.

118. Dekkers JCM. Use of high-density marker genotyping for genetic improvement of livestock by genomic selection. CAB Reviews: Perspectives in Agriculture, Veterinary Science, Nutrition and Natural Resources 2010;5(No. 037).

119. Wiggans GR, VanRaden PM, Cooper TA. The genomic evaluation system in the United States: Past, present, future. Journal of Dairy Science 2011;94:3202–11.

120. Davis SR. Lactational traits of importance in dairy cows and applications for emerging biotechnologies. New Zealand Veterinary Journal 2005;53:400–5.

121. Kadarmideen HN, Reverter A. Combined genetic, genomic and transcriptomic methods in the analysis of animal traits. CAB Reviews: Perspectives in Agriculture, Veterinary Science, Nutrition and Natural Resources 2007;2(042):16.

122. Stoop WM, Schennink A, Visker MHPW, Mullaart E, van Arendonk JAM, Bovenhuis H. Genome-wide scan for bovine milk-fat composition. I. Quantitative trait loci for short- and medium-chain fatty acids. Journal of Dairy Science 2009;92:4664–75.

123. Schennink A, Stoop WM, Visker MHPW, van der Poel JJ, Bovenhuis H, van Arendonk JAM. Short communication: Genome-wide scan for bovine milk-fat composition. II. Quantitative trait loci for long-chain fatty acids. Journal of Dairy Science 2009;92:4676–82.

124. Polineni P, Aragonda P, Xavier SR, Furuta R, Adelson DL. The bovine QTL viewer: A web accessible database of bovine quantitative trait loci. BMC Bioinformatics 2006;7:283.

125. Hu Z-L, Park CA, Fritz ER, Reecy JM. QTLdb: a comprehensive database tool building bridges between genotypes and phenotypes. In Proceedings of the 9th World Congress on Genetics Applied to Livestock Production, Leipzig, Germany, 1–2 August 2010, Communication No. 17; 2010.

126. Van Raden PM, Van Tassell CP, Wiggans GR, Sonstegard TS, Schnabel RD, Taylor JF, *et al.* Reliability of genomic predictions for North American Holstein bulls. Journal of Dairy Science 2009;92:16–24.

127. Kearney JF, Cromie AR, Berry DP. Implementation and uptake of genomic selection in Irish Holstein-Friesian dairy cattle. In Proceedings of the 9th World Congress of Genetics Applied to Livestock Production, 1–2 August 2010, Leipzig, Germany, Communication 0889; 2010.

128. Spelman RJ, Arias JA, Keehan MD, Obolonkin V, Winkelman AM, Johnson DL, *et al.* Application of genomic selection in the New Zealand dairy cattle industry. In Proceedings of the Ninth World Congress of Genetics Applied to Livestock Production, Leipzig, Germany; 2010.

129. Boichard D, Guillaume F, Baur A, Croiseau P, Rossignol MN, Boscher MY, *et al.* Genomic selection in French dairy cattle. In Proceedings of the Ninth World Congress on Genetics Applied to Livestock Production, 1–2 August 2010, Leipzig, Germany; 2010.

130. Su G, Madsen P, Nielsen US, Mäntysaari EA, Aamand GP, Christensen OF, *et al.* Genomic prediction for Nordic Red Cattle using one-step and selection index blending. Journal of Dairy Science 2012;95:909–17.

131. Schaeffer LR. Strategy for applying genome-wide selection in dairy cattle. Journal of Animal Breeding and Genetics 2006;123:218–23.

132. Gianola D, de los Campos G, Hill WG, Manfredi E, Fernando R. Additive genetic variability and the Bayesian alphabet. Genetics 2009;183:347–63.

133. Calus MPL. Genomic breeding value prediction: methods and procedures. Animal 2010;4:157–64.

134. Bovine Genome Sequencing and Analysis Consortium.The genome sequence of taurine cattle: A window to ruminant biology and evolution. Science. 2009;324:522–8.

135. Niemann H, Kuhla B, Flachowsky G. Perspectives for feed-efficient animal production. Journal of Animal Science 2011;89:4344–63.

136. Wickramasinghe S, Hua S, Rincon G, Islas-Trejo A, German JB, Lebrilla CB, *et al.* Transcriptome profiling of bovine milk oligosaccharide metabolism genes using RNA-sequencing. PLoS ONE 2011;6(4):e18895.

137. Bionaz M, Loor JJ. Gene networks driving bovine milk fat synthesis during the lactation cycle. BMC Genetics 2008;9:366–87.

138. Berry DP, Meade KG, Mullen MP, Butler S, Diskin MG, Morris D, *et al.* The integration of 'omic' disciplines and systems biology in cattle breeding. Animal 2011;5:493–505.

139. Flori L, Fritz S, Jaffrézic F, Boussaha M, Gut I, Heath S, *et al.* The genome response to artificial selection: a case study in dairy cattle. PLoS ONE 2009;4:e6595.

140. Karatzas CN. Designer milk from transgenic clones. Nature Biotechnology 2003;21:138–9.

141. Brophy B, Smolenski G, Wheeler T, Wells D, L'Huillier P, Laible G. Cloned transgenic cattle produce milk with higher levels of β-casein and κ-casein. Nature Biotechnology 2003;21:157–62.

142. Piedrahita JA, Olby N. Perspectives on transgenic livestock in agriculture and bsiomedicine: an update. Reproduction, Fertility and Development 2010;23:56–63.

143. Wall RJ, Hawk HW, Nel N. Making transgenic livestock: genetic engineering on a large scale. Journal of Cellular Biochemistry 1992;49:113–20.

144. Yu S, Luo J, Song Z, Ding F, Dai Y, Li N. Highly efficient modification of beta-lactoglobulin (BLG) gene via zinc-finger nucleases in cattle. Cell Research 2011;21:1638–40.

145. Cecchinato A, De Marchi M, Gallo L, Bittante G, Carnier P. Mid-infrared spectroscopy predictions as indicator traits in breeding programs for enhanced coagulations properties of milk. Journal of Dairy Science 2009;92:5304–13.

146. Lush JL. Animal Breeding Plans. Iowa State College Press, Ames, IA; 1945.

Environmental enrichment and practical applications for welfare in production agriculture

L.M. Dixon*

Address: Animal Behaviour and Welfare, Sustainable Livestock Systems, Scottish Agricultural College, Edinburgh, UK.

***Correspondence:** Email: Laura.Dixon@sac.ac.uk

Received: 23 June 2011
Accepted: 17 April 2012

Abstract

Environmental enrichment refers to stimuli that are added to an animal's environment in an attempt to improve biological functioning and animal welfare by increasing the complexity of the environment and providing opportunities for the animal to perform species-appropriate activities. Enrichments are often provided to help decrease or alleviate performance of abnormal behaviour patterns that are often found in barren and static environments. Enrichments can be divided into four categories: foraging opportunities, structural complexity, sensory stimulation/novelty and social stimulation from conspecifics or humans. However, different species have different behavioural requirements and therefore may benefit from different types of enrichment. To summarize the research that has been conducted on environmental enrichment, tables have been created for some of the more common production species, which give examples of enrichments that can be used and the potential benefits and costs found from implementing them. In general, most species seem to benefit from the provision of a foraging substrate/bedding, places to hide/dividers/platforms and access to conspecifics. However, most enrichment studies are performed on a much smaller scale than normally found in production environments; therefore the full benefits and costs of the enrichment may not yet be known. Practically, enrichments should improve animal welfare and satisfy behavioural needs while being inexpensive, available and easy to maintain for the producer. The development of improved welfare brands that require certain enrichment items should help improve the lives of production animals and help guide producers on the environmental enrichments to provide for their animals.

Keywords: Animal welfare, Environmental enrichment, Animal behaviour, Motivation, Abnormal behaviour

Review Methodology: This review was based on searching the following databases: CAB Abstracts, CAB Heritage and ISI World of Knowledge (keyword search terms used: animal, poultry, duck, fowl, pig, swine, dairy cattle, beef cattle, rabbit, mink, fox, sheep, goat, fish, ostrich, farm animal, animal welfare, environmental enrichment; enriched environment; human-animal interactions; social housing; behavioural needs; motivation). In addition, the author used the references from the articles obtained by this method to extend the search for additional relevant material.

What is Environmental Enrichment?

Environmental enrichment refers to stimuli that are added to an animal's environment in an attempt to improve biological functioning (health, reproduction, etc.) [1] and animal welfare (quality of life) [2] by increasing the complexity of the environment and providing opportunities for the animal to perform a larger number of species-appropriate activities. Enrichments are often provided in an attempt to satisfy unfulfilled motivations, increase behavioural repertoires, add sensory stimulation and/or decrease stress with the end result being improved welfare (e.g. [3]). There may not always be conclusive evidence that a given enrichment has achieved its intended goal, i.e. it may not fulfil the intended motivation, but there is a general consensus that any enrichment given to animals (especially those in barren or static environments) improves welfare by giving the animal more to do and allowing a larger number of behaviour patterns to be performed while generally reducing performance of abnormal behaviour patterns (e.g. [4]). In this paper, the term enrichment will refer to an addition to the environment that is thought to improve welfare but without making assumptions about the underlying mechanisms of the improvement.

How does Environmental Enrichment Improve Welfare?

Production animals are often kept in fairly barren and unstimulating environments, especially when compared with the natural environments that the ancestors of these animals have originated from. These captive environments limit the behaviour patterns the animals can perform and are usually unchanging in appearance and regimented in daily routines. The environments were designed to maximize production and decrease incident of disease (which also had a positive effect on animal welfare); however, they usually did not take into account the animals' behavioural needs or the mental states of the animals housed in the environments [5].

Animals kept in production environments tend to have lower mortality and occurrence of disease than wild animals but they also perform higher levels of abnormal behaviours, such as stereotypic or repetitive behaviour or self-injurious behaviour (e.g. [6]). Additionally some animals become unresponsive or apathetic in these environments [7, 8]. It is believed that these abnormal behaviours stem from frustration and/or attempts to cope with captive environments and some animals may even have developed central nervous system (brain) dysfunction because of unnatural and stressful experiences, all of which indicate a problem with the animals' captive environment [9].

Environmental enrichment was developed as a way to abolish or at least decrease occurrence of these abnormal behaviours and to improve public perception of production systems (e.g. [1]). One of the main ways in which enrichments improve animal welfare is allowing the animal behavioural opportunities to exert control over their environments [10]. Animals in enriched environments also tend to spend more time engaged in more species-specific behaviour patterns, which have decreased stress and fear responses and may be easier to handle (e.g. [11, 12]) than animals in barren environments, which again improves the welfare of the animals and additionally improves the safety of those who work with the animals.

What Types of Enrichments are Available?

Most enrichments are provided to attempt to satisfy motivations, such as foraging, dustbathing and exploration in chickens, and allow animals to perform a larger number of species-appropriate behavioural patterns than would be possible in barren conditions (e.g. [4, 13]). For example, laying hens are motivated to use perches and nestboxes and will work by pushing through doors or squeezing through narrow openings to gain access to them [14, 15]. Thus, the provision of these enrichments to laying hens helps to fulfil motivation and decrease the incidence of abnormal behaviour patterns performed compared with birds housed without nestboxes or perches (e.g. [16]). Similarly pigs are motivated to root/forage in a substrate [17] and when that substrate is provided abnormal behaviours, such as tail biting, are decreased [18].

Enrichments are often divided into four categories: foraging opportunities, structural complexity, sensory stimulation/novelty and social stimulation from conspecifics or humans [1, 12]. Many species are motivated to forage and the provision of a substrate can also have a positive effect on gut health through an increase of roughage in the diet (e.g. [19–21]). Structural complexity can take the form of nestboxes, shelters, dividing partitions, water baths or platforms (e.g. [16, 22–25]). Enrichments designed to provide sensory stimulation or novelty are often types of 'toys' made of different materials: balls, hanging chains or ropes, rubber hoses, hanging CDs and so on, which encourage the animal to explore the new object (e.g. [26–28]). These are often rotated to help maintain the animals' interest and effectiveness of the enrichment [29, 30]. Social stimulation from conspecifics can be as simple as group or pair housing. This allows opportunities for increased behavioural complexity and often decreases stress and fearfulness (e.g. [31, 32]). However, some species tend to be solitary for part or all of their lives and in these cases conspecific socialization may be aversive (e.g. [33]). Social stimulation from humans is often used to decrease handling stress and fearfulness and in cases may improve productivity (e.g. [11, 34]). Then again, this can have negative effects for the animal handlers as less fearful animals may be harder to move/manage (e.g. [35]).

Some enrichment types are provided to many species, e.g. forages are given to fowl, pigs, cattle and rabbits [4, 20, 36], whereas other enrichments tend to be species-specific, e.g. perches in poultry [34] and water troughs for ducks [24]. In the following paragraphs, I summarize different types of enrichment available for fowl, pigs, cattle, mink and foxes, rabbits and other production animals and discuss some pros and cons of their usage.

Enrichments available to fowl (Table 1)

Jungle fowl and other types of poultry housed in more extensive environments spend a large proportion of their active time foraging [41] and even commercially produced fowl are motivated and will pay a cost for access to forages, dustbaths and areas to explore (e.g. [4, 37, 73]). Provision of these enrichments can lead to an increase in

Table 1 Examples of environmental enrichments available for commercial fowl (chickens, turkeys, ducks and quail) and the potential benefits and costs associated with their use

Enrichment	Examples	Potential benefits	Potential costs	References
Dustbath	Peat moss, sand (wood shavings were often found to be inappropriate for dust bathing)	Allows natural behaviour Motivated to access appropriate dustbathing substrate Decreases foot and feather damage Decreases feather pecking	Additional cost/work to maintain Increases dust levels	[4, 37–40]
Foraging substrate/ bedding	Outdoor access to grass, hay, grains scattered over the bedding, cabbage, woodshavings, sawdust, suet block with seeds	Allows natural behaviour Motivated to peck and scratch Increase exploration Increases foraging behaviour Decreases pecking at pen fixtures Decreases aggression Decrease pacing Decreases feather pecking/pecking injuries	Additional cost/work to maintain May be incompatible with waste management system Increases dust levels Decreases hygiene which can increase incident of disease	[4, 39, 41–49]
Nestbox	Front rollaway nestboxes, hollow nestboxes	Allows natural behaviour Motivated to access nestbox Decreases pacing near time of egg lay Decreases abnormal behaviour Decreases foot and feather damage	Additional cost/ work to maintain Increases aggression due to nestbox use Increases interruption of lay by other birds Increases downgraded eggs	[13, 14, 16, 38, 50–53]
Perches	Soft, round perches (e.g. polyurethane), wooden perches, rectangular perches, steel perches	Allows natural behaviour Motivated to perch Increases bone strength/volume Improves foot pad health (broilers) Increases activity Decreases incidence of feather pecking and cannibalism	Additional cost/work to maintain Increases keel bone damage Increases foot pad problems (layers)	[13, 15, 38, 43, 54–61]
Conspecific socialization	E.g. in groups of 2 or more birds	Allows natural behaviour Decreases fear and stress Decreases abnormal behaviour	Increases aggression Increases feather wear/damage Increases disease spread	[53, 62–64]
Human socialization	E.g. handled twice daily, stroked, hand feed birds	Improves handling Improve growth, egg production, eggshell quality, feed conversion Increases antibody response Decreases fear Decreases injury Decreases aggression	Additional cost/work to maintain May not be feasible on commercial size farms	[48, 62, 65–68]
Water source (Ducks)	Open water source – trough, bath or overhead shower	Allows natural behaviour Improves nostril hygiene Improves feather condition Improves body weight Decreases injuries and feather pecking Decreases leg problems	Additional cost/work to maintain Decreases hygiene which can increase incident of disease and injury	[24, 62, 69–72]

Table 1 (Cont.)

Enrichment	Examples	Potential benefits	Potential costs	References
Other	String, Peck-a-Block, CD's from ceiling (Freedom foods), barriers in shed, porous concrete block to peck, 'toys' (bells, mirrors, balls – different textures/colours), vertical panels, metal panels, chains, tissue paper, increase space allowance	Allows natural behaviour (?) (e.g. exploration, locomotion) Increases activity Improves reproductive performance (vertical panels) improves egg production, fertility and hatchability (vertical panels) Decreases fear Decreases aggression and feather pecking Decreases abnormal behaviour like polydipsia and spot pecking Improved spatial distribution (vertical panels)	Additional cost/work to maintain Increased disease and mortality ('toys') Decreased weight gain ('toys')	[4, 23, 28, 29, 42, 69, 73–77]

activity and a decrease in aggression and the incidence of abnormal behaviour performed (e.g. [4, 42, 74]). Some enrichments have been shown to improve production factors such as an increase in bone strength when birds had access to perches [38] and improvements in growth, egg production and egg quality when the birds were gently handled each day [62, 65]. Housing birds with conspecifics can lead to decreased fearfulness and stress (e.g. [31]) and positive human handling can help decrease fearfulness, injury and aggression (e.g. [66, 67]). Even just increasing the space allowance by e.g. adding perches or nestboxes can help decrease aggression, feather pecking and fear (e.g. [15, 23, 43]). However, there are costs associated with enrichment provision. Any addition to the environment will incur a financial cost and will take more time and effort to maintain (e.g. [23]). Some substrates may increase dust levels in the air and as enrichments tend to be hard to keep clean, can lead to decreased hygiene and increased disease (e.g. [69, 78]). Some enrichments may even lead to increased aggression, such as nestboxes, owing to competition to access the preferred box (e.g. [50]). Good farm management may help reduce the negative consequences of providing access to environmental enrichments and these are often related to maintenance and cleaning [44, 75]. Also, birds have been shown to be highly motivated to perform some of the behaviours that enrichments allow, such as dustbathing, foraging, perching and accessing a nestbox, which does indicate that their welfare is improved when they are given the opportunity to do this (e.g. [14, 15, 37]). Additionally, certain enrichments will soon be required for laying hens such as perches and dustbaths when traditional or battery cages are banned in the EU in 2012 [79]. Thus, there seems to be evidence that providing fowl with at minimum dustbaths, foraging substrates, perches, nestboxes and conspecifics will improve bird welfare and birds should be housed with these enrichments when possible.

Enrichments available to pigs (Table 2)

Pigs have been shown to be highly motivated to perform certain behaviour patterns such as rooting in substrates, exploration and nest building in pregnant sows (e.g. [20, 83, 84, 101]). The majority of enrichments that are provided to pigs are designed to attempt to fulfil motivations by providing e.g. suitable substrates for rooting and nesting and objects/areas to explore (e.g. [18, 83, 101]). This in turn decreases the incidence of abnormal behaviours shown, such as bar biting and sham chewing, and also helps to decrease stress and aggression (e.g. [18, 80, 96]). This decreased stress and aggression can also have positive effects on meat quality and other enrichments, such as provision of a bedding material, can help minimize pigs slipping on the flooring and their carcasses becoming bruised (e.g. [80, 85]). Pigs housed on appropriate bedding are also thought to exhibit better

Table 2 Examples of environmental enrichment available for commercial pigs and the potential benefits and costs associated with their use

Enrichment	Examples	Potential benefits	Potential costs	References
Bedding	Straw, sand, sawdust, peat, mats	Increases lying comfort Increased meat tenderness Improved spatial memory Decreases slipping Decreases sole bruising	Additional cost/work to maintain Increases skin abrasions May be incompatible with waste management system Decreases hygiene	[5, 20, 62, 80–82]
Forage/rooting material	Straw (long preferred to chopped), woodshavings, peat/earth, grass silage	Allows natural behaviour Motivated to forage/root Increases activity Increases meat quality Decreases abnormal behaviours Decreases aggression and injuries Decreases stress Less rooting of the ground in outdoor systems	Additional cost/work to maintain May be incompatible with waste management system Decreases hygiene Higher gastric lesion score	[18, 20, 80, 83–88]
Nesting material	Straw, sand, sawdust	Allows natural behaviour Motivated to nest build Increases piglet survival Decreases abnormal behaviour Decreases stress	Additional cost/work to maintain May be incompatible with waste management system Decreases hygiene	[88–91]
Conspecific socialization	Group housing, group farrowing	Allows natural behaviour Increases activity Decreases abnormal behaviour	Additional cost/work to maintain Increases aggression	[92, 93]
Human socialization	Talk to pigs Stroke pigs	Decreases fear Decreases stress	Additional cost/work to maintain May not be feasible on commercial size farms Reduced fear of humans may make handling more difficult	[18, 35, 94, 95]
Other	Toys: suspended chains, rope, metal pipe, rubber hose Increased space Mineral blocks, Edinburgh football	Allows natural behaviour (?) (e.g. exploration, locomotion) Motivated to explore Increases activity Increases average daily gain and feed to gain ratio Increases oestrus detection and mating (increased space) Decreases abnormal behaviour Decreases aggression Decreases fear	Additional cost/work to maintain Increased aggression (mineral blocks) and increased injury	[26, 27, 30, 96–100]

lying postures and shift positions less frequently than those on slatted or concrete surfaces, indicating increased comfort [20, 81]. Pigs are a social species and access to conspecifics increases the behavioural repertoire available to perform and decreases abnormal behavioural patterns performed (e.g. [92, 93]). Pigs that are socialized with humans by positive handling, such as being spoken to or stroked also show decreased fear and stress compared with those without this interaction (e.g. [18, 35]). There have also been many 'toys' designed to keep pigs occupied in production environments and to give them new opportunities to explore different objects (e.g. [26, 96, 102]). Generally, these toys can be manipulated by the pigs snout and are durable enough for the pig to chew on, for example hanging chains, ropes, rubber hoses or the Edinburgh football® (e.g. [26, 30, 102]). Provision of these enrichments can help decrease aggression, fear and performance of abnormal behaviour [26, 96]. These enrichments do have downsides: the addition of items adds another production cost and the enrichments are harder to keep clean, which may decrease hygiene and could lead to increased disease. Also pigs' interest in enrichments decreases as the objects become soiled (e.g. [103]). The type of substrate used, e.g. straw, sawdust, sand, peatmoss, silage, etc., needs to be compatible with the waste management system on farm or it could lead to blockages in the system (e.g. [83]). The enrichments need to be properly maintained and sometimes rotated in and out of pens for them to maintain pig interest and do the job they were intended to (e.g. fulfil motivations) [83]. Insufficient access to enrichments may lead to competition for these items and may increase aggression. Group-housed animals also have increased aggression and potentially increased injuries compared with singly housed animals (e.g. [92]). However, overall there appears to be more benefits than costs for adding enrichments to a well-run pig production system. This is backed up by EU directives (2001/88/EC and 2001/93/EC), which require that all ages of pigs have the provision of substrate enrichments, such as straw, hay or sawdust, at all times. There is a growing public concern for improved animal welfare systems and increasing markets for pigs produced in these systems. For example in the UK, pork producers who wish to sell under the Freedom Foods label must provide their pigs with straw or another rooting substrate [104]. Therefore pigs should be housed with enrichment items that at least meet minimum EU requirements and, if possible, with additional items as this appears to benefit the animals and improve their welfare and may also have some commercial benefits.

Enrichments available to dairy and beef cattle (Table 3)

The intake of forages is important to digestive health in cattle and in natural environments foraging and rumination occupies a large part of their day. Production cattle are often fed a diet of concentrates with little forage available. As a result, foraging enrichments can be provided that increase the time spent ruminating and decreases rumen acidosis and the incidence of abnormal behaviour patterns, such as tongue rolling [107]. Another concern for mainly dairy cattle is the high incidence of lameness that can occur in production herds. Dairy cattle have increased in size, weight and milk production but the stalls they are housed in may be outdated and designed for smaller animals. The provision of various beddings and cow comfort mats increases lying and rumination time and decreases foot and leg problems as well as skin lesions [121, 122]. Grooming tools, such as steel drums covered in turf or mounted grooming brushes can be provided and can help decrease performance of abnormal behavioural patterns while improving carcass characteristics [108, 113]. Group housing helps decrease fear reactions, improves feed intake and increases rumination time [32, 114, 115], whereas positive human handling such as brushing and feeding treats has been found to improve handling of cattle, decrease stress and fear and improve milk yield [32]. Cattle can also be given mineral or salt blocks or 'toys' made of plastic hose pipe, pieces of tires or chains, which can increase activity and play behaviour and decrease abnormal behavioural patterns [117–120]. The addition of extra items to the production system increases production costs and decreases hygiene, especially soiled bedding, which can lead to increased incidence of disease [122]. Animals in group housing will have higher aggression levels than those in singly or tie stall environments [32, 115]. For calves on milk replacers, the presence of conspecifics may actually increase abnormal cross-suckling behaviour [119]. There are improved animal welfare schemes or recommendations that require as a minimum the provision of forages and comfortable lying areas that are appropriate for the size and number of animals housed and if abnormal behaviours develop, additional enrichments must be added to improve this (e.g. [123, 124]). The overall benefits of adding enrichment to cattle environments appear to outweigh the costs and cattle should be provided at the least minimum enrichments laid out in improved welfare standards.

Enrichments available to mink and fox (Table 4)

Mink and fox raised for fur production are often housed in fairly barren environments in an attempt to keep their pelts clean and un-damaged. However, these environments may not be adequate for the animals' welfare and a number of enrichments can be and often are added. For example, mink and fox are motivated to access and will work for use of a nestbox or box of substrates containing hay, straw or sand (e.g. [125, 143]). Access to these enrichments decreases stress, fear and incidence of abnormal behaviours; as well, bone strength, kit mortality

Table 3 Examples of environmental enrichments available for commercial dairy and beef cattle and the potential benefits and costs associated with their use

Enrichment	Examples	Potential benefits	Potential costs	References
Bedding/comfort mat	Straw, rubber mat, sand, sawdust	Allows natural behaviour Increases lying Increase rumination Increases milk production Increases oestrous behaviour and fertility Decreases foot and leg problems Decreases skin injuries	Additional cost/work to maintain Decreases hygiene of lying surface and cleanliness of animals	[20, 105, 106]
Forage	Drum can filled with hay, bales of straw, access to pasture	Allows natural behaviour Increased daily gain Increases rumination Decreases abnormal behaviour Increases plasma dopamine Improve carcass characteristics, e.g. belly fat thickness and marbling Decreases lameness (pasture)	Additional cost/work to maintain Increases aggressive interactions/competition	[20, 107–112]
Grooming tool	Drum covered in artificial turf, trees, mounted scratching brushes	Allows natural behaviour Improve carcass characteristics, e.g. belly fat thickness and marbling Decrease abnormal behaviour	Additional cost/work to maintain	[108, 113]
Conspecific socialization	Two or more animals housed together	Allows natural behaviour Increases feed intake Increase rumination Decreases fear	Increases aggression Increases incident of disease Increases some abnormal behaviours (e.g. cross-suckling in calves)	[32, 114, 115]
Human socialization	Brushing and petting cattle, standing in enclosure, talking in soft voice, feeding treats	Improves handling Decreases fear Decreases stress Increased milk yield	Additional cost/work to maintain May not be feasible on commercial size farms	[12, 116]
Other	Salt/mineral blocks, hanging 'toy' made of plastic hose pipe, increase space allowance, piece of tire, chain	Allows natural behaviour (?) (e.g. exploration) Increases activity Increase milk protein yield (space allowance) Increase play behaviour (space allowance) Decreases abnormal behaviour	Additional cost/work to maintain Decreases hygiene	[117–120]

Table 4 Examples of environmental enrichments available for mink and fox used commercially and the potential benefits and costs associated with their use

Enrichment	Examples	Potential benefits	Potential costs	References
Nestbox/bedding	Boxes with hay, straw Sand floor, sand box	Allows natural behaviour Motivated to access Improves fur condition (mink) Increases bone strength Decreases stress Decreases fear Decreases kit death Decreases abnormal behaviour	Additional cost/work to maintain Decreases hygiene Decreases fur condition (fox)	[11, 125–131]
Shelf/platform	Wire mesh shelf, solid wood	Allows natural behaviour Decreases stereotypic behaviour	Additional cost/work to maintain	[25, 130, 131]
Conspecific socialization	Housed in same sex pairs, mixed sex pairs, family groups	Allows natural behaviour (during certain times of life, i.e., mother with kits or during mating season) Better pelt quality (in some studies) Decreases abnormal behaviour Decreased fearfulness	Increased stress Decreased pelt quality (in some studies) Poor teat condition (family groups)	[33, 132, 133]
Human socialization	Handle once a day, feed treats, petted	Increases exploration Decreases fear Decreases stress	Additional cost/work to maintain May not be feasible on commercial size farms	[134–136]
Swimming water (mink)	Trough, bath	Allows natural behaviour Motivated to access water	Additional cost/work to maintain Decreases hygiene – increased kit mortality	[128, 130, 131, 137–139]
Other	Running wheel, increased space, toys: ropes, plastic tunnels, wooden blocks	Allows natural behaviour (?) (e.g. locomotion/ranging, exploration) Motivated to access (wheel) Decreases abnormal behaviour Decreases tail chewing Decreases stress	Additional cost/work to maintain Harder to catch animals (increased space)	[137, 140–142]

levels and fur condition (mink) is improved (e.g. [126, 127]). Although hygiene was decreased as droppings could build up in the boxes and foxes were found to have decreased fur quality when provided with a box of bedding [127]. A simple enrichment that can be provided is a mesh or wooden platform and this helps to decrease the incidence of abnormal behaviour [25]. Although they are often solitary animals, the use of pair, group or family group housing has been tried and can decrease abnormal behaviour and fearfulness (e.g. [33, 132]). However, these animals were also found to have increased stress levels, which may indicate that pair or group housing could be detrimental to welfare [33]. Positive human socialization through gentle handling and petting did increase exploratory behaviour while decreasing stress and fear [134–136]. Mink, as a semi-aquatic species, were motivated to access a water bath or trough that they could submerge themselves or paddle in but there was difficultly in keeping the water clean and as a result maintaining good hygiene and mortality levels (e.g. [137, 138]). 'Toys' such as running wheels, ropes, plastic tunnels and wooden blocks helped decrease stress, abnormal behaviours and tail chewing but these additions to the environment made catching the animals more difficult and time consuming (e.g. [140, 141]). As with any addition to a production system, use of enrichments will be an added cost. Also there may be hygiene issues and possibly decreased pelt quality if the enrichments are not properly maintained; however, these risks can be minimized by changing nestbox or bedding material and water supply (mink) regularly. Group housing may have benefits during certain times of the year, e.g. mating or a mother and her young offspring, but at other times of year the costs do not appear to outweigh the advantages. It is recommended that enrichments that the animals are motivated to access, such as nestboxes/boxes of bedding, and that are easy to implement and maintain, such as platforms and certain toys be provided to farmed mink and fox whenever possible to help reduce stress and abnormal behaviour and improve welfare.

Enrichments available to rabbits (Table 5)

Rabbits kept in natural conditions spend the majority of their active hours foraging, then retreat to an underground burrow to rest and digest the forages [167]. However, rabbits kept for meat or fur production do not always have the opportunity to perform these behaviours. Rabbits provided bedding or digging substrates or forages have decreased abnormal behaviours and stress, and improved dental health and gut function (e.g. [144–146]. The addition of these substrates can decrease hygiene and the bedding/forage needs to be changed regularly [145]. Rabbits are also motivated to access a shelter or a platform on which they can sit, which helps reduce restlessness and stress [22, 150]. Rabbits are a social species and will work by pushing through a weighted door for access to conspecifics [150]. Social housing improves stress and abnormal behaviour levels but has higher risk for increased aggression or injury than singly housed animals (e.g. [151, 152]). Positive human socialization has also been found to decrease fear and can even help improve conception rate [34, 157]. There are 'toys' available for rabbits, such as hanging mirrors and different types of gnawing sticks, and these can help increase movement and activity, which helps improve bone strength and density and decreases stress and incidence of abnormal behaviours (e.g. [160–162]). As always, these enrichments are an additional cost for the producer to add and they require regular cleaning and maintenance to prevent poor hygiene and associated problems. Access to forages, a shelter/platform and conspecifics appear to be important for rabbit welfare and it is recommended that, at the least, these be provided whenever possible.

Enrichments available to other commercially farmed species (Table 6)

There is less literature available on enrichment opportunities for farmed species produced on a smaller scale or those normally produced in more extensive conditions (e.g. hill sheep in the UK). However the goals of the enrichment tend to be the same – to fulfil motivations, decrease abnormal behavioural patterns and stress and provide the animal with a great ranger of species appropriate behaviour patterns to perform (e.g. [1, 9]). For example, fish enclosures can be enriched with a stone or gravel 'bedding', live plants, places to hide and novel objects to explore [168, 169, 188]. Fish can also be group housed and habituated to human presence [170]. This can result in increased brain size, behavioural complexity shown and decreased reactions to stressors [170, 171], but can also result in increased aggression and decreased enclosure hygiene [188]. Goats can be provided larger enclosures, access to conspecifics and be habituated to human handling, which helps improve growth rate and handling and decreases stress and fear reactions [175–179]. Farmed ostrich can be provided bedding, forages and sticks, which increases their activity and decreases fear [180]. Sheep which are normally housed indoors can be given access to outdoor paddocks and increased space allowances, coloured dividing boards, access to conspecifics and positive human handling [181–186]. This can improve handling, increase activity and decrease fear and response to a stressor (e.g. [181, 182, 185]). Improvement in immune response and milk protein in animals given outdoor access was also found [182, 183]. When considering more extensively housed animals, enrichments in the form of shelter from the elements and supplemental food for the animals may be of greatest benefit [187]. However, all of these additions to the environment cost money to implement and require

Table 5 Examples of environmental enrichments available for commercial rabbits and the potential benefits and costs associated with their use

Enrichment	Examples	Potential benefits	Potential costs	References
Bedding/digging substrate	Straw, box of hay, peat moss, sand	Allows natural behaviour Decreases abnormal behaviour	Additional cost/work to maintain Decreased hygiene Bedding consumption	[144, 145]
Foraging substrate	Hay, grass cubes, fresh grass	Allows natural behaviour Improves dental health Increases growth rate Decreased abnormal behaviour Decreased stress Reduces incidence of diarrhoea	Additional cost/work to maintain Decreases hygiene	[36, 146–149]
Shelter/platform	Wooden box, plastic box, drainage tubes	Allows natural behaviour Motivated to access Decreased restlessness Decreased abnormal behaviour Decreased stress	Additional cost/work to maintain	[22, 150]
Conspecific socialization	Pair housing, group housing	Allows natural behaviour Motivated to access Increases movement Decreases abnormal behaviour Decreases stress	Additional cost/work to maintain? Increased risk of injury Increased risk of disease Increased aggression Decreases meat quality	[145, 151–156]
Human socialization	Daily handling of kittens, nest checks, handling does	Increases conception rate Decreases fear Decreases duration of pregnancy	Additional cost/work to maintain May not be feasible on a commercial scale	[157–159]
Other	Mirrors, increased space/larger run, increased height, gnawing sticks	Allows natural behaviour (?) (e.g. exploration, locomotion) Motivated to access (more space) Increases growth rate (hanging gnaw sticks) Decreases bone problems (increased space) Decreases respiratory problems (outdoor run) Increased movement Decreases abnormal behaviour Decreases aggression	Additional cost/work to maintain Decreases hygiene	Drescher and Loeffler 1991 referenced in [160–166]

Table 6 Examples of environmental enrichments available for other commercially farmed species and the potential benefits and costs associated with their use

Animal	Examples	Potential benefits	Potential costs	References
Fish	Stones/gravel, plants, feeding live prey, places to hide, novel objects, access to conspecifics, habituation to humans	Allows natural behaviour Increases brain size Increases behavioural complexity Decreases response to a stressor	Additional cost/work to maintain Human habituation may not be feasible on a commercial scale Increases aggression Decreases hygiene	[168–174]
Goat	Access to conspecifics, habituation to humans, increased space	Allows natural behaviour Improves handling Improves growth Increases movement Decreases fear Decreases stress	Additional cost/work to maintain Human habituation may not be feasible on a commercial scale Decreased hygiene	[175–179]
Ostrich	Bedding, forages, sticks	Allows natural behaviour Increases movement Decreases fear	Additional cost/work to maintain Decreased hygiene	[180]
Sheep	Habituation to humans, increased space, access to an outdoor area, access to conspecifics, coloured boards	Allows natural behaviour Improves handling Improves immune response Increases milk protein (outdoor access) Increases movement Decreases fear Decreases response to a stressor	Additional cost/work to maintain Human habituation may not be feasible on a commercial scale Decreased hygiene	[181–187]

more time and effort to maintain and keep clean. Nevertheless, even small additions to a static or barren environment can have positive welfare consequences and should be encouraged whenever possible.

Are there Problems with Providing Environmental Enrichments?

Enrichments need to be properly maintained and replenished for them to have maximum effect (e.g. [24, 189]). As mentioned above, when bedding or forages become wet and dirty, they need to be replaced as the benefit to the animal is reduced and the decreased hygiene can lead to injury/illness (e.g. [75]). For example, when ducks are provided an open water source, their straw bedding can become wet and dirty from spilled water. If not replaced, ducks on this bedding can develop leg problems, similar to those found in ducks housed on slatted floors (Raud and Faure 1994 in [69]). Similarly, when mink were provided swimming water, kit mortality increased, most likely due to water spillage on the bedding, which decreased hygiene [138]. It is possible for animals to lose interest in enrichments, especially those that the animal is not highly motivated to access and those that do not provide meaningful feedback, after having them available for a period of time [190]. If used, these types of enrichments should only be provided on occasion to help maintain the animals' interest in them.

Whenever animals are pair or group housed, the risk of aggression is increased compared with singly housed animals (e.g. [92]). However, equal distribution of resources and including enrichments, which allow the animals to escape from each other, such as wall partitions, platforms or objects to hide in, can be effective at reducing aggression and decreasing chances of injury (e.g. Erwin *et al.* 1976 in [1, 23]). For example, pigs provided deep straw bedding spent more time foraging and rooting through the straw and less time on the amount of aggression, belly nosing and tail biting performed [18]. Similarly, pigs provided with hide areas had decreased aggression when a new pig was introduced into the pen than pigs without the hide [97].

One of the main criticisms of providing environmental enrichment to decrease the occurrence of abnormal behavioural patterns is that the enrichment may merely be giving the animals more to do, decreasing the time available to perform the abnormal behaviour [191]. However, there may not actually be any improvement in the underlying problem that caused the abnormal behaviour to be performed in the first place. For an enrichment to be practical, from the animals' point of view, it should actually be something the animal wants/needs [5]. If only a limited amount of money and time can be spent on enriching an animal's environment, then we need to ensure the enrichments are the best ones possible for the animal, as well as from a producer's standpoint. If the enrichment is something we know animals' are motivated to obtain and will work hard to access, then we should try to provide this in lieu of less important resources. For example, rabbits are motivated to access conspecifics and will work almost as hard by pushing through a weighed door for access to another rabbit as they will for food [150]. Therefore rabbits should be pair- or group-housed whenever possible.

Finally, one of the most problematic issues with environmental enrichment research is that most of it is done in small-scale studies compared with the size and number of animals involved in a production system. This may affect how well the enrichment works, the ability to maintain the enrichment and the cost of providing enrichment to the animals. Some epidemiological studies can give a good indication of what enrichments provide benefits at an on-farm level but these types of studies cannot explain causation between the enrichment and improved welfare, only correlations between the factors. For example, in a survey study, Green *et al.* [192] found that providing laying hens with loose litter was associated with a decreased risk of feather pecking. This indicates that providing hens with some form of loose substrate will have positive welfare effects on-farm and attempts should be made to provide this to hens whenever possible. This is an area that could use further research and as many systems change to accommodate new EU directives or for food scheme criteria, we should have more opportunities to do so.

Conclusions

Allowing production animals the opportunity to perform a larger repertoire of behavioural patterns by providing increased foraging opportunities, structural complexity, sensory stimulation/novelty and social stimulation from conspecifics or humans generally has a positive effect on their biological functioning and welfare. In addition to benefitting the animals, the enrichment items must be affordable and easy to maintain for them to be incorporated into a production setting and for the animals to get the best use out of them. In conclusion, the best enrichment is one that satisfies behavioural needs and allows performance of species-specific behavioural patterns, while also being readily available and affordable and easy to maintain. The 'best' enrichment may therefore differ, depending on the location of the production system and the resources that are readily available.

Acknowledgements

L.M.D. was funded by a grant from the UK Department for Environment, Food and Rural Affairs (Defra) as part of a larger project (AW1141) investigating the subjective state of food restricted broiler chickens using behavioural and neurochemical measures. Scottish Agricultural College was supported by the Rural and Environment Research and Analysis Directorate of the Scottish Government.

References

1. Newberry RC. Environmental enrichment: increasing the biological relevance of captive environments. Applied Animal Behaviour Science 1995;44:229–43.
2. Fraser D, Weary DM, Pajor EA, Milligan BN. A scientific conception of animal welfare that reflects ethical concerns. Animal Welfare 1997;6:187–205.
3. Swaisgood R, Shepherdson D. Environmental enrichment as a strategy for mitigating stereotypies in zoo animals: a literature review and meta-analysis. In: Mason G, Rushen J, editors. Stereotypic Animal Behaviour: Fundamentals and Applications to Welfare. 2nd edn. CAB International, Wallingford, UK; 2006. p. 256–85.
4. Dixon LM, Duncan IJH, Mason GJ. The effects of four types of enrichment on feather-pecking behaviour in laying hens housed in barren environments. Animal Welfare 2010;19:429–35.
5. Duncan IJH. Behaviour and behavioural needs. Poultry Science 1998;77:1766–72.
6. Millet S, Moons CPH, Van Oeckel MJ, Janssens GPJ. Welfare, performance and meat quality of fattening pigs in alternative housing and management systems: a review. Journal of the Science of Food and Agriculture 2005;85:709–19.
7. Wood-Gush DGM, Vestergaard K. Exploratory behaviour and the welfare of intensively kept animals. Journal of Agricultural Ethics 1989;2:161–9.
8. Wemelsfelder F. Animal boredom: do animals miss being alert and active? In: Appleby MC, Horrell RI, Petherick JC, Rutter SM, editors. Applied Animal Behaviour: Past, Present and Future: Potters Bar. UFAW, UK; 1991. p. 120–3.
9. Mason G. Stereotypic behaviour in captive animals: fundamentals and implications for welfare and beyond. In: Mason G, Rushen J, editors. Stereotypic Animal Behaviour: Fundamentals and Applications to Welfare. 2nd edn. CAB International, Wallingford, UK; 2006. p. 325–56.
10. Wiepkema PR, Koolhaas JM. Stress and animal welfare. Animal Welfare 1993;2:195–218.
11. Nimon AJ, Broom DM. The welfare of farmed mink (*Mustela vison*) in relation to housing and management: a review. Animal Welfare 1999;8:205–28.
12. Hemsworth PH. Human–animal interactions in livestock production. Applied Animal Behaviour Science 2003;81:185–98.
13. Appleby MC, Walker AW, Nicol CJ, Lindberg AC, Freire R, Hughes BO, *et al.* Development of furnished cages for laying hens. British Poultry Science 2002;43:489–500.
14. Cooper JJ, Appleby MC. The effects of experience on motivation: prelaying behaviour in laying hens. Applied Animal Behaviour Science 1995;42:283–95.
15. Olsson IAS, Keeling LJ. The push-door for measuring motivation in hens: laying hens are motivated to perch at night. Animal Welfare 2002;11:11–9.

16. Sherwin CM, Nicol CJ. A descriptive account of the pre-laying behavior of hens housed individually in modified cages with nests. Applied Animal Behaviour Science 1993;38:49–60.

17. Ladewig J, Matthews LR. The role of operant conditioning in animal welfare research. Acta Agriculturae Scandinavica, Section A: Animal Science 1996;27(Suppl.):64–8.

18. Day JEL, Burfoot A, Docking CM, Whittaker X, Spoolder HAM, Edwards SA. The effects of prior experience of straw and the level of straw provision on the behavior of growing pigs. Applied Animal Behaviour Science 2002;76:189–202.

19. Nielsen BL, Thodberg K, Steenfeldt S. Proportion of insoluble fibre in the diet affects behaviour and hunger in broiler breeders growing at similar rates. Animal 2011;5:1247–58.

20. Tuyttens FAM. The importance of straw for pig and cattle welfare. Applied Animal Behaviour Science 2005;92:261–82.

21. Mason G, Mendl M. Do the stereotypies of pigs, mink and chickens reflect adaptive species differences in the control of foraging? Applied Animal Behaviour Science 1997;53:45–58.

22. Hansen LT, Berthelsen H. The effect of environmental enrichment on the behaviour of caged rabbits (*Oryctolagus cuniculus*). Applied Animal Behaviour Science 2000;68:163–78.

23. Cornetto TL, Estevez I. Influence of vertical panels on use of space by domestic fowl. Applied Animal Behaviour Science 2001;71:141–53.

24. Jones TA, Waitt CD, Dawkins MS. Water off a duck's back: showers and troughs match ponds for improving duck welfare. Applied Animal Behaviour Science 2009;116:52–7.

25. Axelsson HMK, Alden E, Lidfors L. Behaviour in female mink housed in enriched standard cages during winter. Applied Animal Behaviour Science 2009;121:222–9.

26. Young RJ, Carruthers J, Lawrence AB. The effect of a foraging device (The 'Edinburgh Foodball') on the behaviour of pigs. Applied Animal Behaviour Science 1994;39:237–47.

27. Hill JD, McGlone JJ, Fullwood SD. Environmental enrichment influences on pig behavior, performance and meat quality. Applied Animal Behaviour Science 1998;57:51–68.

28. Kells A, Dawkins MS, Borja MC. The effect of a 'freedom food' enrichment on the behaviour of broilers on commercial farms. Animal Welfare 2001;10:347–56.

29. Jones RB, Carmichael NL, Rayner E. Pecking preferences and pre-dispositions in domestic chicks: implications for the development of environmental enrichment devices. Applied Animal Behaviour Science 2000;69:291–312.

30. Bracke MBM. Multifactorial testing of enrichment criteria: pigs 'demand' hygiene and destructibility more than sounds. Applied Animal Behaviour Science 2007;107:218–32.

31. Duncan IJH. The pros and cons of cages. World's Poultry Science 2001;57:381–90.

32. Babu LK, Pandey HN, Sahoo A. Effects of individual versus group rearing on ethological and physiological responses of crossbred calves. Applied Animal Behaviour Science 2004;87:177–91.

33. Pedersen V, Jeppesen LL. Effects of family housing on behavior, plasma cortisol and performance in adult female mink (*Mustela vison*). Acta Agriculturae Scandinavica, Section A: Animal Science 2001;51:77–88.

34. Csatádi K, Kustos K, Eiben C, Bilkó Á, Altbäcker V. Even minimal human contact linked to nursing reduces fear responses toward humans in rabbits. Applied Animal Behaviour Science 2005;95:123–8.

35. Hemsworth PH, Barnett JL, Hansen C. The influence of handling by humans on the behavior, growth and corticosteroids in the juvenile female pig. Hormones and Behaviour 1981;15:396–403.

36. Leslie TK, Dalton L, Phillips CJC. Preference of domestic rabbits for grass or coarse mix feeds. Animal Welfare 2003;13:57–62.

37. De Jong IC, Wolthuis-Fillerup M, van Reenen CG. Strength of preference for dustbathing and foraging substrates in laying hens. Applied Animal Behaviour Science 2006;104:24–36.

38. Barnett JL, Tauson R, Downing JA, Janardhana V, Lowenthal JW, Butler KL, *et al.* The effects of a perch, dust bath, and nest box, either alone or in combination as used in furnished cages, on the welfare of laying hens. Poultry Science 2009;88:456–70.

39. Huber-Eicher B, Wechsler B. Feather pecking in domestic chicks: its relation to dustbathing and foraging. Animal Behaviour 1997;54:757–68.

40. Tauson R. Furnished cages and aviaries: production and health. World's Poultry Science Journal 2002;58:49–63.

41. Dawkins MS. Time budgets in red jungle fowl as a baseline for the assessment of welfare in domestic fowl. Applied Animal Behaviour Science 1989;24:77–80.

42. Martrenchar A, Huonnic D, Cotte JP. Influence of environmental enrichment on injurious pecking and perching behaviour in young turkeys. British Poultry Science 2001;42:161–70.

43. Huber-Eicher B, Audige L. Analysis of risk factors for the occurrence of feather pecking in laying hen growers. British Poultry Science 1999;40:599–604.

44. Enstrom B, Schaller G. Experimental studies of the health of laying hens in relation to housing system. In: Savory CJ, Hughes BO, editors. Proceedings of the Fourth European Symposium on Poultry Welfare. Universities Federation for Animal Welfare, Potters' Bar, UK; 1993. pp. 87–96.

45. Hocking PM, Jones EKM, Picard M. Assessing the welfare consequences of providing litter for feed restricted broiler breeders. British Poultry Science 2005;46:545–52.

46. Miller KA, Mench JA. The differential effects of four types of environmental enrichment on the activity budgets, fearfulness and social proximity preference of Japanese quail. Applied Animal Behaviour Science 2005;95:169–87.

47. Sherwin CM, Lewis PD, Perry GC. The effects of environmental enrichment and intermittent lighting on the behavior and welfare of male domestic turkeys. Applied Animal Behaviour Science 1999;62:319–33.

48. Faure JM, Guy G, Guemene D. Is there avoidance of the force feeding procedure in ducks and geese? Animal Research 2001;50:154–7.

49. Blokhuis HJ, van der Haar JW. Effects of floor type during rearing and of beak trimming on ground pecking and feather pecking in laying hens. Applied Animal Behaviour Science 1989;22:359–69.

50. Wood-Gush DGM, Gilbert AB. Observations on the laying behaviour of domestic hens in battery cages. British Poultry Science 1972;10:29–36.

51. Cooper JJ, Appleby MC. Individual variation in prelaying behaviour and the incidence of floor eggs in laying hens. British Poultry Science 1996;37:245–54.

52. Cooper JJ, Appleby MC. The value of environmental resources to domestic hens: a comparison of the work-rate for food and for nests as a function of time. Animal Welfare 2003;12:39–52.

53. Meijsser FM, Hughes BO. Comparative analysis of pre-laying behaviour in battery cages and in three alternative systems. British Poultry Science 1989;30:747–60.

54. Duncan IJH. Designing environments for animals – not for public perceptions. British Veterinary Journal 1992;148:475–77.

55. Gunnarsson S, Yngvesson J, Keeling LJ, Forkman B. Rearing without early access to perches impairs the spatial skills of laying hens. Applied Animal Behaviour Science 2000;67:217–28.

56. Pickel T, Schrader L, Scholz B. Pressure load on keel bone and foot pads in perching laying hens in relation to perch design. Poultry Science 2011;90:715–24.

57. Sandilands V, Moinard C, Sparks NHC. Providing laying hens with perches: fulfilling behavioural needs but causing injury? British Poultry Science 2009;50:395–406.

58. Ventura BA, Siewerdt F, Estevez IE. Effects of barrier perches and density on broiler leg health, fear and performance. Poultry Science 2010;89:1574–83.

59. Tauson R. Health and production in improved cage designs. Poultry Science 1998;77:1820–7.

60. Tauson R. Effects of a perch in conventional cages for laying hens. Acta Agriculturae Scandinavica, Section A: Animal Science 1984;34:193–209.

61. Hughes BO, Wilson S, Appleby MC, Smith SF. Comparison of bone volume and strength as measures of skeletal integrity in caged laying hens with access to perches. Research in Veterinary Science 1993;54:202–6.

62. Gross WB, Siegel PB. Socialization as a factor in resistance to disease, feed efficiency, and response to antigen in chickens. American Journal of Veterinary Research 1982;43:2010–2.

63. Hughes BO. Feather damage in hens caged individually. British Poultry Science 1980;21:149–54.

64. Keiper R. Studies of stereotypy function in the canary (*Serinus cararius*). Animal Behaviour 1970;18:353–7.

65. Hughes BO, Black AJ. Influence of handling on egg production, egg shell quality and avoidance behaviour of hens. British Poultry Science 1976;17:135–44.

66. Jones RB. Reduction of the domestic chick's fear of human beings by regular handling and related treatments. Animal Behaviour 1993;46:991–8.

67. Jones RB. Regular handling and the domestic chick's fear of human beings: generalization of response. Applied Animal Behaviour Science 1994;42:129–43.

68. Collins JW, Siegel PB. Human handling, flock size and responses to an *E. coli* challenge in young chickens. Applied Animal Behaviour Science 1987;19:183–8.

69. Rodenburg TB, Bracke MBM, Berk J, Cooper J, Faure JM, Guemene D, *et al.* Welfare of ducks in European duck husbandry systems. World's Poultry Science 2005;61:633–46.

70. O'Driscoll KKM, Broom DM. Does access to open water affect the health of Pekin ducks (*Anas platyrhynchos*)? Poultry Science 2011;90:299–307.

71. Klemm R, Pingel H, Reiter K, Bierschenk F, Rauch W. Investigations on feather pecking in Muscovy ducks. In World's Poultry Congress, Amsterdam, The Netherlands; 1992. pp. 390–3.

72. Ruis MAW, Lenskens P, Coenen E. Welfare of Pekin-ducks increases when freely accessible open water is provided. In Second World Waterfowl Conference, Alexandria, Egypt; 2003. p. 17.

73. Jones RB, McAdie TM, McCorquodale C, Keeling LJ. Pecking at other birds and at string enrichment devices by adult laying hens. British Poultry Science 2002;43:337–43.

74. Holcman A, Gorjanc G, Stuhec I. Porous concrete block as an environmental enrichment device increases activity of laying hens in cages. Poultry Science 2008;87:1714–9.

75. Lay Jr DC, Fulton RM, Hester PY, Karcher DM, Kjaer JB, Mench JA, *et al.* Hen welfare in different housing systems. Poultry Science 2011;90:278–94.

76. King LA. Environmental enrichment for broiler breeders [Ph.D. Thesis]. University of Oxford, Oxford; 2001.

77. Jones RB. Effects of early environmental enrichment upon open-field behaviour and timidity in the domestic chick. Developmental Psychobiology 1982;15:105–11.

78. Huff GR, Huff WE, Balog JM, Rath NC. The effects of behavior and environmental enrichment on disease resistance of turkeys. Brain, Behaviour and Immunity 2003;17:339–49.

79. Europa. Council Directive 1999/74/EC – laying down the minimum standards for the protection of laying hens. 1999. Available from: URL: http://ec.europa.eu/food/animal/welfare/farm/laying_hens_en.htm

80. Beattie VE, O'Connel NE, Moss BW. Influence of environmental enrichment on the behavior, performance and meat quality of domestic pigs. Livestock Production Science 2000;65:71–9.

81. Boyle LA, Regan D, Leonard FC, Lynch PB, Brophy P. The effect of mats on the welfare of sows and piglets in the farrowing house. Animal Welfare 2000;9:39–48.

82. Mouttotou N, Hatchell FM. The prevalence and risk factors associated with forelimb skin abrasions and sole bruising in preweaning piglets. Preventative Veterinary Medicine 1999;39:231–45.

83. Studnitz M, Jensen MB, Pedersen LJ. Why do pigs root and in what will they root? A review on the exploratory behavior of pigs in relation to environmental enrichment. Applied Animal Behaviour Science 2007;107:183–97.

84. Arey DS. Straw and food as reinforcers for preparital sows. Applied Animal Behaviour Science 1992;33:217–26.

85. Lyons CAP, Bruce JM, Fowler VR, English PR. A comparison of productivity and welfare of growing pigs in four intensive systems. Livestock Production Science 1995;43:265–74.

86. Mench JA. Environmental enrichment and the importance of exploratory behavior. In: Shepherdson DJ, Mellen JD,

Hutchins M, editors. Second Nature: Environmental Enrichment for Captive Animals. Smithsonian Institution Press, Washington, DC; 1998. p. 30–46.

87. Moinard C, Mendl M, Nicol CJ, Green LE. A case control study of on-farm risk factors for tail biting in pigs. Applied Animal Behaviour Science 2003;81:333–55.

88. Oostindjer M, Bolhuis JE, Mendl M, Held S, Gerrits W, van den Brand H, *et al.* Effects of environmental enrichment and loose housing of lactating sows on piglet performance before and after weaning. Journal of Animal Science 2010;88:3554–62.

89. Jarvis S, van der Vegy B, Lawrence AB. The effect of parity and environmental restriction on behavioural and physiological responses of pre-parturient pigs. Applied Animal Behaviour Science 2001;71:203–16.

90. Jarvis S, Calvert SK, Stevenson J, Vanleeuwen N, Lawrence AB. Pituitary-adrenal activation in pre-parturient pigs (*Sus scrofa*) is associated with behavioural restriction due to lack of space rather than nesting substrate. Animal Welfare 2002;11:371–84.

91. Lawrence AB, Petherick JC, Mclean K, Deans L, Chirnside J, Vaughan A, *et al.* The effect of environment on behaviour, plasma cortisol and prolactin in parturient sows. Applied Animal Behaviour Science 1994;39:313–30.

92. Barnett JL, Winfield CG, Cronin GM, Hemsworth PH, Dewar AM. The effect of individual and group housing on behavioural and physiological responses related to the welfare of pregnant pigs. Applied Animal Behaviour Science 1985;14:149–61.

93. Csermely D, Wood-Gush DGM. Agonistic behaviour in grouped sows: I. The influence of feeding. Biology of Behaviour 1986;11:244–52.

94. Hemsworth PH, Barnett JL, Hansen C. The influence of inconsistent handling by humans on the behavior, growth and corticosteroids of young pigs. Applied Animal Behaviour Science 1987;17:245–52.

95. Pedersen V, Barnett JL, Hemsworth PH, Newman EA, Schirmer B. The effects of handling on behavioural and physiological responses to housing in tether-stalls among pregnant pigs. Animal Welfare 1998;7:137–50.

96. Blackshaw JK, Thomas FJ, Lee JA. The effect of a fixed or free toy on the growth rate and aggressive behavior of weaned pigs and the influence of hierarchy on initial investigation of the toys. Applied Animal Behaviour Science 1997;53:203–12.

97. McGlone JJ, Curtis SE. Behaviour and performance of weanling pigs in pens equipped with hide areas. Journal of Animal Science 1985;60:20–4.

98. Grandin T, Curtis SE, Greenough WT. Objective measurement of effects of environmental complexity on young pigs. Swine Research Reports, University of Illinois at Urbana-Champaign, USA; 1983.

99. Hemsworth PH, Barnett JL, Hansen C, Winfield CG. Effects of social environment on welfare status and sexual behavior of female pigs. II. Effects of space allowance. Applied Animal Behaviour Science 1986;1:259–67.

100. Mas Reixach G, Felius S, Coonan C, English PR, Edwards SA. The influence of mineral blocks on the behaviour of newly weaned pigs. In Proceedings of the Winter Meeting of the British Society of Animal Science, Scarborough, UK; 2001. p. 180.

101. Thodberg K, Jensen KH, Herskin MS, Jorgensen E. Influence of environmental stimuli on nest building and farrowing behaviour in domestic sows. Applied Animal Behaviour Science 1999;63:131–44.

102. Mason BM, Blackshaw JK. Toys for dry sows. The Pig Farmer 1985;19:4–5.

103. Van de Weerd HA, Docking CM, Day JEL, Avery PJ, Edwards SA. A systematic approach towards developing environmental enrichment for pigs. Applied Animal Behaviour Science 2003;84:101–18.

104. Anon. RSPCA Welfare Standards for Pigs. 2010. Available from: URL: http://www.rspca.org.uk/ImageLocator/LocateAsset?asset = document&assetId = 1232712363838& mode = prd

105. Ladewig J. Physiological results of welfare research in fattening bulls. In: Schlichting MC, Smidt D, editors. Welfare Aspects of Housing Systems for Veal Calves and Fattening Bulls, CEC Report. CEC, Office for Official Publications, Luxembourg; 1987. p. 123–9.

106. Ninomiya S, Sato S. Effects of 'Five freedoms' environmental enrichment on the welfare of claves reared indoors. Animal Science 2009;80:347–51.

107. Redbo I, Emanuelson M, Lundberg K, Oredsson N. Feeding level and oral stereotypies in dairy cows. Animal Science 1996;62:199–206.

108. Isiwata T, Uetake K, Abe N, Eguchi Y, Tanaka T. Effects of an environmental enrichment using a drum can on behavioural, physiological and productive characteristics in fattening beef cattle. Animal Science Journal 2006;77:352–62.

109. Hernandez-Mendo O, von Keyserlingk MAG, Veira DM, Weary DM. Effects of pasture on lameness in dairy cows. Journal of Dairy Science 2007;90:1209–14.

110. Lindstrom T, Redbo I. Effect of feeding duration and rumen fill on behavior in dairy cows. Applied Animal Behaviour Science 2000;70:83–97.

111. Pelley MC, Lirette A, Tennessen T. Observations on the responses of feedlot cattle to attempted environmental enrichment. Canadian Journal of Animal Science 1995;75:631–2.

112. Redbo I, Nordblad A. Stereotypies in heifers are affected by feeding regime. Applied Animal Behaviour Science 1997;53:193–202.

113. Kohari D, Kosako T, Fukasawa M, Tsukada H. Effect of environmental enrichment by providing trees as rubbing objects in grasslands: grazing cattle need tree-grooming. Animal Science Journal 2007;78:413–6.

114. Albright JL. Nutrition, feeding and calves. Journal of Dairy Science 1993;76:485–98.

115. Arave CW, Albright JL, Sinclair CL. Behavior, milk yield and leucocytes of dairy cows in reduced space and isolation. Journal of Dairy Science 1974;57:1497.

116. Bertenshaw CE, Rowlinson P. Exploring heifers' perception of 'positive' treatment through their motivation to pursue a retreated human. Animal Welfare 2008;17:313–9.

117. Jensen MB, Vestergaard KS, Krohn CC. Play behaviour in domestic calves kept in pens: the effect of social contact and space allowance. Applied Animal Behaviour Science 1998;56:97–108.

118. Linnane M, Horan B, Connolly J, O'Connor PO, Buckley F, Dillon P. The effect of strain of Holstein–Friesian and feeding system on grazing behavior, herbage intake and productivity in the first lactation. Animal Science 2004;78:169–78.

119. Veissier I, Chazal P, Pradel P, Le Neindre P. Providing social contacts and objects for nibbling moderates reactivity and oral behaviours in veal calves. Journal of Animal Science 1997;75:356–65.

120. Westerath HS, Laister S, Winckler C, Knierim U. Exploration as an indicator of good welfare in beef bulls: an attempt to develop a test for on-farm assessment. Applied Animal Behaviour Science 2009;116:126–33.

121. Grant RJ. Incorporating dairy cow behavior into management tools. In Penn State Dairy Cattle Nutrition Workshop 2006; p. 31–41.

122. Hogan JS, Smith KL, Hoblet KL, Todhunter DA, Schoenberger PS, Hueston WD, *et al.* Bacterial counts in bedding materials used on nine commercial dairies. Journal of Dairy Science 1989;72:250–8.

123. Anon. RSPCA Welfare Standards for Beef Cattle. 2010. Available from: URL: http://www.rspca.org.uk/ImageLocator/LocateAsset?asset = document&assetId = 1232712365259&mode = prd

124. Anon. FAWC Opinion on the Welfare of Dairy Cattle. 2009. Available from: URL: http://www.fawc.org.uk/pdf/dcwelfar-091022.pdf

125. Koistinen T, Orjala H, Mononen J, Korhonen HT. Position of operant cost affects blue foxes' time budget between sand floor and mesh floor. Applied Animal Behaviour Science 2009;116:266–72.

126. Hansen SW, Hansen BK, Berg P. The effect of cage environment and *ad libitum* feeding on the circadian rhythm, behaviour and feed intake of farm mink. Acta Agriculturae Scandinavica, Section A: Animal Science 1994;44:120–7.

127. Korhonen H, Jauhianinen L, Niemela P, Harri M, Sauna-aho R. Physiological and behavioural responses in blue foxes (*Alopex lagopus*): comparisons between space quantity and floor material. Animal Science 2001;72:375–87.

128. Cooper JJ, Mason GJ. Environmental requirements in mink. In: Duncan IJH, Widowski TM, Haley DB, editors. Proceedings of the 30th International Congress of the International Society for Applied Ethology, The Colonel K L Campbell Centre for the Study of Animal Welfare, Guelph, Canada; 1996. p. 37.

129. Korhonen HT, Jauhiainen L, Rekila T. In-cage sandbox as a ground substitute for farmed blue foxes (*Alopex lagopus*): effects on digging activity and welfare. Canadian Journal of Animal Science 2003;83:703–12.

130. Cooper JJ, Mason G. Increasing costs of access to resources cause re-scheduling of behaviour in American mink (*Mustela vison*): implications for the assessment of behavioural priorities. Applied Animal Behaviour Science 2000;66:135–51.

131. Mason GJ, Cooper J, Clarebrough C. Frustrations of fur farmed mink. Nature 2001;410:35–6.

132. Ahola L, Mononen J, Pyykonen T, Mohaibes M, Hanninen S. Group size and space allocation in farmed juvenile blue foxes (*Alopex lagopus*). Animal Welfare 2005;14:1–9.

133. Heller KE, Jeppesen LL. Behavioural and eosinophil leukocyte responses to single and repeated immobility stress in mink. Scientifur 1985;9:174–8.

134. Pedersen V. Effects of different post-weaning handling procedures on the later behavior of silver foxes. Applied Animal Behaviour Science 1993;37:239–50.

135. Pedersen V, Jeppesen LL. Effects of early handling on later behavior and stress responses in the silver fox (*Vulpes vulpes*). Applied Animal Behaviour Science 1990;26:383–93.

136. Pedersen V, Moeller NH, Jeppesen LL. Behavioural and physiological effects of post-weaning handling and access to shelters in farmed blue foxes (*Alopex lagopus*). Applied Animal Behaviour Science 2002;77:139–54.

137. Hansen SW, Jensen MB. Quantitative evaluation of the motivation to access a running wheel or a water bath in farm mink. Applied Animal Behaviour Science 2006;98:127–44.

138. Skovgaard K, Jeppesen LL, Hansen CPB. Would you like a swim, madam mink? Scientifur 1997;21:253–9.

139. Hansen CPB, Jeppesen LL. Swimming activity of farmed mink (*Mustela vison*) and its relation to stereotypies. Acta Agriculturae Scandinavica, Section A: Animal Science 2001;51:71–6.

140. Hansen SW, Malmkvist J, Palme R, Damgaard BM. Do double cages and access to occupational materials improve the welfare of farmed mink? Animal Welfare 2007;16:63–76.

141. Korhonen HT, Jauhiainen L, Niemelä P, Sauna-aho R. Wooden blocks and straw as environmental enrichments for juvenile blue foxes (*Alopex lagopus*). Acta Ethologica 2002;5:29–37.

142. Korhonen HT, Orjala H. Effect of cage dimensions on welfare and production of farmed blue fox. Annals of Animal Science 2010;10:311–24.

143. Cooper JJ, Mason G. Increasing costs of access to resources cause re-scheduling of behaviour in American mink (*Mustela vison*): implications for the assessment of behavioural priorities. Applied Animal Behaviour Science 2000;66:135–51.

144. Crowell-Davis SL. Behaviour problems in pet rabbits. Journal of Exotic Pet Medicine 2007;16:38–44.

145. Dal Bosco A, Castellini C, Mugnai C. Rearing rabbits on a wire net floor or straw litter: behaviour, growth and meat qualitative traits. Livestock Production Science 2002;75:149–56.

146. Tawfik ES, Sherif SY, El-Hindawy M, Attia AI, Abou-Ela S. The role of fibre in rabbit nutrition. Der Tropenlan Beitrage zur Tropischen Landwirtschaft und Veternamedizin 1997;98:73–81.

147. Berthelsen H, Hansen LT. The effects of hay on the behavior of caged rabbits (*Oryctolagus cuniculus*). Animal Welfare 1999;8:149–57.

148. Kamphues J. The species-specific feeding of rabbits in pet husbandry. Deutsch Tierarztlich Wochenschrift 2001;108:131–5.

149. Lidfors L. Behavioural effects of environmental enrichment for individually caged rabbits. Applied Animal Behaviour Science 1999;52:157–69.

150. Seaman SC, Waran NK, Mason G, D'Eath RB. Animal economics: assessing the motivation of female laboratory

rabbits to reach a platform, social contact and food. Animal Behaviour 2008;75:31–42.

151. Chu L, Garner JP, Mench JA. A behavioural comparison of New Zealand White rabbits (*Oryctolagus cuniculus*) housed individually or in pairs in conventional laboratory cages. Applied Animal Behaviour Science 2004;85:121–39.

152. Held SDE, Turner RJ, Wootton RJ. Choices of laboratory rabbits for individual or group-housing. Applied Animal Behaviour Science 1995;46:81–91.

153. Bilkò A, Altbacker V. Regular handling early in the nursing period eliminates fear responses toward human beings in wild and domestic rabbits. Developmental Psychobiology 2000;36:78–87.

154. Mugnai C, Dal Bosco A, Castellini C. Effects of different rearing systems and pre-kindling handling on behaviour and performance of rabbit does. Applied Animal Behaviour Science 2009;118:91–100.

155. Podberscek AL, Blackshaw JK, Beattie AW. The behaviour of group penned and individually caged laboratory rabbits. Applied Animal Behaviour Science 1991;28:353–63.

156. Zotte AD, Princz Z, Metzger S, Szabo A, Radnai I, Biro-Nemeth E. Response of fattening rabbits reared under different housing conditions. 2. Carcass and meat quality. Livestock Science 2009;122:39–47.

157. Kersten AMP, Meijsser FM, Metz JHM. Effect of early handling on later open-field behaviour of rabbits. Applied Animal Behaviour Science 1989;24:157–67.

158. Csatádi K, Kustos K, Eiben C, Bilkó Á, Altbäcker V. Even minimal human contact linked to nursing reduces fear responses toward humans in rabbits. Applied Animal Behaviour Science 2005;95:123–8.

159. Pongra'cz P, Altbacker V. Arousal, but not nursing, is necessary to elicit a decreased fear reaction toward humans in rabbit (*Oryctolagus cuniculus*) pups. Developmental Psychobiology 2003;43:192–9.

160. Stauffacher M. Group housing and enrichment cages for breeding, fattening and laboratory rabbits. Animal Welfare 1992;1:105–25.

161. Edgar JL, Seaman SC. The effects of mirrors in the behaviour of singly housed male and female laboratory rabbits. Animal Welfare 2010;19:461–71.

162. Luzi F, Ferrante V, Heinzl E, Verga M, Mariani P, Soperchi P, *et al.* Effect of environmental enrichment on productive performance and welfare aspects in fattening rabbits. Italian Journal of Animal Science 2003;2:438–40.

163. Dixon LM, Hardiman JR, Cooper JJ. The effects of spatial restriction on the behaviour of pet rabbits (*Oryctolagus cuniculus*). Journal of Veterinary Behaviour 2010;5:302–8.

164. Dixon LM, Cooper JJ. How low can they go? The effects of height restriction on the behaviour of pet rabbits (*Oryctolagus cuniculus*). In: Proceedings of the ISAE UK Meeting, Harper Adams University College, Newport, UK; 2010. p. 6.

165. Dixon LM, Cooper JJ. Are traditional rabbit hutches too small? Assessing pet rabbits' motivation to access additional space using consumer demand techniques. In: Recent advances animal welfare science UFAW Animal Welfare Conference, UFAW, York, UK; 2010.

166. Princz Z, Zotte AD, Metzger S, Radnai I, Biro-Nemeth E, Orova Z, *et al.* Response of fattening rabbis reared under different housing conditions. 1. Live performance and health status. Livestock Science 2009;121:86–91.

167. Mykytowycz R, Rowley I. Continuous observations of the activity of the wild rabbit, *Oryctolagus cuniculus*, during 24-hour periods. CSIRO Wildlife Research 1958;3:26–31.

168. Brown C, Davidson T, Laland K. Environmental enrichment and prior experience of live prey improve foraging behaviour in hatchery reared Atlantic salmon. Journal of Fish Biology 2003;63:187–96.

169. Nijman V, Heuts BA. Effect of environmental enrichment upon resource holding power in fish in prior residence situations. Behavioural Processes 2000;49:77–83.

170. Williams TD, Readman GD, Owen SF. Key issues concerning environmental enrichment for laboratory-held fish species. Laboratory Animals 2009;43:107–20.

171. Kihslinger RL, Nevitt GA. Early rearing environment impacts cerebellar growth in juvenile salmon. Journal of Experimental Biology 2006;209:504–9.

172. Barreto RE, Carvalho GGA, Volpato GL. The aggressive behaviour of Nile tilapia introduced into novel environments with variation in enrichment. Zoology 2011;114:53–7.

173. Handley JW. Environmental enrichment for fish used in regulatory toxicity studies. Animal Technology 2001;52:227–31.

174. Salvanes AGV, Braithwaite VA. Exposure to variable spatial information in the early rearing environment generates asymmetries in social interactions in cod (*Gadus morhua*). Behavioural Ecology and Sociobiology 2005;59:250–7.

175. Jackson KMA, Hackett D. A note: The effects of human handling on heart girth, behaviour and milk quality in dairy goats. Applied Animal Behaviour Science 2007;108:332–6.

176. Kor A, Karaca S, Ertugrul M. Effect of different housing systems on fattening performance, slaughter and carcass characteristics of Akkeci (White Goat) male kids. Tropical Animal Health and Production 2011;43:591–6.

177. Lyons DM, Price EO. Relationships between heart rates and behavior of goats in encounters with people. Applied Animal Behavior Science 1986;18:363–9.

178. Lyons DM, Price EO, Moberg GP. Social modulation of pituitary-adrenal responsiveness and individual differences in behaviour of young domestic goats. Physiology and Behaviour 1988;43:451–8.

179. Miranda-de la Lama GC, Mattiello S. The importance of social behaviour for goat welfare in livestock farming. Small Ruminant Research 2010;90:1–10.

180. Christensen JW, Nielsen BL. Environmental enrichment for ostrich, *Struthio camelus*, chicks. Animal Welfare 2004;13:119–24.

181. Bouissou MF, Vandenheede M. Fear reactions of domestic sheep confronted with either a human or a human-like model. Behavioural Processes 1995;34:81–92.

182. Caroprese M, Napolitano F, Albenzio M, Annicciarico G, Musto M, Sevi A. Influence of gentling on lamb immune response and human–lamb interactions. Applied Animal Behaviour Science 2006;99:118–31.

183. Caroprese M, Annicchiarico G, Schena L, Muscio A, Migliore R, Sevi A. Influence of space allowance and housing conditions on the welfare, immune response and production

performance of dairy ewes. Journal of Dairy Research 2009;76:66–73.

184. Lyons DM, Price EO, Moberg GP. Social grouping tendencies and separation-induced distress in juvenile sheep and goats. Developmental Psychobiology 1993;26:251–9.

185. Moberg GP, Wood VA. Effect of differential rearing on the behavioural and adrenocortical response of lambs to a novel environment. Applied Animal Ethology 1982;8:269–79.

186. Vandenheede M, Bouissou MF. Effects of an enriched environment on subsequent fear reactions of lambs and ewes. Developmental Psychobiology 1998;33:33–45.

187. Dwyer CM. Welfare of sheep: providing for welfare in an extensive environment. Small Ruminant Research 2009;86:14–21.

188. Barreto RE, Carvalho GGA, Volpato GL. The aggressive behaviour of Nile tilapia introduced into novel environments with variation in enrichment. Zoology 2011;114:53–7.

189. Trickett SL, Guy JH, Edwards SA. The role of novelty in environmental enrichment for the weaned pig. Applied Animal Behaviour Science 2009;116:45–51.

190. Tarou LR, Bashaw MJ. Maximizing the effectiveness of environmental enrichment: suggestions from the experimental analysis of behaviour. Applied Animal Behaviour Science 2007;102:189–204.

191. Mason G, Clubb R, Latham N, Vickery S. Why and how should we use environmental enrichment to tackle stereotypic behaviour? Applied Animal Behaviour Science 2007;102:163–88.

192. Green LE, Lewis K, Kimpton A, Nicol CJ. Cross-sectional study of the prevalence of feather pecking in laying hens in alternative systems and its associations with management and disease. Veterinary Record 2000;147:233–8.

Selection of growth functions for describing length-at-age relationships in fish species displaying different life history traits: unexpected alternatives to the von Bertalanffy equation and advantages of a pluralistic statistical approach

André Dumas[1]*, Secundino López[2], Ermias Kebreab[3], Marc Gendron[4], John H.M. Thornley[5] and James France[5]

Address: [1] Coastal Zones Research Institute, 232B avenue de l'Église, Shippagan, NB E8S 1J2, Canada.[2] Instituto de Ganadería de Montaña, Universidad de León – Consejo Superior de Investigaciones Científicas, Departamento de Producción Animal, Universidad de León, E-24007 León, Spain.[3] Department of Animal Science, University of California, Davis, CA 95616, USA.[4] Environnement Illimité Inc., 1453 rue Saint-Timothée, Montréal, QC H2L 3N7, Canada.[5] Centre for Nutrition Modelling, Department of Animal and Poultry Science, University of Guelph, Guelph, ON N1 G 2W1, Canada.

***Correspondence**: André Dumas. Fax. 506-336-6601. Email: andre.dumas@irzc.umcs.ca

Received: 23 January 2012
Accepted: 22 May 2012

Abstract

Simple growth models have proven helpful to study and predict growth trajectories of fish with different life histories and from various environments, and have found numerous applications in ecology, fisheries and aquaculture worldwide. The most applied simple models (e.g. von Bertalanffy) convey little information about life history traits, are facing criticism, and their goodness of fit is assessed using a limited number of statistical outputs. This review challenges the von Bertalanffy and certain statistical outputs using real-life data, and proposes more explanatory alternatives. The length-at-age relationship from three different wild populations of four fish species (lake herring, lake whitefish, northern pike and walleye) is described using the monomolecular, Schumacher, Gompertz, logistic, von Bertalanffy and Richards equations. Comparison between models was based on least squares and likelihood theories involving several statistical outputs currently encountered in model comparison and selection studies. The analysis demonstrates that the monomolecular, Schumacher and Richards equations often stood as alternatives to the von Bertalanffy and highlighted residual sum of squares, standard error (SE) of parameter estimates and, to a lesser extent, Akaike weight as statistical outputs facilitating discrimination between candidate models. Based on visual appraisal, different equations fitted similar trajectories across the 12 growth profiles. However, the assumptions and statistical performance served to select the most appropriate models and discern life history traits of the species under study. Results indicate that comparison and selection of models should consider not only the statistical performance of equations but also the purpose of the study along with the biological/physical interpretation of parameters.

Keywords: Fish, Growth functions, Statistical inference, von Bertalanffy, Life history

Introduction

Simple growth models have found numerous applications in the animal and plant sciences, including fisheries, ecology and aquaculture where they have proven helpful to predict biomass gain and assess the growth trajectory of fish, and served as decision-making tools for stock management [1–3]. These models, referred to as growth functions, are generally analytical solutions to differential equations that can be fitted by non-linear regression analysis [4, 5] and have the capability to exhibit a wide array of dynamic behaviour [6]. Comparison and selection

Table 1 General characteristics of the lakes where fish populations originated

Lake	Mercator coordinates	Area (km^2)	Mean depth (m)	Watershed
Matagami	49°53′N, 77°20′W	121	3.2	Nottaway
Poncheville	50°03′N, 77°01′W	121	4.8	Broadback
Evans	50°53′N, 77°05′W	453	3.7	Broadback
Mesgouez	51°25′N, 75°06′W	58	3.9	Rupert
Nemiscau	51°26′N, 76°43′W	43	3.2	Rupert
Jolliet	51°33′N, 76°55′W	46	4.9	Rupert

Note: Lakes Matagami and Nemiscau correspond to Matagami West and Nemiscau North, respectively, in the study of Consortium Groupe de Recherche SEEEQ Ltée – Environnement Illimité Inc. [32].

of models are generally based on physical and biological parameter interpretation, and statistical evaluation of performance [7, 8]. The latter involves several different statistical approaches (e.g. hypothesis testing and information theory), methods (e.g. least squares and maximum likelihood) and outputs (e.g. standard error (SE) of parameters, coefficient of determination, residual sums of squares and information criteria) of varying complexity, which are used to assess goodness of fit of the models [9–12].

To date, fish growth studies have relied on a rather limited set of candidate equations and statistical approaches. The von Bertalanffy equation and its special cases stand as the most studied and applied fish growth model [13–16], although other growth functions have been proposed as alternatives (e.g. [17–22]). Various growth functions encountered in the animal and plant sciences have not been examined in detail in fish studies. Among them, the Schumacher [23] and monomolecular [24] equations deserve further consideration. The Schumacher was originally intended to model stem volume growth in forest plantations [23]. It has been introduced in fisheries apparently by Krüger in the 1960s [17] and has been successfully applied in animal science more recently [25]. The monomolecular equation was first used to describe the effect of chemical fertilizers on crop yields [24] and has found application in animal science ever since, especially to represent weight gain as a function of feed or energy intake [26, 27].

Statistical analysis of growth functions is generally confined to either hypothesis testing or an information theory approach [10, 11, 28] and to a limited amount of statistical outputs [29–31]. Although generally assumed, consistency and reliability of candidate models across freshwater species and populations exhibiting different life histories require deeper examination.

This analytical review was undertaken to compare the Schumacher and monomolecular equations with more established growth functions in fish studies and determine -which assumptions and statistical information are meaningful guidelines to orientate model selection. Furthermore, consistency of candidate models across species and populations was examined. Ultimately, the work aims at helping biologists and managers discern the most appropriate growth function for their intended application.

Comparative Analysis Using Real-Life Data

Data Source

A total of 12 growth profiles (length versus age) from four different fish species were obtained from a study conducted by Hydro-Québec in the James Bay territory (Province of Québec, Canada) in August and September 1991 [32]. Species were selected on the basis of their different life history traits (e.g. growth rate, age or size at sexual maturity and lifespan), wide distribution across North America, as well as ecological and economic importance. The fish species are: lake herring (*Coregonus artedii*), lake whitefish (*Coregonus clupeaformis*), northern pike (*Esox lucius*) and walleye (*Sander vitreus*). For each species, three different populations originating from lakes covering three different watersheds (Nottaway, Broadback and Rupert; see Table 1) were collected using experimental nets (mesh sizes from 2.5 to 10.2 cm). General characteristics of the fish populations are summarized in Table 2. The mean age at sexual maturity and the actual total mortality rate per year were estimated using the methods of Lysack [33] and Ricker [13], respectively.

Age (*t*, year) and total length (*L*, mm) were determined for each individual. The structures to determine the age of fish were the otoliths for lake herring and whitefish, scales and cleithral bones for pike, and scales and opercular bones for walleye. At least two independent age determinations were conducted on each structure (without reference to fish size) by trained technicians. In situations of disagreement between two readings, a third one was performed. The means of *L* for each year class were computed and served afterwards to plot the length-at-age relationship. For fitting of equations, these means were weighted by sample size because of different number of observations between age groups.

Water quality in all lakes did not compromise the growth or survival of fish. The dissolved oxygen concentrations were $\geq$90% saturation and pH varied between 5.2 and 7.3. Water temperature is colder and turbidity is much lower in the Rupert than in the Broadback and Nottaway watersheds. Further details can be found in the report by Consortium Groupe de Recherche SEEEQ Ltée – Environnement Illimité Inc. [32].

Table 2 General characteristics of the fish populations selected for fitting the models

Species	Lake (population)	N	Range of ages (year)	Age_m (year)	A (%)
Lake herring	Matagami	139	1–10	2.6	57
	Evans	145	0–6	1.4	63
	Jolliet	88	1–8	1.9	64
Lake whitefish	Matagami	100	1–15	5.1	31
	Evans	161	0–14	6.0	15
	Nemiscau	75	1–13	4.7	49
Northern Pike	Matagami	77	1–22	3.6	27
	Poncheville	60	3–24	4.0	29
	Nemiscau	29	0–15	5.8	42
Walleye	Matagami	235	1–22	9.4	37
	Evans	231	1–23	11.6	46
	Mesgouez	246	1–23	10.7	20

Note: Number of observations (N), age at sexual maturity (Age_m) and actual total mortality rate per year (A). Lakes Matagami and Nemiscau correspond to Matagami West and Nemiscau North, respectively, in the study of Consortium Groupe de Recherche SEEEQ Ltée – Environnement Illimité Inc. [32].

Mathematical Models

The length-at-age relationship was described for each fish population using six growth functions: monomolecular [24], Schumacher [23], Gompertz [34], logistic [35], von Bertalanffy [36, 37] and Richards [38]. Models were selected on the basis of their capability to describe different curve patterns, wide acceptance in animal or plant science and position of the inflexion point. Differential and integrated (sometimes reparameterized) forms of the equations used in this analysis are described below (for further details, see Thornley and France [5, chapter 5]).

Monomolecular

$$\frac{dL}{dt}=k(L_f-L),$$

$$L=L_f-(L_f-L_0)e^{-kt},$$

$$t=0,\quad L=L_0;\quad t\to\infty,\quad L=L_f,$$

where L_0 and L_f are initial and final values of total length (L, mm) of fish, time t is measured as age (year) and the rate constant k has units per year. The monomolecular equation has no point of inflexion and assumes that growth rate decreases continuously.

Schumacher

$$\frac{dL}{dt}=\mu L,$$

$$\frac{d\mu}{dt}=-\frac{2\mu_0}{t_0}\left(\frac{\mu}{\mu_0}\right)^{3/2},$$

$$L=L_0\exp\left(\frac{\mu_0 t_0 t}{t+t_0}\right),$$

$$t=0,\quad \mu=\mu_0,\quad L=L_0;\quad t\to\infty,\quad L=L_f=L_0\exp(\mu_0 t_0),$$

where μ (per year) is the specific growth rate, μ_0 is the initial value of μ and t_0 (year) is a time parameter. The Schumacher equation assumes that senescence or decay affects μ according to a 3/2 power function [5, 25]. In contrast with the monomolecular, the Schumacher equation describes a sigmoidal curve and has an inflexion point that occurs early in the trajectory at a constant fraction of L_f (i.e. $L_f\, e^{-2}$).

Gompertz

$$\frac{dL}{dt}=\mu L,$$

$$\frac{d\mu}{dt}=-D\mu,$$

$$L=L_0\exp\left[\frac{\mu_0}{D}(1-e^{-Dt})\right],$$

$$t=0,\quad L=L_0,\quad \mu=\mu_0;\quad t\to\infty,\quad L=L_f=L_0 e^{\mu_0/D},$$

where D (per year) describes decay in specific growth rate μ and reflects the effect of ageing. Unlike the Schumacher equation, μ decays exponentially and L at inflexion occurs at a higher position on the growth curve corresponding to L_f/e [5].

Logistic

$$\frac{dL}{dt}=\mu L\left(1-\frac{L}{L_f}\right),$$

$$L=\frac{L_0 L_f}{L_0+(L_f-L_0)e^{-\mu t}},$$

$$t=0,\quad L=L_0;\quad t\to\infty,\quad L=L_f.$$

Unlike the Schumacher and Gompertz, the logistic equation has a constant value of μ and describes a symmetrical sigmoidal curve as the point of inflexion occurs at $L=L_f/2$.

von Bertalanffy

$$\frac{dL}{dt} = \eta L^b - \kappa L,$$

$$L = \left[L_f^v - \left(L_f^v - L_0^v\right) e^{-kt}\right]^{1/v},$$

Parameters: $\eta > k \geq 0$, $0 < b \leq 1$,

$$t=0, \quad L=L_0; \quad t \to \infty, \quad L=L_f=\left(\frac{\eta}{k}\right)^{1/(1-b)},$$

where η and κ are constants for anabolism and catabolism, respectively, k is a rate constant equal to $\kappa(1-b)$, and υ equals $1-b$. The equation has a flexible point of inflexion that can occur between L_0 and $0.368L_f$ [5]. The von Bertalanffy equation encompasses therefore the monomolecular, Schumacher and Gompertz.

Richards

$$\frac{dL}{dt} = \frac{kL(L_f^\eta - L^\eta)}{\eta L_f^\eta},$$

$$L = \frac{L_0 L_f}{\left[L_0^\eta + \left(L_f^\eta - L_0^\eta\right) e^{-kt}\right]^{1/\eta}},$$

Parameters: $k>0$, $-1 < \eta \leq \infty$

$t=0$, $L=L_0$; $t \to \infty$, $L=L_f$.

Richards [38] based the development of his equation on von Bertalanffy [37]. As the inflexion point can occur at any fraction of maximum asymptotic weight [5], the Richards equation encompasses the monomolecular, Schumacher, Gompertz, logistic and von Bertalanffy. The function is thus highly flexible and has been proven very effective in animal science to accommodate various growth patterns [17, 39, 40].

Statistical Procedures for Model Comparison

The integrated forms of the monomolecular, Schumacher, Gompertz, logistic, von Bertalanffy and Richards were fitted to the 12 growth profiles (four species×three populations) using the MODEL procedure of SAS [41]. Parameter estimates for the length-at-age relationship were obtained with the weighted data. Model parameters were estimated using the Marquardt method of iteration. Normal distribution of residuals was assessed for all datasets using the UNIVARIATE procedure of SAS [41].

Models were compared on the basis of their underlying theory and empirical evidence, and using the following statistical measures of goodness of fit: residual sum of squares (RSS), residual mean squares (RMS), significance level and SE of the parameter estimates, bias corrected Akaike's Information Criterion (AIC_c), Akaike weight (AIC_w) and Bayesian Information Criterion (BIC) [4, 9–12]. Degrees of freedom of the model used to compute RMS in the MODEL procedure do not include constrained parameters whose estimates are equal to their boundaries, and thus not significant. This can sometimes lead to slight overestimation of RMS. The coefficient of determination (r^2) did not serve as an indicator of the goodness of fit of the model, because its use is inappropriate in nonlinear regression analysis [9, 28]. Parameters with a P value ≥ 0.05 or high SE (i.e. greater than one-third of the parameter estimate) were considered non-significant or uncertain, and suggest possible over parameterization of the model. Higher values of AIC_w and lower values of RSS, RMS, AIC_c and BIC indicated superior fit within a given population.

The criteria AIC_c and BIC are indicators of the distance from the fitted model to observational reality and are based on information theory [10]:

$$AIC_c = -2\log(J) + 2E + \left[\frac{2E(E+1)}{N-E-1}\right],$$

$$BIC = -2\log(J) + E\log(N),$$

where J is the maximum likelihood, E is the number of estimable parameters in the model and N is the number of observations. Maximum likelihood, which was determined using the FIML option of the MODEL procedure of SAS [41], can be interpreted as a measure of fit of the model. The criteria AIC_c and BIC impose a penalty on models with additional parameters.

The relative weight of evidence in favour of a given model in the context of a set of (R) candidate models was determined using Akaike weight (AIC_w). The AIC_w (%) was determined for each model within a fish population and computed from the equation of Burnham and Anderson [10]:

$$AIC_w = \frac{\exp(-0.5\Delta)}{\sum_{i=0}^{R} \exp(-0.5\Delta_i)} \times 100,$$

where Δ is the difference in AIC_c between that of the model and the model with the lowest AIC_c value, i.e. $\Delta = AIC_c - AIC_{c,min}$, and the Δ are summed for the set of R models.

The statistical difference between models within each dataset was assessed by comparing their RSS values using an F-test, which is based on hypothesis testing theory. The equations proposed by Motulsky and Ransnas [28] served to test the statistical difference:

for models with the same number of parameters

$$F = \frac{RSS_1}{RSS_2},$$

for models with a different number of parameters

$$F = \frac{(RSS_1 - RSS_2)/(df_1 - df_2)}{RSS_2/df_2},$$

where subscripts 1 and 2 correspond to the fits with the larger and smaller RSS values, respectively, and df is the

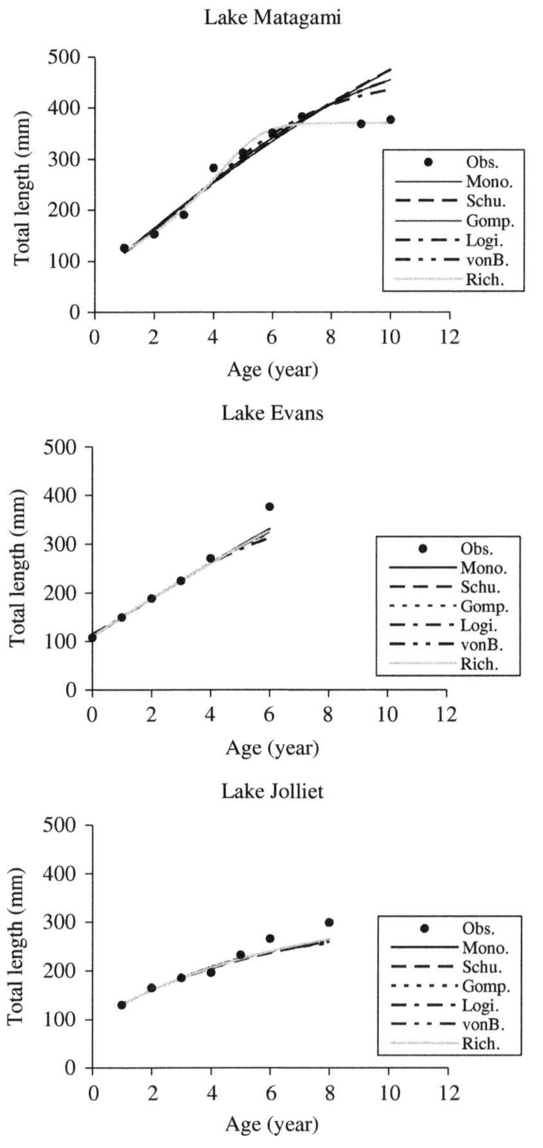

Figure 1 Length-at-age growth profiles for three populations of lake herring (*Coregonus artedii*). The data points are observed means (Obs.) for a given age class. The fitted lines show the monomolecular (Mono.), Schumacher (Schu.), Gompertz (Gomp.), logistic (Logi.), von Bertalanffy (vonB.) and Richards (Rich.) equations. The equations were fitted using means weighted by sample size.

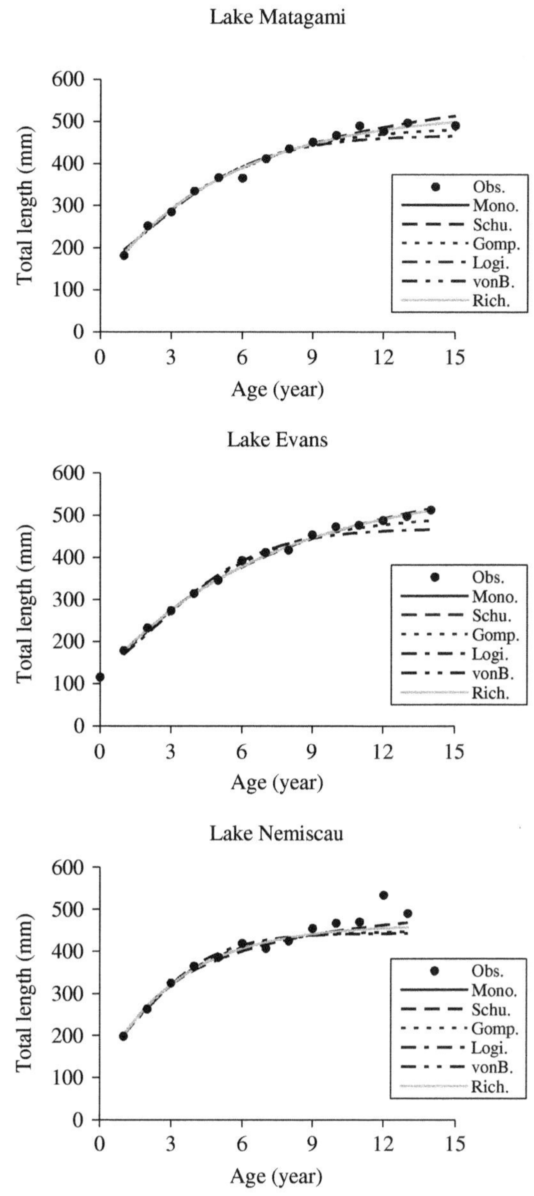

Figure 2 Length-at-age growth profiles for three populations of lake whitefish (*Coregonus clupeaformis*). The data points are observed means (Obs.) for a given age class. The fitted lines show the monomolecular (Mono.), Schumacher (Schu.), Gompertz (Gomp.), logistic (Logi.), von Bertalanffy (vonB.) and Richards (Rich.) equations. The equations were fitted using means weighted by sample size.

degrees of freedom of the error (df=number of observations − number of estimable parameters in model). The number of parameters in a given model included those that failed to be significant ($P \geq 0.05$).

The Outcomes

Model Fitting

Overall, length-at-age relationships tended to display diminishing returns behaviour (Figures 1–4). A more obvious sigmoidal shape for length-at-age relationships might have been observed if the data included more individuals less than 2 years of age. Linear and exponential patterns occurred when datasets covered a narrower age range (e.g. Figure 1, Lake Evans). Based on visual appraisal, different growth functions yielded similar curves across species and populations. Disagreement between growth functions was observed more often in older fish. Variability within age classes was wider for northern pike than for other species (data not shown).

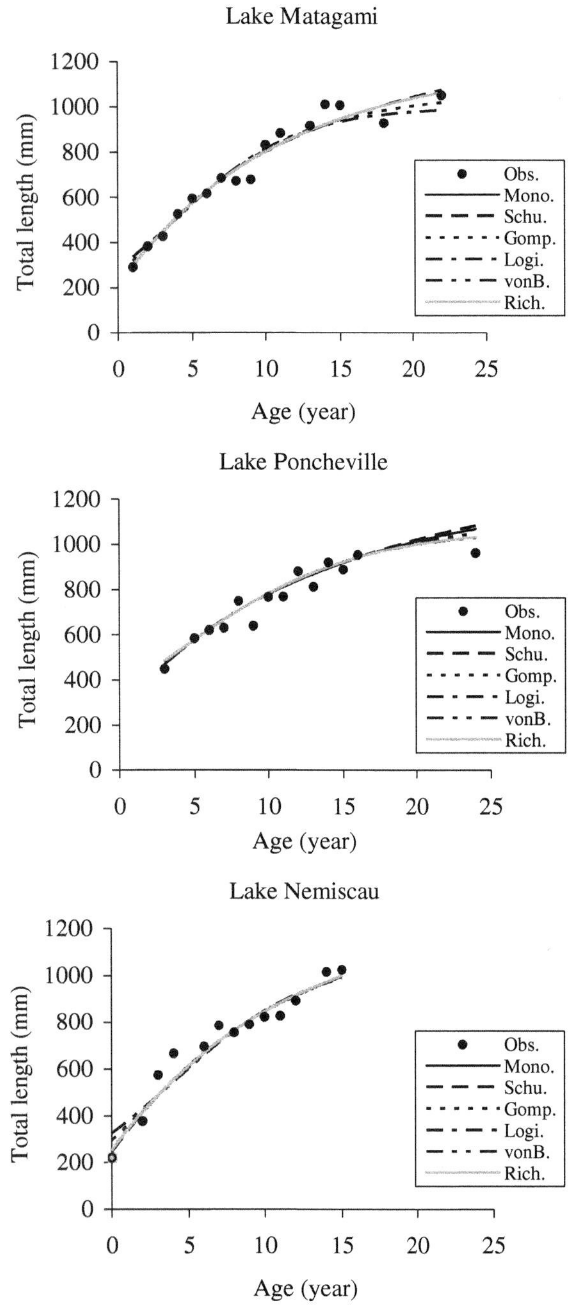

Figure 3 Length-at-age growth profiles for three populations of northern pike (*Esox lucius*). The data points are observed means (Obs.) for a given age class. The fitted lines show the monomolecular (Mono.), Schumacher (Schu.), Gompertz (Gomp.), logistic (Logi.), von Bertalanffy (vonB.) and Richards (Rich.) equations. The equations were fitted using means weighted by sample size.

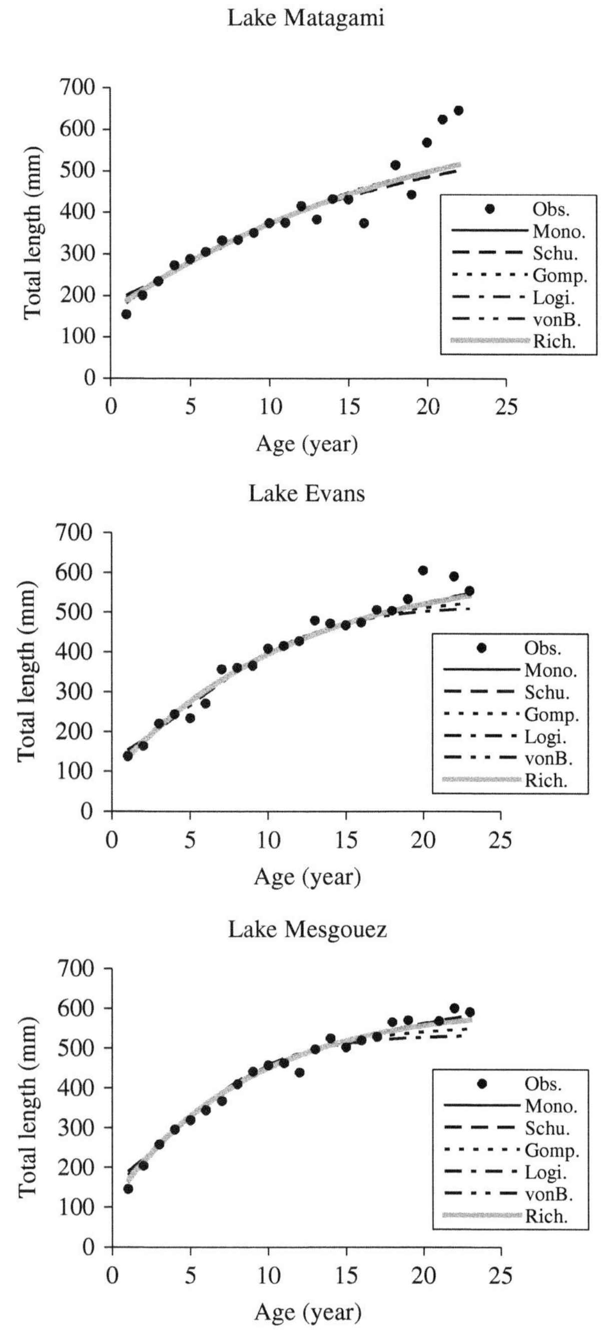

Figure 4 Length-at-age growth profiles for three populations of walleye (*Sander vitreus*). The data points are observed means (Obs.) for a given age class. The fitted lines show the monomolecular (Mono.), Schumacher (Schu.), Gompertz (Gomp.), logistic (Logi.), von Bertalanffy (vonB.) and Richards (Rich.) equations. The equations were fitted using means weighted by sample size.

Systematic deviations were often observed in older fish. For example, very nearly all models over-estimated and under-estimated total length of older lake herring and walleye in Lake Matagami, respectively (Figures 1 and 4). Better fits for older fish were obtained with the Richards equation, which seemed to force its approach to a final asymptotic length in some growth profiles (Figures 1 and 3, Lake Matagami).

The von Bertalanffy equation and, to a lesser extent the Richards, often had one parameter that corresponded to one of its bounds, indicating that this parameter could

Table 3 Performance and goodness of fit of the six growth functions used to describe the length-at-age relationships for three populations of lake herring and lake whitefish

	Lake herring					Lake whitefish				
Models	RSS	RMS	AIC_c	AIC_w	BIC	RSS	RMS	AIC_c	AIC_w	BIC
	Lake Matagami					Lake Matagami				
Monomolecular	25 402	187	1387	0.0	1387	5308	55	921	30.3	921
Schumacher	24 723	182	1381	0.0	1381	5285	55	920	37.6	920
Gompertz	23 601	174	1374	0.0	1375	5432	56	923	9.5	923
Logistic	22 159	163	1366	0.2	1366	5599	58	926	2.1	926
von Bertalanffy	23 602	175	1376	0.0	1377	5308	55	923	10.2	923
Richards	19 944	148	1353	99.8	1353	5308	55	923	10.2	923
	Lake Evans					Lake Evans				
Monomolecular	20 234	143	1309	22.1	1310	3658	23	1366	42.9	1367
Schumacher	20 240	143	1310	21.7	1310	3679	23	1367	27.1	1368
Gompertz	20 248	143	1310	21.0	1310	4080	26	1384	0.0	1384
Logistic	20 265	143	1310	19.8	1310	4956	31	1415	0.0	1416
von Bertalanffy	20 236	144	1312	7.7	1312	3658	23	1368	15.0	1369
Richards	20 236	144	1312	7.7	1312	3658	23	1368	15.0	1369
	Lake Jolliet					Lake Nemiscau				
Monomolecular	1427	17	643	30.2	643	1972	27	634	0.5	633
Schumacher	1418	17	643	41.0	642	1716	24	624	99.1	623
Gompertz	1475	17	646	7.1	646	2187	30	642	0.0	641
Logistic	1527	18	649	1.6	649	2395	33	649	0.0	648
von Bertalanffy	1427	17	645	10.1	645	1972	28	636	0.2	635
Richards	1427	17	645	10.1	645	1972	28	636	0.2	635

Note: Residual sum of squares (RSS), residual mean squares (RMS), bias corrected Akaike's information criterion (AIC_c), Akaike weight (AIC_w) and Bayesian information criterion (BIC).

not be estimated in an unbiased way. Fitted lines and parameter estimates in these cases were almost identical to those of the monomolecular. In spite of this, all models converged and therefore it was decided to include them in the comparison.

Statistical analyses provided more quantitative information, but in general confirmed observations from Figures 1–4. Comparisons based on RSS, RMS, AIC_c and BIC showed that models gave relatively similar fits within populations and for all species (Tables 3 and 4). Differences among models in the values of AIC_c and BIC were small for each growth profile.

Based on RSS, the monomolecular, Schumacher and Richards equations often performed slightly better than the other functions (Tables 3 and 4). The Schumacher was the best model for 7 of the 12 length-at-age profiles followed by the monomolecular and Richards, which each performed the best on one. Among the 12 growth profiles, the von Bertalanffy shared lowest RSS values with other equations, mostly the Richards, in three profiles but never stood out above other functions. The logistic was the worst model for 10 of the 12 profiles. The Schumacher appeared the worst model in one (northern pike, Lake Poncheville). The Gompertz equation gave intermediate values of RSS often closer to the logistic than other equations.

The ranking of models based on AIC_w, came out similar to that based on RSS, although disagreements were observed for certain datasets (e.g. Table 3, whitefish, Evans). The von Bertalanffy never stood above the other models according to AIC_w score. The logistic was the worst model for 5 of the 12 profiles.

Models with relatively similar values of RSS and AIC_c often had quite different AIC_w scores (Tables 3 and 4). For instance, percentage differences in RSS and AIC_c between the logistic and Richards were 10 and 1%, respectively, for the Matagami population of lake herring (Table 3). However, the AIC_w gave scores of 0.2 and 99.8% for the logistic and Richards, respectively. The monomolecular, von Bertalanffy and Richards resulted in the exact same fitted lines (Figure 4) with identical RSS (Table 4) and certain parameter estimates for three populations of walleye. Despite this agreement between models, AIC_w assigned scores above 30% to the monomolecular and penalized the von Bertalanffy and Richards for their extra parameter by assigning an AIC_w value below 12% (Table 4).

Based on *F*-tests, the von Bertalanffy and Richards equations most of the times performed equally well as the other functions, except in comparison with the logistic. There was no statistical difference between models with the following datasets: lake herring (Evans) and northern pike (Matagami, Poncheville and Nemiscau). However, statistical differences were observed particularly between the logistic and both the von Bertalanffy and Richards with the following datasets: lake herring (Jolliet), whitefish (Matagami, Evans and Nemiscau) and walleye (Matagami, Evans and Mesgouez). Over the 12 growth profiles, no significant difference was observed between the following models: monomolecular and Schumacher, monomolecular

Table 4 Performance and goodness of fit of the six growth functions used to describe the length-at-age relationships for three populations of northern pike and walleye

	Northern pike					Walleye				
Models	RSS	RMS	AIC_c	AIC_w	BIC	RSS	RMS	AIC_c	AIC_w	BIC
	Lake Matagami					Lake Matagami				
Monomolecular	29 913	404	869	26.6	869	15 860	68	2309	32.4	2310
Schumacher	29 913	404	869	26.6	869	15 821	68	2308	43.7	2309
Gompertz	30 192	408	870	18.6	869	16 362	71	2316	0.8	2317
Logistic	30 627	414	871	10.7	870	16 820	72	2322	0.0	2323
von Bertalanffy	29 913	410	871	8.7	870	15 860	69	2311	11.5	2312
Richards	29 913	410	871	8.7	870	15 860	69	2311	11.5	2312
	Lake Poncheville					Lake Evans				
Monomolecular	26 936	473	677	21.1	676	24 401	107	2379	33.0	2380
Schumacher	26 984	473	677	20.0	676	24 373	107	2379	37.6	2380
Gompertz	26 891	472	677	22.2	676	24 779	109	2383	5.6	2383
Logistic	26 880	472	677	22.5	676	25 411	112	2388	0.3	2389
von Bertalanffy	26 891	480	679	7.0	677	24 401	107	2381	11.7	2382
Richards	26 880	480	679	7.1	677	24 401	107	2381	11.7	2382
	Lake Nemiscau					Lake Mesgouez				
Monomolecular	17 764	683	345	27.6	342	26 195	108	2413	30.4	2414
Schumacher	17 701	681	345	29.0	342	26 142	108	2412	39.0	2413
Gompertz	18 323	705	346	17.6	343	26 494	109	2415	7.6	2416
Logistic	18 867	726	346	11.5	344	26 871	111	2419	1.3	2420
von Bertalanffy	17 764	711	347	7.1	344	26 195	108	2415	10.8	2416
Richards	17 764	711	347	7.1	344	26 195	108	2415	10.8	2416

Note: Residual sum of squares (RSS), residual mean squares (RMS), bias corrected Akaike's information criterion (AIC_c), Akaike weight (AIC_w) and Bayesian information criterion (BIC).

and Gompertz, Schumacher and Gompertz, Gompertz and logistic, and von Bertalanffy and Richards.

Parameter Estimates

Within populations, the models generally agreed well on values of L_0, whereas estimates of L_f varied (Table 5). The monomolecular equation was not a good predictor of asymptotic lengths (L_f) for the Matagami and Evans populations of lake herring (Table 5). In contrast, the logistic provided the most realistic estimates of L_f in lake herring populations. All growth functions gave realistic values for L_f in pike populations.

Estimates of initial specific growth rate (μ_0) were always higher for the Schumacher than the Gompertz and were relatively similar within species and between populations (Table 5). Values of μ_0 and the decay parameter (D) were usually larger for lake herring and whitefish than for pike and walleye.

The models with three parameters (monomolecular, Schumacher, Gompertz and logistic) usually displayed proportionately fewer non-significant parameters — and therefore lower SEs — than the models with four parameters, i.e. von Bertalanffy and Richards (Table 5). The logistic stood out as the equation where all parameters were significant most of the times. Estimation of parameters was generally not a problem, except for the von Bertalanffy and Richards. The number of parameter estimates that could not be obtained in fitting the equations of von Bertalanffy and Richards totalled 14 and 10, respectively. Fitting problems were encountered mostly with respective shape parameters υ and η (both dimensionless). As mentioned previously, the monomolecular often gave identical or very similar parameter estimates to von Bertalanffy and Richards (e.g. Table 5, Lake Jolliet). These similarities occurred when parameters υ and η could not be estimated (converging to prescribed values) and illustrate that these two equations encompass the monomolecular.

Residuals from each growth function were normally distributed for all pike populations based on the Shapiro–Wilk test. Similarly, residuals for populations Matagami and Jolliet lake herring and Evans and Nemiscau lake whitefish followed a normal distribution. Only the Gompertz and logistic equations generated residuals with a normal distribution in population Matagami walleye. Deviations from normality in other populations were minor based on the orientation of curvature in Q–Q plots and absolute skewness values (<0.6 in most datasets).

Discussion of Outcomes

This comparative review demonstrates that the monomolecular, Schumacher and Richards equations can be alternatives to the von Bertalanffy and identifies which statistical outputs facilitate discrimination among candidate models. These contributions provide guidance and

Table 5 Estimated growth parameters for describing the length-at-age relationship of three populations of lake herring using the six growth functions. Standard errors of the parameters are given in brackets. Significance ($P<0.05$) of the parameters is indicated by the symbol*

Models	Parameters						
	L_0 (mm)	L_f (mm)	k or μ (per year)	υ or η	t_0 (year)	μ_0 (per year)	D (per year)
Lake Matagami							
Monomolecular	66* (13)	1098 (616)	0.0504 (0.0358)				
Schumacher	73* (11)				5.1000* (1.4236)	0.5547* (0.1342)	
Gompertz	76* (8)					0.4753* (0.0753)	0.2418* (0.0411)
Logistic	82* (7)	459* (52)	0.4469* (0.0516)				
Von Bertalanffy	76* (8)	542 (78)	0.2422* (0.0413)	0.0001 (na)			
Richards	92* (7)	371* (93)	1.8404 (2.3293)	6.9660 (9.3350)			
Lake Evans							
Monomolecular	109* (17)	1662 (16 318)	0.0257 (0.2839)				
Schumacher	110* (16)				6.5094 (10.0707)	0.3469 (0.2389)	
Gompertz	112* (13)					0.3189 (0.1691)	0.2224 (0.2775)
Logistic	114* (11)	368 (230)	0.4189 (0.2733)				
Von Bertalanffy	109* (22)	1000 (na)	0.0499 (0.1010)	0.9344 (1.9042)			
Richards	109* (22)	1000 (na)	0.0489 (0.0990)	−0.9530 (1.8962)			
Lake Jolliet							
Monomolecular	96* (5)	333* (48)	0.1565* (0.0476)				
Schumacher	94* (6)				3.5247* (0.9412)	0.4163* (0.0909)	
Gompertz	101* (4)					0.2994* (0.0383)	0.2821* (0.0469)
Logistic	104* (4)	272* (16)	0.4109* (0.0477)				
Von Bertalanffy	96* (5)	333* (48)	0.1564* (0.0476)	1 (na)			
Richards	96* (5)	333* (48)	0.1565* (0.0476)	−1 (na)			

Note: L_0 and L_f are initial and final total lengths, respectively; k and μ are specific growth rate constants; υ and η are shape parameters; t_0 is the age at L_0; μ_0 is the initial specific growth rate at L_0; D is the constant for decay; na means not applicable (i.e. parameter was fixed at constraint limit or not fitted).

supplementary tools to help assess fish production and implement the decision-making process.

Choice of model depends heavily on the purpose and scope of the study. If one is interested in discerning life history traits, the Schumacher and Gompertz, which include estimates of initial growth rate and indirect indications on lifespan, would likely stand as most appropriate candidate models. On the other hand, size of older fish was often better estimated by the Richards.

Model Fitting

Although the growth functions are based on different assumptions and have different mathematical expressions, they resulted in relatively similar fitted lines. Resorting to statistical performance was therefore necessary to compare and select models.

Comparison and selection of models rely on various theoretical approaches and criteria or statistical outputs [4, 28, 42]. Here, the most popular statistical indices were extracted from fish growth datasets and used to discriminate between models. RSS more than RMS facilitated discrimination because the values displayed, to some extent, differences between models. In contrast, AIC_c and BIC values were relatively similar across models within a fish population, which suggests that these information criteria were perhaps ineffective in this particular analysis. Katsanevakis [14] and Helidoniotis *et al.* [43] also obtained similar results for AIC_c with five fish growth functions, especially for large datasets and had to rely on AIC_w to discriminate between them. Scores for AIC_c and BIC were often very similar within models. Wang and Liu [7] compared AIC_w and BIC and also reported similar values for the two criteria. They concluded that both criteria can be applied in non-nested (i.e. one of two competing models is not a sub-model of the other) model selection.

The AIC_w was arguably designed to accentuate any gap between models [10, 42], but tended to be misleading in this analysis because it excessively down-weighted models having biological interpretation and very acceptable goodness of fit such as the von Bertalanffy and Richards. The AIC_w scores were most of the times below 95% and similar between two models, indicating uncertainties associated with model selection based on this criterion [10]. These outcomes suggest that AIC_w has to be used cautiously in model selection.

The *F*-test, in spite of being simple [12, 44, 45], nevertheless allowed differentiation between extreme models (i.e. those with divergent RSS) by considering the significance of the divergence between RSS values.

The tests used to assess statistical difference between models, though useful, have limited application since they overlook other properties of equations such as biological interpretation and position of the inflexion point [21, 43].

Growth profiles and candidate equations examined in this review were useful for discerning variations in life history among freshwater species. Different anatomical structures are used to assess ageing depending on fish species. In this study, the most reliable and useful hard body parts of each species were sampled to determine the age of the fish [46]. Scales were collected from recommended areas on pike and walleye. In the James Bay territory, the climate is cold and annuli become clearly distinct on otoliths, scales, cleithral and opercular bones, and allow precise estimates of age. Therefore, the age results reported in this review are considered reliable. Wide variations in size within and between year classes, relatively young age at sexual maturity and prolonged lifespan of northern pike indicate that this species has a particular life history. Vélez-Espino *et al.* [47] looked at the fecundity, survival rate, age at maturity and longevity of more than 80 freshwater fish species and also concluded that northern pike fell into a different life history category from lake herring, whitefish and walleye. Our results suggest that more explanatory models are required for pike. Conclusions drawn about each equation from the Nemiscau population of pike were in line with those of other populations and species, even though sample size was inferior to 30. A larger sample would be required to conduct a more robust study on length-at-age relationships of this population. Weight-at-age relationships (data not shown) suggest that lake herring, whitefish, northern pike and walleye exhibit indeterminate growth and keep growing well beyond age at sexual maturity. The use of models which relax the constraint on final asymptotic weight should therefore be more appropriate [21, 48–50].

Parameter Estimates

Incorrect estimates of L_f should be expected when indeterminate growth is described using models that assume fixed asymptotic size. In agreement with Day and Taylor [51] and Charnov *et al.* [52], candidate growth functions examined in this paper are not recommended to estimate growth trajectories of fish at the end of their lifespan. Discrepancies were particularly obvious in lake herring, pike and walleye. Parameter estimates of the monomolecular tended not to be good descriptors of length-at-age relationships in profiles where a diminishing returns pattern was not displayed clearly.

Unrealistic asymptotic length may also arise from weighting the samples. Weighting gave less importance to young and old fish because of their usual small sample sizes. Fitted lines were thus less influenced by data points belonging to these classes and this may have sometimes resulted in overestimation of asymptotic size. In such cases, systematic errors were observed in older fish. The Richards was apparently less affected by weighting of samples as reflected by lake herring and pike. The flexible inflexion point of the Richards perhaps contributed to spreading out deviations and attenuating systematic errors.

Higher values of parameters μ_0 and D occurred in species with a younger age at sexual maturity and shorter lifespan. These outcomes agree with other studies where higher initial rates of growth were observed in fish maturing at an earlier age [53–55] and the onset of physiological degeneration or decay has been correlated with rapid individual development and early reproductive maturity in plants, fish and mammals [56–59]. Despite this strong agreement with the literature, valid estimates of parameters μ_0 and D would ideally require that the distribution of stresses and dangers remained uniform across age classes [56], a situation hardly encountered in nature. Nevertheless, the Schumacher and Gompertz equations were apparently robust enough against the effects of varying environmental conditions on life history of fish, since their parameter estimates concurred with current theories of growth and ageing.

Declaration of significance of parameter estimates alone does not stand as a reliable statistical index in selecting a growth function. Darmani Kuhi *et al.* [39] reached the same conclusion in a similar study conducted with chickens. Here, the logistic more often had significant parameters compared with other growth functions but resulted most of the times in the poorest fits based on other statistical criteria. The occurrence of non-significant parameters with the von Bertalanffy and Richards suggests that these equations are over parameterized. Although sometimes informative [11], *P*-values focus on accepting or rejecting parameters rather than measuring the size of a parameter estimate relative to its SE. Here, SEs provided further information on the size of parameter effects and should therefore be assessed in parameter estimation. In datasets where the residuals were not normally distributed, the assumption regarding normality and presumably homoscedasticity can be relaxed because of the large size of our datasets [12].

An additional criterion that could be used to compare models relates to the capability of fitting all parameters in a model. Models providing parameter estimates that correspond to constrained or prescribed values might be considered as over-parameterized or non-ideal.

Models are simplifications of reality and unavoidably overlook factors affecting growth responses of fish. In this review, all models ignored the apparent indeterminate growth of fish species under study and their ectothermic nature. Moreover, fish growth was considered a function of time rather than, for example, food intake, physiological state and sex. The models excluded temperature and sexual maturation as driving variables of the growth trajectory. The effect of gender on model performance, and thus selection, was reported by Darmani Kuhi *et al.* [39] in a growth study on poultry. It would be interesting therefore to assess the effect of sex in fish on goodness of fit of candidate models. Imperfections in simple growth functions enhance their empirical content and impair their ability to represent causality in growth processes. Nevertheless, these models can find useful application provided users are aware of their respective assumptions and familiar with the life history traits of the species involved. Equations examined in this review along with their special cases are, despite their simplicity, currently and successfully applied not only in fisheries, ecology and aquaculture management but also in fish bioenergetics [60, 61], animal production and nutrition [39, 62, 63], and metabolic modelling [64, 65]. Therefore, we contend that models should be rejected on the basis of their inability to describe patterns or satisfy the purposes of a study rather than on their simplicity.

This analytical review compared different growth functions to describe length-at-age relationships with a view to helping scientists discern further information on the life history of freshwater fish and providing managers in fisheries and aquaculture with relevant tools for use in decision-making. All growth functions resulted in relatively good fits but some, such as the monomolecular, Schumacher and Richards, did especially well on the basis of statistical inference. The logistic should not be applied to describe and simulate the growth trajectory of fish species examined in this paper. Comparison and selection of models should consider statistical performances (goodness of fits) of each equations and their parameter estimates along with the purpose and scope of the study. The RSS, SE of parameter estimates and, to a lesser extent, AIC_w were particularly relevant outputs for discriminating between models. Much statistical information on models can be extracted, but not all of it is helpful and of interest in model selection. An approach based on strong theoretical concepts and detailed statistical analyses of parameters can be of little value if, ultimately, it does not facilitate the decision-making process. This analysis also showed that different conclusions can sometimes be reached depending on which statistical outputs are used for inference. Comparison and selection of models should therefore rely on proper understanding of both the system under study and the mathematical tools that are being used to describe it.

Acknowledgements

Sincere thanks to Hydro-Québec and Environnement Illimité Inc. for access to data and to Mrs Michelle Edwards for her assistance with SAS. This review was undertaken, in part, thanks to funding from the Canada Research Chairs Program. André Dumas was the recipient of scholarships from the Natural Sciences and Engineering Research Council of Canada, the Fonds québécois de recherche sur la nature et les technologies and the University of Guelph.

References

1. Stead SM, Laird L. Handbook of Salmon Farming. Springer, Chichester, UK; 2000.

2. Charnov EL, Gillooly JF. Size and temperature in the evolution of fish life histories. Integrative and Comparative Biology 2004;44:494–7.

3. Joung S-J, Liao Y-Y, Chen C-T. Age and growth of sandbar shark, *Carcharhinus plumbeus*, in northeastern Taiwan waters. Fisheries Research 2004;70:83–96.

4. Ratkowski DA. Nonlinear Regression Modelling: a Unified Practical Approach. Marcel Dekker Inc., New York, NY; 1983.

5. Thornley JHM, France J. Mathematical Models in Agriculture: Quantitative Methods for the Plant, Animal and Ecological Sciences. CAB International, Wallingford, UK; 2007.

6. May RM. Simple mathematical models with very complicated dynamics. Nature 1976;261:459–67.

7. Wang Y, Liu Q. Comparison of Akaike information criterion (AIC) and Bayesian information criterion (BIC) in selection of stock-recruitment relationships. Fisheries Research 2006;77:220–5.

8. Zhou S. Discriminating alternative stock-recruitment models and evaluating uncertainty in model structure. Fisheries Research 2007;86:268–79.

9. Ratkowski DA. Handbook of Nonlinear Regression Models. Marcel Dekker Inc., New York, NY; 1990.

10. Burnham KP, Anderson DR. Model Selection and Multimodel Inference: A Practical Information-Theoretic Approach, 2nd edn. Springer, New York, NY; 2002.

11. Stephens PA, Buskirk SW, Hayward GD, Martínez Del Rio C. Information theory and hypothesis testing: a call for pluralism. Journal of Applied Ecology 2005;42:4–12.

12. Vittinghoff E, Glidden DV, Shiboski SC, McCulloch CE. Regression Methods in Biostatistics: Linear, Logistic, Survival, and Repeated Measures Models. Springer, New York, NY; 2005.

13. Ricker WE. Computation and Interpretation of Biological Statistics of Fish Populations. Bulletin of the Fisheries Research Board of Canada No. 191, Ottawa, Canada; 1975.

14. Katsanevakis S. Modelling fish growth: model selection, multi-model inference and model selection uncertainty. Fisheries Research 2006;81:229–35.

15. Gurney WSC, Tyldesley G, Wood SN, Bacon PJ, Heath MR, Youngson A, Ibbotson A. Modelling length-at-age variability under irreversible growth. Canadian Journal of Fisheries and Aquatic Sciences 2007;64:638–53.

16. Shih N-T, Hsu K-C, Ni I-H. Age, growth and reproduction of cutlassfishes *Trichiurus* spp. in the southern East China Sea. Journal of Applied Ichthyology 2011;27:1307–15.

17. Ricker WE. Growth rates and models. Fish Physiology 1979;8:677–743.

18. Iwama G, Tautz AF. A simple growth model for salmonids in hatcheries. Canadian Journal of Fisheries and Aquatic Sciences 1981;38:649–56.

19. Elliott JM, Hurley MA, Fryer RJ. A new, improved growth model for brown trout, *Salmo trutta*. Functional Ecology 1995;9:290–8.

20. Lopez S, France J, Gerrits WJJ, Dhanoa MS, Humphries DJ, Dijkstra J. A generalised Michaelis–Menten equation for the analysis of growth. Journal of Animal Science 2000;78:1816–28.

21. Dumas A, France J. Modelling the ontogeny of ectotherms exhibiting indeterminate growth. Journal of Theoretical Biology 2008;254:76–81.

22. Lee E, Sasi S, Shivaji R. An ecological model with a Σ-shaped bifurcation curve. Nonlinear Analysis: Real World Applications 2012;13:634–42.

23. Schumacher FX. A new growth curve and its applicability to timber yield studies. Journal of Forestry Research 1939;37:819–20.

24. Mitscherlich-Königsberg EI. Das Gesetz des Minimums und das Gesetz des abnehmenden Bodenertrages. Landwirtschaftliches Jahrbücher 1909;38:537–52.

25. Schulin-Zeuthen M, Kebreab E, Dijkstra J, López S, Bannink A, Darmani Kuhi H, Thornley JHM, France J. A comparison of the Schumacher with other functions for describing growth in pigs. Animal Feed Science and Technology 2008;143:314–27.

26. Jull MA, Titus HW. Growth of chickens in relation to feed consumption. Journal of Agricultural Research 1928;36:541–50.

27. Darmani Kuhi H, Kebreab E, López S, France J. A comparative evaluation of functions for the analysis of growth in male broilers. Journal of Agricultural Science Cambridge 2003;140:451–9.

28. Motulsky HJ, Ransnas LA. Fitting curves to data using nonlinear regression: a practical and nonmathematical review. FASEB Journal 1987;1:365–74.

29. Chen Y, Jackson DA, Harvey HH. A comparison of von Bertalanffy and polynomial functions in modelling fish growth data. Canadian Journal of Fisheries and Aquatic Sciences 1992;49:1228–35.

30. Hernandez-Llamas A, Ratkowski DA. Growth of fishes, crustaceans and molluscs: estimation of the von Bertalanffy, logistic, Gompertz and Richards curves and a new growth model. Marine Ecology Progress Series 2004;282:237–44.

31. Katsanevakis S, Maravelias CD. Modelling fish growth: multi-model inference as a better alternative to *a priori* using von Bertalanffy equation. Fish and Fisheries 2008;9:178–87.

32. Consortium Groupe de Recherche SEEEQ Ltée - Environnement Illimité inc. Complexe NBR, faune ichtyenne Volume 2: dynamique de population. Rapport. Hydro-Québec, vice-présidence Environnement, Canada, 1993.

33. Lysack W. 1979 Lake Winnipeg Fish Stock Assessment Program. Report. Manitoba Department of Natural Resources, Research MS Report No. 80-30, Canada; 1980.

34. Gompertz B. On the nature of the function expressive of the law of human mortality, and on a new mode of determining the value of life contingencies. Philosophical Transactions of the Royal Society of London 1825;115:513–83.

35. Verhulst P-F. Notice sur la loi que la population suit dans sa croissance. Correspondance, Mathématiques et Physique 1838;10:113–21.

36. Pütter A. Studien über physiologische Ähnlichkeit. VI. Wachstumsähnlichkeiten. Pflgers Archiv fr die gesamte Physiologie des Menschen und der Tiere 1920;180:298–340.

37. von Bertalanffy L. Quantitative laws in metabolism and growth. Quarterly Review of Biology 1957;32:217–31.

38. Richards FJ. A flexible growth function for empirical use. Journal of Experimental Botany 1959;10:290–300.

39. Darmani Kuhi H, Kebreab E, López S, France J. An evaluation of different growth functions for describing the profile of live weight with time (age) in meat and egg strains of chicken. Poultry Science 2003;82:1536–43.

40. Black JL, Davies GT, Bray HJ, Giles LR, Chapple RP. Modelling the effects of genotype, environment and health on nutrient utilisation. In: Proceeding of the IVth International Workshop on Modelling Nutrient Utilisation in Farm Animals. National Institute of Animal Science, Foulum, Denmark, 1995. p. 85–105.

41. SAS. SAS/STAT user's guide. SAS Institute Inc., Cary, NC; 2004.

42. Buckland ST, Burnham KP, Augustin NH. Model selection: an integral part of inference. Biometrics 1997;53:603–18.

43. Helidoniotis F, Haddon M, Tuck G, Tarbath D. The relative suitability of the von Bertalanffy, Gompertz and inverse logistic models for describing growth in blacklip abalone populations (*Haliotis rubra*) in Tasmania, Australia. Fisheries Research 2011;112:13–21.

44. Rozeboom WM. The fallacy of the null-hypothesis significance test. Psychological Bulletin 1960;57:416–28.

45. Dayton CM. Model comparisons using information measures. Journal of Modern Applied Statistical Methods 2003;2:281–92.

46. Jearld AJR. Age determination. In: Nielsen LA, Johnson DL, editors, Fisheries Techniques. Bethesda, MD, USA: American Fisheries Society; 1983. p. 301–24.

47. Vélez-Espino L, Fox MG, McLaughlin RL. Characterization of elasticity patterns of North American freshwater fishes. Canadian Journal of Fisheries and Aquatic Sciences 2006;63:2050–66.

48. Parker LA, Larkin PA. A concept of growth in fishes. Journal of the Fisheries Research Board of Canada 1959;16:721–45.

49. Knight W. Asymptotic growth: an example of nonsense disguised as mathematics. Journal of the Fisheries Research Board of Canada 1968;25:1303–7.

50. Roff DA. A motion for the retirement of the von Bertalanffy function. Canadian Journal of Fisheries and Aquatic Sciences 1980;37:127–9.

51. Day T, Taylor PD. Von Bertalanffy's growth equation should not be used to model age and size at maturity. American Naturalist 1997;149:381–93.

52. Charnov EL, Turner TF, Winemiller KO. Reproductive constraints and the evolution of life histories with indeterminate growth. Proceedings of the National Academy of Sciences, USA 2001;98:9460–4.

53. Dhillon RS, Fox MG. Growth-independent effects of temperature on age and size at maturity in Japanese medaka (*Oryzias latipes*). Copeia 2004;1:37–45.

54. Shearer K, Parkins P, Gadberry B, Beckman B, Swanson P. Effects of growth rate/body size and a low lipid diet on the incidence of early sexual maturation in juvenile male spring Chinook salmon (*Oncorhynchus tshawytscha*). Aquaculture 2006;252:545–56.

55. Yoneda M, Kurita Y, Kitagawa D, Ito M. Spatial variation in the relationship between growth and maturation rate in male Japanese flounder *Paralichthys olivaceus* off the Pacific coast of northern Japan. Journal of Sea Research 2007;52:171–9.

56. Williams GC. Pleiotropy, natural selection, and the evolution of senescence. Evolution 1957;11:398–411.

57. Harvey PH, Zammuto RM. Patterns of mortality and age at first reproduction in natural populations of mammals. Nature 1985;315:315–20.

58. Valdesalici S, Cellerino A. Extremely short lifespan in the annual fish *Nothobranchius furzeri*. Proceedings of the Royal Society of London B 2003;270(Suppl.):S189–91.

59. de Magalhães JP, Church GM. Genomes optimize reproduction: aging as a consequence of the developmental program. Physiology 2005;20:252–9.

60. Cho CY, Bureau DP. Development of bioenergetics models and the Fish-PrFEQ software to estimate production, feeding ration and waste output in aquaculture. Aquatic Living Resources 1998;11:199–210.

61. Lupatsch I, Kissil GW. Feed formulations based on energy and protein demands in white grouper (*Epinephelus auneus*). Aquaculture 2005;248:83–95.

62. Schinckel AP. Describing the pig. In: Kyriazakis I, editor A Quantitative Biology of the Pig. Wallingford, UK: CABI Publishing; 1999. p. 9–38.

63. Kebreab E, Schulin-Zeuthen M, López S, Soler J, Dias RS, de Lange CFM, France J. Comparative evaluation of mathematical functions to describe growth and efficiency of phosphorus utilization in growing pigs. Journal of Animal Science 2007;85:2498–507.

64. Buyse J, Geypens B, Malheiros RD, Moraes VM, Swennen Q, Decuypere E. Assessment of age-related glucose oxidation rates of broiler chickens by using stable isotopes. Life Sciences 2004;75:2245–55.

65. Boyce SJ, Murray AWA, Peck LS. Digestion rate, gut passage, time and absorption efficiency in the Antarctic spiny plunderfish. Journal of Fish Biology 2006;57:908–29.

The function, social implications and management of barking in dogs

Elsa. L. Flint*

Address: Institute of Veterinary, Animal and Biomedical Sciences, College of Sciences, Massey University, Private Bag 11-222, Palmerston North, New Zealand.

***Correspondence:** Email: e.flint@xtra.co.nz

Received: 9 January 2012
Accepted: 22 May 2012

Abstract

The function and social significance of barking in domestic dogs (*Canis familiaris*) is still not fully understood. Barking is poorly tolerated by some people, is frequently identified by owners as problem behaviour, is cited as a reason for the relinquishment of dogs to animal shelters and is a common cause for complaint to authorities. It is as yet unclear why many humans find barking disturbing. Despite a recent increase in research in this area, further work is needed to achieve a greater understanding of the behaviour itself and its impact on society and to establish humane and effective strategies to manage and control this 'socially unacceptable' canine behaviour. This paper reviews information on the purpose of and factors influencing barking in domestic dogs, the possible reasons for its negative impact on humans and methods of control and management.

Keywords: Barking, Dogs, Communication, Behaviour, Society, Understanding, Management

Review Methodology: Databases used to source information included Web of Science, Medline, CAB Abstracts and Biological Abstracts. Other sources included collegial discussion and personal investigation.

Introduction

Barking in the domestic dog (*Canis familiaris*) is often considered undesirable. It is commonly reported to authorities and may result in dogs being removed from properties by authorities (usually councils) or fines being issued to owners. In Auckland, New Zealand, 35% of annual complaints about dogs are about barking and in one Auckland-based behaviour clinic, 42% of cases presented initially for problem barking. The explanation for this is unclear. There is a need for greater understanding of this behaviour, its form and function, the factors that influence it and the reasons why humans find it so difficult to cope with. More detailed knowledge in these areas may enable appropriate management and prevention strategies to be implemented which will not only address the human need for companionship but will also consider the welfare of dogs kept as suburban pets.

The function of barking

The purpose of barking in dogs has been subject to some controversy. However, recent studies on canine vocalization have helped to elucidate its function. Most recent studies focus on the communicative function of barking in dogs both intra- and interspecific situations, with an emphasis on dog–human communication.

Intraspecific communication

It is now recognized that barking is a significant and variable form of communication used by domestic dogs that is context-specific [1–3] and not, as some have suggested in the past, limited in purpose to territorial guarding [4] or random and with no specific function [5].

While barking may indeed be part of territorial defence, alerting the rest of the pack to the presence of intruders [6, 7]; it may also signify excitement during play, before a walk or in response to the return of pack or family members [8]. It may be an expression of distress, a calling back of pack members when a dog is left alone, or an expression of extreme anxiety caused by isolation [8]. Short-range vocalizations that accompany social interactions in wild canids are important in forming and maintaining bonds between pack members and minimizing conflict [8]. In domestic dogs, growls recorded in different situations (play, response to a stranger and food guarding) evoke quite different responses when played to dogs approaching a bone, indicating that the dogs are receiving and understanding specific information about the reason for the growl [9]. Similarly recent studies emphasize the communicative function of barking [10–12] and have shown that dogs of various breeds use barks that are similar sonographically when exposed to various stimuli or situations [3].

It is believed that canine vocalizations may also disclose information about an individual's sex [10] and identity [12]. Taylor [13] and Farago [14] showed that growling (evoked by food guarding) is correlated to body size and that individuals hearing a growl then appear to form a mental image of the size of the growling animal. It is possible that similar information may be conveyed by territorial barking; however, this has not been tested.

It is to some extent understandable that people may become irritated when subjected to frequent, persistent noise for which they see no purpose. Understanding that barking is a relatively complex behaviour with specific functions may help people to be more tolerant of it. This information can be used in public education.

Interspecific communication

Recent research has confirmed that dogs can in fact use barking to successfully communicate with humans as well as conspecifics; however, there is little information available on efficacy in communicating with other species. The structure of barking in specific situations is significantly different when analysed acoustically [1]. Owners become adept at recognizing the meaning of their dogs' barks [2], being able to distinguish whether their dogs are guarding, playing, welcoming a returning family member or seeking attention. In the domestic situation, barking may be used to draw attention to an empty food dish or water bowl, to inform an owner of the location of an inaccessible toy [15] or to request that a door be opened. This suggests that barking is important in enabling dogs to communicate successfully with humans, although some authors [15] propose that when intentionally directed at the owner, this communication is limited to requesting something that the dog needs or wants rather than disseminating information and so is manipulative rather than truly communicative.

Individual variation in context-specific barking allows owners and other dogs to identify individuals by their barks [10–12]. This individual variation in context-specific barking does not seem to alter human ability to correctly interpret the bark. It has also been shown that people who are not dog owners, with no prior experience of dogs and who have never seen accompanying body language, can correctly identify the situation in which a bark is occurring and the emotionality of the bark [2]. In fact, infants as young as 6 months may be capable of differentiating between aggressive and non-aggressive barking [16]. It is interesting that studies have shown that while humans can accurately interpret barking they are not so successful with growling. Humans repeatedly incorrectly classify growls from large dogs as aggressive [17]. It has been suggested that this is the result of social conditioning resulting in preconceived ideas about large dogs. Whatever the reason, it emphasizes the fact that barking has a unique communicative function as barks also vary with size but seem not to be misinterpreted in the same way.

Some authors suggest that because barking in domestic dogs occurs under a wider range of circumstances than was true of their ancestors, barking could be largely an acquired behaviour, developed as a consequence of domestication, to facilitate communication with humans [18]. This may be true, especially of non-territorial barking. Territorial barking, however, was probably a vocalization developed to warn conspecifics of a real or perceived threat [7] that, through experience, humans learned to recognize as such, rather than one developed by dogs with the intention of communicating with humans. Recognizing the meaning of territorial barking may well have conferred a survival advantage on humans living in close association with dogs.

More studies in this area may help to confirm the relative significance of this inherent human ability to detect emotionality and meaning in canine vocalizations with respect to human reaction to barking in modern day life.

What influences barking behaviour?

To manage a particular behaviour, it is important to know what drives it. It is generally accepted within the literature that barking, like most behaviour, is influenced by genetics, environment, experience and physiological state.

Influence of genes

Barking occurs more frequently and under a wider range of circumstances in the domestic dog than in other canids [19]. In the wolf (*Canis lupus*), barking is rare [6]; however, it is more common in the African wild dog (*Lycaon pictus*). Since repetitive high-frequency harmonic vocalizations have been shown to be more prevalent in young African wild dogs than adults and are associated with soliciting and contesting food, [20], it has been hypothesized that early selective breeding for neotenic (puppy-like) traits in domestic dogs has increased their tendency to bark [8]. Adult domestic dogs show significant etepimelectic (care-seeking) behaviour patterns in the form of submissive gestures and these are often accompanied by vocalizations, which could also be consistent with neoteny [8]. There is probably as yet insufficient scientific data to support or refute this theory [18].

The frequency of barking may be influenced by the breed of dog and although it is unclear whether early selection for neotenic traits did in fact occur, we know that dogs have been selectively bred over centuries for specific purposes and in many cases one of the traits selected for particularly in hunting dogs is a propensity to bark. It is generally accepted that terriers which are bred for chasing and digging bark readily, especially when excited [21, 22] and many working farm dogs are bred for strong persistent barking. Conversely, selection towards other traits can result in dogs that rarely bark such as chow-chows, sharpeis and some sheepdogs.

There is also evidence of individual variation within breeds. Karjalainen [23] found that barking behaviour could be influenced by selective breeding in Finnish spitz. Seksel [24] found individual variability in the vocal response of puppies to novel stimuli irrespective of breed. Gogoleva [25] discovered that specific vocalizations could be selected for in silver foxes (*Vulpes vulpes*) and that these were linked to tameness and aggression. This not only provides further evidence that barking may be selected for or against but should serve to remind us that selection for one trait may inadvertently result in the expression of another.

As yet there have been few studies done to examine the influence of breed on problem barking, although Cross [26] concluded that the greatest risk for nuisance barking in a sample of dogs in Brisbane, Australia occurred with dogs of herding type. This is not surprising, given that with the exception of eye dogs, herding dogs are bred to bark.

The knowledge that barking may be selected against may be helpful in preventing problem barking, allowing us to selectively breed dogs less predisposed to this behaviour that may be more suitable as suburban pets.

Environment

The environment in which a dog is kept can influence barking behaviour. It is known that dogs bark in response to environmental stimuli and that such stimuli may be auditory, olfactory or visual [21].

Few quantitative studies have been made to identify the relative significance of auditory, olfactory and visual stimuli to problem barking. However, the sound of other dogs barking has been identified as a particularly significant stimulus [27, 28]. The intensity of response elicited is related to the factors provoking the barking dogs. For example, if the dog responsible for the stimulus is reacting to an intruder and feels threatened, the intensity of the reacting dog's response will be greater than if the dog providing the stimulus was barking at a passing cat [27].

Other auditory stimuli are also significant in provoking or decreasing barking [29]. When on their home territory, dogs commonly bark at unfamiliar or sudden noises and also at familiar noises associated with things they enjoy chasing or people they are excited to see [21]. Barking triggers identified by owners are many and varied and may be animate or inanimate in origin [30]. Wells [29] found that dogs in an animal welfare centre barked more in response to heavy metal music than any other stimulus including human voices. Visual stimuli are often significant trigger factors too [30]. These include cats, birds, people, passing the property, other dogs and toys. It is logical to assume that reactivity to specific triggers may vary according to the dog's experience and the frequency to which they are exposed to certain stimuli [21, 29].

Olfactory stimuli have not been specifically studied as triggers for problem barking. We know that the scent of prey triggers barking behaviour in hunting dogs but this is unlikely to be of significance in a suburban situation. Olfactory stimuli are, however, used in behavioural therapy. Gaultier [31] showed that chamomile and lavender scent had a calming effect on kennelled dogs and reduced vocalization. Pheromones are used to calm dogs suffering from separation anxiety [31].

Activities such as play and exercise all have the potential to precipitate barking [21], owing to excitement or as a method of continued solicitation. Conversely, it is postulated that lack of environmental stimulation and exercise may result in excessive barking that is born of frustration [32].

Dogs left alone may bark to seek attention [33, 34]. This is often a manifestation of anxiety precipitated by separation from their owners or from human or canine company in general and can escalate to a state of panic. They may also bark in response to anxiety-provoking noises such as thunderstorms and fireworks [21].

Environmental influences and associated management practices are extremely important considerations in the investigation and treatment of problem barking.

Physiological state

Barking is potentially influenced by age and state of health, including hormone fluctuations, levels of anxiety and level of arousal. Dogs are less responsive to auditory triggers when asleep [27] so it may be that exercise and work on a daily basis serves to decrease barking as animals spend more time asleep.

Young dogs and puppies with high energy levels may vocalize more than older animals as they engage in play or react to new stimuli [35]. Old dogs suffering from canine cognitive dysfunction (CCD, senile degeneration equivalent to Alzheimer's in people) may engage in monotonous apparently meaningless and unprovoked barking [36]. This is the result of neurological changes including amyloid deposition and tangling of neurofibrils and a reduction in levels of the neurotransmitter dopamine. Old dogs not suffering from CCD may bark less as a result of age-related impairment in auditory and visual acuity [21].

Kim [4] found an increase in territorial aggression in ovario-hysterectomized bitches and hence in reactive barking. The reason for this is unclear, although earlier research has suggested that some female puppies situated between males *in utero* may be affected by testosterone and show increased levels of aggression. This seems to improve if they are not spayed before their first heat [21]. There is little research on the effects of testosterone on barking behaviour in males, although it is generally accepted that overly aggressive entire male pets should be castrated to reduce reactivity. As a result, barking associated with challenge and guarding behaviour may also be reduced [21].

Dogs that are anxious or fearful may bark or otherwise vocalize as an expression of this anxiety [33–35]. This may occur in dogs that are left at home alone and suffer separation anxiety, or in animals reacting to noxious stimuli such as fireworks or thunderstorms. Barking in these circumstances varies, but is often repetitive and monotonous and may be interspersed with howling [33]. Dogs showing this behaviour tend to be neurologically different from their counterparts resulting in an inability to cope with anxiety-provoking situations caused by insufficient neurotransmitters (especially serotonin) or receptor sites for those transmitters [21].

Hunger and physical discomfort is also likely to contribute to increased vocalization although no studies are available to quantify this because of welfare considerations.

Human response to barking

Barking is cited worldwide as a behavioural problem in dogs [21, 22, 26, 29, 32, 34, 37, 38] and has been recognized as such for many years. In fact, the first dog control act written in the Bahamas in 1941 describes dogs as a nuisance [39]. As early as 1977 in a paper entitled 'The Anti-Social Behaviour of Urban Dogs', Loew and Fryer [40] suggested that society needs a 'toothless silent constipated canine' before it can be socially acceptable.

Despite our awareness of barking as a social problem, there has been very little research undertaken to investigate why humans respond negatively to barking and to elucidate human perception of this canine behaviour or to quantify public attitudes toward it. Although we know people complain about the behaviour, do we really know why?

There are studies in the field of bioacoustics that investigate the effects of suburban noise on sleep and concentration [41], but most tend to focus on traffic and human-created noises. Trimmel and Poelzl [42] included barking in investigating noise disturbance and concluded that background noise, even at low intensity, is associated with energy consumption and therefore can be potentially tiring and may cause impaired performance in spatial attention. Kaiser *et al.* [43] showed that barking caused humans to orient to the source of the sound. This appears to be the result of a direct effect on the brain and is not under voluntary control. Riede *et al.* [44] showed that measures of the harmonic to noise ratio (HNR) were useful predictors of perceptual rating by human listeners. There was a positive correlation between human classification of barks with respect to noisiness and the actual HNR measurements. This suggests that humans differentiate between barks based on harmonic to noise ratios and lends further support to the theory that a high noise component may be the most disturbing feature of barking.

Dogs barking in close proximity to humans may cause significant discomfort as the sound may reach levels greater than 60–75 decibels, which is the level used as the upper threshold by authorities [45] when monitoring workplace noise levels or noise complaints. Noise levels in kennels have been recorded at 125 decibels [46]. However, it is difficult to know if it is purely the physical effect of barking that causes people to react negatively to barking. It may be that the context of the bark has some relevance to the degree of disturbance caused. The relative disturbance of different bark types has not yet been studied in detail.

The question remains as to whether it is the acoustic components of the bark itself or the meaning conveyed by the bark that disturbs humans. Is the degree of disturbance a reflection of social attitude towards dogs and expectations about control and care or is it influenced by individual experience with dogs and lifestyle?

Given the ability of humans to distinguish between bark types [2, 11, 12], it is likely that early experiences and associations with dogs may have some influence on response to and tolerance of barking; however, this has not been studied.

Humans have evolved living in close proximity to canids for many years and it is highly likely that the ability to recognize a bark signifying concern and a compulsion to investigate it conveyed a significant survival advantage in early human civilizations. Lord *et al.* [7] theorizes that barking is primarily designed as a mobbing behaviour that is a response to a perceived or real threat and shows that such 'barking' vocalizations are not confined to Canidae but occur in many species. This may be true of territorial barking in dogs and if so could support the theory that humans are conditioned to respond to barking dogs. If this is true, one would expect humans to find territorial barking more disturbing than play barking or barking associated with separation anxiety.

'Quick Fix' methods used to control problem barking

Several methods to control unwanted barking have been implemented over the years, often without any attempt to understand the cause of the problem behaviour. One of the early solutions was the practice of debarking or ventral laryngotomy [47]. This is still performed by some veterinarians, although the procedure has been prohibited in many countries. More recently, there has been a move towards the use of 'anti-bark' collars.

The use of anti-bark collars is highly controversial. These collars work by delivering a noxious stimulus in response to the barking behaviour. There are different types of collars for this purpose; those that deliver an electric shock, those that release a spray of citronella or cold air into the dog's face and those that emit a high frequency sound. Electric collars may cause significant discomfort [48] and have been banned as inhumane in some countries. Citronella collars do not inflict pain but can cause nausea in some dogs. Scentless spray collars do not cause nausea and work on the principal of interrupting the behaviour with a cold blast of air. Although anti-bark collars have been shown to temporarily stop barking behaviour [21, 49], they do not address the underlying cause of the barking or permanently modify the behaviour. In many, the motivation to bark overrides the negative feedback from the collar [48, 49]. Although punishment is delivered, no alternative behaviour is positively reinforced, which does not comply with the basic concepts of behaviour modification [21]. It may also be unclear to the dog just what has triggered the collar, as dogs rarely bark without performing other associated behaviours such as running or tail wagging.

Barking muzzles have also been evaluated. Cronin *et al.* [50] found that while some dogs initially pawed and rubbed at anti-bark muzzles, most became accustomed to wearing them and were not significantly stressed by them. However, these muzzles as with any of quick-fix strategies, do nothing to address and remedy the underlying cause of the unwanted behaviour.

Recommended methods of diagnosis, management and treatment of problem barking

Councils investigating barking complaints use varying criteria to decide if the complaint is valid. One Australian study defines problem or nuisance barking as 'Barking that is of sufficient frequency to cause distress or interruption of the life of the dog's owners or other residents within the neighbourhood' [26]. In New Zealand, there is no stipulated acceptable level of barking (F. Eames, Environmental Health Officer, North Shore City Council, personal communication). Although, in some instances, recordings are made, assessment is purely subjective. Unfortunately a quantified universally acceptable 'normal' level of barking in dogs that are not problem barkers has not been established. Such a parameter would be useful as a benchmark against which to compare problem barkers. Without an accepted normal level to use as a baseline there is pressure for owners of dogs that have been complained about to reduce the amount of barking their dogs are doing regardless of the level or the cause. As a result owners may resort to inappropriate quick-fix strategies such as anti-bark collars [30].

It is accepted among experts that the diagnosis of barking problems requires proper investigation [21, 22, 33]. After a basic health check, a thorough behavioural history should be taken [21]. It is important to quantify how much barking actually is occurring over a 24-hour period and the type and pattern of barking. This information can be gathered from the observations of cooperative neighbours, the use of video cameras, tape recorders and owner diaries [33]. From this information it is possible to distinguish between barking that is anxiety based, territorial barking, barking in response to other environmental stimuli and possibly barking caused by senility [12].

Where barking is not the result of an underlying pathological state, there are several different strategies that may be employed to modify the behaviour without resorting to quick-fix strategies.

Dogs can be taught to respond to a 'Quiet' command and can be redirected onto other behaviours if they are barking at some environmental stimulus [21, 22]. However, often the motivation to comply with an innate drive to protect may override the training. Owners can be taught to manage dogs that bark for attention by reinforcing quiet behaviour and ignoring barking until the behaviour is extinguished by lack of positive reinforcement. This is effective but takes time and patience, so owner compliance is often difficult, especially if the dog is particularly persistent.

Dogs can also be desensitized to certain auditory and visual stimuli that provoke barking [21, 33]. If there is no opportunity for the owners to work through a desensitization programme, simply putting up a screen to decrease access to visual stimuli is often effective [21]. Leaving the dog inside when it is alone is a simple solution where visual and auditory stimuli pose a problem. Leaving soothing music playing may also be useful in decreasing barking in dogs that are confined [29].

It seems reasonable to assume that dogs that have been well socialized and make positive associations with human strangers and other dogs are likely to be less intense in territorial guarding behaviour and so are likely to settle more quickly after barking in response to the presence of a stranger. Puppy socialization classes and frequent exposure to visitors (both canine and human) in the home environment are probably useful in decreasing the potential for 'problem' barking related to territorial defence, although this has not been quantified.

It is noted by people working in canine behaviour that dogs receiving daily exercise and off-property stimulation tend to sleep more and be less reactive when at home, so owners of problem barkers are usually encouraged to run their dogs for at least 30 min, morning and night.

However, these effects have not been quantified in field studies.

Although there is no absolute data to support efficacy in the field, environmental enrichment strategies in the form of toys that release food have been trialled with some success in kennelled beagles [51] and similar strategies are used successfully in clinical behavioural practice to reduce barking. A dog that is occupied chewing a bone, working out how to extract food from treat dispensing toy, digging up toys containing food buried in a sandpit, or tracking down hidden treats is less likely to spend time barking at extraneous environmental stimuli.

Dogs that vocalize because they are anxious about being alone, i.e. suffer separation anxiety, require specific medical treatment and intensive behavioural therapy [21, 33]. Treatment should include anti-anxiety medication, ideally given over a period of several months, while a behavioural modification programme designed to accustom the dog to remaining alone is instituted. The medication may then be gradually withdrawn [21]. The aim in using this medication is to reduce anxiety in order to facilitate effective behavioural modification and to modify the neuroanatomy of susceptible individuals by increasing receptor sites for the relaxing neurotransmitter serotonin [21]. Permanent changes are achievable in some individuals, although some dogs require life-long medication to control the disorder. Adjunctive therapies may be used in dogs where standard treatments are not having the desired effect [52].

Pheromone therapy is a useful addition to the management of canine separation anxiety. Dog appeasement pheromone (DAP) is an analogue of a pheromone produced by bitches from the mammary sulcus when they have puppies and serves as an anxiolytic. It is usually used in the form of a diffuser, although sprays and impregnated collars are available in some countries [31].

Recommendations for dogs barking excessively as a result of senile degeneration include a mixture of medical therapy, nutritional support and behaviour modification. Drugs to increase available dopamine and drugs that assist brain perfusion are often recommended, along with dietary supplementation with antioxidants [36]. *Ginkgo biloba* is useful [53], and recent studies have shown that feeding medium-chain triglycerides derived from coconut provides a source of nutrition for neurons in the form of ketones which counters the decreased ability of the ageing brain to utilize glucose, so helping to prevent or at least delaying the onset of cognitive dysfunction [54]. Environmental modification and exercises to encourage mental activity are also recommended as part of a treatment protocol [36].

Conclusions

Dogs bark for various reasons during their daily life. Although there are circumstances where it may be excessive, barking is a normal and important component of the repertoire of canine behaviours. It is unreasonable to expect that dogs living in suburbia should not bark at all.

Unfortunately there is no documented universally socially acceptable level of barking as there is for other suburban noise and there is a tendency in some cases for complaints to result in council action with little investigation. Establishing some sort of baseline expected or acceptable level of barking that could be used as a benchmark would be extremely useful, even though there would probably need to be some regional variability depending on the level of environmental stimuli.

Since excessive barking may indicate an inappropriate level of care or the presence of pathology, it is important that cases of problem barking should be properly investigated and appropriate treatment or modification strategies implemented.

Further research is required to investigate why people find barking so obnoxious so that we can better understand how to manage this very significant social problem. It may be that better public education about barking will lead to increased tolerance and encourage owners to seek professional advice where appropriate.

By further understanding the factors that influence barking behaviour and identifying risk factors for problem barking, it may be possible to improve management strategies and implement prevention plans.

Acknowledgements

The assistance of Dr K.J. Stafford and Dr P. Perry of Massey University, Palmerston North, New Zealand is acknowledged.

References

1. Yin S. A new perspective on barking in dogs (*Canis familiaris*). Journal of Comparative Psychology 2002;116(2):189–93.
2. Pongracz P, Molnar C, Miklosi A, Csanyi V. Human listeners are able to classify dog (*Canis familiaris*) barks recorded in different situations. Journal of Comparative Psychology 2005;119(2):136–44.
3. Maros K, Pongracz P, Bardos G, Molnar C, Farago T, Miklosi A. Dogs can discriminate barks from different situations. Applied Animal Behaviour Science 2008;114(1–2):159–67.
4. Kim H, Yeon S. Comparison of the individual characteristics of the vocalization and behavior in adult female German Shepherd with territorial aggression. Journal of Veterinary Clinics 2002;19(1):23–7.
5. Fox M. Behaviour of Wolves, Dogs, and Related Canids. Kreiger Publishers, Malabar, FL; 1971.
6. Schassburger RM. Wolf vocalisation. An integrated model of structure, motivation and ontogeny. In: Frank H, editor. Man

and Wolf. Dr W. Junk, Dordrecht, The Netherlands; 1987. p. 313–46.

7. Lord K, Feinstein M, Coppinger R. Barking and mobbing. Behavioural Processes 2009;81(3):358–68.
8. Robbins. Vocal communication in free ranging African wild dogs (*Lycaon pictus*). Behaviour 2000;137:1271–98.
9. Farago T, Pongracz P, Range F, Viranyi Z, Miklosi A. 'The bone is mine': affective and referential aspects of dog growls. Animal Behaviour 2010;79(4):917–25.
10. Chulkina MM, Volodin IA, Volodina EV. Individual, intersexual, and interbreed variability of barks in the dog *Canis familiaris* (Carnivora, Canidae). Zoologichesky Zhurnal 2006;85(4):544–55.
11. Molnar C, Pongracz P, Doka A, Miklosi A. Can humans discriminate between dogs on the base of the acoustic parameters of barks? Behavioural Processes 2006;73(1):76–83.
12. Yin S, McCowan B. Barking in domestic dogs: context specificity and individual identification. Animal Behaviour 2004;68:343–55.
13. Taylor AM, Reby D, McComb K. Size communication in domestic dog, *Canis familiaris*, growls. Animal Behaviour 2010;79(1):205–10.
14. Farago T, Pongracz P, Miklosi A, Huber L, Viranyi Z, Range F. Dogs' expectation about signalers' body size by virtue of their growls. PLoS One 2010;5(12): e15175.
15. Kaminski J, Neumann M, Brauer J, Call J, Tomasello M. Dogs, *Canis familiaris*, communicate with humans to request but not to inform. Animal Behaviour 2011;82(4):651–8.
16. Flom R, Whipple H, Hyde D. Infants' intermodal perception of canine (*Canis familairis*) facial expressions and vocalizations. Developmental Psychology 2009;45(4):1143–51.
17. Taylor AM, Reby D, McComb K. Why do large dogs sound more aggressive to human listeners: acoustic bases of motivational misattributions. Ethology 2010;116(12):1155–62.
18. Pongracz P, Molnar C, Miklosi A. Barking in family dogs: an ethological approach. Veterinary Journal 2010;183(2):141–7.
19. Cohen JA, Fox MW. Vocalizations in wild canids and possible effects of domestication. Behavioural Processes 1976;1(1):77–92.
20. Robbins RLME. African wild dog pup vocalisations with special reference to Morton's model. Behaviour 2003;140:333–51.
21. Overall KL. Clinical Behavioural Medicine for Small Animals. Mosby, St. Louis, MO; 1997.
22. Juarbe-Diaz SV. Assessment and treatment of excessive barking in the domestic dog. Veterinary Clinics of North America, Small Animal Practice 1997;27(3):515–8.
23. Karjalainen L, Ojala M, Vilva V. Environmental effects and genetic parameters for measurements of hunting performance in the Finnish Spitz. Journal of Animal Breeding and Genetics-Zeitschrift Fur Tierzuchtung Und Zuchtungsbiologie 1996;113(6):525–34.
24. Seksel K, Mazurski EJ, Taylor A. Puppy socialisation programs: short and long term behavioural effects. Applied Animal Behaviour Science 1999;62(4):335–49.
25. Gogoleva SS, Volodin IA, Volodina EV, Kharlamova AV, Trut LN. Vocalization toward conspecifics in silver foxes (*Vulpes vulpes*) selected for tame or aggressive behavior toward humans. Behavioural Processes 2010;84(2):547–54.
26. Cross NJ, Rosenthal K, Phillips CJC. Risk factors for nuisance barking in dogs. Australian Veterinary Journal 2009;87(10):402–8.
27. Adams GJ, Johnson KG. Behavioral-responses to barking and other auditory-stimuli during nighttime sleeping and waking in the domestic dog (*Canis familiaris*). Applied Animal Behaviour Science 1994;39(2):151–62.
28. Askew HR. Treatment of Behaviour Problems in Dogs and Cats. Blackwell Science Inc., Cambridge, UK; 2006.
29. Wells DL, Graham L, Hepper PG. The influence of auditory stimulation on the behaviour of dogs housed in a rescue shelter. Animal Welfare 2002;11(4):385–93.
30. Flint EL. The tragedy of a quick-fix approach to canine behaviour problems: a New Zealand perspective. In Mills D, editor. Fifth International Veterinary Behaviour Meeting 2005, Minnesota. Purdue University Press, West Lafayette, IN; 2005. p. 255–6.
31. Gaultier E, Bonnafous L, Bougrat L, Lafont C, Pageat R. Comparison of the efficacy of a synthetic dog-appeasing pheromone with clomipramine for the treatment of separation-related disorders in dogs. Veterinary Record 2005;156(17): 533–8.
32. Hassan AZ, Emmanuel EG, Awasum CA, Remi-Adewunmi BD, Hassan FB, Mohammed A, *et al.* Behavioural conditions in dogs – a review. Part I – Disagreeable normal behaviours. Nigerian Veterinary Journal 2009;30(2):45–50.
33. Palestrini C, Minero M, Cannas S, Rossi E, Frank D. Video analysis of dogs with separation-related behaviors. Applied Animal Behaviour Science 2010;124(1–2):61–7.
34. Lund JD, Jorgensen MC. Behaviour patterns and time course of activity in dogs with separation problems. Applied Animal Behaviour Science 1999;63(3):219–36.
35. Cannas S, Frank D, Minero M, Godbout M, Palestrini C. Puppy behavior when left home alone: changes during the first few months after adoption. Journal of Veterinary Behavior: Clinical Applications and Research 2010;5(2):94–100.
36. Landsberg GM. Therapeutic agents for the treatment of cognitive dysfunction syndrome in dogs. Progress in Neuropsychopharmacology and Biological Psychiatry 2005;29:471–9.
37. Uchida Y, Yamada K, Nakade T, Otomo K. Owner complaints about canine and feline behaviour. Journal of the Japan Veterinary Medical Association 1996;49(5):337–41.
38. Rafiie SM, Sattari B, Naderi MM, Ale-Davoud SJ, Seif AA, Bokaei S. The prevalence of behaviour problems in domestic dogs in Tehran–Iran. Global Veterinaria 2011;6(3):333–8.
39. Fielding WJ. Dogs: A continuing and common neighborhood nuisance of new providence, The Bahamas. Society and Animals 2008;16(1):61–73.
40. Loew FM, Fraser AF. Anti-social behavior of urban dogs. Applied Animal Ethology 1977;3(2):101–4.
41. Omlin S, Bauer GF, Brink M. Effects of noise from non-traffic-related ambient sources on sleep: review of the literature of 1990–2010. Noise and Health 2011;13(53):299–309.
42. Trimmel M, Poelzl G. Impact of background noise on reaction time and brain DC potential changes of VDT-based spatial attention. Ergonomics 2006;49(2):202–8.

43. Kaiser J, Birbaumer N, Lutzenberger W. Magnetic oscillatory responses to lateralization changes of natural and artificial sounds in humans. European Journal of Neuroscience 2002;15(2):345–54.

44. Riede T, Herzel H, Hammerschmidt K, Brunnberg L, Tembrock G. The harmonic-to-noise ratio applied to dog barks. Journal of the Acoustical Society of America 2001;110(4):2191–7.

45. North Shore City Council District Plan. Section 10. p. 10–13, table 10.1

46. Sales G, Hubrecht R, Peyvandi A, Milligan S, Shield B. Noise in dog kennelling: is barking a welfare problem for dogs? Applied Animal Behaviour Science 1997;52(3–4):321–9.

47. Franklin AD, Fearnside SM, Black AP. Canine de-vocalisation by ventral laryngotomy – a retrospective study of 25 cases (2000–2008). Australian Veterinary Practitioner 2011;41(1):18–23.

48. Schalke E, Stichnoth J, Jones-Baade R. Stress symptoms caused by the use of electric training collars on dogs (*Canis familiaris*) in everyday life situations. In: Mills D, editor. Current Issues and Research in Veterinary Behavioural Medicine. Papers presented at the 5th International Veterinary Behaviour Conference, Minneapolis, U.S.A. Purdue University Press; 2005. p. 139–45.

49. Moffat KS, Landsberg GM, Beaudet R. Effectiveness and comparison of citronella and scentless spray bark collars for the control of barking in a veterinary hospital setting. Journal of the American Animal Hospital Association 2003;39(4):343–8.

50. Cronin GM, Hemsworth PH, Barnett JL, Jongman EC, Newman EA, McCauley I. An anti-barking muzzle for dogs and its short-term effects on behaviour and saliva cortisol concentrations. Applied Animal Behaviour Science 2003;83(3):215–26.

51. Schipper LL, Vinke CA, Schilder MBH, Spruijt BM. The effect of feeding enrichment toys on the behaviour of kennelled dogs (*Canis familiaris*). Applied Animal Behaviour Science 2008;114(1–2):182–95.

52. Gruen ME. Use of Trazodone as an adjunctive agent in the treatment of canine anxiety disorders 56 cases (1995–2007). Journal of the American Veterinary Medical Association 2008;233(12):1902–7.

53. Reichling J, Frater-Schroder M, Herzog K, Bucher S, Saller R. Reduction of behavioural disturbances in elderly dogs supplemented with a standardised Gingo leaf extract. Schweizer Archiv Fur Tierheilkunde 2006;148(5):257–63.

54. Pan YL. Enhancing brain functions in senior dogs: a new nutritional approach. Topics in Companion Animal Medicine 2011;26(1):10–6.

High-affinity and -specificity, bacteriophage-based technologies: beyond phage therapy - Mini Review

Paul Hyman and Stephen T. Abedon

Abstract

Taking advantage of bacteriophage versatility and potential to have little negative impact on human health, several medically oriented, phage-based technologies have been developed in addition to phage therapy. Here, we emphasize bacteriophage capacities for natural and enhanced high-specificity binding in phage display, bacterial detection, bacterial identification and nanotechnology.

Introduction

Bacteriophages, after their discovery, were put to use as antibacterial agents. Subsequently, in much of the world, this 'phage therapy' fell from favour because of inconsistent results and the advent of chemical antibiotics, although phage therapy again is being explored in response to various antibiotic shortcomings. Historically, bacteriophages were also put into use to help elucidate basic molecular biological principles. From this work came diverse phage-based technologies that take advantage of the ease of phage genetic manipulation. Phages, for example, may be used to carry non-phage DNA including as vaccines and gene therapy vectors [1], technologies that are currently under development. A more mature application is the use of bacteriophage to create novel peptides and proteins with defined binding affinities (phage display). The natural binding specificity of bacteriophages for their target bacteria in addition has long been harnessed towards bacterial detection and identification. In this brief overview, we describe pharmacological and other technologies based on phage particle binding specificities and affinities for a diversity of molecules, highlighting both existing and developing applications.

Phage Display

Phage display is used to identify peptides with desired binding specificities. Libraries of gene sequences are inserted into specific phage genes that encode proteins found on virion-particle surfaces. The resulting phage-associated peptide library is exposed to organic or inorganic target material, a procedure called biopanning. After selection for substantial levels of binding, the phages carrying high-affinity peptides are sequenced, which provides the amplified binding sequences. Peptides identified by this procedure can be used in association with phage particles, in isolation, or can provide binding specificity to other molecules.

Phage-display peptides have been used to study enzyme–substrate and protein–protein interactions, enzyme inhibition and towards the development of catalytic antibodies [2]. Peptides have also been used in protein epitope mapping and vaccine design. Although typically used to identify short peptides, phage display can be adapted towards the discovery of larger peptides that can be used as binding sequences in single chain or complete monoclonal antibodies. In 2010, over 35 such monoclonal antibodies were used in clinical development, including 25 in clinical trials [3]. Two of these, adalimumab and belimumab, have been approved in several countries to treat autoimmune disorders.

Bacterial Detection and Identification

As phages utilize bacterial surface molecules as receptors, the presence or absence of these molecules can modify phage–bacterial interactions. The result is a potential for phage-based detection of bacteria within samples at low bacterial concentrations as well as identification of bacterial strains and species once those bacteria are grown in pure culture. Technologies have been developed to detect contaminating bacteria in food and water as well as for diagnosis of patients. Methods to reduce prior bacterial culturing time are accomplished by mixing phages with a sample and looking for phage growth, gene expression, or just bacterial binding. For example, the Keypath MRSA/MSSA test kit (Microphage, Longmont, CO, USA) uses phage growth for detection of *Staphylococcus aureus* and phage growth in the presence of an antibiotic to identify MRSA strains, with amplified phage virions detected using antibodies [4]. The presence of even relatively low levels of antibiotic-resistant bacteria in association with patients consequently can be fairly rapidly ascertained.

Phage typing of pure bacterial cultures was an early phage-based approach to bacterial identification, in this case to the strain level. It was first described in 1938 and remains in use to distinguish particularly among strains of *Clostridium difficile*, *Escherichia coli*, *Listeria monocytogenes*,

Mycobacterium spp., *Salmonella* spp. and *Vibrio cholera*, among others.

The FAST-Plaque TB kit series (Biotec Laboratories, Ipswich, UK) identifies *Mycobacterium tuberculosis* strains including in terms of antibiotic resistance, relying on plaque formation as the detection end point. This method is substantially faster than via standard culturing methods, since phage growth is much faster than *M. tuberculosis* replication and plaques can be detected using fast-growing *Mycobacterium smegmatis* as indicator bacteria.

Phage-based Nanotechnologies

Phages are being used in several nanotechnology applications including as biosensors [5]. Phages, for example, can be attached to materials to create mechanical or electrical biosensors for particular pathogens, with the phages supplying the nanoscale features. In this role, the phage acts as an attachment point for the target bacterium and this binding creates a detectable signal by changing the mechanical or electrical properties of the non-phage material. Although none of this work has moved beyond the laboratory, the approach should offer several advantages over culture-based methods. First, the presence of specific bacteria can be detected in real time rather than requiring phage growth or gene expression. Second, because of the specificity of binding by phage proteins, extensive bacterial purification is not needed. Finally, detection can be automated so that specially trained operators or facilities are not required. The major challenge at this time appears to be developing a system with sufficient sensitivity to detect the target bacteria without prior sample culturing.

Filamentous bacteriophages that carry peptides derived by phage display are also being explored for use in developing materials with novel chemical, magnetic or biological characteristics. See [5] for references. Among the biological materials are those for use in tissue engineering. Bacteriophages carrying peptides that mimic binding epitopes similar to those found on extracellular matrix proteins can be patterned onto a surface or scaffold. These should promote cell attachment and guide cell growth. This has been tested *in vitro* with various cell types including neurons and osteocytes (bone-producing cells). The hope is to improve the *in situ* growth of cultured tissues via phage-mediated binding for replacement of damaged tissue in the body.

Conclusions

Bacteriophages have evolved to be highly specific in their interactions with bacteria, using proteins to identify and then infect their hosts. Biologists have learned to use these interactions to detect pathogens including identifying antibiotic-resistant strains. Taking advantage of the relative ease of genetically engineering bacteriophage, researchers have developed other phage-based technologies to create protein drugs and are exploring the use of phages as vaccines and biosensors. These existing and newer technologies ensure that bacteriophages will continue to have applications to human health that go well beyond their evolved biological roles.

Acknowledgement

We are grateful for the feedback provided by two anonymous reviewers.

References

1. Clark JR, March JB. Bacteriophages and biotechnology: vaccines, gene therapy and antibacterials. Trends in Biotechnology 2006;24(5):212–8.
2. Bratkovic T. Progress in phage display: evolution of the technique and its application. Cellular and Molecular Life Sciences 2010;67(5):749–67.
3. Nelson AL, Dhimolea E, Reichert JM. Development trends for human monoclonal antibody therapeutics. Nature Reviews Drug Discovery 2010;9(10):767–74.
4. Monk AB, Rees CD, Barrow P, Hagens S, Harper DR. Bacteriophage applications: where are we now? Letters in Applied Microbiology 2010;51(4):363–9.
5. Hyman P. Bacteriophages and nanostructured materials. Advances in Applied Microbiology 2012;78:55–73.

Issues and possible intervention strategies relating to early weaning of piglets

Massimo Amadori*, Elisabetta Razzuoli and Claudia Nassuato

Address: Laboratory of Cellular Immunology, Istituto Zooprofilattico Sperimentale della Lombardia e dell'Emilia-Romagna, via A. Bianchi 9, 25124 Brescia, Italy

***Correspondence:** Massimo Amadori. Email: massimo.amadori@izsler.it

Received: 29 March 2012
Accepted: 11 July 2012

Abstract

Early weaning of farmed pigs before 25 days of age is a complex multi-factor stressor, often leading to health and welfare issues. The most obvious and crucial feature is that many piglets do not consume food in this period. On this basis, the final stressing impact is derived from a combination of physical, nutritional, immunological and emotional components. On the whole, there is strong evidence that a reduction in the pigs' weaning age can lead to increased adaptation difficulties in terms of behavioural, neuroendocrine and immune responses, as also observed by means of clinical immunology and chemistry analyses. This is the argument underlying the recommended minimum age of 28 days at weaning in EU Directive 2001/93/EC. Inflammatory, stress and innate immune responses sustain the coping efforts of piglets at weaning. Some responses are weaning-specific, whereas others reflect the related social stress of regrouping (mixing) and separation from the sow. An innate immune response to tissue damage at weaning (shorter, atrophic villi and larger, deeper crypts) and relevant control mechanisms are likely to be of paramount importance. In this respect, type I (α/β) interferons (IFNs) were shown to play a major regulatory role of the inflammatory response at weaning. Also, a low-dose, oral IFN-α treatment was shown to sustain the coping efforts of piglets. These results are of some importance in view of a global assessment of food safety 'from farm to fork', and of the related need to reduce the extensive, diarrhoea-related use of drug-supplemented feeds after weaning.

Keywords: pig, weaning, age, animal health, animal welfare, innate immunity, cytokines, interferon-α

Review Methodology: We searched the following databases: Medline, CAB Abstracts, Scopus and ProQuest. In addition, we used the references of our previous papers and from the obtained articles to check for further relevant material.

Weaning as a Risk Factor for Health and Welfare of Farmed Pigs

Weaning is a physiological event in mammals that occurs after dam milk supply is progressively reduced and the cub has become able to find and utilize feed on its own [1]. The suckling phase of feral pigs gradually comes to an end at 3–4 months of age. Within this term, weaning is a stepwise process, whereby piglets get access to solid food during the sow's lactation and gradually decrease and cease suckling as a function of both milk and feed availability, and increasing age [2]. Instead, in the domestic pig kept under intensive farming conditions, the current management practices cause a dramatic shortening of the lactation phase, aimed at a substantial increase of farrowings per sow per year. This tendency was expressed in its utmost features by segregated early weaning of piglets at about 10 days of life, practiced in USA and France [3], and no longer admitted in the European Union. Substantial limits to this practice are laid down in EU Directive 2001/93/EC, which recommends a minimum age of 28 days at weaning.

Despite the development of specific feeds and the use of procedures for reducing the stressing impact on piglets, weaning undoubtedly remains a critical phase in pig husbandry. In particular, early weaning is a major stressor, sometimes associated with enteric disease onset, increased antibiotic usage and related greater risk for the food chain [4].

It should be stressed that intestinal epithelial cells in the small intestine perform a range of functions: digestion, absorption, secretion and acting as a barrier against noxious agents. In pigs, all of these functions are affected by early weaning following the sudden contact of the piglet's gut mucosa (functionally and immunologically immature) with solid food; this is definitely a novel diet, often rejected by pigs around weaning [5]. As a result of both fasting (anorexia) and exposure to a novel diet, the structure of the intestinal wall is strongly altered: villi get shorter whereas crypts considerably deepen in the first 3–4 days after early weaning [6], and the absorption capacity of liquids and electrolytes is strongly reduced [7].

The intestinal environment exists in a state of controlled inflammation because of the continuous exposure to commensal bacteria; this is possible because different populations of gut dendritic cells (CD103+ and CX_3CR1+) are decision-makers, i.e. they can select a T-cell response appropriate to the nature of challenge [8]. Weaning may affect such a state of controlled inflammation, since the abrupt rupture of the milk-suckling phase has a negative impact on both the local immune status and gut microflora [9]; the situation is further compounded by the new social order and microclimate, which may cause exposure to a new bacterial populations and wounds associated with fighting. In this scenario, antibodies to fed proteins can be detected at weaning, indicating an active immune response by the piglets [10]. Also, early weaning causes a delayed reactivity of intraepithelial lymphocytes to T-cell mitogens and a transient depression of such functions in *lamina propria* lymphocytes [11]. Specific deficits in cell-mediated immunity (skin test reaction to phytohaemagglutinin) and in antibody responses to sheep red blood cells were demonstrated in pigs weaned before 35 days of age [3].

On the whole, intestinal damage at weaning can be accounted for by adverse immune responses to dietary antigens, or by primary intestinal inflammation because of the lack of intestinal stimulation during post-weaning anorexia, immune responses then being secondary [4]. This latter hypothesis is corroborated by a substantial return to the previous values of villus height and crypt depth upon resumption of feed intake [12]. Also, the stress of early weaning causes mast cell activation and loss of epithelial barrier function during low feed intake [13].

Regardless of the actual causes, *Escherichia coli* and other gut pathogens can take advantage of the new permissive intestinal environment, outgrow the usual gut microflora and produce toxins, which underlie the onset of post-weaning diarrhoea and related production inefficiencies [14]. Finally, under intensive farming conditions weaning is performed by commingling piglets of different litters, which thus experience novel hierarchy and housing conditions (transfer to the weaning pens) [5]. As a result, the final stressing impact is derived from a combination of physical, nutritional, immunological and emotional components, which may cause potentially dangerous, additive and/or synergistic effects on animal health and welfare [15]. This is the reason underlying the recommended term of 28 days at weaning in EU Directive 2001/93/EC, in agreement with a previous EU scientific opinion [16].

Risks were also reported for the sow's reproductive performance in the case of short lactation periods (< 3 weeks), as shown by reduced litter sizes, increased wean-to-first insemination intervals, reduced conception rates and reduced number of living embryos [3].

The Issue of the Age at Weaning

Owing to the arguments reported in the above section, early weaning of pigs can be defined as a complex, multi-factor stressor, sometimes affecting health and welfare of farmed pigs with complementation or synergism of its components. The problem is further compounded by the genetic selection for an increased number of piglets per litter, leading to a greater competition for the sow's milk [17]. This can give rise to contradictory effects: it can stimulate earlier feed intake by the smaller piglets, whereas the bigger ones (though more mature) get most of the milk and may not be motivated to eat solid food. In this scenario, the emergence of lean, more stress-susceptible pig phenotypes, the very structure of intensive pig farms and fundamental criteria of profitability prevent a return to the natural timescale of weaning observed in feral pigs. Therefore, once fundamental requirements of farm hygiene and welfare of nursing sows have been met, the crucial point for the herd is actually the age at weaning of piglets. This is in fact a trade-off decision, where the opposite requirements of sows and piglets for shortened versus expanded nursing periods, respectively, must be confronted and reconciled on a rational basis. In particular, fundamental ethical concerns and even farm profitability demand that piglets be weaned when adequate homoeostatic responses can be displayed in the framework of good housing conditions and stockman's control. Since piglets are mostly weaned in intensive farms between 21 and 35 days of life, the coping responses should be compared in this time period for a proper evaluation of the age issue. Such an evaluation implies a multi-factor approach that includes the clinical outcome of weaning, as well as the parameters of clinical immunology and chemistry defining the extent of the coping efforts displayed at weaning. In this respect, the data obtained in experimental studies in isolation facilities need to be confirmed by proper field trials. For an evaluation of the age issue at weaning, the reader is referred to the recent comprehensive review by Hameister *et al.* [3], which also illustrates the main behavioural disturbances of piglets following early weaning (increased vocalization rate, aggressive and exploratory behaviour, belly nosing), as well as the main effects on the neuroendocrine system (Adrenocorticotropic hormone, glucocorticoids, catecholamines and glucocorticoid receptors); the above findings characterize the piglets' efforts to cope with the new conditions at weaning. On the whole, there is strong evidence that a reduction of the pigs' weaning age can lead to increased adaptation difficulties in terms of behavioural, neuroendocrine and immune responses.

Post-Weaning Pathologies

The concentration of maternally derived antibody continuously drops in the first few weeks and the piglet's own production of antibody is not fully developed until the fourth week of life, which outlines an accentuated risk of disease occurrence in this time period [3]. In this respect, gastrointestinal disturbances during the post-weaning period account for large economic losses in pig farms, amounting in the European Union to about 17% of the total losses during the pig production cycle [4]. The common agents of infectious diarrhoea in nursing and nursery pigs are *E. coli*, transmissible gastroenteritis coronavirus (TGEV), rotavirus, *Isospora suis* and *Clostridium perfringens* [18]. Most of them infect the small intestine, inducing various degrees of villous atrophy and malabsorptive, maldigestive diarrhoea. With respect to the layout of the intestine, these agents show preferential colonization of different vertical (villus/crypt) and horizontal (duodenum, jejunum, ileum and colon) sites, and diarrhoea intensity is often related to the extent of pathogen replication and/or toxin production at these sites. Also, concurrent infections with multiple enteric agents can produce synergistic or additive effects leading to more extensive villous atrophy throughout the intestine and more severe and prolonged diarrhoea. Diarrhoea may show up with high morbidity and mortality, and microbial agents cannot be differentiated clinically. Diagnosis depends on examination of newly affected, untreated pigs presented for post-mortem examination, and on tests designed specifically to identify each pathogen [13].

Post-weaning diarrhoea can occur in the framework of post-weaning multisystemic wasting syndrome (PMWS). This has caused important economic losses in the last 20 years. A peculiar aspect of this pathology is lymphocyte depletion together with histiocytic and/or multinucleate giant cell infiltration, associated with alterations of the immune responses. This predisposes to secondary infections and related losses. With the commercialization of successful vaccines, PMWS and other porcine circovirus 2 (PCV2)-associated syndromes have been substantially silenced [19].

Apart from post-weaning diarrhoea, oedema disease (OD) and endotoxin shock (ES) may also occur after weaning. OD mostly occurs in the second and third week after weaning with typical oedema of the submucosa of the cardial gland region of the stomach; different O serogroups of *E. coli* are involved in post-weaning diarrhoea, OD, or both [14]. OD is an enterotoxemia, caused by strains of *E. coli* that colonize the small intestine via F18ab fimbriae and produce Shiga toxin Stx2e [20]. ES is a shock-like syndrome because of the systemic effect of lipopolysaccharides (LPS), which usually occurs 1 week after weaning.

The Homoeostatic Response of Farmed Pigs to the Weaning Stress

In addition to the above European study [16], further scientific contributions highlighted the pig's homoeostatic response to early weaning stress. In particular, the main metabolic responses of pigs at weaning were characterized as diet-dependent and diet-independent; interestingly, major metabolic and immunological responses, i.e. decreased plasma glucose, increased plasma glucagon, fibrinogen and interleukin (IL)-1, were shown to be diet-independent [21]. Also, a low-titred interferon (IFN)-α response was shown in serum samples of piglets some days after weaning, as also observed after transportation to the fattening farms [22, 23]. The same response was also observed in calves after a long-distance journey in trucks (Massimo Amadori, personal communication 2009). Most important, early weaning of pigs causes a dramatic up-regulation of inflammatory cytokine genes in both proximal and distant tracts of the small intestine [24].

On the whole, an orchestrated complex of hormonal and biochemical adjustments, inflammatory and innate immune responses (cytokines) is generated to cope with a combined metabolic, environmental and emotional challenge at weaning, and to prevent weaning-associated health disorders. These responses are also reflected by specific differences of the blood serum proteome of piglets after weaning, which can be correlated with distinct deficits in innate immunity [25].

Innate Immune Responses and Inflammation as General Adaptation Strategy

Beyond the early weaning model, a large body of evidence exists about the strict correlation between innate immune response, inflammation and stress. In particular, the inflammatory, immune and stress responses should be thought of as a unitary, ancestral, overlapping set of responses aimed at the neutralization of stimuli perturbing body homoeostasis [26]. In particular, the inflammatory response is currently viewed as a fundamental adaptation strategy to various infectious and non-infectious stressors; the inflammatory response is started by the host to achieve better fitness in managing adverse environmental conditions (infectious or non-infectious) and then curtailed to avoid major tissue damage and waste of food energy [27]. As a result, adaptation implies a fine tuning of pro- and anti-inflammatory responses aimed at checking outright and potential stressors, thus optimizing the interaction with the environment and the exploitation of the available resources [28]. The inflammatory responses to environmental, non-infectious stressors and microbial agents, respectively, do not substantially differ in terms of basal mediators (cytokines, chemokines, reactive oxygen species, etc.); they differ instead in terms of extent, duration and combinations of inflammatory mediators [27]. The above host's responses are relevant to the concept of 'sterile inflammation', which has been recently introduced to depict a variety of responses ranging from subclinical to even lethal conditions, such as the systemic inflammatory response syndrome (SIRS) [29]. The term 'sterile inflammation' defines a response to cellular stress and tissue damage induced by causes as diverse as traumas, toxic compounds, endocrine destroyers and metabolic stress. 'Sterile inflammation' aims at removing dangerous compounds and metabolites and restoring the previous homoeostatic conditions. In this conceptual framework, 'sterile inflammation' is fully relevant to the physio-pathological conditions of early weaning in farmed pigs, since they usually proceed in the absence of detectable microbial infections.

The Foundation of the Response to Non-infectious Stressors

The rationale behind the innate immune response to a non-infectious stressor such as early weaning has been elusive for a long time. More recently, the 'danger theory' [30] has become the foundation of the present knowledge about the immune recognition of tissue damage and cellular stress, and about the involvement of the innate immune system [31]. Owing to the above condition, it is worth mentioning the molecular and biological basis for such an immune recognition as a general conceptual framework, in which the weaning issue could be adequately set. Animal cells exposed to both infectious and non-infectious stressors can be recognized following the release of metabolites as diverse as ATP, $K+$ and oxidative stress products; these can be perceived by specialized recognition and processing structures named inflammosomes, with the eventual production of caspase-1 and IL-1β [32, 33]. In addition, stressed cells can be recognized by an outright lymphoid cell surveillance system that detects changes of the usual surface phenotype; in particular, a network of lymphocyte populations (mainly $\gamma\delta$ T-cells) recognizes neo-antigens such as MHC Class I chain-related (MIC) on stressed cells [34], i.e. cells exposed to events as diverse as heat shock, infections, DNA damage, etc. MIC proteins are ligands for the activating Natural Killer (NK) cell receptor NKG2D, expressed on NK cells, CD8+$\alpha\beta$ T-cells and $\gamma\delta$ T-cells; stimulation of these cells with recombinant MIC causes an IFN-γ response [35]; the latter can sustain further inflammatory cascades by activating macrophages to release both reactive oxygen metabolites (ROMs) and reactive nitrogen intermediates (RNIs) [36].

The above mechanisms of recognition and response are likely to be implemented at weaning following a major cellular stress exerted on pig enterocytes. The metabolites released and the cytokine responses at weaning indirectly confirm the above tenets (see the following sections).

Intervention Strategies

Poor feed intake and possible enteric disease onset after weaning demand robust and effective farm management strategies within this crucial phase. In this respect, many options have been proposed and tested with varying degrees of success. Among alternative farming schemes, an improvement of the health conditions at weaning can be achieved by means of the Multi-Site-Production technology, whereby nursery piglets are isolated and kept separated from the fattening pens; this causes a reduction of the environmental infectious pressure and related infections; under these strictly controlled conditions, the age at weaning is of less importance, although health and production parameters can still improve on passing from 12 to 21 days of age at weaning; alternative rearing systems (multi-suckling and communal piglet area schemes), or a simple farrow-to-finish scheme with no commingling of litters may also play a positive role; improvements of the sows' conditions in loose-farrowing and controlled housing schemes are also likely to improve feed intake and weight gain of piglets after weaning [3]. In particular, enrichment of the pre-weaning environment is likely to stimulate the development of feeding behaviours and, therefore, the feed intake after weaning [37].

Porcine and bovine spray-dried plasma (SDP) can be used in the diets of early weaned pigs to improve growth rates and feed intake; also, SDP can promote feed conversion efficiency and nitrogen retention, i.e. a greater efficiency of dietary protein utilization, partly mediated by decreased amino acid catabolism [38]. The response to plasma does not diminish with increase in weaning age and it is not strongly affected by the protein source the plasma replaces; hormonal hunger signals in SDP because of the fasting of pigs/cattle prior to slaughter may account for the increased feed intake [39]. SDP can also reduce pro-inflammatory cytokine expression in the gut of early weaned piglets challenged with enterotoxigenic *E. coli* K88 [40], i.e. it can check an excessive amplification of the inflammatory response to bacterial LPS.

Immune egg and dairy products, organic acids, lactose, mannan oligosaccharides, alternative cereal grains, nucleotides, nucleosides, direct-fed microbials (probiotics), prebiotics, herbal extracts, essential oils and fermented liquid feeding can notably improve the diets of weanling pigs and lead to greater performance and disease resistance [4, 39, 41].

Zinc oxide (ZnO) in weaning diets (usually from 1500 to 3000 ppm) can increase growth rates and protect against enteric diseases [39]. Zinc supplementation can reduce the expression of immune response genes concerned with inflammation in piglets orally challenged with *E. coli* ETEC K88 [42], which is also reminiscent of the aforementioned SDP control action.

The possible role of immunoprophylactic treatments should be examined as well, as a strategy for preventing excessive usage of antibiotics in intensive pig farms and meeting fundamental food safety concerns [43].

Specificity of the Homoeostatic Responses to Early Weaning

The previous section has illustrated important findings related to the physiopathology of early weaning in piglets as well as to various intervention protocols on farm. We felt, however, that the global picture was still obscure with regard to one main point, at least: is it possible to dissect the homoeostatic control actions of piglets into those related to anorexia/diet change and social stress, respectively? In our opinion, this could be of some importance for obtaining reliable indicators of the actual coping strategy of piglets with the dietary component of the weaning stress. Therefore, at the beginning of our studies, we set out to compare the responses of piglets weaned at 21 and 35 days of age under the conditions of little, if any, environmental infectious pressure. For this purpose, we employed 14 specific-pathogen-free (SPF) piglets born on the same day in our Experiment Station at IZSLER, Brescia; they belonged to two litters of as many sows of the same (Landrace×Large White) genetic profile. On day 20 after birth, blood samples were collected from all suckling piglets in tubes with and without anticoagulant (lithium heparin). On day 21, three and four piglets of litters 1 and 2 were weaned, and on day 24 blood samples were collected once again from all piglets (weaned and not weaned). The seven suckling piglets left were weaned at 35 days of age, and blood was collected last from all animals on day 39. All the piglets were housed together post-weaning, under SPF conditions.

The results obtained in this trial need to be dissected on the basis of the poor health conditions of some piglets (wasting and scour). These were observed in some suckling piglets on day 20 and in the whole group on day 39. The disease status of piglets went along with a tumour necrosis factor α (TNF-α) response. Therefore, the only meaningful comparisons refer to healthy piglets on day 20 and 24, respectively.

There was no significant increase in cortisol around weaning, whereas zinc significantly decreased in weaned animals (*P* 0.01) as opposed to controls. In both control and weaned piglets, haptoglobin was increased (*P*=0.02 in both control and weaned animals, respectively); separation of littermates and establishment of a new group hierarchy probably underlies such a result. As expected, weaning did not decrease the values (^{3}H-thymidine incorporation) of Concanavalin A-driven blast transformation of lymphocytes, which is likely to be affected in our experience by chronic, rather than acute stressing events. Haematological data also confirmed the complexity of the host's homoeostatic response. Whereas total leukocyte counts did not substantially differ between the two groups, neutrophil percentages significantly increased in weaned animals, only (*P* 0.01), that also showed a very similar time-course of neutrophil counts (*P*=0.05). Further experiments are needed to clarify the origin of the major changes in the granulocyte counts; both a massive influx of young, immature forms from the bone marrow and a major shift of the adherent granulocyte pool from endothelia to circulating blood may account for the above results. The shift might be linked to a transient cortisol response at weaning, no more detectable on day +3, as usually expected [44].

Sera of healthy piglets collected on days 20 and 24 (−1 and +3 days with respect to weaning) were submitted to two-dimensional (2D) electrophoresis and proteome examination as previously described [45]. This study revealed significant differences in the expression of some serum proteins before and after weaning (Figure 1), also in terms of qualitative criteria (post-translational changes, data not shown). The increased expression of the haptoglobin dimer was in line with our clinical immunology data. Intriguingly, the up-regulation of both α1-acid glycoprotein and fetuin is reminiscent of the serum profile of piglets prenatally, since both components are highly expressed in foetal life and decay thereafter [46].

On the whole, these results indicate that some homoeostatic adjustments of piglets are peculiar to the cessation of suckling, whereas others are the outcome of the combined effects of weaning and related social stress. Weaned and unweaned piglets remained in two separate cages of the same room, attended by the same personnel, so that a major influence of the new environment on the different responses can be substantially dismissed. Plasma zinc levels and granulocyte counts were reliable indicators of the weaning-specific inflammatory response of piglets to the sudden diet change. The observed changes in the neutrophil/lymphocyte ratio after early weaning are in line with previous studies [3].

The above results imply that an inflammatory response takes place at weaning, which may underlie the clinical repercussions sometimes observed in this phase. This tenet has been substantially confirmed *ex iuvantibus* by a few immunoprophylaxis studies: the intramuscular injection at 100–200 mg/kg body weight of the IL-1 receptor antagonist (IL-1 RA) can decrease the disease losses of weaned piglets, and it is the basis of a patented procedure

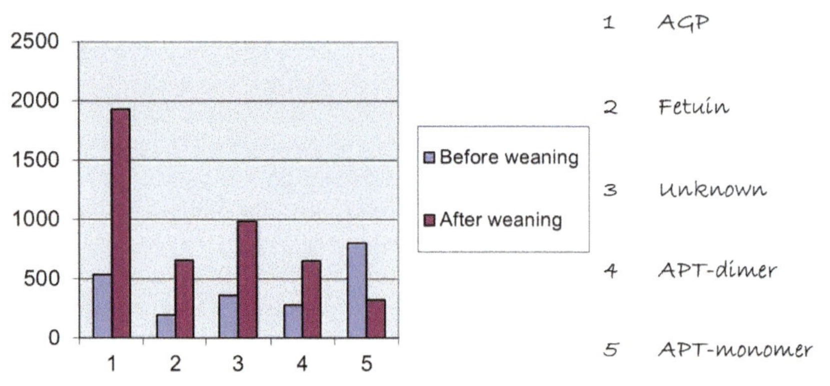

Figure 1 Expression of some serum proteins before and after weaning. Serum samples were collected from five healthy piglets at 20 and 24 days of age (−1 and +3 with respect to weaning) and frozen at −20 °C. After thawing and centrifugation (10′, 4 °C, 14000 rpm) the supernatant was mixed with one volume of 8 M urea, 4% CHAPS, 1% DTT, 15 mM Tris, 2% ampholines, pH 3.5–10 (Amersham Biosciences) and mixed for 2 h. A 300-μl volume of the diluted serum was used to rehydrate 18-cm, 3–10 NL strips (Amersham Biosciences). After 9 h, isoelectrofocusing was carried out with IPGPhor II (Amersham Biosciences) up to a total of 150 kV/h. After equilibration, strips were transferred onto a SDS–PAGE, 9–16% gradient gel for second dimension separation by a Protean II xi 2-D Cell (BIO-RAD). Gels were stained with silver nitrate and colloidal Coomassie G-250. After acquiring digitalized images, image analysis was carried out using Image Master 2D Platinum software (Amersham Biosciences). AGP: α-1-acid glycoprotein; APT: haptoglobin.

[47]. In this respect, we reasoned that the cytokine control circuits could be targeted to prevent excessive amplification of the inflammatory response at weaning and related disease losses. On the basis of previous convincing evidence [48], we considered the use of oral, low-dose IFN-α treatments, which is not stressing for the animal; in particular, it does not show the well-known detrimental side effects of high-dose parenteral injections, also observed in farm animals [49].

Oral low-dose IFN-α treatments are effective in models of both viral infections and immuno-pathological conditions [50]. Even in the model of feline immunodeficiency virus (FIV) infection of cats, dramatic improvements of the health conditions are observed without any decrease in virus concentrations of plasma and leukocytes [51]. Therefore, IFN-α is likely to promote a substantial balance between 'danger' and response [27], and to restore compromised homoeostatic response mechanisms following exposure to both infectious and non-infectious stressors. In a global view, such a possible mechanism would show an obvious synergism with a 'hinge' role of IFN-α (and type I IFNs on the whole) between innate and adaptive immunity, widely recognized by the scientific community [52].

Modulation of the Early Weaning Stress by Oral, Low-Dose IFN-α

In order to verify the above hypotheses, a pilot study was carried out on eight SPF piglets of the same litter [53]. Four piglets were given human lymphoblastoid IFN-α for 10 days in a row in the form of a freeze-dried preparation, mixed in the diet, starting on the day of early weaning at 21 days of age. The amount of IFN was adjusted to provide 10 international units (IU)/kg b.w. on the basis of the expected growth of the piglets. The other four littermate piglets served as untreated controls; they received the same daily mass of placebo (reagent grade lactose) in a separate weaning cage. Treated and control groups were matched in terms of weight at weaning. With respect to weaning day (beginning of oral IFN-α treatment), blood samples in tubes containing heparin or no anticoagulant, respectively, were collected on day −1, +3 and + 21 to determine cytokines and clinical chemistry parameters in serum, as well as cytokine gene expressions in peripheral blood mononuclear cells (PBMC) by reverse transcriptase (RT) real-time PCR. Also, piglets were individually weighed on days −1, +21 and +30 to reckon the corresponding mean daily weight gains (DWG). Expression data, normalized on a housekeeping gene, were scored as Delta cycle threshold (ΔCt) and analysed using a general linear model (GLM) to assess the significance of time and treatment factors. Gene expression data scored as *n*-fold ($\log_2 2^{-\Delta\Delta Ct}$) with respect to the values observed on the day before weaning were submitted to one-way ANOVA [53]. The effects of time and IFN-α treatment on the weight gain of piglets were evaluated by two-way ANOVA (Prism 2.01, GraphPad Software, San Diego, CA).

Even though the sample size was very small and a reasonable caveat can be duly expressed about the actual weight of our findings, the latter were interesting and worth analysing in detail. IFN-α-treated piglets showed, in fact, a greater mean weight gain between days −1 and +21 with respect to weaning day (236 ± 71 versus 164 ± 25 g/day) and also between days +21 and +30 (528 ± 80 versus 383 ± 106 g/day); despite the reduced number of piglets in this study, the effect of the treatment between days −1 and +30 was almost significant ($P < 0.06$). Since piglets actually grew less than

Table 1 Field trial on piglets weaned at different ages

Event	Litters 1 and 2 (group 2, early weaning)	Litters 3 and 4 (group 3, early weaning + IFN-α)	Litters 5 and 6 (group 1, late weaning)
Day 0: farrowing	x	X	x
Day 15: clinical examination	x	X	x
Days 19–21: blood sampling, weight control and ear-tagging	x	X	x
Day 22: weaning	x	X	
Day 22: start of oral, low-dose IFN-α treatment		X	
Day 28: blood sampling and weight control	x	X	
Day 28: weaning and blood sampling			x
Day 31: end of oral, low-dose IFN-α treatment		X	
Day 34: blood sampling and weight control	x	X	x
Day 47: blood sampling and weight control	x	X	x
Days 54–56: final blood samplings	x	X	x

expected in the period under study, the daily dose of oral IFN-α amounted to some 20 IU/kg b.w.

Only one control piglet was IFN-α-positive on the day before weaning, whereas two other control piglets were IFN-α-positive on day 3 after weaning as opposed to none in the treated group. Also, only one control piglet was weakly TNF-α positive on day 3 after weaning. As shown in the previous experiment, plasma zinc levels significantly decreased in both groups between days −1 and +3 (P=0.001 and 0.0048 in control and treated group, respectively). Also, in both groups, urea and bilirubin levels were significantly lower on day +3 compared with day −1. Therefore, the greater weight gain of IFN-α-treated animals cannot be associated with any of the above cytokine and clinical chemistry parameters. Haematological and flow cytometry data revealed instead notable differences between treated and control animals. In control animals, there was in fact a notable decrease in the leukocyte counts between day −1 and +3 (P 0.09, tendency) as opposed to all the IFN-treated ones, which showed a notable stability of such counts up to day +21. Furthermore, flow cytometry experiments on PBMC provided two interest findings:

- Only treated animals showed on day +21 a stable prevalence of both CD4+ T-cells and CD4+/CD8+, double-positive memory T-cells, which points at a better homoeostatic control of the T-cell compartment.
- Only control animals showed a sharp increase in CD14 (LPS receptor)-positive cells on day +3, as opposed to treated subjects. This probably underlies a major influx of immature monocytes from the bone marrow following cytokine/chemokine responses at weaning, probably suppressed by the IFN-α treatment.

RT real-time PCR experiments also helped define the effectiveness of the IFN-α treatment [53]. The expression of IFN-γ was highest on day +3 (P < 0.05) and significantly reduced by the IFN-α treatment on both day +3 and +21 (P < 0.05). The expression of the IL-1β and IL-6 genes was up-regulated on day +3 (both) and +21 (IL-1β only). The expression of the TNF-α gene differed along the time, with lowest and highest expression on days +3 and +21, respectively. These data were substantially in agreement with the *n*-fold data, which confirmed the lower expression of IFN-γ on day +21 (P < 0.08) and the higher expression of IL-1β on day +3 (P < 0.05) in piglets orally treated with IFN-α.

Interestingly, IFN-α was detected by flow cytometry in PBMC of control animals collected on day +3 and not on day −1, in agreement with the tests on serum samples and a general homoeostatic role of this cytokine in the post-weaning phase. On the whole, a number of clinical (growth), immunological and clinical chemistry parameters enabled us to depict a global, favourable impact of a low-dose IFN-α treatment in the context of the post-weaning conditions of farmed pigs.

Field Validation of our Experimental Model

The above findings were observed within an experimental setting, which is not likely to reproduce the huge variety of environmental conditions usually experienced by farmed pigs around weaning. This is the reason why we decided to verify the main tenets of our interpretation scheme in a field scenario.

For this purpose, six litters of Large White × Landrace (sow)/commercial Hypor hybrid (boar) were selected in a farm showing excellent figures of animal health, welfare and post-weaning growth. Animals were allocated to the experimental groups shown in Table 1, weaned at the indicated ages (22 and 28 days of age) and given the same starter feed with antibiotic (amoxicillin 400 ppm + colistin 120 ppm) and ZnO (2000 ppm) supplements over the first 10 days. Litters 3 and 4 were treated with human lymphoblastoid IFN-α (TF-21, Guna, Milan) in drinking water at a dose of 1 IU/kg b.w. over 10 days on the basis of the expected growth rate and related water consumption. Litters 5 and 6 were weaned at 28 days of age in accordance with the official recommendations of Italian Legislative Decree 20/2/2004, n. 53.

Table 2 Weight gain of piglets after weaning at 22 and 28 days of age

Days around weaning	−9	−1	+6	+12	+19	+26	+26
Days of age	**19**	**21**	**28/34**	**34**	**47**	**54**	**48**
Weaned at 28 days	5.974 ± 0.716	ND	10.16 ± 0.898	ND	15.24 ± 1.759	18.68 ± 2.405	
Weaned at 22 days	ND	6.028 ± 0.931	7.667 ± 1.383	8.667 ± 1.562	ND	ND	13.72 ± 2.396
Weaned at 22 days, IFN-α treatment	ND	7.075 ± 1.15	8.800 ± 1.250	9.525 ± 1.23	ND	ND	15.200 ± 2.353

Results are shown in terms of mean weight (kg) ± 1 standard deviation.
ND, not done.

The six litters of piglets showed good clinical conditions (absence of disease symptoms and usual thriftiness) all over the trial, as did the other litters inside the pen. In particular, no clinical difference was observed between subjects weaned at 22 and 28 days, respectively. On the contrary, in terms of DWG, greater growth rates ($P < 0.001$) were shown by IFN-α-treated piglets (litters 3 and 4) compared with the control ones (litters 1 and 2). In practice, litters 3 and 4 showed the same final weight at 47–48 days of age as litters 5 and 6, despite the different age at weaning (Table 2); the one-day difference between the weight checks (47 days of age for litters 5 and 6, as opposed to 48 days for litters 3 and 4; see Table 2) might imply though a residual growth advantage of the late weaning group (litters 5 and 6), in line with an increased DWG in piglets weaned at 28 days of age compared with those weaned at 21 days [54]. The better growth performance of IFN-α-treated piglets could be apparently be the result of the slightly greater mean weight at weaning, since this is reputedly predictive of the following growth rate [55]. However, the mean weights at weaning of all litters were by far beyond 10-lbs, a limit for which a delay of the final market weight was observed [56]. The absence of a significant effect beyond the above limit is also indicated by very similar weights at 47 days of age of litters 5 and 6, weaned later (Table 2). On closer examination, this hypothesis can be also dismissed on the basis of the actual time-course of the weight gains (Figure 2). DWG data were similar in IFN-treated and control animals between days −1 and +6 (234 and 246 g in control and treated animals, respectively), and DWG was even lesser in the treated ones between days +6 and +12 (121 versus 166 g). The overt advantage in treated piglets was observed instead between days +12 and +26 (405 versus 361 g). This latter result is in agreement with the much higher concentrations of TNF-α on day +26 in control piglets weaned at 22 days of age, also showing a much greater prevalence of subjects with a high serum TNF-α concentration (≥300 pg/ml) (Chi-square, $P < 0.05$) (Table 3). The IL-6 titres showed a negative correlation with the TNF-α titres, which underlies a substantial nadir of IL-6 on day +26; the highest titres on the day of weaning were shown instead in groups 5 and 6 (Table 3). The reported values of both TNF-α and IL-6 were determined by bioassays [57, 58]; the absolute values are not usually comparable to those obtained by ELISA or other immuno-enzyme techniques, even though the relative differences would persist.

In line with the previous experiments and our experimental hypothesis, an endogenous low-titred IFN-α response to the stress of weaning was observed by a bioassay in all the litters, despite the very good health and welfare conditions of pigs and the absence of detectable virus infections [59]. The frequency of IFN-α responses in serum was significantly higher (Chi-square, $P < 0.01$) in pigs weaned at 22 days compared with those weaned at 28 days. IFN-α was unambiguously identified by means

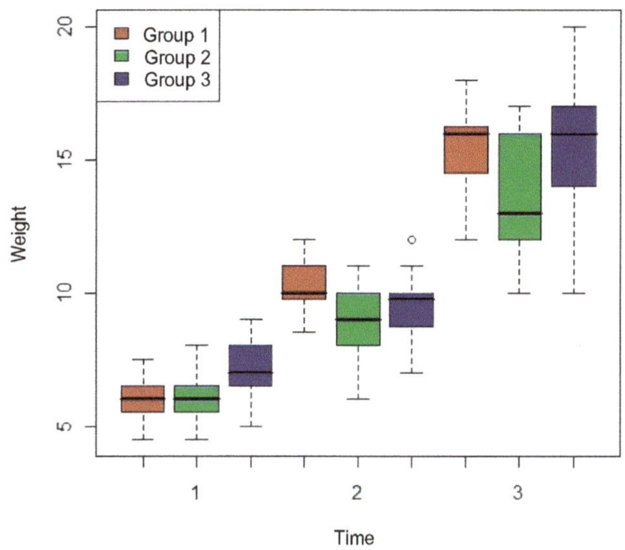

Figure 2 Weight gain of piglets at different times around weaning. Piglets were weaned at 28 days (group 1, litters 5 and 6), or 22 days of age (groups 2 and 3, litters 1, 2, 3 and 4). Group 3 animals received an oral, low-dose IFN-α treatment. Three times/animal (21, 34 and 48 days of life) were compared for weight gain evaluation. Results are shown in terms of kg (median) and interquartile range (box-and-whisker plot).

of a neutralization assay with a monoclonal antibody to porcine IFN-α [59]. An increased prevalence of IFN-α responses was also observed on day +26 in litters 3, 4, 5 and 6, in possible association with the peak of the TNF-α response.

Interestingly, by using a panel of primers for IFN-β and the different IFN-α genes in RT real-time PCR, constitutive expression of IFN-β and some IFN-α genes in PBMC was observed in the piglets under study before weaning. Expression was shown to be shut-off on day 6 after weaning when IFN-α/β was detected in serum. Constitutive gene expression was shown to resume 6 days later, on day 12 after weaning [59, 60].

The different groups of piglets also showed interesting peculiarities with regard to some clinical immunology parameters measured on days −1 and +6 with respect to weaning [61]. The significance of the observed differences was evaluated by one-way ANOVA for repeated measures. Haptoglobin levels significantly increased in all piglets ($P<0.01$ in all groups). Cortisol levels were significantly increased in litters 5 and 6, only (late weaning, $P<0.001$). Total protein levels were stable in litters 1 and 2, as opposed to litters 5 and 6 (late weaning, +3.78 g/l, $P<0.001$) and litters 3 and 4 (IFN treatment, −3.65 g/L, $P<0.001$). Serum albumin levels showed minor changes in litters 1, 2, 5 and 6, whereas a significant decrease ($P<0.001$) was observed in IFN-α-treated litters 3 and 4. A significant increase in α-globulin levels ($P<0.001$) was seen in control litters, as opposed to the IFN-treated ones. β-globulin levels significantly decreased in IFN-α-treated piglets, only ($P<0.001$), whereas levels were substantially stable in control animals. γ-globulin levels were shown to decrease in all groups, with significant differences observed in litters 1, 2, 3 and 4 ($P<0.001$).

On the basis of the above results, our data were further analysed by a linear mixed effects model to evaluate: (i) the effect of treatment with IFN-α and (ii) the effect of different times of withdrawal on biochemical parameters and weight. As for biochemical parameters, two times of withdrawals, weaning time and 6 days later, were considered, while three times/animal (21, 34 and 48 days of life) were compared concerning weight evaluation. Treatment and time were included as fixed effects, while pig was included as a random effect to take into account the possible correlation between repeated measures within pig. The interaction term between treatment and time was added to the model in order to evaluate if treatment effect changed at different times of withdrawal. *P*-values equal to or lower than 0.05 were considered statistically significant. *P*-values were adjusted by *post hoc* Tukey procedure for multiple comparisons. Statistical analyses were performed using R software (release 2.8.1). The group effect on weight gain was shown to change significantly over time: pairwise comparisons between different groups and times were all significant except for those between groups at the same time; as for group 2 at time 3 (48 days of life), this was because of the high variability within this group (see Figure 2). Likewise, the group effect (post-weaning increase) on haptoglobin values was significantly different at the two times; haptoglobin levels significantly increased in all groups between time 1 and 2, whereas no significant difference was observed between groups at time 1. There were on average significantly higher serum albumin levels after weaning in groups 1 and 2 compared with 3. Concerning β-globulin values, a significant decrease was shown in group 3 after weaning as opposed to both groups 1 and 2. A weak positive association between albumin and cortisol values, and a much stronger positive one between β-globulin and cortisol ($P<0.01$) were demonstrated as well.

On the whole, the major findings of the statistical analyses can be summarized as follows:

- Weight gains did not differ between pigs weaned at 4 weeks and pigs weaned at 3 weeks, orally treated with IFN-α (P=1.0 in the pairwise comparison at time 3).
- The variability of weight gains was lower in IFN-α-treated subjects compared with the control ones (see Figure 2).
- A peculiar decrease in both albumin and β-globulin concentrations occurred in IFN-α-treated piglets, only.

Major Findings of the Field Trial

The immunological findings of the above field trial confirmed the outcome of other clinical and zootechnical

Table 3 Inflammatory cytokine responses around weaning

Days around weaning	−9	−1	0	+6	+12	+19	+26	+34
(A) Serum TNFα								
Weaned 28 days	76 (0/19)	ND	178 (2/19)	36 (1/19)	ND	161 (3/19)	736 (13/19)	ND
Weaned 22 days	ND	450 (6/20)	ND	108 (2/20)	155 (3/20)	ND	12.050 (20/20)	160 (2/20)
Weaned 22 days IFN-α-treated	ND	364 (7/20)	ND	65 (2/20)	403 (6/20)	ND	2.613 (13/20)	299 (6/20)
(B) Serum IL-6								
Weaned 28 days	4061 (19/19)	ND	34.620 (19/19)	1200 (6/19)	ND	1200 (13/19)	2.5 (3/19)	ND
Weaned 22 days	ND	450 (13/20)	ND	300 (6/20)	1200 (9/20)	ND	2.5 (1/20)	2430 (11/20)
Weaned 22 days IFN-α-treated	ND	2317 (20/20)	ND	100 (0/20)	1200 (12/20)	ND	2.5 (3/20)	2005 (9/20)

Serum TNFα concentrations (pg/ml) are shown as 75th percentile values. Also, the number of high-responder pigs (≥300 pg/ml) out the total is reported between brackets.
Serum IL-6 concentrations (pg/ml) are shown as 75th percentile values. Also, the number of high-responder pigs (≥100 pg/ml) out the total is reported between brackets.
ND, not done.

studies: weaning is a major stressing event for pigs in intensive farms. This conclusion is based on the acute phase response in terms of serum haptoglobin levels in the three groups under study, which indicates a pronounced inflammatory response in piglets. In this scenario, the adaptation strategies and the relevant coping efforts of the three groups were shown to differ. Only pigs of the late weaning group mounted a robust cortisol response, as the likely result of a more vigorous adrenal activity compared with pigs weaned at 22 days of age; in this respect, our field trial confirmed the findings obtained on pigs weaned at 21 days in our isolation facilities (see above). This finding was rather unexpected since the cortisol response should be only high on the first day or so on weaning [44]. The lower levels of serum albumin and α-globulins in sera of IFN-α-treated piglets can be accounted for by a peculiar negative acute-phase response and stability of concentrations, respectively [61]. The specificity of the negative acute-phase response (decreased serum concentrations of proteins largely produced by liver) is also suggested by the increasing concentrations of serum albumin in pigs until 6 months of age [46]. The decreased β-globulin levels may be also part of the negative acute-phase response of IFN-treated piglets, since transferrin (a major β-globulin) acts as a negative acute-phase reactant [46, 62]. As for the changes in γ-globulin levels in the three groups under study, it should be stressed that γ-globulins usually decrease between days 14 and 28, and level off until 56 days of age following the decay of colostrum-derived antibody [46]. On the whole, an oral, low-dose IFN-α treatment could strengthen the homoeostatic response capacity of piglets, thus easing adaptation to this difficult phase. The better adaptive response of IFN-α-treated pigs was confirmed by the greater DWG, very similar to that of piglets weaned later (see Table 2 and Figure 2).

The Role of Type I (α/β) IFNs

According to the results of our studies, type I (α/β) IFNs play a major homoeostatic role at weaning in farmed pigs. Weaning was shown to affect the constitutive expression of different IFN-α genes by shutting them off at the time of IFN protein release in serum, as a result of a feedback inhibition strategy [59]. What rationale can be envisaged for the type I response at weaning? In our opinion, the regulatory action is mainly directed to a control of IFN-γ [53], because of its notable pro-inflammatory potential in intestinal epithelial cells [63] and the detrimental influence on intestinal permeability [64], also affected by TNF-α [24]; this latter feature underlies an accentuated risk of enteritis and diarrhoea. Since matrix metalloproteinases (MMP) are up-regulated after weaning [12] and IFN-γ is a powerful MMP inhibitor [65], it is tempting to speculate that the IFN-γ response after weaning is part of a wider feedback control of such proteinases in the gut. The IFN-γ response is also in line with the dramatic increase of

IFN-γ-secreting lymphocytes in gut *lamina propria* of mice at weaning [66], and the infiltration of activated T-cells with increased expression of CD25 in gut *lamina propria* of piglets after weaning [67]. Also, the increased prevalence of IL-10 secreting cells in PBMC of pigs at 6 weeks of age [68] might be generated by a control strategy of pro-inflammatory cytokines such as IFN-γ; interestingly, such an increase at 6 weeks of age is in contrast with the stable prevalence in previous and subsequent phases [68].

The correlation between excessive IFN-γ response and onset of enteritis was clearly shown in a murine model, in which severe diarrhoea was caused by co-injection of IL-12 and IL-18, i.e. two IFN-γ inducing cytokines [69]; the observed gut lesions (shortening of villi and increased depth of crypts) were reminiscent of the changes observed in the gut mucosa of piglets after weaning; interestingly, diarrhoea was not induced in IFN-γ KO mice [69].

Having in mind the possible role of IL-1β in the induction of type I IFNs [59] *in vitro* and the notable up-regulation of IL-1β in the gut of piglets after weaning [24], we can hypothesize that a major inflammosome reaction is induced in the gut as a response to tissue damage after weaning, producing caspase-1 and IL-1β as final products [32]. IL-1β would in turn sustain the type I IFN response, which may also occur as release of pre-formed IFN molecules from leukocytes [70]. A contribution of bacterial endotoxin to induction of type I IFN following tissue damage can be also hypothesized on a sound basis [71]. A possible flow chart of some cytokine responses in the gut after weaning is shown in Figure 3; this is the authors' interpretation of the events, which obviously needs confirmation in further experimental studies before a complete understanding of the phenomena occurs.

Discussion of Results

Our studies are aimed at assessing the impact on piglets of weaning performed at different ages. Results have confirmed the strong pro-inflammatory impact of the weaning stress, which is in line with the expression of inflammatory cytokine genes in the gut [24]. Furthermore, we could discriminate between weaning-specific and non-specific responses, the latter being observed in unweaned piglets following the collapse of group hierarchy. In this scenario, we investigated the homoeostatic response strategy of pigs and highlighted the role of some endogenous cytokines. Concerning our field data, it should be stressed that the trial took place in a farm with high hygiene standards, under good health and welfare conditions. Needless to say, results might be quite different in 'problem' farms, in which pigs are confronted with high infectious pressure and unfavourable housing and/or feeding conditions. The very occurrence of an IFN-α response at weaning in healthy subjects, with no evidence of intercurrent viral infections, is a further demonstration of the important homoeostatic role of this cytokine and of

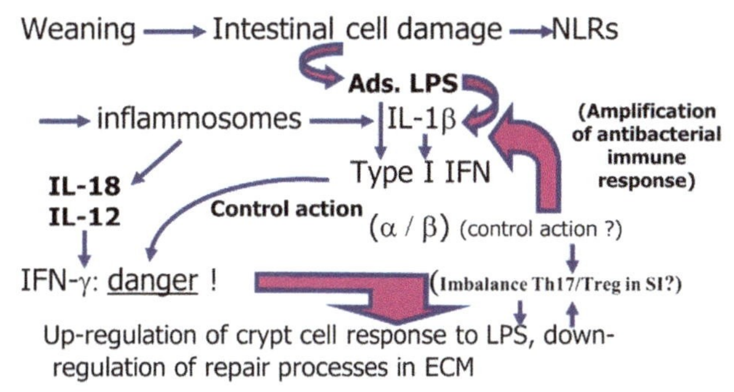

Figure 3 Possible flow chart of some cytokine responses in the gut after weaning. The intestinal damage at weaning is likely to trigger an inflammosome reaction sustained by NOD-like receptors (NLRs) in enterocytes, giving rise to caspase-1 and IL-1β as final products. IL-1β and adsorbed LPS (ads. LPS) could induce type I IFN (α/β), acting as a homoeostatic control factor for the type II IFN-γ response. The latter could be caused by IFN-γ inducing cytokines (IL-12 and IL-18, the latter directly generated by the inflammosome reaction). An excessive IFN-γ response can lead to noxious intestinal inflammation, which may imply a major imbalance between activating (Th17) and suppressive (Treg) lymphocytes in the small intestine (SI). Serious clinical consequences for the host may eventually take place.

type I IFNs on the whole. The homoeostatic functions of IFN-α can be eased by the constitutive expression of this cytokine in swine PBMC [70] and skin dendritic cells [72]. The pattern of constitutive IFN-α expression in swine PBMC is very similar to that of human PBMC [73].

With respect to immune parameters, a limited age difference at weaning was shown to play an important role, actually beyond the authors' expectations. On the whole, piglets weaned at 28 days of age could mount in our experience a more effective homoeostatic response compared with piglets weaned at 22 days of age (see DWG and cortisol data). Furthermore, the lesser coping efforts of piglets weaned at 28 days were confirmed by the lower prevalence of serum IFN-α responses in the first week after weaning. Therefore, there is evidence that weaning of piglets at 3 weeks of age determines the need for a more intense homoeostatic response compared with that of 4-week-old animals. This is possibly associated with a major failure of enterocytes of 3-week-old animals to properly absorb the nutrients of conventional starter diets. In this respect, fundamental parameters of clinical immunology and chemistry can adequately depict the impact of the weaning stress, as well as the different extent of the homoeostatic response in piglets. The lesser impact of weaning at 4 rather than 3 weeks of age is in line with the aforementioned EU Scientific Opinion [16] and the present regulations enacted in Europe (EU Directive 2001/93/EC). These recognize that early weaned piglets show enhanced behavioural problems and neuroendocrine stress reactions, as well as reduced immune responses [3]. Concerning the present farmed pigs, the limit of 4–5 weeks of age should not be deferred though in our opinion, since dams must find a balance in the allocation of resources for nursing both present and future litters, and maximizing their reproductive performance [3]. In this respect, the high IL-6 response at 28 days of age just before weaning (Table 3) is likely to reflect the stress of piglets competing for the sow's milk, having to match a 6- or 7-fold greater body mass with respect to the birth weight. If such a hypothesis should be confirmed, this would make a strong case against weaning beyond 4–5 weeks of age, in agreement with the present practice on farm and also with fundamental sow welfare concerns, regarding in particular the conventional systems with farrowing crates. Beyond that, this is a matter of litters per year per sow, i.e. a genuine financial matter.

Piglets weaned at 22 days and orally treated with IFN-α revealed a better adaptive response compared with age-matched control animals on the basis of their growth curves. Interestingly, the effectiveness of the IFN-α treatment in terms of DWG could be appreciated not earlier than 2 weeks after weaning; it went along with a reduction of the TNF-α response in this phase compared with control piglets, in line with the values observed in pigs weaned later (Table 3). This result is in agreement with our previous study on SPF pigs in our Experiment Station, in which a control action on TNF-α gene expression was exerted by the oral IFN-α treatment on day +21 with respect to weaning. Interestingly, this is reminiscent of the SDP treatment, which reduces the expression of TNF-α, IL-6 and IL-1β genes after weaning [74]. Therefore, low-dose IFN-α may affect the regulatory circuits of inflammatory cytokines after weaning in a time-related manner. In this conceptual framework, an oral IFN-α treatment could partly replace the need for the corresponding endogenous cytokine in the post-weaning phase. The described treatment should be carried out in the framework of correct farm management and disease control procedures. In this scenario, the alternative farming schemes reported in the section 'Intervention strategies' could enable pigs to better express their behavioural repertoire, and effectively complement a sound immunoprophylaxis strategy. Most important, the intervention strategy based on oral IFN-α is in line with a global need for a major reduction of antibiotic usage in pig farms and their qualification in terms of Good Farming Practices (EU Regulation 852/2004), and 'Welfare-friendly' labelling of products of animal origin in Europe [75].

Conclusion

Adaptation of pigs to the weaning stress may be of concern in farms where lean pig phenotypes, their pressing needs and inability to cope with sub-optimal environmental conditions are confronted with poor management and infrastructure. Our results indicate that weaning at 4 weeks of age under good management and hygiene conditions can be compatible with the piglets' coping capacity, and it is likely to represent a decent compromise in terms of health and welfare of both piglets and sows. On the basis of this tenet, validated schemes of Good Farming Practices at weaning should be developed and accepted by farmers. This could help reduce the despicable practice of drug-supplemented feeds after weaning, often meant as an 'insurance policy' against possible disease losses in this crucial phase, in the framework of a sustainable pig husbandry and reduction of antibiotic usage. An oral, low-dose IFN-α treatment at weaning could be an important side measure on farm, in line with major concerns about food safety 'from farm to fork' [76]. Such a strategy can be adequately supported by a timely assessment of clinical immunology parameters. These can be seen as a practical and accurate reporter system of environmental adaptation, and suitable readout of interventions for better health and welfare conditions, in association with fundamental clinical and productive parameters of farmed pigs.

Acknowledgements

The authors thank Ivonne Laura Archetti and Paolo Candotti (IZSLER, Brescia, Italy) for their co-operation in clinical chemistry analyses and field activities, respectively. We are also indebted to Monica Colitti (Faculty of Veterinary Medicine, Udine, Italy) and Paola Roncada, Francesca Deriu (Faculty of Veterinary Medicine, Milan, Italy) for real-time PCR on leukocytes and serum proteome analyses, respectively. The skilful technical assistance of Antonio Cristiano, Cinzia Mantovani and Giovanna Ferrari (IZSLER, Brescia) is gratefully acknowledged.

References

1. Jensen P, Stangel G, Algers B. Nursing and suckling behaviour of semi-naturally kept pigs during the first 10 days postpartum. Applied Animal Behaviour Science 1991;31(3–4):195–209.
2. Manners MJ. The development of digestive function in the pig. Proceedings of the Nutrition Society 1976;35(1):49–55.
3. Hameister T, Puppe B, Tuchscherer M, Kanitz E. Effects of weaning age on behavioural and physiological responses of domestic piglets–a review. Berliner und Munchener Tierarztliche Wochenschrift 2010;123(1–2):11–9.
4. Lalles JP, Bosi P, Smidt H, Stokes CR. Nutritional management of gut health in pigs around weaning. Proceedings of the Nutrition Society 2007;66(2):260–8.
5. Moeser AJ, Blikslager AT. Mechanisms of porcine diarrheal disease. Journal of the American Veterinary Medical Association 2007;231(1):56–67.
6. Nabuurs MJ, Hoogendoorn A, van der Molen EJ, van Osta AL. Villus height and crypt depth in weaned and unweaned pigs, reared under various circumstances in The Netherlands. Research in Veterinary Science 1993;55(1):78–84.

7. Nabuurs MJ, Hoogendoorn A, van Zijderveld FG. Effects of weaning and enterotoxigenic *Escherichia coli* on net absorption in the small intestine of pigs. Research in Veterinary Science 1994;56(3):379–85.

8. Laffont S, Powrie F. Immunology: dendritic-cell genealogy. Nature 2009;462(7274):732–3.

9. Barnett KL, Kornegay ET, Risley CR, Lindemann MD, Schurig GG. Characterization of creep feed consumption and its subsequent effects on immune response, scouring index and performance of weanling pigs. Journal of Animal Science 1989;67(10):2698–708.

10. Bailey M, Haverson K, Inman C, Harris C, Jones P, Corfield G, *et al.* The development of the mucosal immune system pre- and post-weaning: balancing regulatory and effector function. Proceedings of the Nutrition Society 2005;64(4): 451–7.

11. Stokes CR, Bailey M, Wilson AD. Immunology of the porcine gastrointestinal tract. Veterinary Immunology and Immunopathology 1994;43(1–3):143–50.

12. McCracken BA, Spurlock ME, Roos MA, Zuckermann FA, Gaskins HR. Weaning anorexia may contribute to local inflammation in the piglet small intestine. Journal of Nutrition 1999;129(3):613–9.

13. Wijtten PJ, van der Meulen J, Verstegen MW. Intestinal barrier function and absorption in pigs after weaning: a review. British Journal of Nutrition 2011;105(7):967–81.

14. van Beers-Schreurs HM, Vellenga L, Wensing T, Breukink HJ. The pathogenesis of the post-weaning syndrome in weaned piglets: a review. Veterinary Quarterly 1992;14(1):29–34.

15. D'Eath RB, Roehe R, Turner SP, Ison SH, Farish M, Jack MC, *et al.* Genetics of animal temperament: aggressive behaviour at mixing is genetically associated with the response to handling in pigs. Animal 2009;3(11):1544–54.

16. Anonymous. The welfare of intensively kept pigs. 1997, Doc XXIV/B3/ScVC/0005/1997 final.

17. Dostaikova I, Spinka M, Kindlmann P. Begging for milk: evolution of teat massaging in suckling pigs. Journal of Theoretical Biology 2002;215(3):321–32.

18. Dewey C, Carman S, Pasma T, Josephson G, McEwen B. Relationship between group A porcine rotavirus and management practices in swine herds in Ontario. Canadian Veterinary Journal 2003;44(8):649–53.

19. Darwich L, Mateu E. Immunology of porcine circovirus type 2 (PCV2). Virus Research 2012;164(1–2):61–7.

20. Moxley RA. Edema disease. Veterinary Clinics of North America: Food Animal Practice 2000;16(1):175–85.

21. McCracken BA, Gaskins HR, Ruwe-Kaiser PJ, Klasing KC, Jewell DE. Diet-dependent and diet-independent metabolic responses underlie growth stasis of pigs at weaning. Journal of Nutrition 1995;125(11):2838–45.

22. Wattrang E, Wallgren P, Lindberg A, Fossum C. Signs of infections and reduced immune functions at weaning of conventionally reared and specific pathogen free pigs. Zentralblatt fur Veterinarmedizin B 1998;45(1):7–17.

23. Artursson K, Wallgren P, Alm GV. Appearance of interferon-alpha in serum and signs of reduced immune function in pigs after transport and installation in a fattening farm. Veterinary Immunology and Immunopathology 1989;23(3–4):345–53.

24. Pie S, Lalles JP, Blazy F, Laffitte J, Seve B, Oswald IP. Weaning is associated with an upregulation of expression of inflammatory cytokines in the intestine of piglets. Journal of Nutrition 2004;134(3):641–7.

25. Roncada P, Begni B, Amadori M, Cristoni S, Archetti IL, Boldetti C, *et al.* Blood serum proteome for welfare evaluation in pigs. Veterinary Research Communications 2007;31 Suppl 1:321–5.

26. Ottaviani E, Franceschi C. A new theory on the common evolutionary origin of natural immunity, inflammation and stress response: the invertebrate phagocytic immunocyte as an eye-witness. Domestic Animal Endocrinology 1998;15(5):291–6.

27. Amadori M. The role of IFN-alpha as homeostatic agent in the inflammatory response: a balance between danger and response? Journal of Interferon and Cytokine Research: The Official Journal of the International Society for Interferon and Cytokine Research 2007;27(3):181–9.

28. Broom DM. Adaptation. Berliner und Munchener Tierärztliche Wochenschrift 2006;119(1–2):1–6.

29. Koo EG, Lai LM, Choi GY, Chan MT. Systemic inflammation in the elderly. Best Practice and Research Clinical Anaesthesiology 2011;25(3):413–25.

30. Matzinger P. An innate sense of danger. Annals of the New York Academy of Sciences 2002;961:341–2.

31. Hansen JD, Vojtech LN, Laing KJ. Sensing disease and danger: a survey of vertebrate PRRs and their origins. Developmental and Comparative Immunology 2011;35(9):886–97.

32. Martinon F, Mayor A, Tschopp J. The inflammasomes: guardians of the body. Annual Review of Immunology 2009;27:229–65.

33. Zeiser R, Penack O, Holler E, Idzko M. Danger signals activating innate immunity in graft-versus-host disease. Journal of Molecular Medicine (Berlin, Germany) 2011;89(9):833–45.

34. Hayday AC. Gammadelta T cells and the lymphoid stress-surveillance response. Immunity 2009;31(2):184–96.

35. Guzman E, Birch JR, Ellis SA. Cattle MIC is a ligand for the activating NK cell receptor NKG2D. Veterinary Immunology and Immunopathology 2010;136(3–4):227–34.

36. Schreiber G.HSRD. Interferon-gamma. In Thomson A.W.Lotze M.T., editors. The Cytokine Handbook. 4th ed. Academic Press, London; 2003. p. 567–601.

37. Oostindjer M, Bolhuis JE, Mendl M, Held S, Gerrits W, van den Brand H, *et al.* Effects of environmental enrichment and loose housing of lactating sows on piglet performance before and after weaning. Journal of Animal Science 2010;88(11):3554–62.

38. Jiang R, Chang X, Stoll B, Ellis KJ, Shypailo RJ, Weaver E, *et al.* Dietary plasma protein is used more efficiently than extruded soy protein for lean tissue growth in early-weaned pigs. Journal of Nutrition 2000;130(8):2016–9.

39. Pettigrew JE. Reduced use of antibiotic growth promoters in diets fed to weanling pigs: dietary tools, part 1. Animal Biotechnology 2006;17(2):207–15.

40. Bosi P, Casini L, Finamore A, Cremokolini C, Merialdi G, Trevisi P, *et al.* Spray-dried plasma improves growth performance and reduces inflammatory status of weaned

pigs challenged with enterotoxigenic Escherichia coli K88. Journal of Animal Science 2004;82(6):1764–72.

41. Stein HH, Kil DY. Reduced use of antibiotic growth promoters in diets fed to weanling pigs: dietary tools, part 2. Animal Biotechnology 2006;17(2):217–31.

42. Sargeant HR, McDowall KJ, Miller HM, Shaw MA. Dietary zinc oxide affects the expression of genes associated with inflammation: transcriptome analysis in piglets challenged with ETEC K88. Veterinary Immunology and Immunopathology 2010;137(1–2):120–9.

43. Graveland H, Duim B, van Duijkeren E, Heederik D, Wagenaar JA. Livestock-associated methicillin-resistant *Staphylococcus aureus* in animals and humans. International Journal of Medical Microbiology 2011;301(8):630–4.

44. Merlot E, Meunier-Salaün M-C, Prunier A. Behavioural, endocrine and immune consequences of mixing in weaned piglets. Applied Animal Behaviour Science 2004;85(3–4):247–57.

45. Roncada P, Deriu F, Archetti IL, Greppi GF, Amadori M Proteomica del siero per la valutazione degli effetti flogistici dello svezzamento precoce nel suino. In *Atti Società Italiana delle Scienze Veterinarie*, 26–29 Settembre 2007, Salsomaggiore TermeTipografia Antonio Trischitta; 2007. p. 113–4.

46. Martin M, Tesouro MA, Gonz Am N, Pi A, Lampreave F. Major plasma proteins in pig serum during postnatal development. Reproduction, Fertility and Development 2005;17(4):439–45.

47. Strom ADG, Knowles AG, Husband A, Inventors. A method of improving the growth performance of an animal. WO 02/096216; 2002.

48. Dec M, Puchalski A. Use of oromucosally administered interferon-alpha in the prevention and treatment of animal diseases. Polish Journal of Veterinary Sciences 2008;11(2):175–86.

49. Straub OC. Studies on the suitability of alpha-hybrid interferon application in cattle. Comparative Immunology, Microbiology and Infectious Diseases 1995;18(4):239–43.

50. Cummins JM, Krakowka GS, Thompson CG. Systemic effects of interferons after oral administration in animals and humans. American Journal of Veterinary Research 2005;66(1):164–76.

51. Pedretti E, Passeri B, Amadori M, Isola P, Di Pede P, Telera A, *et al.* Low-dose interferon-alpha treatment for feline immunodeficiency virus infection. Veterinary Immunology and Immunopathology 2006;109(3–4):245–54.

52. Kaser A, Tilg H. Interferon-alpha in inflammation and immunity. Cell and Molecular Biology (Noisy-le-grand, France) 2001;47(4):609–17.

53. Amadori M, Farinacci M, Begni B, Faita R, Podavini D, Colitti M. Effects of interferon-alpha on the inflammatory response of swine peripheral blood mononuclear cells. Journal of Interferon and Cytokine Research: The Official Journal of the International Society for Interferon and Cytokine Research 2009;29(4):241–7.

54. Niekamp SR, Sutherland MA, Dahl GE, Salak-Johnson JL. Immune responses of piglets to weaning stress: impacts of photoperiod. Journal of Animal Science 2007;85(1):93–100.

55. Wolter BF, Ellis M, Corrigan BP, DeDecker JM. The effect of birth weight and feeding of supplemental milk replacer to piglets during lactation on preweaning and postweaning growth performance and carcass characteristics. Journal of Animal Science 2002;80(2):301–8.

56. Azain MJ, Tomkins T, Sowinski JS, Arentson RA, Jewell DE. Effect of supplemental pig milk replacer on litter performance: seasonal variation in response. Journal of Animal Science 1996;74(9):2195–202.

57. Grenett HE, Danley DE, Strick CA, James LC, Otterness IG, Fuentes N, *et al.* Isolation and characterization of biologically active murine interleukin-6 produced in *Escherichia coli.* Gene 1991;101(2):267–71.

58. Asai T, Okada M, Ono M, Irisawa T, Mori Y, Yokomizo Y, *et al.* Increased levels of tumor necrosis factor and interleukin 1 in bronchoalveolar lavage fluids from pigs infected with *Mycoplasma hyopneumoniae*. Veterinary Immunology and Immunopathology 1993;38(3–4):253–60.

59. Razzuoli E, Villa R, Sossi E, Amadori M. Characterization of the interferon-alpha response of pigs to the weaning stress. Journal of Interferon and Cytokine Research: The Official Journal of the International Society for Interferon and Cytokine Research 2011;31(2):237–47.

60. Razzuoli E, Villa R, Sossi E, Amadori M. Reverse transcription real-time PCR for detection of porcine interferon alpha and beta genes. Scandinavian Journal of Immunology 2011;74(4):412–8.

61. Razzuoli E, Dotti S, Archetti IL, Amadori M. Clinical chemistry parameters of piglets at weaning are modulated by an oral, low-dose interferon-alpha treatment. Veterinary Research Communications 2010;34 Suppl 1:S189–92.

62. Petersen HH, Nielsen JP, Heegaard PM. Application of acute phase protein measurements in veterinary clinical chemistry. Veterinary Research 2004;35(2):163–87.

63. Alvarado J, Taylor P, Castillo JR, Thomas LE. Interferon gamma bound to extracellular matrix changes the hyporesponsiveness to LPS in crypt but not villous intestinal epithelial cells. Immunology Letters 2005;99(1):109–12.

64. Beaurepaire C, Smyth D, McKay DM. Interferon-gamma regulation of intestinal epithelial permeability. Journal of Interferon and Cytokine Research: The Official Journal of the International Society for Interferon and Cytokine Research 2009;29(3):133–44.

65. Alvarado J, del Castillo JR, Thomas LE. Modulation of membrane type 1 matrix metalloproteinase by LPS and gamma interferon bound to extracellular matrix in intestinal crypt cells. Cytokine 2008;41(2):155–61.

66. Vazquez E, Gil A, Garcia-Olivares E, Rueda R. Weaning induces an increase in the number of specific cytokine-secreting intestinal lymphocytes in mice. Cytokine 2000;12(8):1267–70.

67. Harris C, Haverson K, Jones P, Miller B, Stokes C, Bailey M. A comparison of the development of lamina propria lymphocytes in pre-weaning and post-weaning piglets. In Seventh International Veterinary Immunology Symposium, 25–30 July, Quèbec City; 2004 p. 198.

68. Diaz I, Mateu E. Use of ELISPOT and ELISA to evaluate IFN-gamma, IL-10 and IL-4 responses in conventional pigs. Veterinary Immunology and Immunopathology 2005;106(1–2):107–12.

69. Chikano S, Sawada K, Shimoyama T, Kashiwamura SI, Sugihara A, Sekikawa K, *et al.* IL-18 and IL-12 induce intestinal inflammation and fatty liver in mice in an IFN-gamma dependent manner. Gut 2000;47(6):779–86.

70. Amadori M, Cristiano A, Ferrari M. Constitutive expression of interferons in swine leukocytes. Research in Veterinary Science 2010;88(1):64–71.

71. Pietras EM, Saha SK, Cheng G. The interferon response to bacterial and viral infections. Journal of Endotoxin Research 2006;12(4):246–50.

72. Bautista EM, Ferman GS, Gregg D, Brum MC, Grubman MJ, Golde WT. Constitutive expression of alpha interferon by skin dendritic cells confers resistance to infection by foot-and-mouth disease virus. Journal of Virology 2005;79(8):4838–47.

73. Greenway AL, Hertzog PJ, Devenish RJ, Linnane AW. Constitutive and virus-induced interferon production by peripheral blood leukocytes. Experimental Hematology 1995;23(3):229–35.

74. Touchette KJ, Carroll JA, Allee GL, Matteri RL, Dyer CJ, Beausang LA, *et al.* Effect of spray-dried plasma and lipopolysaccharide exposure on weaned pigs: I. Effects on the immune axis of weaned pigs. Journal of Animal Science 2002;80(2):494–501.

75. Anonymous. Options for animal welfare labelling and the establishment of a European Network of Reference Centres for the protection and welfare of animals. 2009, COM(2009) 584 final.

76. Anonymous. White Paper on Food Safety of 12 January 2000. COM/99/0719; 2000.

Major bacterial diseases in aquaculture and their vaccine development

Julia W. Pridgeon* and Phillip H. Klesius

Address: Aquatic Animal Health Research Unit, USDA-ARS, 990 Wire Road, Auburn, AL 36832, USA.

***Correspondence:** Julia W. Pridgeon. Email: Julia.Pridgeon@ars.usda.gov

Received: 7 May 2012
Accepted: 27 July 2012

Abstract

Aquaculture is emerging as the fastest growing food-producing industry in the world because of the increasing demand for food fish consumption. However, the intensive culture of food fish has led to outbreaks of various bacterial diseases, resulting in annual economic losses to the aquaculture industry estimated at billions of dollars worldwide. Feeding infected fish with antibiotic-medicated food is a general practice but has led to antibiotic resistance development in bacterial pathogen, resulting in a higher dose requirement for effective control, a matter of increasing public concern. Therefore, alternatives to antibiotics that give similar or enhanced protection to aquatic animals are urgently needed. Various vaccines have been developed recently to combat bacterial diseases in aquaculture. The purpose of this review is to summarize the major bacterial pathogens in aquaculture and the development of vaccines as alternatives to antibiotics to protect aquatic animals from these bacterial diseases.

Keywords: Aquaculture, Attenuated, Bacteria, Disease, Vaccine

Review Methodology: United States Department of Agriculture National Agricultural Library's DigiTop Navigator platform was used to search the following databases: AGRICOLA, AGRIS, BIOSIS, CAB Abstracts, Fish, Fisheries & Aquatic Biodiversity Worldwide, Food Science and Technology Abstracts, MEDLINE, Wildlife & Ecology Studies Worldwide and Zoological Record (keyword search terms used: vaccine, bacteria, attenuated, aquaculture). In addition, references from the articles obtained by this method were used to check for additional relevant material.

Aquaculture as the Fastest Growing Food-Producing Industry

Aquaculture is also called 'underwater agriculture' [1]. There is an ever-increasing human population and consequent demand for food, but significant expansion of the traditional land-food production systems (such as plant agriculture, poultry, cattle/goat/pig farming, etc.) is limited by the fact that the earth has only 30% land coverage. To provide enough food for the rapidly growing human population, we need to utilize the remaining 70% area under water for food production [1]. Worldwide, the aquaculture industry has grown at an average rate of 8.9% per year since 1970, compared with a 1.2% growth rate of capture fisheries and 2.8% growth rate of terrestrial farmed meat production over the same period [2]. According to the 2010 FAO review on world fishery production [3], the capture production system has maintained similar level at 90 million tonnes for more than a decade [4], whereas the aquaculture production of fisheries has increased from 34.6 million tonnes in 2001 to 55.7 million tonnes in 2009 [3]. The value of aquaculture production was estimated at $105.3 billion in 2009 [2]. China remains as the leading force contributing to the increase of world total fishery production [4].

Major Bacterial Diseases Affecting Aquaculture

Notwithstanding the fact that aquaculture is the fastest growing food-production industry in the world the sector is plagued by diseases. The annual economic loss to the aquaculture industry through diseases is estimated to be billions of US dollars worldwide [5]. Major pathogens that are affecting the aquaculture industry include: bacteria [6–9], fungi [10, 11], viruses [12–15] and parasites

[16, 17]. Given that bacteria can survive well in aquatic environment independently of their hosts, bacterial diseases have become major impediments to aquaculture, especially when water temperature is warm [18]. Thus far, bacterial species belonging to at least 13 genera have been reported to be pathogenic to aquatic animals, including: (1) Gram-negative bacteria such as *Aeromonas*, *Edwardsiella*, *Flavobacterium*, *Francisella*, *Photobacterium*, *Piscirickettsia*, *Pseudomonas*, *Tenacibaculum*, *Vibrio* and *Yersinia*; and (2) Gram-positive bacteria such as *Lactococcus*, *Renibacterium* and *Streptococcus* [18]. Major pathogenic Gram-negative and Gram-positive bacterial species reported in disease outbreaks are summarized in Tables 1 and 2, respectively.

Fresh-water cultured fish as well as marine fish are devastated by bacterial diseases, including tilapia, catfish, carp, trout, salmon, bass, perch, sturgeon and eel. The annual economic loss associated with three bacterial fish pathogens (*Aeromonas hydrophila*, *Yersinia ruckeri* and *Vibrio fluvialis*) between 1990 and 1992 in China was estimated at more than $120 million [168]. Since the value of aquaculture production was estimated at $105.3 billion in 2009 [2], the global economic impact of bacterial diseases on aquaculture production would be within the range of hundreds of millions to billions of dollars annually.

Control of Bacterial Diseases in Aquaculture

To control bacterial diseases, feeding infected fish with antibiotic-medicated food is a general practice. However, this is usually expensive and may be ineffective because sick fish may remain off feed. In addition, frequent use of antimicrobial compounds has led to the development of resistance to antimicrobial compounds in pathogens, posing serious challenges to health and national security [169]. Resistance can arise in disease-causing bacteria, rendering life-saving antimicrobial compounds ineffective. Furthermore, diseases caused by antibiotic-resistant bacteria are difficult to treat, and there are only a few new antimicrobial compounds in the drug development pipeline. Therefore, alternatives to antimicrobial compounds that give similar or enhanced protection to aquatic animals are urgently needed. An integrated approach that considers not only the pathogen but also the host and the environment will be the most effective method in the long-term to improve aquatic animal health. From the pathogen perspective, besides using antimicrobial compounds, alternative methods in integrated management include: (1) using pathogen-specific bacteriophages; (2) using short-chain fatty acids and polyhydroxyalkanoates to inhibit the growth of bacteria; and (3) using compounds to inhibit virulence gene expression or interrupt the signal transduction pathways of the pathogens [170, 171]. From the environmental perspective of integrated management, good hygiene such as quarantine and disinfection to optimize water quality is important [171]. From the host perspective, it is important to: (1) provide quality feed; (2) prevent stress; (3) improve the breeding stock's disease resistance ability; and (4) adopt the use of immunostimulants or vaccines to improve host immunity [171]. The use of vaccine has been proven to be highly effective in controlling diseases in the salmon industry in Europe, America and Japan [172]. Vaccines have been used in fish, in particular the salmon industry, for approximately 30 years. Vaccination of salmon has dropped the industrial use of antibiotics to a mere fraction of its original use. For example, in 1987, approximately 50,000 kg of antibiotics was used in Norway to control fish diseases (mostly *Vibrio* spp.) in salmon prior to the widespread use of vaccines [173]. However, by 1997, when an efficacious oil-adjuvant vaccine was extensively used, antibiotic usage had dropped to less than 1000–2000 kg, concurrent with a threefold increase in fish production (Figure 1) [173]. In the following paragraphs, the recent development of vaccines to combat the major bacterial diseases in aquaculture will be reviewed.

Types of Bacterial Vaccines

In general, there are six types of bacterial vaccines: (1) bacterins (killed bacteria) such as the *Aeromonas salmonicida* bacterin vaccine that is currently available in the USA; (2) attenuated live bacterial vaccine such as the attenuated live *Edwardsiella ictaluri* vaccine that is commercially available in the USA; (3) toxoid vaccine containing inactivated bacterial toxins rather than the whole bacteria; (4) subunit vaccine such as a protein subunit of the bacterium, where instead of an inactivated or attenuated whole bacterium to stimulate the immune system, a fragment or a subunit of the bacterium is used to stimulate the immune response; (5) conjugate vaccine by conjugating certain membrane proteins of bacteria to other proteins to make the bacterial proteins recognizable, thereby inducing immune response; and (6) experimental vaccines such as DNA vaccines or recombinant vector vaccine. Of the six types of bacterial vaccines, the first two types are commonly used in aquaculture. Modified live vaccines containing live bacteria usually survives and replicates within the host, therefore stimulating a stronger cellular immune response and conferring a longer protection compared with killed vaccines.

Vaccine Development Status for Major Bacterial Diseases in Aquaculture

Currently, eight bacterial fish vaccines are licensed for use in the USA (http://www.aphis.usda.gov/animal_health/vet_biologics):

(1) *A. salmonicida* vaccine (bacterin);
(2) *Arthrobacter* vaccine (live culture);

Table 1 Summary of pathogenic Gram-negative bacteria and their hosts reported in literature

Pathogen	Disease	Host of pathogen
Aeromonas hydrophila	Motile aeromonads septicaemia	Catfish [19], carp [20, 21], trout [22, 23], eel [24], sturgeon [25], tilapia [26] and bass [27]
Aeromonas salmonicida	Furunculosis	Salmon [28, 29], trout [30], flounder [31], turbot [32], carp [33], tilapia [33] and sole [31]
Chryseobacterium sp.	Chryseobacteriosis	Salmon [34, 35] and trout [35]
Edwardsiella ictaluri	Enteric septicaemia of catfish	Catfish [36–39]
Edwardsiella tarda	Edwardsiellosis or putrefactive disease	Turbot [40, 41], flounder [42, 43], carp [44], catfish [45], eel [45, 46] and tilapia [47, 48]
Flavobacterium columnare	Columnaris	Carp [49, 50], trout [51, 52], perch [53], tilapia [54, 55], catfish [56] and salmon [39]
Flavobacterium johnsonae	False columnaris	Barramundi [57]
Flavobacterium psychrophilum	Flavobacteriosis or rainbow trout fry syndrome	Trout [58, 59]
Flavobacterium branchiophilum	Bacterial gill disease	Trout [60]
Francisella spp.	Francisellosis	Tilapia [61] and hybrid striped bass [62]
Moritella viscosa	Winter ulcer disease	Salmon [63, 64]
Photobacterium spp. (formerly *Pasteurella* spp.)	Pasteurellosis	Sturgeon [65], hybrid striped bass [66], seabream [67–69], yellowtail [70], sea bass [71], snakehead [72], tuna [73] and cobia [74]
Piscirickettsia salmonis and *Piscirickettsia*-like organism	Piscirickettsiosis or rickettsial septicaemia	Salmon [75], trout [75] and tilapia [76, 77]
Pseudomonas spp.	Pseudomonads septicaemia or red spot disease	Seabream [78], trout [79], eel [80], rabbitfish [81], catfish [82], eel [83], shrimp [84] and salmon [85]
Tenacibaculum maritimum	Tenacibaculosis	Sole [86] and turbot [87]
Vibrio spp.	Vibrosis	Croaker fish [88], puffer fish [89], grouper [90], cod [91], shrimp [92, 93], big-scale sand smelt [94], flounder [95], abalone [96], seabream [97, 98], turbot [99, 100], sole [101], red drum [102], cobia [103], eel [104, 105], salmon [106], sweetfish [107], sheatfish [108] and catfish [109]
Yersinia ruckeri	Yersiniosis or enteric redmouth disease	Trout [110–112], tilapia [113] and salmon [114, 115]

Table 2 List of pathogenic Gram-positive bacteria and their hosts reported in literature

Pathogen	Disease	Host of pathogen
Lactococcus garvieae (formerly *Enterococcus seriolicida*)	Lactococcosis	Yellowtail [116, 117], trout [118–120], rockfish [121] and mullet [122]
Nocardia sp.	Nocardiosis	Tigerfish [123], snakehead [124, 125], croaker [126, 127], mullet [128], seabass [129], largemouth bass [130, 131], gourami [132] and yellowtail [133]
Renibacterium salmoninarum	Bacterial kidney disease	Trout [134] and salmon [135, 136]
Staphylococcus spp.	Staphylococcosis	Trout [137], tilapia [138], carp [139], perch [140], seabream [141, 142] and yellowtail [142]
Streptococcus agalactiae	Streptococcosis	Tilapia [143, 144] Grouper [145], mullet [146] and pomfret [147]
Streptococcus ictaluri	Streptococcosis	Catfish [148]
Streptococcus iniae	Streptococcosis	Tilapia [149], seabream [150], red porgy [151], trout [151], flounder [152], barramundi [153], rabbitfish [154], hybrid striped bass [155], yellowtail [156] and catfish [157]
Streptococcus dysgalactiae	Streptococcosis	Sturgeon [158], amberjack [159, 160] and yellowtail [160]
Streptococcus parauberis	Streptococcosis	Turbot [161] and flounder [162]
Streptococcus phocae	Streptococcosis	Salmon [163]
Vagococcus salmoninarum	Coldwater 'streptococcosis'	Trout [164, 165]
Weissella sp.	Haemorrhagic septicaemia	Trout [166, 167]

(3) *A. salmonicida–Vibrio anguillarum–Vibrio ordalii–Vibrio salmonicida* (*Aliivibrio salmonicida*) vaccine (bacterin);
(4) *Flavobacterium columnare* vaccine (bacterin);
(5) *Y. ruckeri* vaccine (bacterin);
(6) *V. anguillarum–V. ordalii* vaccine (bacterin);
(7) *E. ictaluri* vaccine (avirulent live culture);
(8) *F. columnare* vaccine (avirulent live culture).

The eight vaccines are licensed for use in aquaculture practices for protection against bacterial diseases. However, vaccines are usually licensed for specific fish species. Therefore, license of certain vaccine for one fish species may not be used to protect another fish species, although the pathogen might be the same. For example, the *F. columnare* bacterium vaccine is currently licensed for use only in catfish.

Worldwide, vaccines have been developed and commercially available to treat at least 18 bacterial infections [173, 175, 176]:

(1) vibriosis caused by *V. anguillarum* (formerly *Listonella anguillarum*) or *V. ordalii* in salmonids, cod, halibut, seabass, seabream, amberjack and yellowtail;
(2) coldwater vibriosis caused by *V. salmonicida* in salmonids;
(3) winter ulcer disease or wound disease caused by *Moritella viscosa* in salmonids;
(4) furunculosis caused by *A. salmonicida* subsp. *salmonicida* in salmonids;
(5) atypical *A. salmonicida* caused by *A. salmonicida* in salmonids;
(6) yersiniosis caused by *Y. ruckeri* in salmonids;
(7) piscircickettsiosis caused by *Piscirickettsia salmonis* in salmonids;
(8) bacteria gill disease caused by *Flavobacterium branchiophilum* in salmonids and carp;
(9) flavobacteriosis caused by *Flavobacterium psychrophilum* in salmonids;
(10) columnaris caused by *F. columnare* in channel catfish and salmonids;
(11) enteric septicaemia of catfish caused by *E. ictaluri* in catfish;
(12) Edwardsiellosis or putrefactive disease caused by *Edwardsiella tarda* in tilapia;
(13) bacterial kidney disease caused by *Renibacterium salmoninarum* in salmonids;
(14) lactococcosis caused by *Lactococcus garvieae* in rainbow trout, amberjack and yellowtail;
(15) Pasteurellosis caused by *Photobacterium damselae* subsp. *piscicida* in seabream, seabass, amberjack and yellowtail;
(16) streptococcosis caused by *Streptococcus iniae* or *Streptococcus phocae* in tilapia, seabass and salmonids;
(17) wound disease or winter ulcer disease caused by *M. viscosa* in salmon; and
(18) streptococcosis/lactococcosis caused by *S. iniae* and *L. garvieae* in trout.

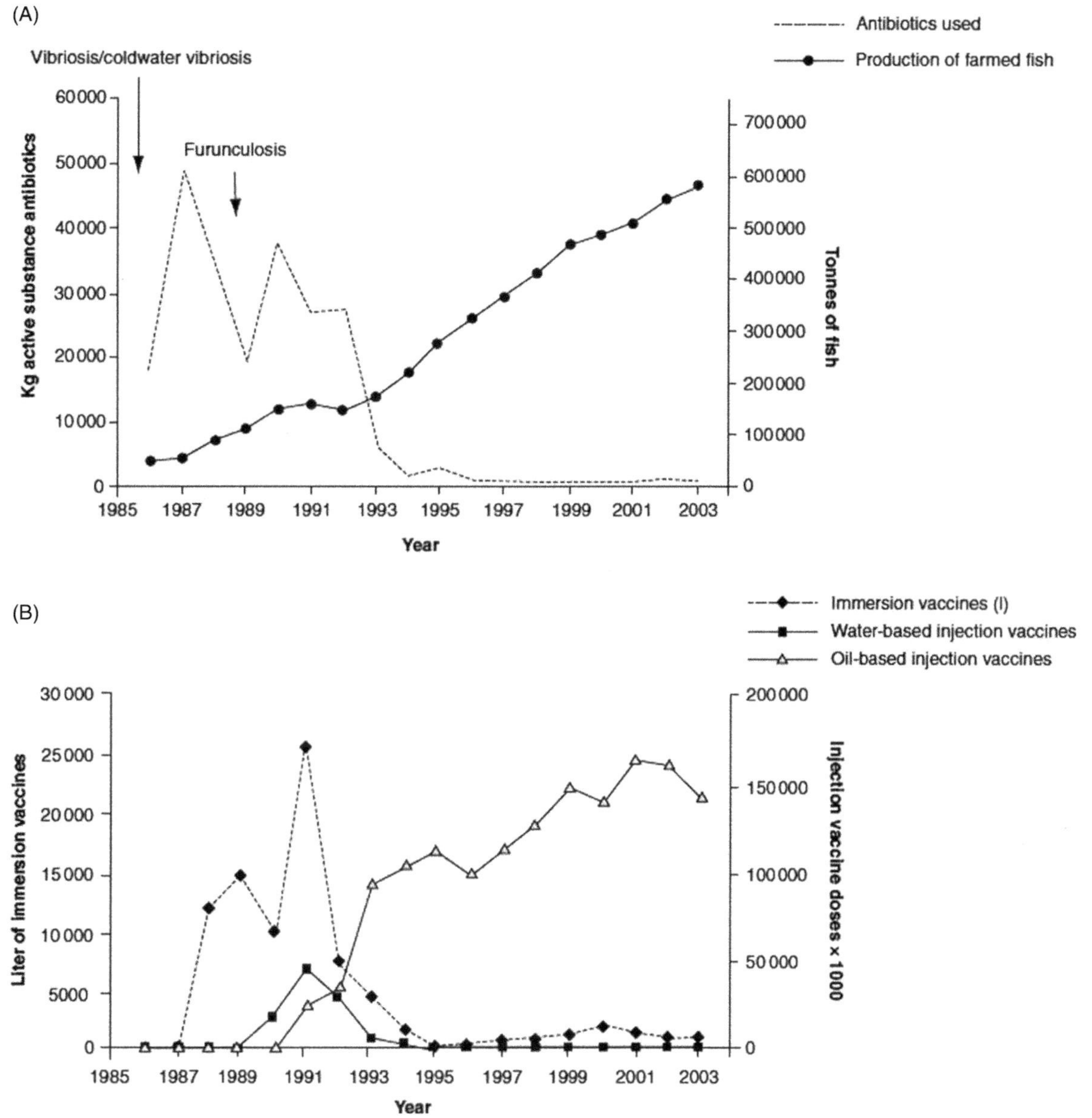

Figure 1 The use of the antibiotics (A) and different types of vaccines (B) during the growth of Norweigian aquaculture industry from 1986 to 2003. Information obtained from Sommerset *et al.* [173]

Although vaccines are commercially available for protection against 18 diseases mentioned above, there are other important diseases in aquaculture, including motile aeromonads septicaemia, chryseobacteriosis, Francisellosis, pseudomonads septicaemia, nocardiosis, staphylococcosis, streptococcosis caused by *Streptococcus agalactiae*, and other emerging diseases, for which vaccines are not yet available. The causative agents of these diseases and their vaccine development status are summarized below:

(1) Motile aeromonad septicaemia (MAS) caused by *A. hydrophila*. Although usually considered as a secondary pathogen associated with disease outbreaks, *A. hydrophila* could also become a primary pathogen, causing outbreaks in fish farms with high mortality rates and severe economic losses to the aquaculture industry [174, 177–181]. In West Alabama of USA, disease outbreaks caused by *A. hydrophila* in 2009 and 2010 caused an estimated loss of more than \$3 million annually [182, 183]. Virulence studies have revealed that the 2009 West Alabama isolates of *A. hydrophila* are highly virulent to channel catfish, with LD_{50} values as low as $2{\times}10^2$ CFU/fish by intraperitoneal injection [183]. MAS disease caused by *A. hydrophila* such as the 2009 West Alabama isolate could be very acute, causing high mortality within 24 h [182, 183]. Therefore, vaccines are urgently needed to prevent MAS. The most extensively studied *A. hydrophila* vaccines are bacterins consisting of formalin or heat-killed

bacteria of pathogenic *A. hydrophila* strains [184–186]. In addition, recombinant protein vaccines such as *A. hydrophila* outer membrane proteins and bacterial lysate have been demonstrated to elicit protection against *A. hydrophila* challenges in laboratory studies [187–190]. Furthermore, live attenuated vaccines such as *aroA* mutant and transposon Tn916-generated mutant have been reported to confer significant protection against homologous *A. hydrophila* challenge [191, 192]. However, it is well known that *A. hydrophila* is very heterogeneous biochemically and serologically, thus presenting the biggest obstacle in developing effective commercial vaccine against *A. hydrophila* [188, 193]. To prevent future disease outbreaks caused by *A. hydrophila*, a vaccine that could offer protection against multiple serotypes from various regions is urgently needed.

(2) Chryseobacteriosis or salmon skin syndrome caused by *Chryseobacterium* sp. Although members of *Chryseobacterium* are not relevant pathogens for domestic animal species, they have been associated with diseases of freshwater and marine fish [34, 35, 194]. In addition, *Chryseobacterium* sp. is reported to be the causative agent of salmon skin syndrome [195]. Therefore, vaccines are needed to protect fish from virulent isolates of *Chryseobacterium*.

(3) Francisellosis caused by *Francisella* sp. Within the last decade, *Francisella* sp. have been reported to cause high mortality in a wide range of diseased fish, including three-line grunt (*Parapristipoma trilineatum*) in Japan [196], hybrid striped bass (*Morone chrysops* × *Morone saxatilis*) in USA [197], cod (*Gadus morhua*) in Norway [198–200], tilapia (*Oreochromis* spp.) in Latin America, Costa Rica, UK and USA [61, 201–203], salmon (*Salmo salar*) in Chile [204] and giant abalone (*Haliotis gigantean*) in Japan [205]. More aquatic animals are possibly affected by *Francisella* bacteria because of their fastidious growth condition requiring specialized solid and liquid media containing cysteine and a source of iron, resulting in difficult culture and isolation. Francisellosis in fish develops in a similar fashion independent of host species. The common disease characteristic is the development of multi-organ granuloma and high morbidity [206]. *Francisella* can be transferred through live fish movement [207], making Francisellosis a significant threat to aquaculture operations. Experimentally, one deletion mutant of *Francisella asiatica* iglC has been evaluated as a live attenuated vaccine, which is able to offer significant protection against virulent *F. asiatica* infections in tilapia [208]. However, no vaccine is currently commercially available to protect fish against infections by *Francisella* bacteria.

(4) Pseudomonad septicaemia or red spot disease caused by *Pseudomonas* spp. *Pseudomonas* spp. are emerging fish pathogens responsible for high mortality and disease outbreaks in various fish species. *Pseuodomonas anguilliseptica* was first reported in Japanese eels (*Anguilla japonica*) affected by haemorrhagic septicaemia in Japan [209]. Subsequently, *Pseudomonas* spp. was recognized as the causative agents of red spot diseases in different aquatic animals, including European eel (*Anguilla anguilla*) in UK and the Netherlands [81, 85], Atlantic salmon (*S. salar*) [210], sea trout (*Salmo trutta*) [210], rainbow trout (*Salmo gairdneri*), whitefish (*Coregonus* sp.) in Finland [210], sea bream (*Sparus aurata*) in Spain [211, 212] and rainbow trout (*Oncorhynchus mykiss*) in Turkey [80]. In addition, when combined with parasite, *Pseudomonas* is responsible for over 70% mortality of sea bream in Egypt [79]. Pseudomonad septicaemia is characterized by high mortality related to a decrease in the water temperatures below 11–12°C [210]. The typical clinical sign is ulceration on the dorsal surface. To prevent diseases caused by *Pseudomonas* spp., different experimental vaccines have been used. Formalin-killed and heat-inactivated cells of *P. anguilliseptica* have been reported to offer protection in eels against virulent *Pseudomonas* infection [213, 214]. Bacterin against *P. anguilliseptica* was also able to provide protection to turbot (*Scophthalmus maximus*) for 12 weeks when tested in an experimental challenge trial [215]. However, no vaccine is currently commercially available to protect aquatic animals against *Pseudomonas* bacterial infections.

(5) Nocardiosis caused by *Nocardia* sp. *Nocardia* bacteria are aerobic, Gram-positive, branching, filamentous and rod-shaped. Although many *Nocardia* species are commonly found in soil and plants [216–218], which may be innocuous to human and animals, some *Nocardia* species are primary pathogen of human and animals [219–222]. Recently, *Nocardia* sp. has been isolated from various diseased aquatic animals, including three striped tigerfish (*Terapon jarbua*) [123], snakehead (*Ophiocephalus argus*) [124, 125], large croaker (*Pseudosciaena crocea*) [126], yellow croaker (*Larimichthys crocea*) [127], striped mullet (*Mugil cephalus*) [128], seabass (*Lateolabrax japonicus*) [129], largemouth bass (*Micropterus salmoides*) [130, 131], giant gourami (*Osphronemus goramy*) [132] and yellowtail (*Seriola quinqueradiata*) [133]. Nocardiosis is a systemic disease of fish where lesions are localized in the skin and several internal organs, with the presence of nodular lesions typical of granulomata [127]. An epizootic disease in large-mouth bass (*M. salmoides*) in a fresh water pond was reported to be caused by *Nocardia asteroides* in Taiwan in 1989, with an accumulative mortality of 30% (15 000 out of total 50 000 fish) within 1 month [131]. *Nocardia* sp. also caused 17.5% mortality (3500 out of 20 000 fish) within a month in pond cultured sea bass (*L. japonicus*) in Taiwan in 1997 [129]. In Zhejiang China, *Nocardia* sp. caused 35% mortality (180 000 out of 500 000) in pond-cultured snakehead (*Ophiocephalus argus*)

in 2006 [125]. Chemotherapy for nocardiosis has been ineffective partly because the tubercles formed in diseased fish prevents access and sequestrate the organisms from the drug effect [133]. Killed vaccines have been evaluated experimentally in yellowtail and largemouth bass. Although yellowtail immunized with formalin-killed cells (FKC) of *Nocardia seriolae* or FKC with Freund's Incomplete Adjuvant (FKC-FIA) showed humoural responses, no protective effect was observed in immunized yellowtail [223]. Similarly, when largemouth bass (*M. salmoides*) were immunized with FKC of four different strains of *N. seriolae*, specific antibody titres were increased significantly at 4 weeks after immunization. However, immunization with different strains with or without booster immunization failed to protect largemouth bass against challenges by virulent strains of *N. seriolae* [224]. On the other hand, when yellowtails (*S. quinqueradiata*) were vaccinated with live bacteria of *Nocardia soli* and *Nocardia fluminea* cells, slight protection against an artificial challenge with *N. seriolae* was observed [23]. In addition, fish that survived the *N. seriolae* infection showed complete resistance to the *N. seriolae* challenge [225], suggesting that live vaccines should be developed to protect aquatic animals against nocardiosis.

(6) Staphylococcosis caused by *Staphylococcus* spp. *Staphylococcus* bacteria are important opportunistic human pathogens and leading cause of a wide variety of diseases in humans and animals [226]. For example, *Staphylococcus aureus* is the aetiological agent responsible for a large extent of morbidity and mortality globally, in both hospital and community settings [227, 228]. Antibiotic resistance in *S. aureus* is a major clinical problem, in particular in infections caused by methicillin-resistant *S. aureus* (MRSA) [226]. In aquatic animals, three *Staphylococcus* species (*Staphylococcus epidermidis* [142], *S. aureus* [141] and *Staphylococcus warneri* [137]) have been reported as disease-causing agents. *S. epidermidis* was reported to be the causative agent of the epizootics in cultured yellowtail (*S. quinqueradiata*) and red sea bream (*Chrysophrys major*) in Japan between 1976 and 1977 [142]. *S. epidermidis* was responsible for the mass mortality of tilapia between 1992 and 1996 in Taiwan [138]. *S. aureus* was reported to be the causative agent of an eye disease during 1982 and 1983 in India, causing mortalities in silver carp, *Hypophthalmichthys molitrix* [227]. *S. warneri* is reported to be the causative agent of a disease outbreak in a rainbow trout farm during the spring of 1997 (May–June) in Spain [137]. The typical symptoms of staphylococcosis are exophthalmia and swollen lesion on the tail [142]. Currently no vaccine against *Staphylococcus* spp. is commercially available or experimentally tested in aquatic animals.

(7) Streptococcosis caused by *Streptococcus* sp. *Streptococcus* bacteria are important pathogens of both human and other animals. For example, *S. agalactiae* (group B *streptococcus*) is an important cause of disease in infants, pregnant women, elderly and immune-suppressed adults [229]. In addition, *S. agalactiae* is responsible for many cases of acute clinical mastitis in dairy animals [230, 231]. *S. agalactiae* also affects a variety of cultured and wild fish species in freshwater, estuarine and marine environments. These include tilapia [143, 144], grouper [145], mullet [146] and pomfret [147]. In August and September of 2001, *S. agalactiae* was responsible for a massive fish kill at the Kuwait Bay, killing over 2500 metric tonnes of wild mullet (*Liza klunzingeri*) [146]. In addition, *S. agalactiae* was reported to be the causative agent of disease outbreaks from 2007 to 2011 in Queensland, Australia in giant Queensland grouper (*Epinephelus lanceolatus*) [145]. No commercial vaccine against *S. agalactiae* is currently available in aquaculture. Experimentally, formalin-killed cells along with concentrated extracellular products (>3 kDa) of a single isolate of *S. agalactiae* have been reported to offer significant protection to 30 g tilapia, but not to 5 g tilapia, with a relative percent of survival (RPS) rate of 80% in vaccinated fish compared with non-vaccinated fish at 30 days post vaccination [232, 233]. However, the RPS dropped to 60, 55 and 46% at 47, 90 and 180 days post vaccination, respectively [234]. It was also reported that a formalin-killed *S. agalactiae* vaccine was able to increase the antibody titre in vaccinated fish compared with control fish. However, vaccination by either intraperitoneal injection or by a combination of intraperitoneal injection with oral booster vaccination (one and two doses per month) failed to protect Nile tilapia from *S. agalactiae* infections [235]. In addition, an experimental *S. agalactiae* biotype 1 (β-haemolytic) vaccine was reported to be able to offer protection against lethal challenges with both biotype 1 and biotype 2 (non-haemolytic) strains, whereas biotype II failed to offer protection against challenges with biotype 1 [236]. Recently, Merck Animal Health Inc. has obtained regulatory approval in Brazil to begin marketing AQUAVAC® Strep Sa, an inactivated oil-adjuvanted vaccine that provides protection against *S. agalactiae* infections in tilapia and other susceptible fish species. However, AQUAVAC® Strep Sa is only available by injection to fish weighing more than 15 g, which is rather labour-intensive. Besides *S. agalactiae*, *S. iniae* is also an important fish pathogen in aquaculture. Although bacterin vaccine is currently available in Asia to protect tilapia from *S. iniae* infection, it is not available for use in other parts of the world. *S. iniae* is responsible for a disease outbreak that killed 75% of hybrid tilapias (*Tilapia nilotica*×*Tilapia aurea*) in a Texas fish farm in USA in the early 1990s [237]. *S. iniae* is also the causative agent of disease outbreaks in seabream [150], red porgy [151], trout [151], flounder [152], barramundi

[153], rabbitfish [154], yellowtail [156] and catfish [157]. The most extensively studied *S. iniae* vaccines are bacterins [238, 239]. These formalin-killed *S. iniae* have been previously successfully used as vaccines to protect rainbow trout in Israel. However, recently it has been reported that these bacterins are unable to protect fish from infection by variant isolates (serotypes) of *S. iniae* [240, 241]. Other vaccines consisting of both formalin killed *S. iniae and* concentrated extracellular products of *S. iniae* have been reported to partially protect Nile tilapia (*Oreochromis niloticus* L.) from *S. iniae* infection [242–245]. Bacterins against *S. iniae* infection in tilapia are commercially available in Asia [179]. In addition to bacterins, live attenuated *S. iniae* strains defective in phosphoglucomutase and M-like protein have been reported to offer protection against homologous *S. iniae* challenge [246, 247]. However, it is not clear whether they offer protection against heterologous *S. iniae*. Recently, a highly efficacious attenuated *S. iniae* vaccine (ISNO) has been reported to protect tilapia at least 6 months against a virulent *S. iniae* strain [248]. In addition, ISNO is able to offer protection against heterologous virulent strains of *S. iniae* (F3CB, 102F1 K, 405F1 K, IF6 and ARS60) at 2-month post vaccination, with relative percent of survival of 78, 90, 100, 100 and 100%, respectively [248]. Furthermore, ISNO vaccine has a wide range of efficacious immunization dosage, with RPS values of 81, 94, 100, 100, 100 and 100 at vaccination dosage of 1×10^2, 1×10^3, 1×10^4, 1×10^5, 1×10^6 and 1×10^7 CFU/fish, respectively [248]. In addition to *S. agalactiae* and *S. iniae*, other *Streptococcus* bacteria are also emerging as important pathogens in aquaculture. For example, *Streptococcus dysgalactiae* is responsible for disease outbreaks in trout, sturgeon, amberjack and yellowtail [157–159], and *S. phocae* is the causative agent of repeated disease outbreaks in cage-farmed Atlantic salmon in Chile since 1999, with mortality up to 25% [160–162]. Currently, no vaccine is commercially available to protect fish from infections by *S. dysgalactiae* or *S. phocae*.

(8) Other emerging diseases. Besides the major diseases mentioned above, other emerging diseases can also be detrimental to the aquaculture industry, including the coldwater 'streptococcosis' caused by *Vagococcus salmoninarum* [163–165], and the haemorrhagic septicaemia caused by *Weissella* sp. [166, 167]. Vaccines have not yet been developed to protect fish from these emerging diseases either commercially or experimentally.

Conclusion

Aquaculture is the fastest growing food-producing industry. To provide sufficient food for the ever-increasing human population, protecting the aquaculture industry from diseases is a priority. To prevent bacterial diseases, using vaccines instead of antibiotics has been proven to be effective and beneficial. Although various vaccines have been developed to protect aquatic animals against various bacterial diseases, vaccines against multiple emerging diseases are still urgently needed for the aquaculture industry. In addition, majority of the vaccines available are bacterins which can only provide partial protection against certain strains for a limited time frame. Furthermore, most vaccines have to be delivered by injection, which is labour-intensive. Therefore, user-friendly (immersion or oral delivery) efficacious vaccines that can offer broader protection for a longer duration are urgently needed for the aquaculture industry.

Acknowledgements

We thank Drs Dehai Xu (USDA-ARS) and Victor Panangala (USDA collaborator) for critical reviews of the manuscript. This study was supported by the USDA/ARS CRIS project no. 6420-32000-024-00D. The use of trade, firm or corporate names in this publication is for the information and convenience of the reader. Such use does not constitute an official endorsement or approval by the United States Department of Agriculture or the Agricultural Research Service of any product or service to the exclusion of others that may be suitable.

References

1. Khan MA, Khan S, Miyan K. Aquaculture as a food production system: a review. Biology and Medicine 2011;3:291–302.
2. Subasinghe RP. Epidemiological approach to aquatic animal health management: opportunities and challenges for developing countries to increase aquatic production through aquaculture. Preventive Veterinary Medicine 2005;67:117–24.
3. Food and Agriculture Organization of the United Nations. FAO Yearbook: Fishery and Aquaculture Statistics. FAO, Rome; 2009. p. 12–3.
4. Food and Agriculture Organization of the United Nations. The State of World Fisheries and Aquaculture. FAO, Rome; 2010. p. 4–5.
5. Subasinghe RP, Bondad-Reantaso MG, McGladdery SE. Aquaculture development, health and wealth. In: Subasinghe RP, Bueno P, Phillips MJ, Hough C, McGladdery SE, Arthur JR, editors. Aquaculture in the Third Millennium. Technical Proceedings of the Conference on Aquaculture in the Third Millennium Bangkok: NACA and FAO; 2001. p. 167–91.
6. Frans I, Michiels CW, Bossier P, Willems KA, Lievens B, Rediers H. *Vibrio anguillarum* as a fish pathogen: virulence factors, diagnosis and prevention. Journal of Fish Diseases 2011;34:643–61.
7. Wang W. Bacterial diseases of crabs: a review. Journal of Invertebrate Pathology 2011;106:18–26.

8. Birkbeck TH, Feist SW, Verner-Jeffreys DW. *Francisella* infections in fish and shellfish. Journal of Fish Diseases 2011;34:173–87.

9. Jacobs JM, Stine CB, Baya AM, Kent ML. A review of mycobacteriosis in marine fish. Journal of Fish Diseases 2009;32:119–30.

10. Ramaiah N. A review on fungal diseases of algae, marine fishes, shrimps and corals. Indian Journal of Marine Sciences 2006;35:380–7.

11. Khoo L. Fungal diseases in fish. Seminars in Avian and Exotic Pet Medicine 2000;9:102–11.

12. Vega-Heredia S, Mendoza-Cano F, Sanchez-Paz A. The infectious hypodermal and haematopoietic necrosis virus: a brief review of what we do and do not know. Transboundary and Emerging Diseases 2012;59:95–105.

13. Ransangan J, Manin BO, Abdullah A, Roli Z, Sharudin EF. Betanodavirus infection in golden pompano, *Trachinotus blochii*, fingerlings cultured in deep-sea cage culture facility in Langkawi, Malaysia. Aquaculture 2011;315:327–34.

14. Oidtmann B, Stentiford GD. White spot syndrome virus (WSSV) concentrations in crustacean tissues – a review of data relevant to assess the risk associated with commodity trade. Transboundary and Emerging Diseases 2011;58:469–82.

15. Gomez-Casado E, Estepa A, Coll JM. A comparative review on European-farmed finfish RNA viruses and their vaccines. Vaccine 2011;29:2657–71.

16. Guo FC, Woo PT. Selected parasitosis in cultured and wild fish. Veterinary Parasitology 2009;163(3, Special Issue):207–16.

17. Brooker AJ, Shinn AP, Bron JE. A review of the biology of the parasitic copepod *Lernaeocera branchialis* (L., 1767) (Copepoda: Pennellidae). Advances in Parasitology 2007;65:297–341.

18. Klesius PH, Pridgeon JW. Live attenuated bacterial vaccines in aquaculture. In: Proceedings of the 9th International Symposium on Tilapia in Aquaculture; 2011. p. 18–26.

19. Pridgeon JW, Klesius PH. Molecular identification and virulence of three *Aeromonas hydrophila* isolates cultured from infected channel catfish during a disease outbreak in west Alabama (USA) in 2009. Diseases of Aquatic Organisms 2011;94:249–53.

20. Shome R, Shome BR, Mazumder Y, Das A, Kumar A, Rahman H, *et al.* Abdominal dropsy disease in major carps of Meghalaya: isolation and characterization of *Aeromonas hydrophila*. Current Science (Bangalore) 2005;88:1897–900.

21. Coz-Rakovac R, Popovic NT, Strunjak-Perovic I, Krca S, Grubacevic N. *Aeromonas hydrophila* outburst on carp (*Cyprinus carpio*, L.) after transfer to aquarium. Periodicum Biologorum 2000;102:293–6.

22. Ogara WO, Mbuthia PG, Kaburia HF, Sorum H, Kagunya DK, Nduthu DI, *et al.* Motile aeromonads associated with rainbow trout (*Onchorhynchus mykiss*) mortality in Kenya. Bulletin of the European Association of Fish Pathologists 1998;18:7–9.

23. Fuentes RJ, Pérez HJ. Isolation of *Aeromonas hydrophila* in the rainbow trout (*Oncorhynchus mykiss*). Veterinaria México 1998;29:117–9.

24. Timur M. An outbreak of a disease of farmed eel (*Anguilla anguilla*) due to *Aeromonas hydrophila* in turkey: histopathological and bacteriological studies. Veteriner Fakultesi Dergisi Ankara Universitesi 1983;30:361–7.

25. Zhao F, Cao J, Liu Q. Study on pathology and etiology of hemorrhagic septicemia in *Acipenser baerii*. Acta Hydrobiologica Sinica 2009;33:316–23.

26. Jiménez AP, Iregui CA, Figueroa J. *In vitro*/*in vivo* characterization and evaluation of *Aeromonas hydrophila* lipopolysacharides (LPS). Acta Biologica Colombiana 2008;13(2):147–62.

27. Plumb JA. Major diseases of striped bass and redfish. Veterinary and Human Toxicology 1991;33(Suppl 1):34–9.

28. St JB. Practical experience with furunculosis in an Atlantic salmon hatchery in Washington State. Bulletin of the Aquaculture Association of Canada 1992;(1):47–8.

29. Rottereng J, Stenmark A, Malmberg G. Control of furunculosis in fresh- and seawater farms in Norway. Parasites and Diseases in Natural Waters and Aquaculture in Nordic Countries 1987;129–30.

30. Balcázar JL, Vendrell D, Blas ID, Ruiz-Zarzuela I, Múzquiz JL. Effect of *Lactococcus lactis* CLFP 100 and *Leuconostoc mesenteroides* CLFP 196 on *Aeromonas salmonicida* infection in brown trout (*Salmo trutta*). Journal of Molecular Microbiology and Miotechnology 2009;17:153–7.

31. Kumagai A, Sugimoto K, Itou D, Kamaishi T, Miwa S, Iida T. *Atypical Aeromonas salmonicida* infection in cultured marbled sole *Pleuronectes yokohamae*. Fish Pathology 2006;41:7–12.

32. Pedersen K, Larsen JL. First report on outbreak of furunculosis in turbot *Scophthalmus maximus* caused by *Aeromonas salmonicida* subsp. *salmonicida* in Denmark. Bulletin of the European Association of Fish Pathologists 1996;16:129–33.

33. Reddy TV, Ravindranath K, Sreeraman PK, Rao MV. *Aeromonas salmonicida* associated with mass mortality of *Cyprinus carpio* and *Oreochromis mossambicus* in a freshwater reservoir in Andhra Pradesh, India. Journal of Aquaculture in the Tropics 1994;9:259–68.

34. Ilardi P, Avendaño-Herrera R. Isolation of *Flavobacterium*-like bacteria from diseased salmonids cultured in Chile. Bulletin of the European Association of Fish Pathologists 2008;28:176–85.

35. Ilardi P, Fernandez J, Avendano-Herrera R. *Chryseobacterium piscicola* sp. nov. isolated from diseased salmonid fish. International Journal of Systematic and Evolutionary Microbiology 2009;59:3001–5.

36. Chen MF, Kumlin ME. Enteric septicemia of channel catfish in California USA. California Fish and Game 1989;75:141–7.

37. Jarboe HH, Bowser PR, Robinette HR. Pathology associated with a natural *Edwardsiella ictaluri* infection in channel catfish (*Ictalurus punctatus* Rafinesque). Journal of Wildlife Diseases 1984;20:352–4.

38. Wagner BA, Wise DJ, Khoo LH, Terhune JS. The epidemiology of bacterial diseases in food-size channel catfish. Journal of Aquatic Animal Health 2002;14:263–72.

39. Ye S, Li H, Qiao G, Li Z. First case of *Edwardsiella ictaluri* infection in China farmed yellow catfish *Pelteobagrus fulvidraco*. Aquaculture 2009;292:6–10.

40. Xiao J, Wang Q, Liu Q, Wang X, Liu H, Zhang Y. Isolation and identification of fish pathogen *Edwardsiella tarda* from mariculture in China. Aquaculture Research 2008;40:13–7.

41. Nougayrede P, Vuillaume A, Vigneulle M, Faivre B, Luengo S, Delprat J. First isolation of *Edwardsiella tarda* from diseased turbot (*Scophthalmus maximus*) reared in a sea farm in the Bay of Biscay. Bulletin of the European Association of Fish Pathologists 1994;14:128–9.

42. Matsuoka S, Nakai T. Seasonal appearance of *Edwardsiella tarda* and its bacteriophages in the culture farms of Japanese flounder. Fish Pathology 2004;39:145–52.

43. Kanai K, Tawaki S, Uchida Y. An ecological study of *Edwardsiella tarda* on a flounder farm. Fish Pathology 1988;23:41–7.

44. Shome R, Shome BR, Sarangi N, Bandyopadhyay AK. Etiological characterization of acute infectious abdominal dropsy outbreak affecting Indian major carp, *Cirrhinus mrigala* in South Andaman. Current Science (Bangalore) 1996;70:744–7.

45. Fuhrmann H, Bohm KH, Savvidis G, Schlotfeldt H. *Edwardsiella tarda* as a possible causative agent of a disease in catfish (*Silurus glanis*) and eel (*Anguilla anguilla*). Bulletin of the European Association of Fish Pathologists 1985;5:57–60.

46. Alcaide E, Herraiz S, Esteve C. Occurrence of *Edwardsiella tarda* in wild European eels *Anguilla anguilla* from Mediterranean Spain. Diseases of Aquatic Organisms 2006;73:77–81.

47. Iregui CA, Guarín M, Tibatá VM, Ferguson HW. Novel brain lesions caused by *Edwardsiella tarda* in a red tilapia (*Oreochromis* spp.). Journal of Veterinary Diagnostic Investigation 2012;24:446–9.

48. Saleh WD. Isolation and identification of *Edwardsiella tarda* from infected Nile tilapia fish "*Oreochromis niloticus*". Bulletin of Faculty of Agriculture, Cairo University 2005;56:839–46.

49. Li N, Guo H, Jiao R, Zhang S, Liu Z, Yao W, *et al.* Identification and pathogenicity of bacterial pathogens isolated in an outbreak on bacterial disease of *Ctenopharyngodon idellus*. Acta Hydrobiologica Sinica 2011;35:980–7.

50. Decostere A, Ducatelle R, Haesebrouck F. *Flavobacterium columnare* (*Flexibacter columnaris*) associated with severe gill necrosis in koi carp (*Cyprinus carpio* L). Veterinary Record 2002;150:694–5.

51. Decei P, Dascalescu P, Macoveschi I. Outbreak of columnaris disease in trout on a fish farm. Revista de Cresterea Animalelor 1985;35:46–51.

52. Rehulka J, Mraz O. Columnarosis in rainbow trout (*Salmo gairdneri* R.) in Czechoslovakia. Acta Veterinaria Brno 1982;51:125–37.

53. Morley NJ, Lewis JW. Consequences of an outbreak of columnaris disease (*Flavobacterium columnare*) to the helminth fauna of perch (*Perca fluviatilis*) in the Queen Mary reservoir, south-east England. Journal of Helminthology 2010;84:186–92.

54. Mohamed SG, Saleh WD. *Flavobacterium columnare* infection in cultured *Oreochromis niloticus*. Assiut Veterinary Medical Journal 2010;56:15–30.

55. Figueiredo HC, Klesius PH, Arias CR, Evans J, Shoemaker CA, Pereira DJ, *et al.* Isolation and characterization of strains of *Flavobacterium columnare* from Brazil. Journal of Fish Diseases 2005;28:199–204.

56. Wagner BA, Wise DJ, Khoo LH, Terhune JS. The epidemiology of bacterial diseases in food-size channel catfish. Journal of Aquatic Animal Health 2002;14:263–72.

57. Soltani M, Munday B, Carson J. Susceptibility of some freshwater species of fish to infection by *Cytophaga johnsonae*. Bulletin of the European Association of Fish Pathologists 1994;14:133–5.

58. Dalsgaard I, Madsen L. Bacterial pathogens in rainbow trout, *Oncorhynchus mykiss* (Walbaum), reared at Danish freshwater farms. Journal of Fish Diseases 2000;23:199–209.

59. Bustos PA, Calbuyahue J, Montaña J, Opazo B, Entrala P, Solervicens R. First isolation of *Flexibacter psychrophilus*, as causative agent of rainbow trout fry syndrome (RTFS), producing rainbow trout mortality in Chile. Bulletin of the European Association of Fish Pathologists 1995;15:162–4.

60. Ostland VE, Lumsden JS, MacPhee DD, Ferguson HW. Characteristics of *Flavobacterium branchiophilum*, the cause of salmonid bacterial gill disease in Ontario. Journal of Aquatic Animal Health 1994;6:13–26.

61. Jeffery KR, Stone D, Feist SW, Verner-Jeffreys DW. An outbreak of disease caused by *Francisella* sp. in Nile tilapia *Oreochromis niloticus* at a recirculation fish farm in the UK. Diseases of Aquatic Organisms 2010;91:161–5.

62. Ostland VE, Stannard JA, Creek JJ, Hedrick RP, Ferguson HW, Carlberg JM, *et al.* Aquatic *Francisella*-like bacterium associated with mortality of intensively cultured hybrid striped bass *Morone chrysops*×*M. saxatilis*. Diseases of Aquatic Organisms 2006;72:135–45.

63. Coyne R, Smith P, Dalsgaard I, Nilsen H, Kongshaug H, Bergh O, *et al.* Winter ulcer disease of post-smolt Atlantic salmon: an unsuitable case for treatment? Aquaculture 2006;253:171–8.

64. Coyne R, Bergh O, Samuelsen O, Andersen K, Lunestad BT, Nilsen H, *et al.* Attempt to validate breakpoint MIC values estimated from pharmacokinetic data obtained during oxolinic acid therapy of winter ulcer disease in Atlantic salmon (*Salmo salar*). Aquaculture 2004;238:51–66.

65. Marcu A, Marcu A, Nichita I, Herman V, Pascu C, Costinar L, *et al.* Studies in the first outbreak of vibriosis associated with pasteurelosis at Siberian sturgeon (*Acipenser baerii*) in the south-west region of Romania. Preliminary Report. Lucrari Stiintifice – Zootehnie si Biotehnologii, Universitatea de Stiinte Agricole si Medicina Veterinara a Banatului Timisoara 2010;43:36–9.

66. Huang H, Tu C, Chuang S, Hung H, Su J, Lin S. Pasteurellosis in reared hybrid striped bass (*Morone saxatilis*×*Morone chrysops*). Fish Pathology 2001;36:35–7.

67. Bakopoulos V, Peric Z, Rodger H, Adams A, Richards R. First report of fish pasteurellosis from Malta. Journal of Aquatic Animal Health 1997;9:26–33.

68. Baptista T, Romalde JL, Toranzo AE. First occurrence of pasteurellosis in Portugal affecting cultured gilthead seabream (*Sparus aurata* [*Pagrus aurata*]). Bulletin of the European Association of Fish Pathologists 1996;16:92–5.

69. Toranzo AE, Barreiro S, Casal JF, Figueras A, Magarinos B, Barja JL. Pasteurellosis in cultured gilthead seabream (*Sparus aurata*): first report in Spain. Aquaculture 1991;99:1–15.

70. Fukuda Y. Studies on humoral immune response of yellowtail [*Seriola quinqueradiata*] against pseudotuberculosis. Bulletin

of Oita Institute of Marine and Fisheries Science (Japan) 1997;1:15–87.

71. Candan A, Ang Kucker M, Karatas S. Pasteurellosis in cultured sea bass (*Dicentrarchus labrax*) in Turkey. Bulletin of the European Association of Fish Pathologists 1996;16:150–3.

72. Tung M, Tsai S, Ho L, Huang S, Chen S. An acute septicemic infection of *Pasteurella* organism in pond-cultured Formosa snake-head fish (*Channa maculata* Lacepede) in Taiwan. Fish Pathology 1985;20:143–8.

73. Mladineo I, Miletic I, Bocina I. *Photobacterium damselae* subsp. *piscicida* outbreak in cage-reared Atlantic bluefin tuna *Thunnus thynnus*. Journal of Aquatic Animal Health 2006;18:51–4.

74. Liu P, Lin J, Lee K. Virulence of *Photobacterium damselae* subsp. *piscicida* in cultured cobia *Rachycentron canadum*. Journal of Basic Microbiology 2003;43:499–507.

75. Gaggero A, Castro H, Sandino AM. First isolation of *Piscirickettsia salmonis* from coho salmon, *Oncorhynchus kisutch* (Walbaum), and rainbow trout, *Oncorhynchus mykiss* (Walbaum), during the freshwater stage of their life cycle. Journal of Fish Diseases 1995;18:277–9.

76. Mauel MJ, Miller DL, Frazier K, Liggett AD, Styer L, Montgomery-Brock D, *et al.* Characterization of a piscirickettsiosis-like disease in Hawaiian tilapia. Diseases of Aquatic Organisms 2003;53:249–55.

77. Iregui CA, Vasquez GM, Rey AL, Verjan N. Piscirickettsia-like organisms as a cause of acute necrotic lesions in Colombian tilapia larvae. Journal of Veterinary Diagnostic Investigation 2011;23:147–51.

78. Ibrahem MD, Hatem ME. Mass mortality caused by *Pseudomonas fluorescens* and *Cryptocaryon irritans* in gilthead sea bream (*Sparus aurata* L.) in Egyptian mariculture. In: Proceedings of the Second Global Fisheries and Aquaculture Research Conference; 2009. p. 275–89.

79. Altinok I, Balta F, Capkin E, Kayis S. Disease of rainbow trout caused by *Pseudomonas luteola*. Aquaculture 2007;273:393–7.

80. Haenen OL, Davidse A. First isolation and pathogenicity studies with *Pseudomonas anguilliseptica* from diseased European eel *Anguilla anguilla* (L.) in The Netherlands. Aquaculture 2001;196:27–36.

81. Saeed MO, Alamoudi MM, Al-harbi AH. A *Pseudomonas* associated with disease in cultured rabbitfish *Siganus rivulatus* in the Red Sea. Diseases of Aquatic Organisms 1987;3:177–80.

82. Gatti R, Nigrelli DA. Haemorrhagic septicaemia in catfish. Pathogenicity of the strains isolated and reproducibility of the disease. Obiettivi e Documenti Veterinari 1984;5:61–3.

83. Woldemariam K, Dear G, Mochaba FM, Stewart DJ. An outbreak of 'Sekiten-byo' among cultured European eels, *Anguilla anguilla* L., in Scotland. Journal of Fish Diseases 1983;6:75–6.

84. Soltani M, Ahmadi MR, Yavari H, Mirzargar SS. Red-pink colony-producing *Pseudomonas sp.* is the causative agent of mass mortality in larvae and post-larvae of *Litopenaeus vannamei* raised in hatcheries in south Iran. International Journal of Veterinary Research 2010;4:89–94.

85. Hatai K, Egusa S, Nakajima M, Chikahata H. *Pseudomonas chlororaphis* as a fish pathogen. Bulletin of the Japanese Society of Scientific Fisheries 1975;41:1203.

86. Vilar P, Failde LD, Bermudez R, Vigliano F, Riaza A, Silva R, *et al.* Morphopathological features of a severe ulcerative disease outbreak associated with *Tenacibaculum maritimum* in cultivated sole, *Solea senegalensis* (L.). Journal of Fish Diseases 2012;35:437–45.

87. Avendano-Herrera R, Magarinos B, Irgang R, Toranzo AE. Use of hydrogen peroxide against the fish pathogen *Tenacibaculum maritimum* and its effect on infected turbot (*Scophthalmus maximus*). Aquaculture 2006;257:104–10.

88. Zhang F, Peng Z, Zhang J, Liu M, Fu R, Luo H. Isolation and identification of the pathogenic strain of *Vibrio harveyi* from *Miichthys miiuy*. Weishengwu Xuebao 2010;50:304–9.

89. Won KM, Kim KH, Park SI. *Vibrio harveyi* associated with deep skin ulcer in river puffer, *Takifugu obscures*, exhibited in a public aquarium. Aquaculture Research 2009;40:963–7.

90. Lin Y, Lin H. Analyses the biology and pathogenesis of virulence *Vibrio* spp. in grouper's aquaculture and the development vaccines for application. Journal of the Fisheries Society of Taiwan 2009;36:147–8.

91. Rodger HD, Colquhoun DJ. Clinical vibriosis in farmed Atlantic cod (*Gadus morhua*) in Ireland. Veterinary Record 2008;162:94–5.

92. Soumya H, Shruti C, Asakura M, Manambrakat V, Yamasaki S. Isolation of *Vibrio parahaemolyticus* and *Vibrio cholerae* (Non-O1 and O139) from moribund shrimp (*Penaeus monodon*) and experimental challenge study against post larvae and juveniles. Annals of Microbiology 2007;57:55–60.

93. Sakai T, Hirae T, Yuasa K, Kamaishi T, Matsuyama T, Miwa S, *et al.* Mass mortality of cultured kuruma prawn *Penaeus japonicus* caused by *Vibrio nigripulchritudo*. Fish Pathology 2007;42:141–7.

94. Yiagnisis M, Vatsos IN, Kyriakou C, Alexis M. First report of *Vibrio anguillarum* isolation from diseased big scale sand smelt, *Atherina boyeri* Risso 1810, in Limnos, Greece. Bulletin of the European Association of Fish Pathologists 2007;27:61–9.

95. Sakai T, Yamada H, Shimizu H, Yuasa K, Kamaishi T, Oseko N, *et al.* Characteristics and pathogenicity of brown pigment-producing *Vibrio anguillarum* isolated from Japanese flounder [*Paralichthys olivaceus*]. Fish pathology (Japan) 2006;41:77–9.

96. Cai J, Han Y, Wang Z. Isolation of *Vibrio parahaemolyticus* from abalone (*Haliotis diversicolor supertexta* L.) post larvae associated with mass mortalities. Aquaculture 2006;257:161–6.

97. Tanrikul TT. Outbreaks of *Vibrio alginolyticus* in gilthead sea bream in Turkey. Indian Veterinary Journal 2006;83:599–601.

98. Rasheed V. Vibriosis outbreak among cultured seabream (*Acanthopagrus cuvieri*) broodstock in Kuwait. Aquaculture 1989;76:189–98.

99. Fan W, Huang J, Wang X, Shi C, Liu L. Identification and phylogenetic study of pathogenic bacteria causing ulcer disease of cultured turbot (*Scophthalmus maximus*). Weishengwu Xuebao 2005;45:665–70.

100. Austin B, Stobie M, Robertson PA, Glass HG, Stark JR, Mudarris M. *Vibrio alginolyticus*: the cause of gill disease leading to progressive low-level mortalities among juvenile

turbot, *Scophthalmus maximus* L., in a Scottish aquarium. Journal of Fish Diseases 1993;16:277–80.

101. Zorrilla I, Arijo S, Chabrillon M, Diaz P, Martinez-Manzanares E, Balebona MC, *et al.* *Vibrio* species isolated from diseased farmed sole, *Solea senegalensis* (Kaup), and evaluation of the potential virulence role of their extracellular products. Journal of Fish Diseases 2003;26:103–8.

102. Liu P, Chuang W, Lee K. Infectious gastroenteritis caused by *Vibrio harveyi* (*V. carchariae*) in cultured red drum, *Sciaenops ocellatus*. Journal of Applied Ichthyology 2003;19:59–61.

103. Rajan PR, Lopez C, Lin JH, Yang H. *Vibrio alginolyticus* infection in cobia (*Rachycentron canadum*) cultured in Taiwan. Bulletin of the European Association of Fish Pathologists 2001;21:228–34.

104. Dalsgaard I, Hoi L, Siebeling RJ, Dalsgaard A. Indole-positive *Vibrio vulnificus* isolated from disease outbreaks on a Danish eel farm. Diseases of Aquatic Organisms 1999;35:187–94.

105. Biosca EG, Amaro C, Esteve C, Alcaide E, Garay E. First record of *Vibrio vulnificus* biotype 2 from diseased European eel, *Anguilla anguilla* L. Journal of Fish Diseases 1991;14:103–10.

106. Nicholson JT, Dodge P. Cold water vibriosis in Shetland salmon. Veterinary Record 1995;137:444.

107. Yashima T, Odajima T. An outbreak of vibriosis in wild ayu (*Plecoglossus altivelis*) and characteristics of the isolates. Journal of the Japan Veterinary Medical Association 1992;45:203–6.

108. Farkas J, Malik SE. Vibrio disease of sheatfish (*Silurus glanis* L.) fry. Aquaculture 1986;51:81–8.

109. Lewis DH. Vibriosis in channel catfish, *Ictalurus punctatus* (Rafinesque). Journal of Fish Diseases 1985;8:539–45.

110. Guguianu E, Vulpe V, Lazar M, Rîmbu C. Yersiniosis outbreak in rainbow trout (*Oncorhynchus mykis*) at a fish farm from northern Romania. Cercetari Agronomice în Moldova 2009;42:75–80.

111. Oraic D, Zrnčic S, Šoštaric B, Bazulic D, Lipej Z. Occurrence of enteric redmouth disease in rainbow trout (*Oncorhynchus mykiss*) in farms in Croatia. Acta Veterinaria Hungarica 2002;50:283–91.

112. Rodríguez LA, Castillo A, Gallardo CS, Nieto TP. Outbreaks of *Yersinia ruckeri* in rainbow trout in north west of Spain. Bulletin of the European Association of Fish Pathologists 1999;19:130–2.

113. Eissa AE, Moustafa M, Abdelaziz M, Ezzeldeen NA. *Yersinia ruckeri* infection in cultured Nile tilapia, *Oreochromis niloticus*, at a semi-intensive fish farm in lower Egypt. African Journal of Aquatic Science 2008;33:283–6.

114. Dear G. *Yersinia ruckeri* isolated from Atlantic salmon in Scotland. Bulletin of the European Association of Fish Pathologists 1988;8:18–20.

115. Sparboe O, Håsteom T, Poppe TT, Koren C, Stenwig H. *Yersinia ruckeri* infection in salmon demonstrated for the first time in Norway. Norsk Veterinaertidsskrift 1986;98:189–92.

116. Park K, Matsuoka S, Nakai T, Muroga K. A virulent bacteriophage of *Lactococcus garvieae* (formerly *Enterococcus seriolicida*) isolated from yellowtail *Seriola quinqueradiata*. Diseases of Aquatic Organisms 1997;29:145–9.

117. Schmidtke LM, Carson J. *Lactococcus garvieae* strains isolated from rainbow trout and yellowtail in Australia, South Africa and Japan differentiated by repetitive sequence markers. Bulletin of the European Association of Fish Pathologists 2003;23:206–12.

118. Soltani M, Nikbakht G, Mousavi HA, Ahmadzadeh N. Epizootic outbreaks of lactococcosis caused by *Lactococcus garvieae* in farmed rainbow trout (*Oncorhynchus mykiss*) in Iran. Bulletin of the European Association of Fish Pathologists 2008;28:207–12.

119. Chang PH, Lin CW, Lee YC. *Lactococcus garvieae* infection of cultured rainbow trout, *Oncorhynchus mykiss*, in Taiwan and associated biophysical characteristics and histopathology. Bulletin of the European Association of Fish Pathologists 2002;22:319–27.

120. Aguado-Urda M, Lopez-Campos GH, Gibello A, Teresa Cutuli M, Lopez-Alonso V, Fernandez-Garayzabal JF, *et al.* Genome sequence of *Lactococcus garvieae* 8831, isolated from rainbow trout lactococcosis outbreaks in Spain. Journal of Bacteriology 2011;193:4263–4.

121. Kobayashi T, Ishitaka Y, Imai M, Kawaguchi Y. Pathological studies on *Lactococcus garvieae* infection of cultured rockfish *Sebastes schlegeli*. Suisanzoshoku (Japan) 2004;52:359–64.

122. Lee J, Chu S, Yu Z, Lee C, Lai C, Hung C, *et al.* Identification and survey on gram-positive *Streptococcus*-like organisms isolated from grey mullet (*Mugil cephalus* Linnaeus). Taiwan Veterinary Journal 2006;32:136–45.

123. Wang PC, Chen SD, Tsai MA, Weng YJ, Chu SY, Chern RS, *et al.* *Nocardia seriolae* infection in the three striped tigerfish, *Terapon jarbua* (Forsskål). Journal of Fish Diseases 2009;32:301–10.

124. Wang G, Xu Y, Jin S, Zhu J, Zhu W. Research on the nocardiosis and pathogen in reared snakehead, *Ophiocephalus argus* Cantor. Acta Hydrobiologica Sinica 2009;33:277–83.

125. Wang G, Xu Y, Jin S, Zhu J, Yuan S. Nocardiosis in snakehead, *Ophiocephalus argus* Cantor. Aquaculture 2007;271:54–60.

126. Wang G, Yuan S, Jin S. Preliminary study on nocardiosis in cage-reared large croaker, *Pseudosciaena crocea* (Richardson). Journal of Fisheries of China 2006;30:103–7.

127. Wang G, Yuan S, Jin S. Nocardiosis in large yellow croaker, *Larimichthys crocea* (Richardson). Journal of Fish Diseases 2005;28:339–45.

128. Huang S, Lai K, Su S, Shei M, Chen S. Isolation and characterization of the pathogenic bacterium, *Nocardia seriolae*, from female broodstock of striped mullet (*Mugil cephalus*). Journal of Taiwan Fisheries Research 2004;12:61–9.

129. Chen S, Lee J, Lai C, Gu Y, Wang C, Chang H, *et al.* Nocardiosis in seabass, *Lateolabrax japonicus*, in Taiwan. Journal of Fish Diseases 2000;23:299–307.

130. Chen S. The study on the pathogenicity of *Nocardia asteroides* to largemouth bass *Micropterus salmoides* Lacepede. Fish Pathology 1992;27:1–5.

131. Chen SC, Tung MC. An epizootic in large mouth bass (*Micropterus salmoides* Lacepede), caused by *Nocardia asteroides* in fresh water pond in southern Taiwan. Journal of the Chinese Society of Veterinary Science 1991;17:15–22.

132. Kitao T, Ruangpan L, Fukudome M. Isolation and classification of a *Nocardia* species from diseased giant gourami *Osphronemus goramy*. Journal of Aquatic Animal Health 1989;1:154–62.

133. Kusuda R, Nakagawa A. Nocardial infection of cultured yellowtail. Fish Pathology 1978;13:25–31.

134. Lorenzen E, Olesen NJ, Korsholm H, Heuer OE, Evensen O. First demonstration of *Renibacterium salmoninarum*/BKD in Denmark. Bulletin of the European Association of Fish Pathologists 1997;17:140–4.

135. Evensen O, Dale OB, Nilsen A. Immunohistochemical identification of *Renibacterium salmoninarum* by monoclonal antibodies in paraffin-embedded tissues of Atlantic salmon (*Salmo salar* L.), using paired immunoenzyme and paired immunofluorescence techniques. Journal of Veterinary Diagnostic Investigation 1994;6:48–55.

136. Hatakeyama M, Miura K. Seasonal change in serum agglutination titer against *Renibacterium salmoninarum* in farmed masu salmon *Oncorhynchus masou*. Scientific Reports of the Hokkaido Fish Hatchery (Japan) 2006;60:91–7.

137. Gil P, Vivas J, Gallardo CS, Rodriguez LA. First isolation of *Staphylococcus warneri*, from diseased rainbow trout, *Oncorhynchus mykiss* (Walbaum), in northwest Spain. Journal of Fish Diseases 2000;23:295–8.

138. Huang S, Chen W, Shei M, Liao I, Chen S. Studies on epizootiology and pathogenicity of *Staphylococcus epidermidis* in Tilapia (*Oreochromis* spp.) cultured in Taiwan. Zoological Studies 1999;38:178–88.

139. Wang W, Chang Y, Shieh M, Lin C. *Staphylococcus epidermidis* and cestode infection of cultured grass carp (*Ctenopharyngodon idella*) in Taiwan. Reports on Fish Disease Research 1996;0:57–63.

140. Wang W, Shieh M, Lee J, Liu C. Case report: *Staphylococcus epidermidis* isolated from diseased cultured sea perch (*Lateolabrax japonicus*) in Taiwan. Taiwan Journal of Veterinary Medicine and Animal Husbandry 1995;65:305–9.

141. Baxa DV, Kawai K, Ando H, Kusuda R. *Edwardsiella tarda* and *Staphylococcus aureus* isolated from cultured red sea bream. Reports of the USA Marine Biological Institute Kochi University 1985;7:1–8.

142. Kusuda R, Sugiyama A. Studies on the characters of *Staphylococcus epidermidis* isolated from diseased fish. I. Morphological, biological and biochemical properties. Fish Pathology 1981;16:15–24.

143. Ye X, Li J, Lu M, Deng G, Jiang X, Tian Y, *et al.* Identification and molecular typing of *Streptococcus agalactiae* isolated from pond-cultured tilapia in China. Fisheries Science 2011;77:623–32.

144. Abuseliana AF, Daud HH, Aziz SA, Bejo SK, Alsaid M. Pathogenicity of *Streptococcus agalactiae* isolated from a fish farm in Selangor to juvenile red tilapia (*Oreochromis* sp.). Journal of Animal and Veterinary Advances 2011;10:914–9.

145. Bowater RO, Forbes-Faulkner J, Anderson IG, Condon K, Robinson B, Kong F, *et al.* Natural outbreak of *Streptococcus agalactiae* (GBS) infection in wild giant Queensland grouper, *Epinephelus lanceolatus* (Bloch), and other wild fish in northern Queensland, Australia. Journal of Fish Diseases 2012;35:173–86.

146. Glibert PM, Landsberg JH, Evans JJ, Al-Sarawi MA, Faraj M, Al-Jarallah MA, *et al.* A fish kill of massive proportion in Kuwait Bay, Arabian Gulf, 2001: the roles of bacterial disease, harmful algae, and eutrophication. Harmful Algae 2002;1:215–31.

147. Al-Marzouk A, Duremdez R, Al-Gharabally H. Efforts to control outbreaks of diseases among cultured silver pomfret *Pampus argenteus* in Kuwait. Journal of Aquaculture in the Tropics 2004;19:103–10.

148. Pasnik DJ, Evans JJ, Klesius PH, Shoemaker CA, Yeh, H. Pathogenicity of *Streptococcus ictaluri* to channel catfish. Journal of Aquatic Animal Health 2009;21:184–8.

149. Al Harbi AH. Molecular characterization of *Streptococcus iniae* isolated from hybrid tilapia [*Oreochromis niloticus*×*Oreochromis aureus*]. Aquaculture 2011;312:15–8.

150. El Aamri F, Padilla D, Acosta F, Caballero MJ, Roo J, Bravo J, *et al.* First report of *Streptococcus iniae* in red porgy (*Pagrus pagrus* L.). Journal of Fish Diseases 2010;33:901–5.

151. Haghighi Karsidani S, Soltani M, Nikbakhat-Brojeni G, Ghasemi M, Skall H. Molecular epidemiology of zoonotic streptococcosis/lactococcosis in rainbow trout (*Oncorhynchus mykiss*) aquaculture in Iran. Iranian Journal of Microbiology 2010;2:198–209.

152. Ozaki A, Okamoto H, Yamada T, Matuyama T, Sakai T, Fuji K, *et al.* Linkage analysis of resistance to *Streptococcus iniae* infection in Japanese flounder (*Paralichthys olivaceus*). Aquaculture 2010;308:S62–7.

153. Creeper JH, Buller NB. An outbreak of *Streptococcus iniae* in barramundi (*Lates calcarifera*) in freshwater cage culture. Australian Veterinary Journal 2006;84:408–11.

154. Yuasa K, Kitancharoen N, Kataoka Y, Al-Murbaty FA. *Streptococcus iniae*, the causative agent of mass mortality in rabbitfish *Siganus canaliculatus* in Bahrain. Journal of Aquatic Animal Health 1999;11:87–93.

155. Stoffregen DA, Backman SC, Perham RE, Bowser PR, Babish JG. Initial disease report of *Streptococcus iniae* infection in hybrid striped (Sunshine) bass and successful therapeutic intervention with the fluoroquinolone antibacterial enrofloxacin. Journal of the World Aquaculture Society 1996;27:420–34.

156. Kaige N, Miyazaki T, Kubota SS. The pathogen and histopathoglogy of vertebral deformaity in cultured yellowtail *Seriola quinqueradiata*. Fish Pathology 1984;19:173–80.

157. Chen D, Wang K, Geng Y, Wang J, Huang L, Li J. *Streptococcus iniae* isolated from channel catfish (*Ictalurus punctatus*) in China. Israeli Journal of Aquaculture Bamidgeh 2011;63:593.

158. Yang W, Li A. Isolation and characterization of *Streptococcus dysgalactiae* from diseased Acipenser schrenckii. Aquaculture 2009;294:14–7.

159. Nomoto R, Unose N, Shimahara Y, Nakamura A, Hirae T, Maebuchi K, *et al.* Characterization of Lancefield group C *Streptococcus dysgalactiae* isolated from farmed fish. Journal of Fish Diseases 2006;29:673–82.

160. Nomoto R, Munasinghe LI, Jin D, Shimahara Y, Yasuda H, Nakamura A, *et al.* Lancefield group C *Streptococcus dysgalactiae* infection responsible for fish mortalities in Japan. Journal of Fish Diseases 2004;27:679–86.

161. Domenech A, Fernandez-Garayzabal JF, Pascual C, Garcia JA, Cutuli MT, Moreno MA, *et al. Streptococcosis in cultured turbot, Scophthalmus maximus* (L.), associated with

Streptococcus parauberis. Journal of Fish Diseases 1996;19:33–8.

162. Han SY, Kang BK, Kang BJ, Kim JM, Han JE, Kim JH, *et al.* Protective efficacy of a combined vaccine against *Edwardsiella tarda*, *Streptococcus iniae* and *Streptococcus parauberis* in farmed olive flounder *Paralichthys olivaceus*. Fish Pathology 2011;46:108–11.

163. Romalde JL, Ravelo C, Valdés I, Magariños B, Fuente ED, San MC, *et al. Streptococcus phocae*, an emerging pathogen for salmonid culture. Veterinary Microbiology 2008;130:198–207.

164. Ruiz-Zarzuela I, de Blas I, Girones O, Ghittino C, Muzquiz JL. Isolation of *Vagococcus salmoninarum* in rainbow trout, *Oncorhynchus mykiss* (Walbaum), broodstocks: characterization of the pathogen. Veterinary Research Communications 2005;29:553–62.

165. Michel C, Nougayrede P, Eldar A, Sochon E, De Kinkelin P. *Vagococcus salmoninarum*, a bacterium of pathological significance in rainbow trout *Oncorhynchus mykiss* farming. Diseases of Aquatic Organisms 1997;30:199–208.

166. Figueiredo HC, Costa FA, Leal CA, Carvalho-Castro GA, Leite RC. *Weissella* sp. outbreaks in commercial rainbow trout (*Oncorhynchus mykiss*) farms in Brazil. Veterinary Microbiology 2012;156:359–66.

167. Liu J, Li A, Ji C, Yang W. First description of a novel *Weissella* species as an opportunistic pathogen for rainbow trout *Oncorhynchus mykiss* (Walbaum) in China. Veterinary Microbiology 2009;136:314–20.

168. Wei Q. Social and economic impacts of aquatic animal health problems in aquaculture in China. FAO Fisheries Technical Paper 2002;406:55–61.

169. Cunha BA. Antibiotic resistance. Medical Clinics of North America 2009;84:1407–29.

170. Defoirdt T, Boon N, Sorgeloos P, Verstraete W, Bossier P. Alternatives to antibiotics to control bacterial infections: luminescent vibriosis in aquaculture as an example. Trends in Biotechnology 2007;25:472–9.

171. Defoirdt T, Sorgeloos P, Bossier P. Alternatives to antibiotics for the control of bacterial disease in aquaculture. Current Opinion in Microbiology 2011;14:251–8.

172. Chinabut S, Puttinaowarat S. The choice of disease control strategies to secure international market access for aquaculture products. Developments in Biologicals Progress in Fish Vaccinology 2005;121:255–61.

173. Sommerset I, Krossøy B, Biering E, Frost P. Vaccines for fish in aquaculture. Expert Review of Vaccines 2005;4:89–101.

174. Fang HM, Ge R, Sin YM. Cloning, characterisation and expression of *Aeromonas hydrophila* major adhesin. Fish and Shellfish Immunology 2004;16:645–58.

175. Håstein T, Gudding R, Evensen O. Bacterial vaccines for fish – an update of the current situation worldwide. Developments in Biologicals (Basel) 2005;121:55–74.

176. Park SI. Disease control in Korean aquaculture. Gyobyo Kenkyu=Fish Pathology 2009;44:19–23.

177. Faisal M, Popp W, Refai M. *Aeromonas hydrophila*-related septicemia in the Nile tilapia *Oreochromis niloticus*. Berliner und Münchener Tierärztliche Wochenschrift 1989;102:87–93.

178. Pathiratne A, Widanapathirana GS, Chandrakanthi WHS. Association of *Aeromonas hydrophila* with epizootic ulcerative syndrome (EUS) of freshwater fish in Sri Lanka. Journal of Applied Ichthyology, 1994;10:204–8.

179. Yambot AV. Isolation of *Aeromonas hydrophila* from *Oreochromis niloticus* during fish disease outbreaks in the Philippines. Asian Fisheries Science 1998;10:347–54.

180. Nielsen ME, Høi L, Schmidt AS, Qian D, Shimada T, Shen JY, *et al.* Is *Aeromonas hydrophila* the dominant motile *Aeromonas* species that causes disease outbreaks in aquaculture production in the Zhejiang Province of China? Diseases of Aquatic Organisms 2001;46:23–9.

181. Xia C, Ma Z, Rahman H, Wu Z. PCR Cloning and identification of the b-hemolysin gene of *Aeromonas hydrophila* from freshwater fishes in China. Aquaculture 2004;229:45–53.

182. Pridgeon JW, Klesius PH. Molecular identification and virulence of three *Aeromonas hydrophila* isolates cultured from infected channel catfish during a disease outbreak in West Alabama in 2009. Diseases of Aquatic Organisms 2011;94:249–53.

183. Pridgeon JW, Klesius PH. Virulence of *Aeromonas hydrophila* in the presence or absence of extracellular products to channel catfish fingerlings. Diseases of Aquatic Organisms 2011;95:209–15.

184. Ruangpan L, Kitao T, Yoshida T. Protective efficacy of *Aeromonas hydrophila* vaccines in Nile tilapia. Veterinary Immunology and Immunopathology 1986;12:345–50.

185. Chandran MR, Aruna BV, Logambal SM, Michael RD. Immunisation of Indian major carps against *Aeromonas hydrophila* by intraperitoneal injection. Fish and Shellfish Immunology 2002;13:1–9.

186. John MB, Chandran MR, Aruna BV, Anbarasu K. Production of superoxide anion by head-kidney leucocytes of Indian major carps immunised with bacterins of *Aeromonas hydrophila*. Fish and Shellfish Immunology 2002;12:201–7.

187. Guan R, Xiong J, Huang W, Guo S. Enhancement of protective immunity in European eel (*Anguilla anguilla*) against *Aeromonas hydrophila* and *Aeromonas sobria* by a recombinant *Aeromonas* outer membrane protein. Acta Biochimica et Biophysica Sinica 2011;43:79–88.

188. Poobalane S, Thompson KD, Ardó L, Verjan N, Han HJ, Jeney G, *et al.* Production and efficacy of an *Aeromonas hydrophila* recombinant S-layer protein vaccine for fish. Vaccine 2010;28:3540–7.

189. Khushiramani R, Girisha SK, Karunasagar I, Karunasagar I. Protective efficacy of recombinant OmpTS protein of *Aeromonas hydrophila* in Indian major carp. Vaccine 2007;25:1157–8.

190. LaPatra SE, Plant KP, Alcorn S, Ostland V, Winton J. An experimental vaccine against *Aeromonas hydrophila* can induce protection in rainbow trout, *Oncorhynchus mykiss* (Walbaum). Journal of Fish Diseases 2010;33:143–51.

191. Hernanz Moral C, Flaño del Castillo E, López Fierro P, Villena Cortés A, Anguita Castillo J, Cascón Soriano A, *et al.* Molecular characterization of the *Aeromonas hydrophila aroA* gene and potential use of an auxotrophic *aroA* mutant as a live attenuated vaccine. Infection and Immunity 1998;66:1813–21.

192. Liu Y, Bi Z. Potential use of a transposon Tn916-generated mutant of *Aeromonas hydrophila* J-1 defective in some exoproducts as a live attenuated vaccine. Preventive Veterinary Medicine 2007;78:79–84.

193. Khashe S, Hill W, Janda JM. Characterization of *Aeromonas hydrophila* strains of clinical, animal, and environmental origin expressing the O:34 antigen. Current Microbiology 1996;33:101–8.

194. Bernardet J, Vancanneyt M, Matte-Tailliez O, Grisez L, Tailliez P, Bizet C, *et al.* Polyphasic study of *Chryseobacterium* strains isolated from diseased aquatic animals. Systematic and Applied Microbiology 2005;28:640–60.

195. Algoet M. Project FC1151 – Risks to public health and the aquaculture industry associated with bacterial fish diseases and their treatment. Trout News 2004;38:23–3.

196. Kamaishi T, Fukuda Y, Nishiyama M, Kawakami H, Matsuyama T, Yoshinaga T, *et al.* Identification and pathogenicity of intracellular *Francisella* bacterium in three-line grunt *Parapristipoma trilineatum*. Fish Pathology 2005;40:67–71.

197. Ostland VE, Stannard JA, Creek JJ, Hedrick RP, Ferguson HW, Carlberg JM, *et al.* Aquatic *Francisella*-like bacterium associated with mortality of intensively cultured hybrid striped bass *Morone chrysops*×*M.* saxatilis. Diseases of Aquatic Organisms 2006;72:135–45.

198. Olsen AB, Mikalsen J, Rode M, Alfjorden A, Hoel E, Straum-Lie K, *et al.* A novel systemic granulomatous inflammatory disease in farmed Atlantic cod, *Gadus morhua* L., associated with a bacterium belonging to the genus *Francisella*. Journal of Fish Diseases 2006;29:307–11.

199. Nylund A, Ottem KF, Watanabe K, Karlsbakk E, Krossoy B. *Francisella* sp. (Family *Francisellaceae*) causing mortality in Norwegian cod (*Gadus morhua*) farming. Archives of Microbiology 2006;185:383–92.

200. Ottem KF, Nylund A, Isaksen TE, Karlsbakk E, Bergh O. Occurrence of *Francisella piscicida* in farmed and wild Atlantic cod, *Gadus morhua* L., in Norway. Journal of Fish Diseases 2008;31:525–34.

201. Mauel MJ, Soto E, Moralis JA, Hawke J. A piscirickettsiosis-like syndrome in cultured Nile tilapia in Latin America with *Francisella* spp. as the pathogenic agent. Journal of Aquatic Animal Health 2007;19:27–34.

202. Soto E, Hawke JP, Fernandez D, Morales JA. *Francisella* sp., an emerging pathogen of tilapia, *Oreochromis niloticus* (L.), in Costa Rica. Journal of Fish Diseases 2009;32:713–22.

203. Soto E, Baumgartner W, Wiles J, Hawke JP. *Francisella asiatica* as the causative agent of piscine francisellosis in cultured tilapia (*Oreochromis* sp.) in the United States. Journal of Veterinary Diagnostic Investigation 2011;23:821–5.

204. Birkbeck TH, Bordevik M, Froystad MK, Baklien A. Identification of *Francisella* sp. from Atlantic salmon, *Salmo salar* L., in Chile. Journal of Fish Diseases 2007;30:505–7.

205. Kamaishi T, Miwa S, Goto E, Matsuyama T, Oseko N. Mass mortality of giant abalone *Haliotis gigantea* caused by a *Francisella* sp. bacterium. Diseases of Aquatic Organisms 2010;89:145–54.

206. Hsieh CY, Tung MC, Tu C, Chang CD, Tsai SS. Enzootics of visceral granulomas associated with *Francisella*-like organism infection in tilapia (*Oreochromis* spp.). Aquaculture 2006;254:129–38.

207. Ellingsen T, Inami M, Gjessing MC, Van Nieuwenhove K, Larsen R, Seppola M, *et al.* *Francisella noatunensis* in Atlantic cod (*Gadus morhua* L.); waterborne transmission and immune responses. Fish and Shellfish Immunology 2011;31:326–33.

208. Soto E, Wiles J, Elzer P, Macaluso K, Hawke JP. Attenuated *Francisella asiatica* iglC mutant induces protective immunity to francisellosis in tilapia. Vaccine 2011;29:593–8.

209. Wakabayashi H, Egusa S. Characteristics of a *Pseudomonas* sp. from an epizootic of pond-cultured eels (*Anguilla japonica*). Bulletin of the Japanese Society of Scientific Fisheries 1972;38:577–87.

210. Wiklund T, Bylund G. *Pseudomonas anguilliseptica* as a pathogen of salmonid fish in Finland. Diseases of Aquatic Organisms 1990;8:13–20.

211. Domenech A, Fernandez-Garayzabal JF, Lawson P, Garcia JA, Cutuli MT, Blanco M, *et al.* Winter disease outbreak in sea-bream (*Sparus aurata*) associated with *Pseudomonas anguilliseptica* infection. Aquaculture 1997;156:317–26.

212. Doménech A, Fernández-Garayzábal JF, García JA, Cutuli MT, Blanco M, Gibello A, *et al.* Association of *Pseudomonas anguilliseptica* infection with 'winter disease' in sea bream, *Sparus aurata* [*Pagrus aurata*] L. Journal of Fish Diseases 1999;22:69–71.

213. Nakai T, Muroga K. Studies on red spot disease of pond-cultured eels. V. Immune response of the Japanese eel to the causative bacterium *Pseudomonas anguilliseptica*. Bulletin of the Japanese Society of Scientific Fisheries 1979;45:817–21.

214. Muroga K, Ohnishi K, Jo Y, Tanimoto H, Nakai T. Studies on red spot disease of pond-cultured eels. IX. A field vaccination trial. Aquaculture 1982;30:131–5.

215. Romalde JL, Ravelo C, Lopez-Romalde S, Avendalo-Herrera R, Magarinos B, Toranzo AE. Vaccination strategies to prevent emerging diseases for Spanish aquaculture. Developments in Biologicals Progress in Fish Vaccinology 2005;121:85–95.

216. Cui Q, Wang L, Huang Y, Liu Z, Goodfellow M. *Nocardia jiangxiensis* sp. nov. and *Nocardia miyunensis* sp. nov., isolated from acidic soils. International Journal of Systematic and Evolutionary Microbiology 2005;55:1921–5.

217. Yamamura H, Tamura T, Sakiyama Y, Harayama S. *Nocardia amamiensis* sp. nov., isolated from a sugar-cane field in Japan. International Journal of Systematic and Evolutionary Microbiology 2007;57:1599–602.

218. Xing K, Qin S, Fei SM, Lin Q, Bian GK, Miao Q, *et al.* *Nocardia endophytica* sp. nov., an endophytic actinomycete isolated from the oil-seed plant *Jatropha curcas* L. International Journal of Systematic and Evolutionary Microbiology 2011;61:1854–8.

219. Olubunmi PA, Ayeni AO. A description of an outbreak of bovine nocardiosis in Western Nigeria. Journal of Animal Production Research 1983;3:127–38.

220. Javadi M, Kanavi MR, Zarei-Ghanavati S, Zarei S, Mirbabaei F, Jamali H, *et al.* Outbreak of *Nocardia keratitis* after photorefractive keratectomy: clinical, microbiological, histopathological, and confocal scan study. Medical trials and studies. Journal of Cataract and Refractive Surgery 2009;35:393–8.

221. Darzi MM, Mir MS, Nashiruddullah N, Kamil SA. Nocardiosis in domestic pigeons (*Columba livia*). Veterinary Record 2006;158:834–6.

222. Sahathevan M, Harvey FA, Forbes G, O'grady J, Gimson A, Bragman S, *et al.* Epidemiology, bacteriology and control of

an outbreak of *Nocardia asteroides* infection on a liver unit. Journal of Hospital Infection 1991;18:473–80.

223. Shimahara Y, Yasuda H, Nakamura A, Itami T, Yoshida T. Detection of antibody response against *Nocardia seriolae* by enzyme-linked immunosorbent assay (ELISA) and a preliminary vaccine trial in yellowtail *Seriola quinqueradiata*. Bulletin of the European Association of Fish Pathologists 2005;25:270–5.

224. Shimahara Y, Huang Y, Tsai M, Wang P, Chen S. Immune response of largemouth bass, *Micropterus salmoides*, to whole cells of different *Nocardia seriolae* strains. Fisheries Science (Tokyo) 2010;76:489–94.

225. Itano T, Kawakami H, Kono T, Sakai M. Live vaccine trials against nocardiosis in yellowtail *Seriola quinqueradiata*. Aquaculture 2006;261:1175–80.

226. Edwards AM, Massey RC, Clarke SR. Molecular mechanisms of *Staphylococcus aureus* nasopharyngeal colonization. Molecular Oral Microbiology 2012;27:1–10.

227. Diekema DJ, Pfaller MA, Schmitz FJ, Smayevsky J, Bell J, Jones RN, *et al.* Beach, M. Survey of infections due to *Staphylococcus* species: frequency of occurrence and antimicrobial susceptibility of isolates collected in the United States, Canada, Latin America, Europe, and the Western Pacific region for the SENTRY Antimicrobial Surveillance Program, 1997–1999. Clinical Infectious Diseases 2001;32:S114–32.

228. Shah KL, Tyagy BC. An eye disease in silver carp, *Hypophthalmichthys molitrix*, held in tropical ponds, associated with the bacterium *Staphylococcus aureus*. Aquaculture 1986;55:1–4.

229. Heath PT. An update on vaccination against group B streptococcus. Expert Review of Vaccines 2011;10:685–94.

230. Zdragas A, Tsakos P, Kotzamanidis C, Anatoliotis K, Tsaknakis I. Outbreak of mastitis in ewes caused by *Streptococcus agalactiae*. Deltion tes Ellenikes Kteniatrikes Etaireias [Journal of the Hellenic Veterinary Medical Society] 2005;56:114–21.

231. Watts JL, Nickerson SC, Pankey JW. A case study of *Streptococcus* group G infection in a dairy herd. Veterinary Microbiology 1984;9:571–80.

232. Evans JJ, Klesius PH, Gilbert PM, Shoemaker CA, Al-Sarawi MA, Landsberg J, *et al.* Characterization of β-haemolytic Group B *Streptococcus agalactiae* in cultured seabream, *Sparus auratus* L., and wild mullet, *Liza klunzingeri* (Day), in Kuwait. Journal of Fish Diseases 2002;25:505–13.

233. Evans JJ, Klesius PH, Shoemaker CA. Efficacy of *Streptococcus agalactiae* (group B) vaccine in tilapia (*Oreochromis niloticus*) by intraperitoneal and bath immersion administration. Vaccine 2004;22:3769–73.

234. Pasnik DJ, Evans JJ, Klesius PH. Duration of protective antibodies and correlation with survival in Nile tilapia *Oreochromis niloticus* following *Streptococcus agalactiae* vaccination. Diseases of Aquatic Organisms 2005;66:129–34.

235. Manjaiang P, Areechon N, Srisapoome P, Mahasawas S. Application of vaccine to prevent disease caused by *Streptococcus agalactiae* in Nile tilapia (*Oreochromis niloticus*). Proceedings of the 45th Kasetsart University Annual Conference 2007;174–82.

236. Sheehan B, Labrie L, Lee Y, Lim W, Wong F, Chan J, *et al.* Streptococcal diseases in farmed tilapia. Aquaculture Asia Pacific 2009;5:26–9.

237. Perera RP, Johnson SK, Collins MD, Lewis DH. *Streptococcus iniae* associated with mortality of *Tilapia nilotica*×*T. aurea* hybrids. Journal of Aquatic Animal Health 1994;6:335–40.

238. Eldar A, Horovitcz A, Bercovier H. Development and efficacy of a vaccine against *Streptococcus iniae* infection in farmed rainbow trout. Veterinary Immunology and Immunopathology 1997;56:175–83.

239. Bercovier H, Ghittino C, Eldar A. Immunization with bacterial antigens: infections with streptococci and related organisms. Developments in Biological Standardization 1997;90:153–60.

240. Eyngor M, Tekoah Y, Shapira R, Hurvitz A, Zlotkin A, Lublin A, Eldar A. Emergence of novel *Streptococcus iniae* exopolysaccharide-producing strains following vaccination with nonproducing strains. Applied and Environmental Microbiology 2008;74:6892–7.

241. Bachrach G, Zlotkin A, Hurvitz A, Evans DL, Eldar A. Recovery of *Streptococcus iniae* from diseased fish previously vaccinated with a streptococcus vaccine. Applied and Environmental Microbiology 2001;67:3756–8.

242. Klesius PH, Shoemaker CA, Evans JJ. Efficacy of a killed *Streptococcus iniae* vaccine in tilapia (*Oreochromis niloticus*). Bulletin of the European Association of Fish Pathologists 1999;19:39–41.

243. Klesius PH, Shoemaker CA, Evans JJ. Efficacy of single and combined *Streptococcus iniae* isolate vaccine administered by intraperitoneal and intramuscular routes in tilapia (*Oreochromis niloticus*). Aquaculture 2000;188:237–46.

244. Klesius PH, Evans JJ, Shoemaker CA, Pasnik DJ. A vaccination and challenge model using calcein marked fish. Fish and Shellfish Immunology 2006;20:20–8.

245. Shoemaker CA, Vandenberg GW, Désormeaux A, Klesius PH, Evans JJ. Efficacy of a *Streptococcus iniae* modified bacterin delivered using Oralject(TM) technology in Nile tilapia (*Oreochromis niloticus*). Aquaculture 2006;255:151–6.

246. Locke JB, Aziz RK, Vicknair MR, Nizet V, Buchanan JT. *Streptococcus iniae* M-like protein contributes to virulence in fish and is a target for live attenuated vaccine development. Public Library of Science One 2008;3:e2824.

247. Buchanan JT, Stannard JA, Lauth X, Ostland VE, Powell HC, Westerman ME, Nizet V. *Streptococcus iniae* phosphoglucomutase is a virulence factor and a target for vaccine development. Infection and Immunity 2005;73:6935–44.

248. Pridgeon JW, Klesius PH. Development and efficacy of a novobiocin-resistant *Streptococcus iniae* as a novel vaccine in Nile tilapia (*Oreochromis niloticus*). Vaccine 2011;29:5986–93.

Intensive versus extensive management systems in commercial egg production

J.B. Golden* and D.V. Arbona

Address: Department of Poultry Science, North Carolina State University, Scott Hall, Room 215, Raleigh, NC 27695, USA.

***Correspondence**: J.B. Golden. Email: jackie_golden@ncsu.edu

Received: 8 November 2011
Accepted: 29 July 2012

Abstract

Despite the degree of environmental protection and economic feasibility offered by traditional poultry production systems, the practice of housing laying hens in cages has been widely criticized worldwide because of the public perception that caged housing systems provide poor welfare for laying hens. As a result of increasing public interest in laying hen welfare, alternative management systems such as free-range environments have been resurrected and new legislation such as the California Proposition Two is expediting a shift back towards extensive practices. The transition from intensive back to extensive practices coupled with the advent and increasing popularity of novel alternative layer housing management methods in recent years necessitates an examination of the influence of environmental stressors associated with different housing methods on layer egg production, immunological response and reproductive well-being. In this review, the evolution of intensive and extensive management practices is comparatively examined based on how these different management practices impact layer hen wellbeing and productivity. Different welfare assessment techniques are also examined in an effort to provide insight into the progression of research concerning layer hen welfare in different management systems.

Keywords: Layers, Cage, Free-range, Growth, Production, Welfare assessments

Review Methodology: In this review, the main search engines used were PubMed and Google Scholar, and databases such as CAB Abstracts and Agricola were used through resources at the NC State University Libraries system. Main key terms used in the search included, layer hens, commercial egg production, alternative management systems, intensive layer production, and extensive layer production.

Part I: Intensive versus extensive commercial egg production management systems stressors impacting egg productivity and egg quality

In commercial egg production, a variety of stressors have been found to significantly impact the laying hen's adrenocortical responses, thereby influencing reproduction and egg productivity. Common stressors include: physiological stressors caused by rapid growth resulting from genetic selection [1], feeding and nutritional stressors caused by shortage of nutrients or feed intake issues [2, 3], social and psychological stressors resulting from pecking order and associated fear behaviours [4, 5], physical stressors from handling, transportation and injections [6, 7], and environmental stressors such as changes in photoperiod [8] and environmental management methods [9, 10].

Of particular recent concern is the impact of environmental management methods and living conditions on population viability, productivity and overall welfare of layer hens [9, 11]. Stress-induced ovarian dysfunction is of major concern to the poultry industry because of the economic costs that can ensue through poor productivity and quality in addition to a significant increase in public scrutiny concerning the welfare status of poultry. As a result of such scrutiny, numerous studies have been conducted to assess layer hen welfare in different housing environments in relation to poultry health [12–14]. It is necessary, however, to first understand the development and history of the most commonly used environmental management methods in commercial egg production in addition to how they came into practice and their impact on the poultry industry.

Development of intensive commercial egg production practices

The transition from backyard coops to intensive production practices occurred as the human population and the demand for food increased and small family farms utilizing small-scale, extensive production techniques were unable to meet these demands in a cost-effective manner [11]. Backyard flocks remained in use throughout the 1940s, where it was customary to use laying chickens for meat after they had declined in egg production levels. Eggs were not sold fresh in the winter since backyard hens went into moult during the cold months and ceased laying. Production numbers were relatively low, partially because of higher mortality rates among backyard flocks through exposure to disease and the elements, as well as high rates of contamination of produce as a result of lacking microbial control methods [11].

As urbanization took over, flock managers began to find quicker and more economical ways to supply eggs to growing cities. Commercial poultry houses were developed to protect birds from environmental extremes that adversely affected growth rates, immune function, and fertility and egg production. With the advent of the modern cage system and large-scale, intensive production, came optimized bird welfare and labour through mechanization, which greatly reduced labour costs, improved stocking density of birds and allowed for larger production units [11, 12]. Commercial layer poultry houses with caged systems were increasingly cost-effective because of a lower incidence of cannibalism [13], disease, dust and ammonia levels [14, 15], as well as better management of flock nutrition and more effective waste management, which provided cleaner eggs for safer consumption [11].

Intensive commercial egg production management systems in the USA

Intensive practices were developed, in part, to save land resources and to manage birds in the most efficient manner possible. In 2010, 95% of commercial egg production in the USA and 90% of the world's egg production was obtained from caged layers, while the remaining 5% of eggs produced in the USA were derived from alternative systems of management, which include cage-free and free-range systems [11].

Conventional battery style and stair step cages are made of wire or plastic mesh and are stacked in three to five tiers. Under 'UEP certified' guidelines for cage production, cages are required to allow the bird to fully stand upright and on a floor, which is not sloped more than 8°. Each cage should allow for 67–86 square inches of usable space per bird in order to allow for natural behaviours to occur. Birds should also have continuous access to water and be able to feed at the same time in order to minimize aggression [11].

Cages are classified based on two different methods of manure collection. In battery style cages, manure is collected via a belt system and must be configured such that manure from birds in upper tiers does not come in contact with lower tier cages. After collection, the manure is transported out of the barn [11, 16]. In conventional stair-step cages, there is a high-rise system where manure falls into a collection and storage area beneath the cages where it is removed one or more times per year [16].

With respect to egg collection, two primary methods exist for collecting eggs: in-line systems and off-line systems. In both systems, when the egg is laid, it rolls to the front of the cage as a result of the slight incline in the cage and lands on a nylon belt. The belt then transports the egg out of the hen house directly to an egg processing facility (in-line) or to a storage cooler facility (off-line). At the egg processing facility, eggs are cleaned in a high temperature solution, visually inspected and graded for packaging. Off-line eggs are stored in a cool room for 2–3 days and then transported to a similar egg processing facility via a refrigerated truck [11]. Feedlines and watering lines are automated and deliver feed to birds along the front of cages via a chain system at specific times of the day and water *ad libitum* via nipple drinkers in each cage, which are supplied by a pipe spanning the length of the house.

Challenges associated with intensive commercial egg production management systems

Despite the stringent welfare guidelines that must be met in order to maintain caged production, numerous critiques of intensive practices have led to the development of alternative management systems for commercial egg production. Public perception of the conventional laying cage system raises concerns that this system provides inadequate welfare for laying hens by preventing hens from carrying out natural behaviours, including dust-bathing and nest building [17] and restricting hens' general freedom of movement [18]. Other areas of concern include issues involving overcrowding and increased susceptibility to osteoporosis and cage-layer fatigue [19] as well as the effects of cages on the spread of disease and deterioration of humoural immune function [20].

These concerns have led to the rising popularity of extensive systems such as free-range, cage-free, and aviary-style environments and an increased desire to alter conventional battery-style cage management methods and regulations. Legislation such as the California Proposition 2: Prevention of Farm Animal Cruelty Act has contributed to the transition from intensive production practices in commercial egg production back to extensive methods of production. Found in Division 20, Chapter 13.8 of the California Health and Safety Code, new regulations for confinement of cattle, swine and poultry raised in the state of California stipulate that animals must be able to lie down, stand up, fully extend their limbs and turn around freely within their confinement. With regard to changes in layer hen housing that are to be implemented by the year 2015, the specific wording of the initiative is imprecise, but when passed the resulting regulations will eliminate the use of cage systems for laying hens in California [21]. Financial concerns with the new initiative include higher economic costs associated with shifts to non-cage barn housing systems, which have been shown to incur farm-level cost increases that are at least 40% higher per dozen eggs than the common cage housing systems [22]. Added costs can be attributed to higher feed costs, housing and labour costs, as well as increased layer hen mortality [21]. These concerns have led to projections that the California egg laying industry will relocate during the 6-year adjustment period between 2009 and 2015 [21].

Development of extensive alternative management systems for US commercial egg production

Alternative layer management systems such as cage free, free-range and aviary systems differ significantly from conventional battery-style cages in their layout of physical facilities, labour costs, stocking densities and risk of parasitic infections among a multitude of other factors.

Cage-free systems, also called barn systems or perchery systems, house birds indoors without cages. A cage-free system consists of a single-level, indoor system [16] that may have litter or slatted floors with feed and water located over the slats. In addition, hens have access to areas for perching and roosting and to nest boxes, which are integrated throughout the house. Provision of feed and water and egg collection are usually automated. Manure is collected through the partially slatted or mesh floor and there is a removal system underneath [16]. Cage-free systems may be further sub-divided into floor-raised systems and aviary style housing which consists of multiple levels in one house.

In contrast, free-range systems require outdoor access to a veranda or range, which may include a covered forage area, paddocks for rotational grazing and a range hut for cover. The range hut is constructed with floors consisting of litter, slats or a combination of the two, and normally contains feeders, waterers and nest boxes. Free-range housing systems are designed to provide a period of time each day when hens can access the outdoor area. When they are not outdoors, the hens remain in a barn similar to the aforementioned cage-free floor-raised barns. Generally, the barn has small doors along the sides called pop-holes that are opened mid-day after egg-lay and closed near dusk when the hens have returned to roost for the evening. The free-range system may vary in size from a few hens to thousands of hens and is quite labour-intensive if examined on a per-hen basis [23]. As of 2010, in the USA, neither the USDA nor the FDA has set prescriptive standards for use of the term 'free-range' for egg production or packaging.

Challenges associated with alternative management systems for commercial egg production

Controversy surrounding implementation of alternative management systems is largely attributed to both decreases in profitability as well as the possibility that these management systems may have negative impacts on welfare, health, hygiene and food safety. As a result, numerous studies [12–14, 18, 20, 24–26], including one eminent large-scale project, the LayWel Project [27] have been conducted to compare the welfare status of laying hens in different housing environments. The LayWel Project was funded by the European Commission's Sixth Framework Programme in addition to national funding from several EU countries. The objective of this project was to produce an evaluation of laying hen welfare in various systems and to disseminate the information gathered to all member states of the EU as well as associated countries. This project developed a database that resulted from the compilation of data from different housing systems allowing for data comparison between systems with respect to risk factors, advantages and disadvantages of various housing systems [27]. The project concluded that all systems have the potential to provide satisfactory welfare for laying hens, with the exception of conventional cages [27]. Additionally, the project developed a feather scoring and integument scoring system that farms can carry out to assess damage caused by pecking [27]. The LayWel project also developed an on-farm auditing procedure in a manual for self-assessment that allow for on-farm assessment of welfare through regular checks of a range of indicators of laying hen welfare that may be carried out in a systematic fashion [27]. Recommendations were made for housing systems on aspects ranging from behaviour to health and mortality as a result of the comprehensive LayWel project [27].

In a review comparing layer hen welfare in alternative management systems, with traditional caged housing, Rodenburg found that the main welfare issues presented by alternative management systems relate to three main differences in group size, freedom of movement and complexity of the environment that exists in alternative management systems as compared with traditional caged housing [28].

Impact of group size

One of the most significant concerns regarding alternative management systems is larger group size and its impact on social behaviour and welfare. Group size affects laying hens' abilities to recognize each other and establish dominance relationships [28]. Different group sizes affect the incidence of aggressive behaviours such as feather pecking (pecking at or pulling out of feathers of conspecifics) and vent-pecking, which can both lead to the development of severe wounds and/or cannibalism [28]. Given the difficulty in identifying a feather-pecker in a large group of laying hens, an outbreak of cannibalism or feather-pecking in an alternative management set-up would be difficult to stop [28]. Group size can also affect fearfulness in laying hens [28]. Studies have shown that larger groups (120 birds) [29] and birds with feather damage [30] were more fearful, having longer, tonic immobility responses than birds from groups of 15 birds or birds without feather damage.

Freedom of movement

In addition to being housed in larger groups, birds in alternative systems are provided with more space than caged birds in order to perform comfort behaviours such as wing flapping, stretching, body shaking and tail wagging. Additionally, alternative systems provide hens with the ability to walk, run or fly [28]. Perches of different heights are also provided in alternative systems, because studies have shown that laying hens are highly motivated to perch, and if offered a choice, will choose a high perch (63 cm) over a low perch (23 cm) [31]. While there are studies showing that this increase in ability to exercise provides hens with stronger bones than those of birds housed in conventional cages [32], a negative side of high perches is that birds are more prone to injury and bone breakage if they should fall or land inaccurately [28]. Additionally, studies have shown that when laying hens are reared without early access to perches, their spatial skills can be impaired and this affects their ability to transition to an environment with perches [33].

The use of perches in general can be hazardous on a hygienic level as well; depending on what material they are constructed from. Wooden perches (used in traditional housing systems) are difficult to clean and can host red mites (*Dermanyssus gallinae*), an infestation that can lead to reduced weight gain, anaemia and even increased mortality [34].

Complexity of the environment

Animal behaviourists argue that to fulfil a laying hen's behavioural requirements, a complex environment is required that provides opportunities for feeding and drinking, pecking and scratching, nesting, perching, hiding and performing comfort behaviours [28]. In an alternative management setting, birds have access to a large floor space where they can perform these behaviours, as well as perches, litter and nest boxes [28]. One issue with many alternative management systems is an inconsistent distribution of birds over the available nests at the beginning of the laying period [28]. Specifically, when some nests are more crowded than others, birds tend to pile up on top of each other, which can lead to increased mortality [28].

Hygiene

The aforementioned issues coupled with the large amount of litter used in alternative management systems are considered potential risk factors for maintaining hygiene, which is essential for health, welfare and food safety [28]. Increased bird movement and large amounts of litter can contribute to higher levels of bacteria and fungi in the air, and higher dust levels compared with conventional and furnished cages [28].

When comparing eggshell contamination in alternative systems with conventional cages, there was a higher rate of contamination of eggshells by total aerobic mesophilic bacteria in alternative systems [35]. In alternative systems, where the birds move more freely in their environment, a significant amount of dust is created from the litter, resulting in air contamination by micro-organisms and endotoxins [35]. Proper litter management appears to be the key in decreasing the amount of bacterial eggshell contamination as a direct correlation between bacterial eggshell contamination and aerobic mesophilic bacteria in the air of poultry houses has been determined [35].

Specifically, facilities with litter have ten times more airborne bacteria and 20–30 times more bacteria on eggshells compared with a caged housing system [35].

Air quality

General management practices emphasize air quality as playing an important role in contributing to respiratory disease in poultry [36]. Accumulation of aerosolized dust and bacteria in the air contribute to poor air quality, and can have a significantly negative effect on animal health [37]. When compared with litter-based systems, improved air quality was found in furnished cages as opposed to litter-based systems (floor housing and aviary) [38]. This may be because of the fact that cage systems have been found to produce significantly lower aerosolized inhalable (particle size: 1–100 μm in diameter) and respirable (particle size: $<8.5\ \mu$m) dust particles when compared with litter-based systems [36]. Other air quality issues are associated with the respiratory health of individuals working in alternative, especially aviary, management systems. These individuals were found to have stronger inflammatory reactions and increased bronchial responsiveness than individuals working in houses with conventional cage systems [28].

High levels of ammonia in the air of alternative systems compared with conventional cage systems are also a cause for concern [39]. Increased ammonia levels can impact both animals and humans by increasing their susceptibility to respiratory pathogens, impairing overall performance, causing eye problems [40] and chronic respiratory symptoms [41].

Environmental exposure to bacteria, viruses and parasites

Questions about the effects of management systems on a hen's general health have also arisen and are mostly centred on foot, bone, feather health and incidence of bacterial infection. There is a significant difference in mortality rates when comparing non-cage, outdoor, conventional cages and furnished cages, with furnished cages being numerically lower than any other system, [42]. In general, mortality has been found to be greater in laying hens raised in litter-based housing compared with furnished cages [38].

In a study assessing production and egg quality in the aviary environment, the primary causes of mortality were salpingitis (inflammation of the oviducts) and cannibalism [12]. Other problems were caused by coccidiosis, lymphoid leucosis (big liver disease) and red mite infestations; along with foot abscesses and keel bone deviations.

Another extensive study, which examined layer hen welfare in alternative management systems, found that bacterial diseases were the most common causes of mortality in birds raised on litter-based systems [43]. Colibacillosis infections, caused by the bacteria *Escherichia coli*, found in flocks raised on litter-based systems resulted in acute or subacute salpingitis, oophoritis and peritonitis [43]. Of the cases of colibacillosis examined, 50% presented with cloacal cannibalism or vent pecking during necropsy, suggesting cannibalism and pecking as an important predisposing factor for bacterial infection to occur [43]. Other common bacterial infections noted in this study were outbreaks of erysipelas, caused by *Erysipelothrix rhusiopathiae*. Hens raised on litter and in free-range environments also had greater mortality associated with viral diseases such as Marek's disease and Newcastle disease [43]. In addition to litter contaminations, infectious disease can also be the result of contact with soil, fomites (e.g. rodents, beetles and equipment) known to carry the agents of those diseases [36]. Infection by soil-borne micro-organisms such as *E. rhusiopathiae* occurs more often in hens exposed to litter-based and free-range systems [43].

With regard to parasitic exposure, a higher percentage of hens (73%) with access to free-range excreted a greater number of coccidial oocysts in faeces compared with hens (58%) without free-range access [44]. Intestinal parasites such as the helminth have been found to have a greater incidence in layer hens raised in free-range systems when compared with hens raised in cages [44, 45]. Parasites and pests common in poultry include flies or beetles that develop in the accumulated manure or feed and several ectoparasites [36, 46]. The house fly, *Musca domestica,* and beetles such as the lesser mealworm, *Alphitobius diaperinus,* thrive in manure feed and feed mixtures especially in conventional and furnished cages [36], and harbour pathogens that can affect poultry and possibly human health [47]. Birds can come in contact with fly larvae and beetles through access to manure in floor pens or outdoors [36]. Lower bird density allows for more efficient manure-drying and lessens problems with flies and with beetles [36], suggesting another reason why group size is an important consideration.

Commonly found ectoparasites can be classified into 1 of 2 categories: nest-dwelling or permanent [36]. Permanent blood-feeding mites, *Dermanyssus* and *Ornithonyssus,* can cause direct economic damage through irritation, blood loss and forcing the hens to mount a metabolically costly immune response [34]. Examples of nest-dwelling mites include the red-mite (*D. gallinae*), soft ticks such as *Argas,* bedbugs (*Cimex* spp.), and certain fleas such as the sticktight flea (*Echidnophaga*) [36]. Nest-dwellers inhabit the birds' environment and travel to the birds (usually to blood feed) for part of their life cycle [36]. Complex environments such as free-range and cage-free systems provide ample small hiding spaces for nest dwellers and other arthropod pests to thrive relative to suspended wire cage systems [36]. As the transition to alternative management systems progresses in the USA, incidence of problems with nest-dwelling ectoparasites will most likely increase [36]. In addition to increased exposure to ectoparasites, hens with access to a free-range environment are also at greater risk for predation and disease because of possible exposure from contact with wild animals [36]. Wild birds carry diseases such as avian influenza and Newcastle disease, in addition to ectoparasites, which can easily spread to domestic poultry [36].

Part I: Summary

Given a significant amount of worldwide criticism and increasing public scrutiny pertaining to the practice of housing laying hens in caged housing systems, a variety of alternative layer hen management systems including free-range, cage free and aviary style housing systems have been developed for commercial egg production. This transition from intensive management methods to extensive management methods has led to a thorough examination of differences in layer hen welfare and egg production. Assessment of differences in egg production in alternative versus conventional caged management systems has brought about a significant amount of controversy as it relates to the pros and cons of commercial egg production in each of the aforementioned management systems. Challenges associated with intensive commercial egg production management systems include the prevention of hens from carrying out natural behaviours such as dust-bathing and nest building and restriction of hens' general freedom of movement. Overcrowding, increased susceptibility to osteoporosis and cage-layer fatigue are additional concerns pertaining to caged layer management systems. In contrast, caged systems provide for better hygiene, more affordable production costs, and lower incidences of cannibalism and egg contamination. Alternative management systems such as cage free, free-range and aviary systems provide more opportunities for hens to carry out natural behaviours, greater degree of freedom of movement and access to outdoors in range environments. However, these systems may also lead to increased incidences of aggression between hens, increased bone breakage in aviaries, increased incidence of egg shell contamination and a significant increase in on-farm production costs. In summary, there are pros and cons to both traditional caged systems as well as alternative management systems. Ultimately, producers must maximize production and layer hen welfare in each environment that they utilize. Producers must choose the rearing environment that best suits their needs based on their consumers, the market they are targeting and the production resources available to them.

Part II: assessment of layer hen welfare in intensive versus extensive management systems

The challenges associated with assessing the physiological and cognitive welfare of laying hens in alternative management systems are numerous. Specifically, the difficulties associated with making welfare measurements without actually influencing the measurements being taken can prove to be a daunting task. Another obstacle to accurate welfare assessments of laying hens includes the influence of the social pecking order on hens' access to resources and space within a given environment. Despite the fact that a flock of hens inhabit the same physical environment, the opportunities or limitations encountered by individual variation that exists in biomarkers of stress and welfare between individual hens makes it difficult to glean meaningful information from comparisons between groups of birds in different environments. Current methods utilized to assess the response of hens to a particular system have been limited to housing groups of hens in different environmental management systems and comparing biological or behavioural measurements that are collected utilizing invasive, post-mortem, or novel non-invasive methodologies.

Environment-based versus animal-based parameters

To date, a majority of research conducted to assess environmental health and layer hen welfare in alternative management systems has focused on environment-based assessments or animal-based parameters [48]. Environment-based assessments such as space allowance, air quality and litter quality focus on the system, management and stockmanship of the farmer and can be assessed with relative ease and good repeatability.

In contrast, animal-based parameters assess the response of the hens to a particular system by measuring parameters such as: levels of stress hormones, abnormal behaviours, or incidence of diseases. The act of obtaining, recording and interpreting animal-based parameters without influencing the data collected has proven to be quite challenging [48]. Other animal-based assessments include the use of hen production characteristics, such as egg production, feed conversion and mortality [23, 49], which if the premise holds true, hens with the best production characteristics would have better welfare status.

Biological sampling methods

Biological sampling methods have centred on collecting blood samples from groups of hens housed in different environments to measure plasma corticosterone levels, humoural immunocompetence in response to various vaccine challenges, and heterophil:lymphocyte ratios to assess stress levels in hens [50]. Minimally invasive research techniques, which can be indicators of welfare issues in a given environment, include assessment of physical condition by scoring integument, i.e. plumage and foot condition.

Immune function assessment

In general, studies have shown that environmental stressors reduce immune responses and cause immunomodulation, which is initiated by the hypothalamus–pituitary–adrenal cortical pathways [51]. The relationship between immune function and disease risk may be greatly influenced by an individual's response to chronic stressors including those that are environmentally induced [52]. Prolonged stress responses have been observed to alter an animal's immune function [53] by increasing their risk for a wide range of adverse health outcomes. Ultimately, stress heightens the risk for adverse health outcomes by suppressing the immune response thus leaving the host vulnerable to opportunistic disease [54]. Measurements of stress-induced immune alterations have been conducted in poultry utilizing haematological and immunological indices. Specifically, heterophil:lymphocyte (H:L) ratios have been used as a sensitive haematological indicator of stress responses in chicken populations [55]. The ratio of heterophils to lymphocytes is a good indicator of stress because stressors induce an elevation in plasma corticosterone concentrations, which, in turn, stimulates the rise in heterophil concentration. Additionally, antibody responses to commonly used vaccines are used to assess alterations in humoural immune function in a variety of species [56], including poultry where decreased immune antibody responses have been observed in hens placed in stressful environments [57].

One such study in laying hens examined the humoural immune response in caged versus non-caged birds after an immunization challenge with human serum immunoglobulin G [58] with or without adjuvant. The results indicated that the mean egg yield and mean antibody titres were significantly higher in caged hens regardless of the presence or absence of adjuvant [59]. Another study examining the physiological responses of laying hens to alternative housing systems found that hens kept in battery cages had greater heterophil levels and decreased lymphocytes in comparison with hens kept in modified cages and free-range housing systems [20]. These findings exemplify the complexity and variation of current findings from studies examining the relationship between stress and immune function in laying hens housed in caged versus non-caged environments.

Blood plasma corticosterone measurements

Plasma corticosterone levels are also utilized to compare levels of stress and determine if there are significant differences in corticosterone production between hens housed in caged versus non-caged environments. Blood samples from birds in each cage type are collected at varying intervals, and the samples are analysed for plasma corticosterone in order to assess corticosterone production. In order to minimize the effects of sampling on plasma corticosterone, which is one of the biggest challenges in executing this invasive approach, a standardized capture and bleeding method [60] can be used, and all samples must be taken within a matter of minutes of the bird being out of the cage [50]. By measuring corticosterone production throughout the laying cycle, it is also possible to juxtapose the potential direct relationship between corticosterone production and H:L ratios.

Scoring of integument

In addition to the effects of environment on the immune system and on corticosterone production, researchers can also use different methods for scoring birds' integument in order to compare welfare in different management systems [61]. The main parts of the integument that are scored are the plumage, foot condition and pecking wounds on different parts of the body [61]. Plumage condition is a key indicator of welfare status as well as a determinant in overall production economics [50]. Birds with poor feather coverage often lose heat and in turn take in increased amounts of feed, resulting in a higher maintenance energy requirement [62]. Feather damage and loss is often an indicator of feather pecking [50], which is highly influenced by group size [30] and stocking density [61]. Foot condition and pecking wounds may describe problems of litter condition, perch design and cannibalistic or aggressive behaviours, respectively [61]. Thus, researchers can use scoring of these traits as an indicator of the effects of a caged versus free-range environment on the physical well-being of the hen. An example scoring system involved a 4-point scoring system [61] in which six different areas of the body (neck, breast, back, wings, tail and vent) were examined and given a score of 1 to 4. Similar systems have been used to compare welfare and general health of birds in conventional cages versus birds in enriched cages [50]. In this study, a score of 4 indicated very good feathering and few broken or deformed feathers. A score of 3 was used when there was complete feather coverage, but some feathers showed deterioration. A score of 2 indicated incomplete coverage with some areas of the body being without feathers, and some areas having broken or worn feathers. Finally, a score of 1 indicated a majority of the body being denuded and the few feathers present were severely deteriorated. Average plumage condition was calculated by adding scores over all six areas totalling for a score ranging from 6 to 24 points. Birds in conventional cages had higher feather scores in the wing region when compared with birds reared in enriched cages; however, the mean score for each of the other five individual body parts that were examined did not differ between cages [50]. This may be an indicator that increased space in enriched cages allows for more wing flapping and therefore increased incidence of wing abrasion [50].

Post-Mortem studies

While integument evaluation may act as an indicator of possible problems in the environment (feather pecking, cannibalism and aggression), using a total body score does not completely explain the possible reasons for deterioration of the integument, and cannot specify an internal physiological issue caused by the environment if there is one [61]. Post-mortem analyses are conducted to analyse hen health at the physiological level by performing tracheal biopsies to examine the influence of air quality on respiratory health, intestinal homogenates for identification of microbial flora [63], and assessment of bone quality traits such as bone mineral density and content [50]. Examining bone strength allows researchers to address one of the main issues with conventional cages, a lack of exercise in laying hens which can lead to bone weakness and during peak production can easily result in a pathological condition termed cage layer fatigue [10]. Hens with cage layer fatigue are described as being able to eat and drink, but are unwilling to stand [64, 65]. When examined, many hens diagnosed with cage layer fatigue have been found with paralysis caused by a compressed spinal cord from fractured thoracic vertebrae [19] and broken bones caused by chronic bone brittleness [10]. Osteoporosis, which is increased porosity and reduced thickness of bone cortex resulting in bone fragility [10], is also an issue in layer hens because of the correlation between high egg production and reduced bone condition. Femur bone mineral content and tibial bone strength have been shown to decline during the first few weeks of egg production [66], which can be attributed to factors such as insufficient dietary intake of calcium, phosphorus and vitamin D, and lack of exercise during peak production periods [10].

In a study that examined the bone mineral density and bone mineral content of post-mortem laying hens from caged and enriched cage environments, bone mineral density of the tibia and humerus of birds housed in enriched cages was significantly higher than birds in conventional cages [50]. Bone mineral content, ash weight, and percentage of calcium and phosphorus of tibia and humerus bones were also measured and found to be higher in enriched cages as well, but resulted in a non-significant pattern. The results of this particular study indicate that the hen's ability to exercise in a particular management system may play a significant role in bone development and maintenance, especially during peak production [50].

Non-Invasive assessments of layer hen welfare

It has been well documented in both mammalian and avian species that environmental stressors elevate concentrations of glucocorticoids in the blood in order to maintain homoeostasis and initiate physiological and behavioural reactions that promote survival [60]. Thus, invasive methods such as plasma corticosterone concentrations have been widely used to monitor stress responses in a variety of species, including egg-laying hens [67]. However, capture, restraint and frequent blood sampling are procedures well known to cause a rapid surge of corticosteroids, specifically corticosterones, in birds [68, 69]. Alternative non-invasive sources utilized to quantify corticosterone levels such as saliva and faeces are problematic because of their non-homogenous composition and strong likelihood of enzymatic/bacterial degradation of large molecules such as corticosterone [70].

Egg corticosterone concentrations

In response to challenges associated with the previously mentioned sampling techniques, several investigations have been conducted to assess the use of corticosterone concentration measurements in egg yolk and albumen as a non-invasive indicator of stress in laying hens. Results from corticosterone analyses in egg yolk and albumen have varied because of the fact that hormones are not homogenously distributed within the egg and the extraction methods for hormones from the yolk and chromatographic separation of different hormones for immunoassays is likely to influence results [71]. Studies have shown a great deal of variability with regard to sensitivity of corticosterone measurements in eggs; one study reported low transfer of corticosterones from plasma to egg after the administration of radiolabelled corticosterone and ACTH to the hen [72], while other studies have found corticosterone measurements sensitive enough to enable the detection of significantly higher levels of corticosterone concentrations in homogenized eggs from free-range hens compared with eggs laid by caged hens or in eggs laid by hens housed in barn and caged environments [73]. On the other hand, another study [74] did not detect any significant differences in the white and yolk corticosterone concentrations or total egg corticosterone between singly housed and group-housed egg-laying hens. Although corticosterone data in eggs appear to vary based upon context, corticosterone concentrations measured in the egg albumen from eggs laid by hens injected with linear concentrations of corticosterone subcutaneously were found to be directly related to plasma corticosterone concentrations [75]. Furthermore, when hens were exposed to various stressors including handling, high ambient temperatures and new cages, the concentrations of corticosterone in albumen increased [75]. In order to thoroughly understand the potential of non-invasive egg analyses for stress assessment in layer hens, further studies are needed to standardize the methodology used to measure and characterize differences in corticosterone concentrations in eggs laid by hens in a variety of alternative management systems.

Behaviour, space and resource utilization, and electronic monitoring

Behavioural observations of hens, typically considered to be more subjective, have concentrated on identifying and quantifying normal behaviours and maladaptive behaviours, in an effort to determine the influence of group size on the formation of dominance hierarchies [17], aggression [30], vent-pecking [30], feather-pecking [76], cannibalism [76], fearfulness [28], freedom of movement [29] and complexity of the environment allowing for opportunities to perform comfort behaviours such as: feeding and drinking, pecking and scratching, nesting, perching, hiding and comfort behaviours [77].

In evaluation of hen well-being in various alternative management systems, behavioural observations of layer hens have frequently been used to assess welfare [78].

Specifically, these observations have focused on analysing hens' innate desires and essential needs, which have led to the modification of traditional caged environments with furnishings, such as perches and nest boxes [77, 79–81]. Ethograms have been developed to assess time budgets under varying housing conditions and social situations in order to assess priorities of hens. Despite the fact that behavioural observations have been used extensively to assess layer hen welfare, gathering behavioural information can be challenging and complex for a variety of reasons. Specifically, the presence of a human observer can influence hens' responses, video can be tedious to analyse, limited angles and views may obscure the observation of certain behaviours, and equipment and data storage can be expensive. Additionally, in order to collect individual behavioural information over time, it is necessary to visually mark each bird or manually track the bird through all of the video recordings. While scan sampling may help to circumvent some of these problems, selecting the interval that best represents the complete behavioural repertoire of a hen is a challenge and opens the analysis to increased risk of missing important pieces of data.

Alternatively, electronic monitoring, if implemented effectively, can yield information 100% of the time at no additional cost or effort. Electronic monitoring is generally non-invasive and removes the influence of humans in the data collection process. Electronic methodologies have been shown to be a viable option for assessing behavioural parameters. Some examples of electronic monitoring include: the development of an algorithm to automatically recognize chicken and turkey behaviour from video [82–85], the use of radio frequency identification (RFID) to assess feeder usage in beef cattle [86], the assessment of geographic space utilization in beef cattle through global positioning systems (GPS) [87, 88], the use of precision scales to assess chick and hen feeding behaviour [89], and the location determination of chickens and mice in enclosed spaces via infrared sensors [90]. Currently, electronic monitoring methods have only begun to enter the realm of layer hen welfare analyses in alternative management systems and require further development and validation.

Part II: Summary and Conclusions

The 60-year transition of the laying industry from backyard coops to large-scale commercialized egg-production facilities has become riddled with controversy regarding housing management methods and animal welfare. Initially, poultry houses were developed and utilized as a means to optimize labour and protect birds from environmental extremes, which could be detrimental to mortality rates, growth rates, immune function, fertility and egg production. The advent of cages allowed for the development of larger production units and the implementation of mechanization in production, which improved stocking densities and required less investment and labour costs per bird and per weight of egg produced [12]. Despite the advantages of extensive cage systems, public perception and scrutiny of hen welfare in conventional cages has triggered a transition back to extensive production methods, and the passing of new legislation such as the California Proposition 2 has pushed the implementation of alternative layer housing management methods.

However, research examining welfare issues for laying hens in alternative management systems versus caged systems raises significant variation in interpretation of data produced from these studies due to the fact that quantifiable biomarkers of stress, distress and pain are difficult to use without influencing the measurements being taken. New non-invasive methods include examining the egg as a more efficient means of assessing immune function and measuring stress, as well as non-invasive forms of behavioural observation may alleviate the current variation in data interpretation that surrounds layer hen welfare studies of birds reared in alternative management systems. It is imperative that this research continues because of the fact that insight into the influence of housing methods on layers' welfare from a behavioural and physiological perspective will enable producers to maximize egg productivity and quality while concurrently promoting animal well-being and environmental adaptability.

References

1. Freeman BM. The stress syndrome. World's Poultry Science Journal 1987;43:15–9.
2. Ben-Nathan B, Drabkin N, Heller D. The effect of starvation on the immune response of chickens. Avian Diseases 1981;25(1):214–7.
3. Etches RJ, WIliams JB, Rzasa J. Effects of corticosterone and dietary changes in the hen on ovarian function, plasma LH and steroids and the response to exogenous LH–RH. Journal of Reproduction and Fertility 1984;70:121–30.
4. Gross WB, Siegel PB. Long-term exposure of chickens to three levels of social stress. Avian Diseases 1981;25(2):312–25.
5. Beuving G, Vonder GMA. The influence of ovulation and oviposition on corticosterone levels in the plasma of laying hens. General and Comparative Endocrinology 1981;44:382–8.
6. Jones RB, Beuving G, Blokhuis HJ. Tonic immobility and heterophil/lymphocyte responses in the domestic fowl to corticosterone infusion. Physiology and Behavior 1988;42:249–53.
7. Gregory NG, Wilkins LJ, Austin SD, Belyavin CG, Alvey DM, Tucker SA. Effect of catching method on the prevalence of broken bones in the end of lay hens. Avian Pathology 1992;21:717–22.
8. Bédécarrats GY, McFarlane H, Maddineni SR, Ramachandran R. Gonadotropin-inhibitory hormone receptor

signaling and its impact on reproduction in chickens. General and Comparative Endocrinology 2009;163(1–2):7–11.

9. Saino N, Romano M, Ferrari RP, Martinelli R, Moller AP. Stressed mothers lay eggs with high corticosterone levels which produce low-quality offspring. Journal of Experimental Zoology 2005;303A:998–1006.

10. Webster A. Welfare implications of avian osteoporosis. Poultry Science 2004(83):184–92.

11. United Egg Producers. UEP Animal Husbandry Guidelines for U.S. Egg Laying Flocks. 2010. Available from: URL: http://www.uepcertified.com/pdf/2010-uep-animal-welfare-guidelines.pdf

12. Abrahamsson PER, Tauson R. Performance and egg quality of laying hens in an aviary system. Journal of Applied Poultry Research 1998;7:225–32.

13. Hilbrich P. History of poultry health. In: Wegner R, editor. Proceedings, Second European Symposium on Poultry Welfare, World's Poultry Science Association, Celle, Germany; 1985. p. 48–54.

14. Engstrom B, Schaller G. Environmental studies of the health of laying hens in relation to housing system. In: Savory CJ, Hughes BO, editors. Proceedings of the Fourth European Symposium on Poultry Welfare. Universities Federation for Animal Welfare, Potters Bar, UK; 1993. p. 87–96.

15. Hauser R. Lufthygeiene in Geflugelstallen. Auswirking aug Mensch und Tier ETH, Zurich, Switzerland; 1988.

16. Chore-Time. Modular Manure Belt Cage System. Product Information Publication CT-2364-0107. 2007.

17. Weeks C, Nicol C. Behavioural needs, preferences and priorities of laying hens. World's Poultry Science Journal 2006;62:297–308.

18. Appleby MC, Hughes BO, Elson HA. Poultry Production Systems: Behavior, Management and Welfare. CAB International, Oxfordshire, UK; 1992.

19. Riddell C, Helmboldt CF, Singsen EP, Matterson LD. Bone pathology of birds affected with cage layer fatigue. Avian Diseases 1968;12(2):285–97.

20. Shini S. Physiological responses of laying hens to the alternative housing systems. International Journal of Poultry Science 2003;2(5):357–60.

21. Sumner DA, Thomas Rosen-Molina J, Matthews WA, Mench JA, Richter KR. Economic Effects of Proposed Restrictions on Egg-laying Hen Housing in California. University of California Agricultural Issues Center, UC Davis, Davis; July 2008.

22. Sumner DA, Gow H, Hayes D, Matthews W, Norwood B, Rosen-Molina JT. *et al.* Economic and market issues on the sustainability of egg production in the United States: analysis of alternative production systems. Poultry Science 2011;90:241–50.

23. Anderson K. Single Production Cycle Report of the 37th North Carolina Layer Performance and Management Test (February 2009). North Carolina Cooperative Extension; 2009. p. 37.

24. Elson HA, Croxall RA. European study on the comparative welfare of laying hens in cage and non-cage systems. European Poultry Science 2006;70:194–8.

25. Elson HA. Poultry housing and husbandry. British Poultry Science 2010;51(Suppl. 1):23–4.

26. Sossidou EN, Elson HA. Hens' welfare to egg quality: a European perspective. World's Poultry Science Journal 2009;65(4):709–18.

27. Blokhuis HJ, Fiks Van Niekerk T, Bessei W, Elson A, Guémené D, Kjaer JB, *et al.* The LayWel project: welfare implications of changes in production systems for laying hens. World's Poultry Science Journal 2007; 63(1): 101–14.

28. Rodenburg TB, Tuyttens FAM, Sonck B. Welfare, health, and hygiene of laying hens housed in furnished cages and in alternative housing systems. Journal of Applied Animal Welfare Science 2005;8(3):211–26.

29. Bilcik B, Keeling LJ, Newberry RC. Effect of group size on tonic immobility in laying hens. Behavioural Processes 1998;43:53–9.

30. Hughes BO, Duncan IJH. The influence of strain and environmental factors upon feather pecking and cannibalism in fowls. British Poultry Science 1972;13:525–47.

31. Olsson IAS, Keeling LJ. Night-time roosting in laying hens and the effect of thwarting access to perches. Applied Animal Behaviour Science 2000;68:243–56.

32. Norgaard-Nielsen G. Bone strength of laying hens kept in an alternative system, compared with hens in cages and on deep litter. British Poultry Science 1990;31:81–9.

33. Gunnarsson. Rearing without early access to perches impairs the spatial skills of laying hens. Applied Animal Science 2000;67:217–28.

34. Kilpinen O, Roepstorff A, Permin A, Nørgaard-Nielsen G, Lawson LG, Simonsen HB. Influence of *Dermanyssus gallinae* and *Ascaridia galli* infections on behaviour and health of laying hens (*Gallus gallus domesticus*). British Poultry Science 2005;46(1):26–34.

35. De Reu K, Grijspeerdt K, Heyndrickx M, Zoons J, De Baere K, Uyttendaele M, *et al.* Bacterial eggshell contamination in conventional cages, furnished cages and aviary housing systems for laying hens. British Poultry Science 2005;46(2):149–55.

36. Lay DC Jr, Fulton RM, Hester PY, Karcher DM, Kjaer JB, Mench JA, *et al.* Hen welfare in different housing systems. Poultry Science 2011;90(1):278–94.

37. Pedersen. Dust in pig buildings. Journal of Agricultural Safety and Health 2000;6(4):261–74.

38. Rodenburg TB, Tuyttens FAM, De Reu K, Herman L, Zoons J, Sonck B. Welfare asessment of laying hens in furnished cages and non-cage systems: an on-farm comparison. Animal Welfare 2008;17:363–73.

39. Groot Koerkamp PWG, Metz JHM, Uenk GH, Phillips VR. Concentrations and emissions of ammonia in livestock buildings in Northern Europe. Journal of Agriculture Engineering Research 1998;70:79–95.

40. Wathes CM, Jones EKM, Kristensen HH, McKeegan DEF. Ammonia and poultry production: biological responses, welfare and environmental impact. In: Yalcin S, editor. Book of Abstracts/Proceedings Published by the World's Poultry Science Association of the 22nd World's Poultry Science Congress, Worlstanbul, Turkey; 2004.

41. Von Essen S, Donham K. Illness and injury in animal confinement workers. Occupational Medicine 1999;14(2):337–350.

42. Sherwin CM, Richards GJ, Nicol CJ. Comparison of the welfare of layer hens in 4 housing systems in the UK. British Poultry Science 2010;51(4):488–99.

43. Fossum O, Jansson DS, Etterlin PE, Vågsholm I. Causes of mortality in laying hens in different housing systems in 2001 to 2004. Acta Veterinaria Scandinavica 2009;51(1):3.

44. Hane M. Legehennenhaltung in der Schweiz. Schlussbericht BVET Forschungsprojekt 1999;2.97.1:164.

45. Permin A, Bisgaard M, Frandsen F, Pearman M, Kold J, Nansen P. Prevalence of gastrointestinal helminths in different poultry production systems. British Poultry Science 1999;40:439–43.

46. Axtell RC, Arends JJ. Ecology and management of arthropod pests of poultry. Annual Review of Entomology 1990;35:101–26.

47. McAllister JC, Steelman CD, Skeeles JK. Reservoir competence of the lesser mealworm (Coleoptera: Tenebrionidae) for *Salmonella typhimurium* (Eubacteriales: Enterobacteriaceae). Journal of Medical Entomology 1994;31(3):369–72.

48. Johnsen PF, Johannesson T, Sandoe P. Assessment of farm animal welfare at herd level: many goals, many methods. Acta Agriculturae Scandinavica Section A Animal Science Supplement 2001;30:26–33.

49. Anderson KE. Overview of natural and organic egg production: looking back to the future. Journal of Applied Poultry Research 2009;18(2):348–54.

50. Tactacan GB, Guenter W, Lewis NJ, Rodriguez-Lecompte JC, House JD. Performance and welfare of laying hens in conventional and enriched cages. Poultry Science 2009;88(4):698–707.

51. Dohms JE, Metz A. Stress-mechanisms of immunosuppression. Veterinary Immunology and Immunophysiology 1991;30:89–109.

52. Sapolsky RM. Individual difference and the stress response. Seminars in the Neurosciences 1994;6:261–9.

53. McEwen B. The concept of allostasis in biology and biomedicine. Hormones and Behavior 2003;43(1):2–15.

54. Miller DB, O'Callaghan JP. Neuroendocrine aspects of the response to stress. Metabolism 2002;51(6):5–10.

55. Gross WB, Siegel HS. Evaluation of the heterophil/lymphocyte ratio as a measure of stress in chickens. Avian Diseases 1983;27(4):972–9.

56. Dohms JE, Saif YM. Criteria for evaluating immunosuppression. Avian Diseases 1983;28(2):305–10.

57. Siegel H. Immunological responses as indicators of stress. World's Poultry Science J 1985;41:36–44.

58. Ballard PL, Baxter J, Higgins SJ, Rousseau GG, Tomkins GM. General prescence of glucocorticoid receptors in mammalian tissues. Endocrinology 1974;94:998–1002.

59. Erhard MH, Ozpinar H, Bilal T, Abas I, Kutay C, Eseceli H, *et al.* The humoral immune response and the productivity of laying hens kept on ground or in cages. Alternatives to Laboratory Animals 2000;28(5):699–705.

60. Silverin B. Stress responses in Birds. Poult Avian Biol Rev 1998;9:153–68.

61. Tauson R, Ambrosen T, Elwinger K. Evaluating procedures for scoring the integument of laying hens – Independent scoring of plumage condition. Acta Agriculturae Scandinavica 1984;34:400–8.

62. Tauson R, Svensson SA. Influence of plumage condition on the hen's feed requirement. Swedish Journal of Agricultural Research 1980;10:35–9.

63. Green AR, Wesley I, Trampel DW, Xin H. Air quality and bird health status in three types of commercial egg layer houses. Journal of Applied Poultry Research 2009;18:605–21.

64. Couch JR. Cage layer fatigue. Feed Age 1955;5:55–7.

65. Grumbles LC. Cage layer fatigue (cage paralysis). Avian Diseases 1959;3:122–5.

66. Cox AC, Balloun SL. Depletion of femur bone mineral after the onset of egg production in a commercial strain of leghorns and in broiler-type pullets. Poultry Science 1971;50:1429–33.

67. Beuving G. Corticosteroids in welfare research of laying hens. In: Smidt D, editor. Indicators Relevant to Farm Animal Welfare. Martinus Nijhoff, Boston, MA; 1983. p. 47–53.

68. Mench JA, Tienhoven AV, Marsh JA, McCormick CC, Cunningham DL, Baker RC. Effects of cage and floor pen managment on behavior, production, and physiological stress responses of laying hens. Poultry Science 1986;65:1058–69.

69. Davis GS, Anderson KE, Carroll AS. The effects of long-term caging and molt on single comb white leghorn hens on heterophil to lymphocyte ratios, corticosterone and thyroid hormones. Poultry Science 2000;79:514–8.

70. Khan MZ, Altmann J, Isani SS, Yu J. A matter of time: evaluating the storage of fecal samples for steroid analysis. Gen Comp Endocrinol 2002;128(1):57–64.

71. Von Engelhardt N, Groothius TGG. Measuring steroid hormones in avian eggs. Annals of the New York Academy of Science 2005;1046:181–92.

72. Rettenbacher S, Mostl E, Hackl R, Palme R. Corticosterone in chicken eggs. Annals of the New York Academy of Science 2005;1046:193–203.

73. Bulmer E, Gil D. Chronic stress in battery hens: measuring corticosterone in laying hen eggs. International Journal of Science of Poultry Science 2008;7(9):880–3.

74. Royo R, Mayo S, Hans-Erik C, Hau J. Egg corticosterone: a noninvasive measure of stress in egg-laying birds. Journal of Avian Medicine and Surgery 2008;22(4):310–4.

75. Downing JA, Bryden WL. Determination of corticosterone concentration in egg albumen: a non-invasive indicator of stress in laying hens. Physiology and Behaviors 2008;95:381–7.

76. Savory CJ. Fether pecking and cannibalism. World's Poultry Science Journal 1995;51:215–9.

77. Olsson IA, Keeling LJ. The push-door for measuring motivation in hens: laying hens are motivated to perch at night. Animal Welfare 2002;11:11–9.

78. Dawkins M. The role of behaviour in the assessment of poultry welfare. World's Poultry Science Journal 1999;55:295–303.

79. Freire R, Appleby MC, Hughes BO. Effects of nest quality and other cues for exploration on pre-laying behaviour. Applied Animal Behaviour Science 1996;48:37–46.

80. Freire R, Appleby MC, Hughes BO. Assessment of pre-laying motivation in the domestic hen using social interaction. Animal behaviour 1997;54:313–9.

81. Cooper JJ, Appleby MC. Motivational aspects of individual variation in response to nestboxes by laying hens. Animal Behaviour 1997;54:1245–53.

82. Leroy T, Vranken E, Struelens E, Sonck B, Berckmans D. Computer Vision Based Recognition of Behavior Phenotypes of Laying Hens. ASABE, Tampa, FL; 2005.

83. Leroy T, Vranken E, Struelens E, Sonck B, Berckmans D, editor. Modeling and recognizing behavior patterns of laying hens in furnished cages. In Proceedings of the Seventh International Symposium of Livestock Environment. ASABE, Beijing, China; 2005.

84. Leroy T, Vranken E, Van Brecht A, Struelens E, Sonck B, Berckmans D. A computer vision method for on-line behavioral quantification of individually caged poultry. Transactions of the ASABE 2006;49(3):795–802.

85. Pousaberi A, Wichman A, Bahr C, Hanninen L, Pastell M, Berckmans D, editors. Automatic monitoring of turkeys: a vision-based approach to detect adn analyse the behaviour of Turkeys in transport cages based on ellipse fitting. In Proceedings of the Seventh World Congress on Computers in Agriculture and Natural Resources. ASABE, Reno, NV; 2009. p. 38.

86. Eigenberg RA, Brown-Brandl JA, editors. Development of a Cattle ID Monitoring System. In Proceedings of the Seventh International Symposium of Livestock Environment. ASABE, Beijing, China; 2005. p. 600.

87. Davis JD, Darr MJ, Xin H, Harmon JD, Brown-Brandl TM. Development of a low-cost GPS Herd Activity and Welfare Kit (HAWK) for livestock monitoring. 607–612. In Proceedings of the Seventh International Symposium of Livestock Environment. ASABE, Beijing, China; 2005. p. 685–93.

88. Green AR, Rodriguez LF. Implementation of a GPS Monitoring System for Cows on Pasture. ASABE, Pittsburgh, PA; 2010.

89. Puma MC, Xin H, Gates RS, Burnham DJ. An instrumentation system for studying feeding and drinking behavior of individual poultry. Applied Engineering in Agriculture 2001;17(3):365–74.

90. Green AR, Wathes CM, Demmers TGM, MacArthur-Clark J, Xin H. Design of a novel preference chamber for laboratory mice and its application for testing tolerance of atmospheric ammonia. JAALAS 2008;47(2):49–56.

Lamb mulesing: impact on welfare and alternatives

Lauren E. Edwards

Address: Animal Welfare and Biodiversity Research Group, Department of Natural Sciences, Unitec Institute of Technology, Carrington Road, Mt Albert, Private Bag 92025, Auckland, New Zealand.

Correspondence: Email: ledwards@unitec.ac.nz

Received: 11 June 2012
Accepted: 20 September 2012

Abstract

Mulesing is a surgical procedure that is routinely performed without pain relief on Merino and Merino cross sheep to prevent flystrike in the breech region. Pain is apparent immediately following the procedure and persists for at least 2–3 days, and possibly up to 2 weeks, as reflected by changes in behaviour. A substantial stress response occurs for 1–2 days following mulesing, and an acute phase response occurs for up to 7 days. Lambs consistently lose weight in the week after mulesing, and this has been associated with increased mortality rates under field conditions. Using these measures, it has been ascertained that mulesing poses a severe and sustained stressor for lambs, resulting in very poor welfare for a prolonged period of time. Pain relief can be provided for a short period (8–12 h) after the procedure, but a pain-free experience is not possible with current methods of pain relief. Alternative methods of breech modification involve the application of plastic occlusive clips or the intradermal delivery of sodium lauryl sulphate (SLS) to the breech. Both of these alternatives cause necrosis of the breech tissue without creating an open wound, and pose a much smaller challenge to welfare than surgical mulesing. Selection for sheep with low susceptibility to breech strike is considered to be the best long-term option, as this will provide a permanent, pain-free solution in which the sheep do not require any interventions. These changes will require time to implement, possibly 10–15 years, and alternative methods and management practices (such as those reviewed here) that provide protection must be used in the interim.

Keywords: mulesing, sheep welfare, clips, sodium lauryl sulphate, Tri-Solfen.

Review Methodology: The CAB Abstract database and the Australian Veterinary Journal were searched for the terms 'mulesing', 'cutaneous myiasis and sheep', 'flystrike and sheep' and 'sodium lauryl sulphate and sheep'. The reference lists from the resulting articles were also used as a source of relevant literature, and the website for Australian Wool Innovation Limited (www.wool.com) was used to gain industry reports and newsletters.

Introduction

Mulesing is a surgical procedure that is routinely performed on Merino and Merino cross sheep to reduce the incidence of flystrike. Flystrike occurs when gravid female blowflies, generally of the species *Lucilia cuprina* [1], lay eggs in the moist fleece of a sheep, resulting in an infestation of larvae (maggots) that feed on the sheep's tissues. During flystrike, sheep show obvious signs of distress such as restlessness and biting of the affected area, as well as a sustained physiological stress response that continues for as long as the infestation is present [2–4], indicating substantial compromise to the welfare of struck sheep. In extreme cases when the sheep is not treated, this condition results in mortality. Flystrike has been estimated to cost the Australian wool industry $280 million per year [5] and causes the death of 3 million sheep annually [6]. Current mortality rates are likely to be lower than this 1990 estimate due to the development of long-acting insecticides such as dicyclanil and cyromazine; however, the prevention of flystrike is still a high priority in terms of both animal welfare and profitability for producers.

A sheep becomes susceptible to flystrike when the fleece becomes wet for prolonged periods, and the odour of the microbial environment within the wet fleece attracts gravid female blowflies [7]. Prolonged rainfall [6] or urine staining can lead to persistently wet fleece, but it is the accumulation of faecal matter around the perineum (dags) that poses one of the greatest risk of flystrike in the tail and breech area [8, 1]. Merino sheep and their crosses are particularly susceptible to breech strike due to the wrinkly nature of their skin. The wrinkles present around the breech not only accumulate more dags and urine, but also trap moisture for longer, creating an optimal environment for oviposition by the blowfly and promoting larvae survival. The amount of wool cover surrounding the perineum also influences the amount of dags and urine that accumulate; however, it is the degree of breech wrinkle and dagginess that are the main predisposing factors for flystrike [8].

To combat breech strike, a surgical procedure called mulesing was developed in 1931 by an Australian woolgrower named John Mules [9, 10]. The procedure is described as 'the removal of wool-bearing skin from the tail and breech area of the sheep' and is performed on Merino or Merino cross lambs between the ages of 2 and 12 weeks old [11], without analgesia or pre-operative anaesthesia. As these wounds heal, the skin around the breech is stretched, creating a bare, wrinkle-free area around the breech and tail that is resistant to the accumulation of urine and faeces [12]. Correct tail docking is also an important component of managing breech strike, as tails that are too short are associated with increased dag accumulation and tail strike [1, 10, 13].

Despite the considerable pain and welfare compromise that this procedure causes, mulesing remains a standard component of Merino husbandry because of the high efficacy of the procedure in combating breech strike. Any viable alternative must offer clear welfare benefits, not only at the time of mulesing, but also throughout the animal's lifetime in terms of flystrike protection. This review will examine the impact of mulesing and the current alternatives to mulesing on sheep welfare.

The Impact of Mulesing on Sheep Welfare

Animal welfare science uses two main approaches to assess the welfare status of animals in response to a stressor: the nature of the affective states that the stressor elicits, such as fear and pain (inferred through changes in behaviour and physiology), and the degree of biological dysfunction that occurs in response to the stressor, such as changes in physiology, health and fitness parameters. In both respects, the magnitude of the response and the duration of the response are used to assess the severity of the welfare challenge. An integrated approach that incorporates both affective states and biological function as the assessment criteria is the optimal method for assessing welfare, and this review will address the indicators of both.

Impact on Affective States

The immediate and intense changes in behaviour and physiology that occur following mulesing indicate that this is a painful procedure. Fell and Shutt [14] reported that neither mulesed nor sham-mulesed lambs vocalized during the procedure, but that the mulesed lambs struggled more. Plasma cortisol concentrations are elevated approximately 70–200% above those of sham-mulesed control lambs within 15 min of mulesing [14–16], and from 10 to 300% above by 30 min [17–21].

Shortly after mulesing, sheep assume the characteristic 'statue standing' response in which they stand with their heads down and back arched for long periods. They also make unsuccessful attempts to lie down [17, 18, 22]. This behavioural inhibition appears to be an attempt by the lambs to reduce movement of the damaged tissues around the breech, and is strongly indicative of pain. When the time budgets of mulesed lambs are compared with those of control lambs, mulesed lambs spend a greater portion of their time standing and less time lying for up to 3 days following the treatment [14–18]. Mulesed lambs also show stiff walking behaviour for 2 days after treatment [14, 18] and in one study, gait score was significantly poorer than that of controls at 18 days post-treatment [15]. Mulesed lambs show an increase in total abnormal behaviours, such as hunched standing, stiff walking and failed attempts at lying, for 2 days following mulesing [17, 18], and a reduction in feeding behaviour for approximately 2 weeks [15, 17]. These studies provide reasonably strong evidence that mulesed lambs undergo an acute pain response indicated by severe behavioural inhibition for 2 to 3 days following mulesing, with some behavioural inhibition for up to 2 weeks following the procedure that is likely to indicate pain.

In addition to pain, there is evidence that mulesing results in increased avoidance of humans [14, 16], with some indication that this response may apply to the specific operators who performed the procedure [14]. This avoidance behaviour persists for 3–5 weeks after the procedure, providing an indication of how aversive the sheep perceive mulesing to be. Fear is an aversive affective state that is detrimental to animal welfare, and while it does not possess the same gravity as pain, it must be acknowledged as a negative welfare outcome of mulesing.

Impacts on Biological Function

The biological dysfunction that occurs in response to mulesing indicates a substantial and sustained challenge to sheep welfare, based on measures of the general stress response, acute phase response to tissue damage and impact on fitness parameters.

Following the initial surgical insult, plasma cortisol concentrations show a biphasic response at 30 min and 12 h [20] or 24 h [14]. The first peak is a response to the pain, fear and novelty, as described above, and the second peak is related to the inflammatory response to the tissue trauma incurred [23]. Cortisol remains significantly elevated above that measured in control lambs for at least 24 h [20, 21] and often 48 h or more [14–19]. The neutrophil:lymphocyte (N:L) ratio is a commonly accepted indicator of stress that is significantly elevated in mulesed lambs and generally peaks at 24 h post-mulesing [18–20]. From this information, it can be concluded that the combined stressors involved in the mulesing procedure and the subsequent acute phase response result in a substantial stress response in lambs for at least 2 days.

Haptoglobin forms part of the acute phase response and is used as an indicator of the severity of tissue damage and inflammation. Following mulesing, haptoglobin concentrations peak on the second or third day after mulesing and remain significantly elevated above controls for 1 week [15, 17–20, 24]. Beta-endorphins are an endogenous opioid released in response to emotional or surgical stress [4] and have been shown to increase at 30 min and 12 h after mulesing [14, 17–20], although in two of these studies the control lambs also showed a similar increase in this protein [17, 19], presumably as a result of the stress associated with handling.

Indicators of fitness in farmed species include measures of survival and changes in bodyweight, rather than the rate of reproduction, as farming practices often limit opportunities for reproduction in these species. One recent study found that mulesed lambs had a 4% lower survival rates than non-mulesed lambs at 90 days post-mulesing [25]. The authors attributed the reduced survival rate to the weight loss that occurred following mulesing, as body weight and growth rates are strongly associated with survival rates in lambs following weaning [26].

Most studies that measured changes in bodyweight following mulesing have found a reduction in weight during the first week. Lambs lost between 0.5 and 1.5 kg during the week after mulesing [15, 16, 20, 24, 25] or showed reduced average daily gains [17, 19, 21], and took 1 [17, 19, 20, 21] to 4 [15, 16, 24] weeks to recover. This weight loss may be attributed to reduced feeding behaviour and reduced availability of resources for growth due to healing. The wounds incurred by mulesing can take up to 47 days to heal [12].

In conclusion, the literature shows that mulesing poses a severe and sustained stressor for sheep that have undergone this procedure, resulting in very poor welfare for a prolonged period of time.

Alternatives to Current Mulesing Practice

For an alternative to surgical mulesing to be acceptable in commercial use it must not only alleviate the acute impact of mulesing on sheep welfare, but also protect the lifetime welfare of the sheep by preventing breech strike. Various alternatives to surgical mulesing had been investigated prior to the anti-mulesing campaign in 2004, such as the topical application of caustic chemicals, freezing the skin of the breech, radiation treatment, light treatment and intradermal injection of the enzyme collagenase to modify the amount of wrinkle or wool cover present on the breech [27]. The intradermal injection of cetrimide has been the subject of a substantial amount of research. However, this method of breech modification has not been shown to offer welfare benefits over surgical mulesing [12, 18, 19, 24] and none of these options have been developed for commercial use by woolgrowers. Only those methods that have been developed as viable options for commercial use will be reviewed, as these are the most relevant to the welfare of the Australian flock. These methods can be categorized into: alternative methods of breech modification; surgical mulesing with pain relief; selection against flystrike susceptibility and alternative management methods.

Alternative Methods of Breech Modification

Clips

Plastic occlusive clips that partially replicate the effects of mulesing without creating an open wound became commercially available in 2009 [5]. The loose skin on the tail and breech is occluded through the clips, laterally stretching the perineal and tail areas to reduce breech wrinkle. The pressure from the clips causes ischaemic necrosis and sloughing of the skin approximately 2–4 weeks after application without creating an open wound [12, 25]. Alternately, the clips can be removed after 4–6 days to allow reuse and avoid contaminating paddocks with sloughed clips [28]

The behavioural indicators of pain were minimal in lambs receiving the clips treatment, although only two published studies have measured the behavioural response of sheep to this procedure. During the 2 h immediately following treatment, lambs that had received the clips treatment spent more time standing with the head down than control lambs [22]. In the following 21 days, the same lambs showed no other differences in behaviour or gait compared with the control lambs [15].

The physiological stress response to the clips was substantially smaller in comparison with that of surgical mulesing. The cortisol concentration of lambs receiving the clips treatment was significantly greater than that of the controls on the day of treatment only [15], compared with 1–2 days for mulesed lambs. The neutrophil:lymphocyte ratio was significantly higher than that of the controls on the day after treatment in one study [15] and for 5 days after treatment in another [24]. In addition, the concentrations of selected acute phase proteins showed a relatively mild increase for approximately 1 week after treatment [15, 24]. This may be related to a lesser degree of tissue damage occurring in comparison with surgical mulesing, or the fact that the necrotizing tissue and its associated toxins were isolated from the live tissue by the clips [19].

The fitness of lambs receiving the clips treatment is substantially higher than that of mulesed lambs, as measured by weight gain, survival rate and rate of wound healing. The clips caused lamb growth to slow for 2–4 weeks after treatment [24, 25, 28], but did not cause weight loss, as seen in mulesed lambs. In fact, lambs receiving the clips treatment on-farm weighed significantly more than mulesed lambs for 6 months post-treatment [25]. The 4% increase in survival rates of the clipped lambs compared with the mulesed lambs in this study was attributed to the increased bodyweight of the clipped lambs following treatment [25]. In a study comparing the rates of wound healing, the wound beds created by clips were completely re-epithelialized by 32 days post-treatment, compared with 47 days in surgically mulesed lambs [12].

In conclusion, clips had a small to moderate impact on acute welfare, and all studies that compared clips against mulesing invariably concluded that clips provided substantial benefits over mulesing in the short term. The behavioural indicators of pain were briefer and of a smaller magnitude than those induced by mulesing (1 day clips versus 2–3 days mules), as was the cortisol response (1 day clips versus 1–2 days mules). The acute phase response to tissue damage was minimal, and while growth was slowed for a short period after treatment the clipped lambs did not lose weight. Clipped lambs also had higher survival rates than mulesed lambs. Based on these welfare indicators, the use of plastic clips appears to offer a more humane method of breech modification than mulesing.

Despite these welfare benefits, recent research has indicated that the use of plastic clips cannot provide the same degree of protection from breech strike as surgical mulesing [29]. Larsen *et al.* [29] reported that the incidence of breech strike was significantly higher for clipped sheep in comparison with surgically mulesed sheep (7.7 and 1.1%, respectively). While further research into the lifetime effectiveness of the plastic clips in preventing breech strike is required, it currently appears that if the plastic clips are to be used, then they must be supported by additional protective measures, such as the use of insecticides [29].

Intradermal Sodium Lauryl Sulphate (SLS)

A second method for modifying the breech involves intradermal injection of SLS into the skin surrounding the breech and on the tail using a needle-less applicator. The applicator forces the SLS compound through the dermis causing immediate (2 min) local necrosis of the tissues, followed by swelling, eschar formation and complete sloughing by 28 days post-treatment [30]. As the wound heals it causes contraction of the skin on the perineum and tail, reducing wrinkle and increasing the size of the bare area. This method is currently being developed for commercial use [5].

The behaviour of lambs receiving intradermal injections of SLS was indicative of discomfort during the first 1–2 days after application. During the 2 h immediately following injections, lambs treated with SLS showed increased kneeling behaviour and spent longer standing in a hunched posture than controls [17, 22]. There was also a tendency for an increase in abnormal behaviours on the day of treatment, as well as more total time standing and less time lying than the control lambs during the 48 h post-treatment [17]. Another study reported a reduction in feeding compared with controls on the day after treatment [15]. Lee and Rothwell [30] reported no behavioural change in lambs receiving SLS on the day after treatment; however, these authors were looking for conspicuous pain behaviours (restlessness, biting/rubbing the area, inhibited movement) and did not measure time budgets.

The physiology of lambs receiving intradermal SLS indicates a moderate stress response to the procedure for 1–2 days. Cortisol was significantly elevated by 15 min post treatment [15], and remained elevated until the day after treatment [15, 17]. The cortisol concentrations were comparable with that of mulesed lambs at both 12 h post-treatment and on the following day [15, 17]. In addition, the N:L ratio and the white cell counts were both significantly elevated above controls on the day after treatment [15, 17]; and again on day 4 in the case of the white blood cell counts, indicating an immunological response to the tissue damage incurred. Haptoglobin concentrations were elevated at 1–2 days after the procedure, and remained elevated for approximately 4 days [15, 17].

Little information has been published on the effects of intradermal SLS on the fitness of the animals, but there was no impact on growth following treatment for 25 [15] or 42 days [17] and this effect would appear to be minimal. There is also a paucity of data relating to the lifetime efficacy of this procedure in protecting sheep from breech strike.

In conclusion, the use of intradermal SLS provides a promising alternative to surgical mulesing from an animal welfare perspective. Behavioural indicators of pain were present for 1–2 days, compared with 2–3 days in mulesing. The cortisol response was comparable between lambs receiving SLS and mulesing, indicating a similar stress response. However, the acute phase response to the SLS treatment was approximately half that of the mulesing treatments indicating a reduced amount of tissue damage incurred. In addition, SLS appears to have little impact on growth of lambs. Based on these welfare indicators, intradermal SLS appears to offer a more humane method of breech modification than mulesing, but still results in a moderate welfare challenge.

Surgical Mulesing with Pain Relief

The pain experienced by sheep during mulesing is the greatest welfare concern resulting from this procedure, and an immediate solution to this problem would at first appear to be the use of pain relief. However, while some relief can be provided immediately following the mulesing procedure, providing total pain relief for a sufficient duration following mulesing is a complex problem that cannot be wholly dealt with using current pain relief methods [23].

The mulesing procedure results in two distinct pain responses. The first is the immediate response to the surgical incisions, tissue removal and the resulting open wound bed. The second response relates to the inflammation of the tissues that occurs during the acute phase response to tissue trauma [23, 31]. As discussed above, this pain is evident during the first 2–3 days following mulesing, and may persist for at least 2 weeks.

The large area of tissue involved in the mulesing procedure makes the pre-administration of local anaesthetic to the breech impractical [21, 23]. A topical anaesthetic (Tri-Solfen) has been developed for commercial use (Bayer Industries), which consists of a gel containing long-acting local anaesthetics combined with adrenaline and an antiseptic that is sprayed directly onto the open wound immediately post-mulesing [32]. This topical anaesthetic was made commercially available in 2005 [33], and its use has been steadily increasing since 2010 (http://www.awex.com.au/market-information/mulesing-status.html). Tri-Solfen has been shown to improve wound healing rates, reduce the behavioural response to direct stimulation of the open wound bed and surrounding skin for 4–8 h following mulesing, and reduce the incidence of abnormal postures and gait during the 4-h observation period following mulesing [32]. A study by Paull *et al.* [21] found that treating mulesed lambs with Tri-Solfen had little effect on behaviour during the 4 h following mulesing, although the duration of hunched standing was reduced and the duration of lying behaviour was increased in comparison with that of mulesed lambs. There were very few behavioural differences between the mulesed lambs and those treated with Tri-Solfen during the subsequent 4–12 h period, and the authors described the effect of Tri-Solfen as 'rather modest' [21]. It should be noted that a placebo gel containing only the adrenaline and antiseptic produced a behavioural response that was intermediate to that of the Tri-Solfen and control treatments, indicating that at least part of the relief could be attributed to the gel barrier covering the open wound, rather than the action of the local anaesthetics alone [32].

Non-steroidal anti-inflammatories (NSAIDs) have been investigated as longer-lasting pain relief options [31]. The use of NSAIDs alone did not reduce the behavioural or physiological response to surgical mulesing [20, 21]. However, when the NSAID carprofen was combined with the topical anaesthetic Tri-Solfen, the behavioural and physiological pain responses of the treated lambs did not differ from control lambs for 6 h following mulesing, and showed only mild differences from controls at 8–12 h [21]. Despite this short-term pain relief, the cortisol response of the treated lambs matched that of the mulesed lambs by 24 h, and both mulesed and treated lambs lost weight following mulesing, indicating that the analgesic benefits of the NSAID and Tri-Solfen combination could not fully address the welfare challenge imposed by surgical mulesing. The re-administration of pain relief at regular intervals following mulesing has not been investigated due to the recommendation that mulesed lambs should not be handled in the 4 weeks following mulesing to aid wound healing [11].

In conclusion, providing pain relief following surgical mulesing does offer benefits to welfare in the short-term (up to 12 h), and any reduction in the severity of pain experienced by lambs is an improvement on traditional surgical mulesing [23]. However, the pain of the surgical incisions and tissue removal is still experienced by the lambs, as is the longer-term pain that is present when the analgesics wear off, and there are no longer-term studies assessing the impact of pain relief on welfare and fitness parameters. In addition, no research has been directed at readministering pain relief options and what impact this may have on wound healing rates, but Fisher [23] warns against developments that may lead to a dependence on pain relief measures rather than moving towards a permanent solution to the mulesing issue through genetic selection.

Selection against Flystrike Susceptibility

The traits that have the greatest impact on susceptibility to breech strike (breech wrinkle, breech cover and dags) are found to vary naturally within the Australian Merino population, and are moderate to highly heritable [5, 34–37]. In addition, whiter fleece is associated with decreased strike susceptibility in areas with summer rainfall [5]. The heritability of these traits offers the Australian wool industry the opportunity to select for sheep with low breech wrinkle, bare breeches, whiter wool and less dags, thus removing the need for mulesing.

A genetic solution to the mulesing issue is an attractive option, as it offers a permanent, pain-free method of combating breech strike. However, it is not an immediate solution for the wool industry. The traits associated with strike susceptibility are also associated to varying degrees with fleece quality, and direct selection against these could result in an unacceptable reduction in productivity [34, 38]. Breech wrinkle is of particular concern, as wrinkle score is correlated unfavourably with both fleece weight and fibre diameter [34, 38]. This issue is overcome by using a selection index that incorporates both strike susceptibility and production traits as criteria for selecting breeding stock. A selection index can improve breech morphology while limiting the impact on production traits, but markedly increases the amount of time taken to achieve selection goals. Edwards *et al.* [39] estimate that breech bareness (measured on a scale of 1–5), derived from a rare mutation, could be reduced by 0.2 units/year, and that it would take 10–15 years to obtain an average flock score of one (completely bare). This trait was correlated with lower dagginess [40] and a reduced incidence of flystrike [41] in New Zealand breeds of sheep. Brown *et al.* [34] estimate that breech wrinkle score (on a scale of 1–5) could be reduced by 0.4–0.9 units over 10 years, and a reduction in breech wrinkle score of 0.5 units/10 years was recommended by Richards and Atkins [38] if financial losses were to be minimized to 20–30%. Recent unpublished results of a breeding programme have determined that a breech wrinkle score, breech cover score and dag score of ≤ 2 (on a scale of 1–5) will provide the same degree of protection from strike as surgical mulesing [42]. Thus, the average score of the flock for each of these traits at the start of the selection programme combined with the strength of the selection pressure will determine the length of time required to reduce the risk of breech strike to an acceptable level.

Another area of genetic research involves selection for improved parasite resistance. The presence of dags and scouring is often the result of intestinal worm burdens [2], and resistance to these parasites is heritable [43]. Sheep also show variation in the attractiveness of their fleece to gravid blowflies [44], presumably because of variation in fleece characteristics and the microbial environment of the fleece. In addition, variation in the ability of the host's immune response to inhibit larval growth has been observed [44]. The effects of fleece characteristics and the immune response to both worm burdens and fly strike may offer further opportunities for selecting improved strike resistance in sheep. It may also help researchers understand the observed variation in breech strike rates among research animals with similar breech morphology [45].

Selection against susceptibility to breech strike is recognized as the optimal solution to the mulesing issue [23, 35, 46]. The selection process does not negatively impact on the welfare of the sheep, nor do the breech modifications being selected for. In fact, there is some evidence that selection for improved breech morphology is also associated with improved body weight and possibly lamb survival [42], suggesting welfare and production improvements above and beyond those related to reducing the need for surgical mulesing. However, the selection process is a lengthy one, and protection from breech strike is required in the interim.

Alternative Management Methods

While mulesing provides the most effective protection against breech strike, alternative management methods that do not involve breech modification are also available. These involve reducing the attractiveness of the fleece to gravid blowflies through the use of insecticides, optimal timing of shearing and crutching and management of scouring [8, 46]. Increasing the frequency of inspections is also required during periods of high fly pressure to minimize the impact of strike on the welfare and productivity of struck animals.

The life cycle of the blowfly *L. cuprina* means that adult flies are only present during the warmer months, and therefore flystrike is a seasonal problem for producers. In addition, the risk of breech strike increases with increasing fleece length [8]. Thus, it is beneficial to shear or crutch sheep at the beginning of the warmer months to reduce the fleece length surrounding the breech, minimizing the availability of oviposition sites for gravid female blowflies. Crutching is an easier procedure than shearing to time correctly, and provides protection from breech strike for approximately 2 months [8].

Insecticides can be applied to the wool of the breech area to deter flies, and these chemicals can be very effective. For example, dicyclanil provided effective protection from flystrike in unmulesed sheep for approximately 5 months [47]. However, the heavy reliance on insecticides is discouraged, as this practice has previously led to chemical resistance in blowflies during the 1950s and 1960 s [10].

Improved management of scouring will help to reduce the risk of flystrike by decreasing the dagginess of the breech area. Typically, this will involve improved management of internal parasite burdens through drenching, correct pasture management and selecting against dagginess in breeding stock [8, 46].

The alternative management options briefly described here all require additional inputs when compared with mulesing. Crutching, jetting insecticides and drenching all involve additional costs and labour, and these are ongoing costs that need to be repeated every year. In relation to the welfare implications of this approach, no research has been conducted on the cumulative effects of the additional handling involved on the welfare of sheep. In addition, the risk of breech strike must be included in these assessments: any increase in the risk of flystrike over the lifetime of the sheep will detract from the welfare benefits obtained by not mulesing. The attitudes of farmers to these husbandry practices are positive, but they are viewed as complementary rather than primary methods of controlling breech strike [48], and it appears unlikely that breech strike will be successfully managed without any form of breech modification.

Conclusion

From this review of the literature, it is evident that mulesing poses a substantial and sustained challenge to sheep welfare. Alternative methods of protecting sheep from breech strike involve non-surgical methods of breech modification (clips or intradermal SLS), the provision of pain relief during surgical mulesing, selection for low susceptibility sheep, and improved management and monitoring of risk factors in the flock. The efficacy of the plastic clips and intradermal SLS in providing lifetime protection from breech strike have not yet been compared with that of surgical mulesing, but warrant further investigation.

Selection of sheep with low susceptibility to breech strike is considered to be the best long-term option, as this will provide a permanent, pain-free solution in which the sheep do not require any interventions. These changes will require time to implement, possibly 10–15 years, and alternative measures (such as those reviewed here) that provide protection must be used in the interim.

Acknowledgements

The author thanks Dr Adele Arnold for her helpful comments on the manuscript.

References

1. Watts JE, Murray MD, Graham NPH. The blowfly strike problem of sheep in New South Wales. Australian Veterinary Journal 1979;55:325–34.
2. Roger PA. Chapter 5: The impact of disease and disease prevention on welfare in sheep. In: Dwyer C, editor. The Welfare of Sheep. Springer Science + Business Media; Dordrecht, The Netherlands; 2008. p. 159–212.
3. Colditz IG, Walkden-Brown SW, Daly BL, Crook BJ. Some physiological responses associated with reduced wool growth during blowfly strike in Merino sheep. Australian Veterinary Journal 2005;83:695–9.
4. Shutt DA, Smith AI, Wallace CA, Connell R, Fell LR. Effect of myiasis and acute restraint stress on plasma levels of immunoreactive beta-endorphin, adrenocorticotrophin (ACTH) and cortisol in sheep. Australian Journal of Biological Sciences 1988;41:297–301.
5. Australian Wool Innovation Limited. Managing Breech Flystrike. 2011.
6. Wardhaugh KG, Morton R. The incidence of flystrike in sheep in relation to weather conditions, sheep husbandry, and the abundance of the Australian sheep blowfly, *Lucilia cuprina* (Wiedemann) (Diptera: Calliphoridae). Australian Journal of Agricultural Research 1990;41:1155–67.
7. Emmens RL, Murray MD. The role of bacterial odours of oviposition by *Lucilia cuprina* (Wiedemann) (Diptera: Calliphoridae), the Australian sheep blowfly. Bulletin of Entomological Research 1982;72:367–75.
8. Horton B, Hogan L. FlyBoss: a web-based flystrike information and decision support system. Animal Production Science 2010;50:1069–76.
9. Beveridge WIB. The origin and early history of the Mules operation. Australian Veterinary Journal 1984;61:161–3.
10. Morley FHW, Johnstone IL. Development and use of the mules operation. Journal of the Australian Institute of Agricultural Science 1984;50:86–97.
11. Primary Industries Standing Committee. Model Code of Practice for the Welfare of Animals. The Sheep. 2nd ed. PISC Report 89. 2006.
12. Lepherd ML, Canfield PJ, Hunt GB, Thomson PC, Bosward KL. Wound healing after mulesing and other options for controlling breech flystrike in Merino lambs: observations on gross and microscopic wound healing. Australian Veterinary Journal 2011;89:27–37.
13. Watts JE, Marchant RS. The effects of diarrhoea, tail length and sex on the incidence of breech strike in modified mulesed Merino sheep. Australian Veterinary Journal 1977;54:118–123.
14. Fell LR, Shutt DA. Behavioural and hormonal responses to acute surgical stress in sheep. Applied Animal Behaviour Science 1989;22:283–94.
15. Hemsworth PH, Barnett JL, Karlen GM, Fisher AD, Butler KL, Arnold NA. Effects of mulesing and alternative procedures to mulesing on the behavior and physiology of lambs. Applied Animal Behaviour Science 2009;117:20–7.
16. Chapman RE, Fell LR, Shutt DA. A comparison of stress in surgically and non-surgically mulesed sheep. Australian Veterinary Journal 1994;71:243–7.
17. Colditz IG, Paull DR, Fisher AD. Physiological and behavioural effects of intradermal injection of sodium lauryl sulfate as an alternative to mulesing in lambs. Australian Veterinary Journal 2010;88:483–9.
18. Colditz IG, Lloyd JB, Paull DR, Lee C, Giraudo A, Pizzato C, *et al.* Assessment of welfare of suckling lambs following intradermal injection of cetrimide as a non-surgical alternative to conventional mulesing. Australian Veterinary Journal 2009;87:12–8.
19. Colditz IG, Lloyd JB, Paull DR, Lee C, Giraudo A, Pizzato C, *et al.* Effect of non-steroidal anti-inflammatory drug, carprofen, on weaned sheep following non-surgical mulesing by intradermal injection of cetrimide. Australian Veterinary Journal 2009;87:19–26.
20. Paull DR, Lee C, Atkinson SJ, Fisher AD. Effects of meloxicam or tolfenamic acid administration on the pain and stress responses of Merino lambs to mulesing. Australian Veterinary Journal 2008;86:303–11.
21. Paull DR, Lee C, Colditz IG, Atkinson SJ, Fisher AD. The effect of a topical anaesthetic formulation, systemic flunixin and carprofen, singly or in combination, on cortisol and behavioural responses of Merino lambs to mulesing. Australian Veterinary Journal 2007;85:98–106.
22. Edwards LE, Arnold NA, Butler KL, Hemsworth PH. Acute effects of mulesing and alternative procedures to mulesing on lamb behavior. Applied Animal Behaviour Science 2011;133:169–74.
23. Fisher AD. Addressing pain caused by mulesing in sheep. Applied Animal Behaviour Science 2011;135:232–40.

24. Lepherd ML, Canfield PJ, Hunt GB, Thomson PC, Bosward KL. Assessment of the short-term systemic effect of and acute phase response to mulesing and other options for controlling breech flystrike in Merino lambs. Australian Veterinary Journal 2011;89:19–26.
25. Evans IB, Lawton PW, Sergeant E, Lloyd JB. Effect of plastic occlusive clips used as an alternative to mulesing on breech conformation, body weight and survival of Merino lambs. Australian Veterinary Journal 2012;90:88–96.
26. Hatcher S, Eppleston J, Graham RP, McDonald J, Schlunke S, Watt B, *et al.* Higher weaning weight improves postweaning growth and survival in young Merino sheep. Australian Journal of Experimental Agriculture 2008;48:966–973.
27. Rothwell J, Hynd P, Brownlee A, Dolling M, Williams S. Research into alternatives to mulesing. Australian Veterinary Journal 2007;85:94–7.
28. Evans IB, Lawton PW, Sergeant E, Lloyd JB. Effect on the breech and tail characteristics of Merino lambs of varying the duration of application of occlusive plastic clips as an alternative to mulesing. Australian Veterinary Journal 2012;90:308–314.
29. Larsen JWA, Tyrell LD, Anderson N. Prevalence of breech strike in mulesed, clipped and unmulesed Merino hoggets in south-eastern Australia. Australian Veterinary Journal 2012;90:158–166.
30. Lee EJ, Rothwell JT. Histological changes to the skin of Merino sheep following deep dermal and subcutaneous injections of sodium lauryl sulfate. Australian Veterinary Journal 2010;88:146–50.
31. Fitzpatrick J, Scott M, Nolan A. Assessment of pain and welfare in sheep. Small Ruminant Research 2006;62:55–61.
32. Lomax S, Sheil M, Windsor PA. Impact of topical anaesthesia on pain alleviation and wound healing in lambs after mulesing. Australian Veterinary Journal 2008;86:159–68.
33. Lomax S, Sheil M, Windsor PA. Use of local anaesthesia for pain management during husbandry procedures in Australian sheep flocks. Small Ruminant Research 2009;86:56–58.
34. Brown DJ, Swan AA, Gill JS. Within- and across-flock genetic relationships for breech flystrike resistance indicator traits. Animal Production Science 2010;50:1060–8.
35. James PJ. Genetic alternatives to mulesing and tail docking in sheep: a review. Australian Journal of Experimental Agriculture 2006;46:1–18.
36. Greeff JC, Karlsson LJE. Opportunities to breed for resistance to breech strike in Merino sheep in a Mediterranean environment. Proceedings of the 18th Conference of the Association for the Advancement of Animal Breeding and Genetics; 2009 September 28–October 1; Barossa Valley, Australia; 2009. p. 272–278.
37. Smith JL, Brewer HG, Dyall T. Heritability and phenotypic correlations for breech strike and breech strike resistant indicators in Merinos. Proceedings of the 18th Conference of the Association for the Advancement of Animal Breeding and Genetics; 2009 September 28–October 1; Barossa Valley, Australia; 2009. p. 334–337.
38. Richards JS, Atkins KD. Will genetics offer a permanent solution to breech strike? Animal Production Science 2010;50:1053–9.
39. Edwards NM, Hebart M, Hynd PI. Phenotypic and genotypic analysis of a barebreech trait in Merino sheep as a potential replacement for surgical mulesing. Animal Production Science 2009;49:56–64.
40. Scobie DR, O'Connell DO, Morris CA, Hickey SM. Dag score is negatively correlated with breech bareness score of sheep. Australian Journal of Experimental Agriculture 2008;48:999–1003.
41. Scobie DR, O'Connell DO. Breech bareness reduces flystrike in New Zealand crossbred sheep. Animal Production Science 2010;50:599–602.
42. Commonwealth Scientific and Industrial Research Organisation. Breech Strike Genetics. Issue 4, June 2011 Newsletter.
43. Eady SJ, Woolaston RR, Mortimer SI, Lewer RP, Raadsma HW, Swan AA, *et al.* Resistance to nematode parasites in Merino sheep: sources of variation. Australian Journal of Agricultural Research 1996;47:895–915.
44. Smith JL, Colditz IG, Piper LR, Sandeman RM, Dominik S. Genetic resistance to growth of *Lucilia cuprina* larvae in Merino sheep. Australian Journal of Experimental Agriculture 2008;48:1210–6.
45. Department of Agriculture and Food Western Australia. Breeding for Breech Strike Resistance Project. Issue 4, October 2011 Newsletter.
46. Phillips CJC. A review of mulesing and other methods to control flystrike (cutaneous myiasis) in sheep. Animal Welfare 2009;18:113–21.
47. Hosking BC, George AJ. Efficacy of CLiK® Spray-On in preventing naturally occurring blowfly strike when applied to unmulesed sheep. International Journal of Applied Research in Veterinary Medicine 2009;7:76–82.
48. Wells AED, Sneddon J, Lee JA, Blache D. Farmer's response to societal concerns about farm animal welfare: The case of mulesing. Journal of Agricultural and Environmental Ethics 2011;24:645–58.

Laminitis in horses

Nicola J. Menzies-Gow*

Address: Department of Veterinary Clinical Sciences, Royal Veterinary College, Hawkshead Lane, North Mymms, Herts AL9 7TA, UK.

*Correspondence: Email: nmenziesgow@rvc.ac.uk

Received: 24 July 2012
Accepted: 4 October 2012

Abstract

Laminitis is characterized by failure of the attachment of the laminar epidermal cells to the underlying basement membrane of the dermal laminae. It can arise in association with gastrointestinal disease, diseases associated with sepsis such as pneumonia and metritis, mechanical overload, endocrine disorders associated with excess glucocorticoids such as pituitary pars intermedia dysfunction (PPID) and endocrine disorders associated with insulin resistance such as equine metabolic syndrome (EMS). However, most commonly, laminitis arises in association with ingestion of non-structural carbohydrates from pasture. Estimates of laminitis frequency range from 1.5 to 34% depending on the population evaluated. There are several different theories concerning the underlying pathophysiological mechanisms involved in laminitis with evidence to support a role for inflammation, metabolic dysfunction, extracellular matrix degradation and vascular and endothelial dysfunction. However, further research is required to fully elucidate the pathways involved and perhaps determine a unifying concept. A diagnosis of laminitis is based on history, clinical signs and radiographs. Furthermore, diagnostic tests are performed if an underlying endocrinopathy is suspected. Treatment involves the use of analgesic drugs, foot support and possibly modification of the digital blood supply. In addition, treatments specific to an underlying endocrinopathy may be employed. Various clinical and radiographic parameters can be used to help determine the prognosis. Prevention is extremely important and is based on management changes focused on minimizing ingestion of non-structural carbohydrates in order to prevent pasture-associated laminitis and on promoting weight loss in obese animals and cryotherapy and foot support in animals with diseases associated with laminitis.

Keywords: Laminitis, Horse, Inflammation, Vascular, Insulin, Extracellular matrix

Review Methodology: The following databases were searched: CAB Abstracts and Medline (keyword search terms used: equine, laminitis, horse). In addition, the references from the articles obtained by this method were checked for additional relevant material. Colleagues were also consulted to seek any upcoming studies not yet published.

Introduction

Laminitis is a common and painful condition of horses characterized by failure of the attachment of the epidermal cells of the epidermal laminae to the underlying basement membrane (BM) of the dermal laminae [1]. It can arise in association with gastrointestinal disease, diseases associated with sepsis such as pneumonia and metritis, mechanical overload (supporting limb laminitis (SLL)), endocrine disorders associated with excess glucocorticoids such as pituitary pars intermedia dysfunction (PPID) [2] and endocrine disorders associated with insulin resistance (IR) such as equine metabolic syndrome (EMS) [3]. However, most commonly, laminitis arises in association with ingestion of pasture [4, 5].

A recent systematic review of laminitis epidemiology reported estimates of laminitis frequency to range from 1.5 to 34% [6]. The prevalence and incidence of laminitis varied depending on the population investigated (general practice or referral institutions), proportion of ponies versus horses, intercurrent diseases (particularly gastrointestinal disease) and the location (generally USA, UK or Australia) [6].

Acute laminitis can be divided into three stages. First, there is a developmental or prodromal phase that begins with contact with the pathophysiological triggers and ends with the onset of lameness 72 h later [2]. This is followed by the acute phase during which the clinical signs are seen [2]. Thus, the clinical signs only become apparent once the lamellar tissues have already been subjected to activation of inflammatory, metabolic and degenerative changes. The acute phase is followed by either resolution of the disease or the chronic phase [2].

Various different experimental models for the induction of laminitis exist including administration of a carbohydrate overload, typically starch [7] or oligofructose [8], administration of black walnut shavings extract (BWE) [9] and infusion of insulin to induce hyperinsulinaemia [10]. Much of the published research is based on observations from these experimental models of laminitis, rather than the naturally acquired disease. There is evidence from these models to support a role for inflammation, extracellular matrix (ECM) degradation, metabolic disease and vascular and endothelial dysfunction in the pathogenesis of laminitis. However, the exact cascade of events and the interactions between all these processes are yet to be elucidated.

Pathogenesis

Inflammatory theory

Although previously the role of inflammation in laminitis was questioned because of the minimal neutrophil infiltration present histologically [2], application of more sensitive research tools has produced abundant evidence of inflammatory changes during laminitis. There is evidence of endothelial activation [11], leucocyte adhesion and emigration [12], pro-inflammatory cytokine expression [13, 14] and oxidative injury occurring in the laminae [15] and it would appear that this is a local manifestation of a systemic disease, as there is also evidence of inflammation occurring concurrently in the skin [16], liver and lungs [17]. Thus, some researchers have hypothesized that there is a systemic inflammatory response syndrome (SIRS) and the inflammatory mechanisms responsible for laminar failure in horses are similar to those responsible for organ failure in human sepsis and multiple organ dysfunction syndrome (MODS) [13].

Endothelial activation, indicated by increased concentration of interleukin-8 (IL-8) and the adhesion molecules E-selectin and intercellular adhesion molecule-1 (ICAM-1), peaked at 1.5 h after BWE administration [11]. Adhesion of leucocytes to the endothelium is a critical step in leucocyte emigration. Emigration of leucocytes into laminar tissue, typically neutrophils and monocytes, occurred during the developmental stage of BWE-induced laminitis [12] as early as 1.5 h after administration [11]. In addition, laminar immunohistochemical staining of calprotectin, a marker of remote tissue inflammation and damage during the SIRS in people, is increased at 12 h after the onset of BWE-induced laminitis [18]. Immunohistochemical staining of calprotectin was perivascular, near the epithelial BM, and mostly associated with neutrophil emigration. Furthermore, calprotectin signalling of epithelial damage occurred 12 h after administration of BWE, which is 9 h after the initial onset of neutrophil emigration, suggesting that leucocyte emigration is a primary event that is not initiated by epithelial damage [18]. Neutrophil emigration from circulation is accompanied by a reduction in the number of circulating neutrophils and monocytes. These leucocytes then produce cytokines, reactive oxygen species (ROS) and matrix metalloproteinases [11–13, 15, 19], which can result in organ dysfunction in human sepsis [20].

Leucocyte infiltration of the laminae has also been well characterized in the carbohydrate-induced model of laminitis but occurs later than with BWE [21]. At the development of Obel Grade 1 laminitis after carbohydrate overload, laminar mRNA expressions of pro-inflammatory cytokines such as IL-1β, IL-6, cyclo-oxygenase-2 (COX-2), E-selectin and ICAM-1 were increased [14]. Plasma concentration of the pro-inflammatory mediator tumour necrosis factor-α (TNF-α) increased 12–24 h after oligofructose administration [22]. Laminar leucocyte immigration does occur in the endocrinopathic model of laminitis, but it is less intense in comparison with carbohydrate overload models [23, 24]. No up-regulation of anti-inflammatory cytokines such as IL-10 or IL-4 occurred during carbohydrate-induced laminitis, suggesting that the laminae do not mount a prominent anti-inflammatory response [14].

Microarray analysis has demonstrated that generation of inflammatory mediators within the lamellar tissues occurred before the development of substantial dermal–epidermal separation, inflammatory infiltrate, or vascular changes, and before the horses began showing signs of foot pain [25]. It has also demonstrated that genes associated with inflammation, activation and extravasation of leucocytes, antimicrobial activities and destruction of the lamellar BM are induced during the early stages of development of laminitis in response to the administration of BWE [26]. However, the role of cytokines in the development of endocrinopathic laminitis remains less clear. Insulin infusion in healthy horses significantly altered plasma concentrations of TNF and IL-6 [27], but peripheral blood cells of obese hyperinsulinaemic horses showed decreased endogenous pro-inflammatory cytokine gene expression (IL-1 and IL-6) and a similar cytokine response following immune stimulation compared with that of control horses [28].

Chemokines play a vital role in leucocyte activation and emigration. Increased laminar CXC chemokine expression consistently precedes peak leucocyte accumulation and onset of lameness in the carbohydrate overload laminitis models [29]. The CXC chemokine CXCL1 plays an early and possibly initiating role in neutrophil accumulation in the BWE laminitis model and laminar keratinocytes are an important source of this chemokine [30].

Oxidative stress reportedly plays a role in sepsis-induced organ dysfunction and failure in many species [31]. Evidence of laminar oxidative stress has been reported experimentally in the black walnut extract (BWE) model, but not in the carbohydrate-overload model [32] or in experimental insulin-induced laminitis [33]. Evidence of systemic oxidative stress has been reported in horses with PPID-associated laminitis [34] and a marked increase in neutrophil ROS production upon phagocytosis was observed in horses with EMS that was strongly correlated to the blood insulin concentration [28].

While in gastrointestinal diseases and diseases associated with sepsis, the triggers for inflammation presumably include endotoxins and exotoxins absorbed into the circulation, the mechanisms linking development of lamellar inflammation in other forms of the disease have not been fully elucidated.

ECM degradation

The ECM is a diverse structure formed of structural proteins, proteoglycans, regulatory proteins, proteases and protease inhibitors that are responsible for the maintenance of structural support, movement, growth, remodelling and healing along with the modulation of cytokines, inflammation, healing and cell migration [35]. The BM is a thin layer of ECM and attachment of the epithelial cell to the underlying BM is mediated through hemidesmosomes. Failure of the hemidesmosomes is a key event in the pathogenesis of laminitis [36, 37]. Previously, it was thought that dysregulation of the protease enzymes known as matrix metalloproteases (MMPs) was fundamental to hemidesmosome failure in laminitis [38–40]. However, more recent research has shown that while MMP activation does occur, it is not an initiating event [41]. Instead, other host-derived proteases and degradation of alternative lamellar matrix components that could cause hemidesmosome failure need to be considered.

The primary components of the BM of epithelial tissues are laminin-332 (Ln-332) and collagen type IV [179]. Immunohistochemistry was used to characterize the distribution of Ln-332 and collagen type IV in the organs of normal horses and these proteins were found to be widespread; Ln-332 and collagen type IV degradation occur in the hoof lamellae of horses with experimentally induced laminitis [42]. Changes in lamellar collagen type IV and Ln-332 were first observed at 12 h post-oligofructose administration [43], i.e. much earlier than MMP activation.

In addition, MMPs, metalloproteases comprise two other classes of metalloproteases including the transmembrane metalloproteases of a disintegrin and metalloprotease domain (ADAM) family known to process cytokines critical for inflammation (ADAM-10 and ADAM-17) [44] and the secreted metalloproteases containing the thrombospondin repeats (ADAMTS) that are responsible for the degradation of proteoglycan in cartilage and contribute to osteoarthritis [45]. All these metalloproteases are inhibited *in vivo* by tissue-specific inhibitors of metalloproteases (TIMPs). TIMPs are produced endogenously and maintain the homoeostasis of proteolytic activity of a tissue. Expression of genes coding for proteins containing an ADAM domain and genes encoding TIMP in horses with carbohydrate overload, BWE and naturally occurring laminitis were studied [46]. Genes encoding ADAM-10 and ADAM-17 (involved in inflammation) were not induced in most laminitic animals, whereas ADAMTS-4 gene expression and expression of ADAMTS-5 (implicated in arthritis) were strongly up-regulated in nearly all horses with experimentally induced and naturally acquired laminitis. In addition, TIMP-2 gene expression decreased in most laminitic horses. In a second study, gene expression of the enzyme ADAMTS-4 was increased and that of its substrate versican decreased in laminitic laminae compared with healthy laminae [47]. In addition, aggrecanase gene expression increased initially at 12–18 h post oligofructose dosing, similar to the BM changes, and gene expression of TIMP-2 decreased during laminitis development [41]. Versican is a large polysulphated proteoglycan providing tissues with resistance to compression deformation; in addition, aggrecan and versican have signalling properties that affect cell proliferation, differentiation, adhesion and intercellular communication and hence have additional non-mechanical roles in regulating cell function [48]. On the basis of their biochemical and biological properties and their tissue distribution, it is possible for the proteoglycans to affect the development and maintenance of the basal epithelial cell layer as well as cushion basal epithelial cells against severe biomechanical stresses associated with their anatomic location [49]. Therefore, elevated gene expression of ADAMTS-4 and aggrecanase in laminitic laminae, if it is accompanied by elevated enzyme expression and degradation of its proteoglycan substrates, and decreased expression of the regulator TIMP-2 may play a critical role in failure of the laminae [49].

The trigger for increased expression of ADAMTS and aggrecanase and decreased TIMP expression remains unknown. It would seem most logical to assume that this dysregulation of cell adhesion is caused by inflammatory and/or hypoxic cellular injury.

However, this may not be applicable to all forms of laminitis as in the hyperinsulinaemic model of laminitis, gene expression of the members of the metalloproteinase family MMP-2, MT1-MMP and ADAMTS-4 and the regulator TIMP-3 were not increased in the lamellar tissue during the developmental stage or acute phase of the disease [50]. Instead, increased MMP-9 gene expression and increased pro-MMP-9 protein concentrations were found during the acute phase of laminitis. The authors postulated that the increased expression of MMP-9 may be associated with the infiltration of inflammatory leucocytes, or may be a direct result of hyperinsulinaemia, but the exact role of MMP-9 in BM degradation in laminitis is uncertain, as it appears to be present largely in the inactive form.

Metabolic theory

Endocrinopathic laminitis has been defined as laminitis developing from hormonal influences [51]. Conditions associated with endocrinopathic laminitis can be divided into two main categories, namely those associated with glucocorticoids including pituitary pars intermedia dysfunction (PPID) [52] and iatrogenic corticosteroid administration [53], and those associated with IR including EMS [52]. In some studies of horses with laminitis, an underlying endocrinopathy appears to be common with 70–89% having a diagnosis of either PPID or basal hyperinsulinaemia indicative of IR [54, 55]. In another study, evidence of an underlying endocrinopathy was only evident during the summer, but not during the winter [56]. However, in yet other studies, an underlying endocrinopathy has not been shown to be a risk factor for the development of acute laminitis [57, 58].

The pathogenesis of laminitis associated with IR is not fully understood, but common to all these endocrinopathic conditions is disturbed glucose and insulin regulation and the development of IR and hyperinsulinaemia; thus, the initial hypothesis was that decreased uptake of glucose by the lamellae could result in laminitis. The normal equine hoof consumes glucose, and glucose consumption at rest exceeds that of the head [59]. Initial research using hoof explants found that decreased glucose availability resulted in separation of the hoof lamellae between the basal epidermal cells and their BM *in vitro,* characteristic of the hoof separation that occurs in naturally occurring laminitis [60]. The authors suggested that the integrity of the hoof explants was dependent on consumption of glucose and that this provided a possible explanation for the development of laminitis caused by conditions such as carbohydrate overload, acute inflammatory conditions, corticosteroid therapy and hyperlipaemia as these conditions would induce a major insulin-mediated metabolic shift away from glucose consumption by many peripheral tissues and if the metabolic change occurred faster than the hoof tissue could adapt to an alternative energy substrate, then hoof separation and laminitis would occur [60]. A second study revealed that the separation occurred as a result of the lack of glucose causing a reduction in hemidesmosome numbers until they disappeared and the basal cytoskeleton collapsed [61]. Thus, even a small decrease in the rate of glucose uptake within the lamellar tissue could be extremely damaging to lamellar integrity, resulting in laminitis.

Glucose transport across the cell membrane is mediated by glucose transport proteins (GLUTS). Equine research has focused on lamellar expression of the insulin-independent GLUT-1 and the insulin-dependent GLUT-4 transporters. GLUT-1 has been consistently demonstrated within equine lamellar keratinocytes [59, 62]. In one study, GLUT-1 was reduced in laminitic tissue [62] and the authors suggested that the lamellae are highly metabolic tissues, and possibly suffer from impaired glucose metabolism during the laminitis disease process. Conversely, in a second study, GLUT-1 was not different between normal and laminitic ponies [63] and was found to have increased during the prodromal stages of insulin-induced laminitis [33]. GLUT-4 was shown to be present in normal lamellae and reduced in laminitic tissue in one study [62], while another study reported poor specificity for the GLUT-4 antibody [59]. However, a recent study revealed that glucose uptake by hoof explants was not affected by insulin and that the relationship between glucose concentration and glucose uptake was consistent with an insulin-independent glucose transport system [64]. In addition, while there was a strong expression of the insulin-independent glucose transporter GLUT-1 mRNA in lamellar tissue, expression of the insulin-dependent glucose transporter GLUT-4 mRNA was absent or barely detectable [64]. Thus, these results do not support a glucose deprivation model for laminitis, in which glucose uptake in the hoof is impaired by reduced insulin sensitivity.

Equine IR is characterized by hyperinsulinaemia and normoglycaemia in the majority of animals. Further investigation into the role of insulin in the pathogenesis of laminitis resulted in the effects of hyperinsulinaemia on the lamellae being evaluated in a number of studies. Prolonged administration of insulin to induce hyperinsulinaemia while maintaining normal glucose concentrations resulted in the development of clinical and histological laminitis in all four feet of healthy ponies within 72 h [65] and of healthy Standard bred horses within 48 h [10] suggesting that IR and the associated hyperinsulinaemia place horses and ponies at risk of developing laminitis through direct insulin toxicity. In addition, infusion of glucose in healthy Standard-bred horses resulted in hyperglycaemia (10.7 ± 0.78 mmol/l), hyperinsulinaemia (208 ± 26.1 μIU/ml) and histological evidence of laminitis in at least one foot in the absence of lameness. The authors suggested that the potential threshold for insulin toxicity (i.e. laminitis) in horses may be at or below a serum concentration of $\sim$200 μIU/ml [66]. However, it must be acknowledged that the serum insulin concentrations achieved in these studies were much higher than those commonly associated with PPID [67] and those reported in some cases of pasture-associated, endocrinopathic laminitis [68] and similarly high serum insulin concentrations have been documented in recurrently laminitic ponies in the absence of clinical signs of laminitis [56] such that very high serum insulin concentrations alone are not necessarily laminitis-inducing. In addition, subsequent histological examination of the laminitis lesions from the ponies with experimental insulin-induced laminitis revealed a lack of widespread BM disintegration and instead the predominant lesion was one of apoptosis and mitosis in the axial regions [24]. Thus, the histological changes induced by insulin are not the same as those seen in various carbohydrate overload models of laminitis or in the naturally occurring disease.

Despite this histological discrepancy, if it is not through glucose deprivation, insulin may cause laminitis through effects on blood flow, inflammatory effects (discussed in the previous section), excess glucose and MMP activation (discussed in the previous sections). Infusion of insulin into healthy Standard-breds resulted in the hoof wall surface temperature being higher and less variable once hyperinsulinaemia was established compared with control horses suggestive of increased digital perfusion [10]. A second study showed that vasodilation without hyperinsulinaemia was not sufficient to induce laminitis [69]. In contrast, in other species, insulin activates endothelial vasodilatory and vasoconstrictive pathways; insulin stimulates nitric oxide (NO) synthase via the phosphatidylinositol 3-kinase (PI-3K) pathway and endothelin-1 (ET-1) release through the mitogen-activated protein kinase (MAPK) pathway [70]; however, in IR, PI-3K activation is blocked and the resultant NO : ET-1 imbalance favours vasoconstriction [70]. This IR-associated vasoconstriction is supported by the finding that following induction of IR, the normal relaxation responses of isolated rings of equine palmar digital arteries and veins were converted to contractions [71]. Thus, whether equine IR is associated with increased or decreased digital perfusion remains unclear.

In humans, GLUT-1 is more sensitive to increased blood flow, which leads to increased flow-mediated glucose uptake. Since the GLUT-1 receptor predominates in the lamellar vasculature it was postulated that increased digital blood flow may result in glucose excess [33]. This, in turn, results in the increased production of advanced glycation end-products (AGEs) and ROS which cause vascular dysfunction and possibly laminitis. However, in a study by de Laat *et al.* [33] involving lamellar tissue from horses with experimental insulin-induced laminitis, there was no increase in ROS production or AGE receptor gene expression, and accumulation of AGE appeared only after 48 h (i.e. at the onset of clinical signs), rather than being an initiating event [33]. Thus, it would appear that excess glucose does not play a role in insulin-induced laminitis.

Vascular and endothelial dysfunction

It has been proposed that laminitis is primarily a vascular disease. Digital blood flow has been shown to decrease during the developmental phase of laminitis in some studies [2, 72, 73]. The initiating vascular event may be digital vasoconstriction, preferential venoconstriction, opening of arteriovenous anastomoses redistributing flow away from the nutrient capillaries, formation of microthrombi or a combination of these factors. All these could lead to hypoperfusion of the laminar region, ischaemia and impaired oxygen and glucose delivery to the dermal laminae, followed by reperfusion injury and activation of the inflammatory cascade triggering the pathological events ultimately leading to collapse of the digit.

Digital vasoconstriction is mediated by α-adrenoceptors and 5-hydroxytryptamine (5-HT; serotonin) receptors, particularly 5-$HT_{1B/1D}$ receptors [74]. 5-HT is a potent digital vasoconstrictor and circulating concentrations of 5-HT have been measured in normal horses which are sufficient to cause digital vasoconstriction *in vitro* [75]. Caecally derived vasoactive amines, such as tryptamine, tyramine and phenylethylamine, can displace 5-HT from platelets, inhibit uptake of 5-HT by digital vein endothelial cells and also have a direct effect on 5-HT and α-adrenergic receptors resulting in potent vasoconstriction [76–78]. Thus, vasoactive amines released from the caecum into circulation can potentiate vascular responses to 5-HT, resulting in digital ischaemia during the developmental phase of laminitis [76]. 5-HT is also released when platelets aggregate [79] and platelet aggregation has been reported in the equine digit during the onset of laminitis [80, 81]. Furthermore, the equine digital vessels are significantly more sensitive to 5-HT than other vessels [75], resulting in selective constriction of the digital vessels. Digital venoconstriction results in increased venous resistance and capillary hydrostatic pressure and promotes the development of tissue oedema because of efflux of fluid across the capillary bed [82, 83]. As the hoof capsule is relatively inexpandable, perivascular oedema increases interstitial pressure and can collapse digital capillaries and further compromise perfusion [79]. Digital hypoperfusion in the prodromal period may also be the result of microvascular thrombi formation in the laminar dermal vessels [2, 80, 81].

On reperfusion there is reactive hyperaemia and an increased digital blood flow [2]. ROS cause reperfusion injury, resulting in lipid and protein peroxidation and cytotoxicity [2]. Ischaemia-reperfusion injury and ROS also increase microvascular permeability and tissue oedema [84, 85]. Superoxide dismutase, an anti-oxidant enzyme responsible for scavenging superoxide ions, is absent from laminar tissue, which may make it more susceptible to damage induced by ROS [86]. However, in the prodromal stages of BWE-induced laminitis, activity of xanthine oxidase, an enzyme which can generate ROS, was not increased suggesting that this pathway is not up-regulated in the pathogenesis of laminitis [86]. This does not preclude the involvement of other enzymatic pathways generating ROS in the pathogenesis of BWE-induced laminitis or for the involvement of ROS in other forms of laminitis.

ROS catalyse arachidonic acid peroxidation to form isoprostanes that have potent vasoactive effects [87]. Concentrations of iso-prostaglandin-2α (iso-$PGF_{2\alpha}$) were increased in laminar tissue in horses with BWE-induced Obel Grade 1 laminitis [87]. In addition, isolated small equine laminar veins were more sensitive to iso-$PGF_{2\alpha}$ after BWE administration, potentially contributing to selective venoconstriction [87].

The endothelium is a layer of cells lining the vascular system which plays a key role in regulating microvascular blood flow, haemostasis and inflammation [88], all of which may be involved in the pathogenesis of laminitis. Thus, it is logical to propose endothelial dysfunction as a key event in laminitis development. The endothelium normally produces vasodilator mediators including NO, prostacyclin (PGI_2) and endothelium-derived hyperpolarizing factor (EDHF) [89] and the potent vasoconstrictor ET-1 [88]. Endothelial dysfunction results in an imbalance between the production of vasoconstrictory and vasodilatory mediators, predominantly because of reduced endothelium-dependent NO bioavailability [90], hence favouring vasoconstriction. Other effects of endothelial dysfunction include leucocyte adherence, platelet activation, coagulation disturbances and vascular inflammation [91]. Endothelial dysfunction may occur as a result of IR and inflammation [70].

The Link between the Gastrointestinal Tract and the Foot

Although in gastrointestinal diseases and diseases associated with sepsis, the trigger for laminitis presumably includes endotoxins and exotoxins absorbed in circulation, the mechanisms linking the consumption of pasture forage with the development of lamellar failure have not been fully elucidated. Clinical cases of laminitis most often occur in conditions that favour accumulation of rapidly fermentable NSC in grass. Therefore, most believe that pasture-associated laminitis is triggered by carbohydrate overload of the hindgut and the subsequent systemic absorption of laminitis trigger factors [92].

Carbohydrate overload of the hindgut results in increased bacterial carbohydrate fermentation and profound changes in the hindgut bacterial population. There is a rapid proliferation of bacteria that preferentially ferment oligofructose and produce lactic acid, including streptococcal species, and death of other bacterial species. The end result is a lactic acid-induced sharp decrease in hindgut pH, death and lysis of large numbers of bacteria releasing endotoxins and exotoxins and hence damage to the hindgut epithelial cells with a resultant increase in intestinal permeability [93]. Using nuclear magnetic resonance-based metabonomics, horses showed increased phosphatidylcholine and/or low density lipoprotein levels, indicating a change in blood-lipid composition following the development of oligofructose-induced laminitis, which could point to a compromise of the intestinal wall during laminitis development [94]. There is also an increase in bacterial synthesis of vasoactive dietary amines [95]. This increased intestinal permeability will allow increased absorption of endotoxins, exotoxins and vasoactive amines into the circulation where they will have inflammatory and vascular effects.

Individual Animal Predisposition to Pasture-Associated Laminitis

Certain individual animals appear to be predisposed to recurrent episodes of pasture-associated laminitis. The mechanisms underlying this predisposition remain a fundamental question in laminitis research. Multiple variables have been evaluated as risk factors for laminitis with the findings generally being inconsistent between studies. An association between the occurrence of laminitis and being a pony, the spring and summer months [4], being female, increasing age [96] and obesity [51, 97] have been demonstrated in some studies.

Animals at the greatest risk of pasture-associated laminitis have a metabolic phenotype including obesity and IR similar to that seen in human metabolic syndrome (HMS) [98]. Thus, the same pathological mechanisms that underlie the cardiovascular disease associated with HMS, including changes in insulin signalling, inflammatory cytokines and endothelial dysfunction, could contribute to laminitis.

IR may predispose to laminitis via vascular function disturbances. Insulin activates endothelial vasodilatory and vasoconstrictive pathways. Insulin stimulates NO synthase via the PI-3K pathway and ET-1 release through the MAPK pathway [70]. In IR, PI-3K activation is blocked; the resultant NO:ET-1 imbalance favours vasoconstriction and contributes to platelet activation and leucocyte adhesion to the endothelium, which are all implicated in laminitis pathophysiology [99].

Increased carbohydrate consumption has been shown to exacerbate IR in horses [100, 101]; feeding a fructan-type carbohydrate (inulin) similar to that present in spring grass to ponies produces an exaggerated insulin response in animals predisposed to laminitis [102]; and reducing feed intake in obese animals results in modulation of obesity-associated hormones including insulin [103]. Thus, it may be exacerbation of IR by increased carbohydrate consumption that results in the development of laminitis in certain individual animals during the spring and summer months.

Obesity may predispose to laminitis through IR or inflammation. Adipocytes are endocrinologically active producing a variety of mediators, some of which antagonise insulin resulting in IR and some of which are pro-inflammatory [104]. TNF-α is one such pro-inflammatory cytokine secreted by adipose tissue which plays an important role in HMS [105] and is implicated in IR development via a blockade of the PI-3K pathway [106, 107]. Within a herd of inbred obese ponies, laminitis-prone animals had significantly higher plasma TNF-α concentrations compared with normal ponies, thus associating laminitis predisposition with increased circulating inflammatory cytokines [108]. An inter-relationship was found between obesity, inflammatory cytokine including TNF-α and insulin sensitivity in the horse [109] and insulin infusion results in increased plasma TNF-α concentrations [110]. In addition, it stimulates apoptotic cell death in many tissues, including equine laminar epithelial cells (S. Bailey, personal communication, 2012) and endothelial dysfunction in other species as a result of stimulation of superoxide radical (ROS) production [105]. Therefore, higher circulating (TNF-α) may enhance laminar epithelial cell apoptosis and endothelial dysfunction, exacerbating the metabolic dysfunction related to IR in laminitis prone animals and, in part, accounting for individual animal predisposition to the disease.

Having a cresty neck, i.e. regional adiposity, has been identified as a risk factor for laminitis [96, 111]. Fat from the nuchal ligament had a more inflammatory profile compared with other fat depots in insulin sensitive and IR horses [112]. Thus, it was initially postulated that this fat depot was the main source of the pro-inflammatory cytokines that may predispose an animal to laminitis. However, subsequent studies have shown that this depot is not more inflammatory than other depots (P. Harris, personal communication, 2011), and that there is no evidence of systemic or skeletal muscle inflammation in IR obese horses [113] and that there is no correlation between plasma TNF-α and body condition score [114]. Instead, it would appear that age is a confounding factor.

It must also be remembered that not all obese animals are IR, not all obese and not all IR animals are prone to recurrent pasture-associated laminitis and not all animals prone to recurrent pasture-associated laminitis are obese and/or IR. Another possibility is that susceptible horses and ponies have differences in their gut flora compared with animals less prone with increased fermentative responses to a given load of NSC and increased production of laminitis trigger factors [115]. The caecal contents from animals maintained on spring/summer pasture had 2–3-fold higher amine concentrations compared with that from animals fed hay or winter grass [116], and the plasma amine concentrations are lower in winter compared with summer [117]. However, while faecal amine concentrations increased following inulin feeding, there was no difference between normal and laminitis prone ponies [118]. Thus, further research is required.

Whether there is a genetic basis to laminitis predisposition remains to be determined. Some horses and ponies appear to be genetically predisposed to EMS, and this is the focus of ongoing study at the University of Minnesota (www.cvm.umn.edu/equinegenetics/ems) and a dominant inheritance pattern for laminitis in ponies with EMS was detected within an inbred herd of ponies [98]. However, further research is required.

SLL

Laminitis in the unaffected supporting limb is a common and devastating consequence of severe lameness. The actual incidence has not been documented, but it was 10% in a retrospective review of horses treated with half limb, full limb or transfixation pin casts [119]. There has been little or no primary research focused on the mechanisms that result in SLL. It is commonly believed that the pathogenesis of SLL has little in common with other forms of the disease until structural failure of the dermal–epidermal junction occurs, at which time secondary inflammation, enzymatic activation and vascular derangements are common [120]. Currently, there are two theories as to the precise causal mechanism, namely direct mechanical overload [121] and inadequate perfusion of the lamellar tissue because of reduced digital blood flow associated with excessive and continuous weight-bearing [122]. However, currently there is little evidence to support either theory.

Diagnosis

A diagnosis of laminitis is usually based on the characteristic clinical signs, which include lameness affecting two or more limbs, characteristic stance of leaning back on the heels and taking weight off toes, bounding digital pulses, increased hoof wall temperature, pain on hoof tester pressure at the region of the point of the frog and possibly a palpable depression at the coronary band. The lameness can vary in severity from that only perceptible at the trot through to spending prolonged periods recumbent. The severity of the laminitis can be classified in many ways, but the commonest is to use the Obel grading system [123]:

Grade 1 – At rest, the horse shifts its weight between the forelimbs; the horse is sound at the walk, but the gait is stilted at the trot in a straight line and on turning.

Grade 2 – The gait is stilted at the walk and the horse turns with great difficulty, but one forelimb can be lifted.

Grade 3 – The horse is reluctant to walk and one forelimb can only be lifted with great difficulty.

Grade 4 – The horse will only move with great coercion and it is impossible to lift one forelimb.

Further tests are performed in those cases where an underlying endocrinological abnormality is suspected such as PPID or EMS and radiographs are taken in those cases where movement of the pedal bone is suspected.

Magnetic resonance imaging (MRI) of cadaver hooves in the initial active stage of laminitis revealed it to be both more sensitive and specific than digital radiology in predicting laminitis [124]. A second study has shown that a magnetic resonance venogram allows the venous deficits of the laminitis patient, as well as the other structures within the hoof capsule in the live horse, to be evaluated [125]. In addition, laminar blood flow and vascular permeability have been measured in healthy horses by using dynamic contrast-enhanced computed tomography (CT) [126]. Thus, the use of advanced imaging techniques in the diagnosis of laminitis may increase in the future.

Diagnosis of Endocrine Disease

PPID

A diagnosis of PPID is based on the signalment (older, pony versus horse), clinical signs and further diagnostic test results. Often the signalment and clinical signs are highly suggestive of a diagnosis of PPID. Diagnosis is more of a challenge when animals are exhibiting only one clinical sign, e.g. recurrent laminitis. Further diagnostic tests include measurement of basal plasma ACTH concentrations, the cortisol response to a standard 19 h overnight dexamethasone suppression test (DST) and the ACTH response to thyrotropin releasing hormone (TRH) administration [127]. Plasma ACTH concentration may be increased in horses with PPID; however, it is significantly affected by season with increases occurring in both normal and PPID animals in autumn [67], thus seasonally adjusted reference ranges must be used to interpret results. The DST is based on the lack of suppression of plasma cortisol concentration in horses with PPID after administration of dexamethasone. The gold standard method of PPID diagnosis was considered, as it was reported to have a sensitivity and specificity of 100% [128]. However, recently it has been shown that this is a significant over-estimate and both false positives and false negatives have been reported [129]. In addition, DST results vary with season with most false positives occurring in autumn [130]. TRH is believed to be a physiologic releasing factor for the pituitary pars intermedia. Thus, exogenous TRH administration should result in increased release of pars-intermedia-derived hormones. Plasma ACTH concentration increased in healthy horses and in horses with PPID following intravenous administration of TRH, but the magnitude and duration of the increase was greater in horses with PPID [131].

EMS

A diagnosis of EMS is based on the history of being an easy keeper; clinical signs of generalized or regional obesity and laminitis that have developed in the absence of recognized causes such as grain overload, colic, colitis or retained placenta; and demonstration of IR [132]. Fasting hyperinsulinaemia is suggestive of IR, however, in some cases dynamic tests such as minimal modelling of the frequently sampled intravenously glucose tolerance test, the combined glucose insulin tests or the oral sugar (or glucose) test may be required [132].

Treatment

Acute laminitis is a medical emergency and treatment should be initiated as soon as possible after the onset of the clinical signs. By the time the clinical signs become apparent, the lamellar tissues have already been subjected to activation of metabolic and degenerative changes. Thus, therapy should be aimed at providing analgesia and foot support.

Non-steroidal anti-inflammatory drugs (NSAIDs) given either orally or intravenously are the first choice for analgesia as they also inhibit part of the inflammatory cascade. There is no evidence to suggest that any one specific NSAID is superior to the next [133]. The use of acetaminophen (paracetamol) as a short-term adjunctive analgesic in a laminitic pony has recently been described [134]. If NSAIDs do not provide sufficient pain relief, then other analgesic drugs such as opiates, gabapentin, lignocaine infusion, ketamine infusion or an infusion containing a combination of drugs can be used.

Supporting the foot is an essential part of the management of acute laminitis. The horse naturally adopts a stance that bears most of the weight over the palmer/plantar part of the foot rather than the painful toe region. Additional support should be supplied to this region of the foot in order to provide pain relief and minimize the mechanical forces on the laminae and hence laminar tearing and pedal bone movement. There is no evidence to suggest that any one foot support method is superior [133].

The use of therapies to alter digital perfusion remains controversial. Vasodilator therapy is frequently used based on laminitis being a consequence of digital hypoperfusion. However, it should be acknowledged that even if the pathophysiology of laminitis involves vasoconstriction, this has normally resolved once the clinical signs become apparent. Acepromazine (ACP) is the most effective digital vasodilator available [135]. The beneficial effect of ACP has only been evaluated in one study of pasture-associated laminitis treated in first opinion practice and there was a trend towards treatment with ACP being associated with survival [133]. Cryotherapy has only been shown to prevent experimental oligofructose-induced laminitis; it is yet to be shown to be beneficial in the treatment of the acute disease.

Treatment of Underlying Endocrinopathy

PPID

Where specific medical therapy is used, there are three types of drugs available namely dopamine agonists, 5-HT antagonists and cortisol synthesis inhibitors. The drug of choice is the dopamine agonist pergolide mesylate. The efficacy of pergolide at improving clinical signs of disease and diagnostic test response has been documented in several studies [136]. The 5-HT antagonist cyproheptadine has been suggested as a second-line drug used in combination with pergolide when maximal doses of pergolide alone are insufficient to achieve resolution of clinical signs since as a monotherapeutic agent, cyproheptadine has limited efficacy [136, 137]. Trilostane is a cortisol synthesis inhibitor reported to improve clinical signs of PPID but not the DST results [138]. It may be beneficial to those horses with PPID, adrenal gland hyperplasia and hypercortisolaemia; however, it would have no effect on the excessive production of pituitary-derived hormones.

EMS

Treatment of EMS relies on management changes to promote weight loss and increase exercise, which, in turn, will improve insulin sensitivity [101, 139, 140] and suppress inflammation [141]. Pharmacological intervention should be reserved for the short term (3–6 months), while these management changes are being instituted or refractory cases. The drugs that have been used in the horses to date have been largely based on drugs used to improve insulin sensitivity in humans with metabolic syndrome. Levothyroxine has been used with some success; it is assumed that it induces weight loss by raising circulating thyroxine concentrations and so stimulating the basal metabolic rate [142]. Metformin is a biguanide drug that is administered to control hyperglycaemia and increase tissue insulin sensitivity in humans with type-2 diabetes mellitus [143]. Durham *et al.* reported that resulting insulin concentrations and proxies of insulin sensitivity improved in horses and ponies with presumed IR when treated with metformin (15 mg/kg po BID) [144]. However, the bioavailability of metformin in the horses has been reported to be as low as 7% [145, 146] and another study demonstrated that metformin (15 and 30 mg/kg BID) had no effect on insulin sensitivity in IR horses [147]. Instead, the beneficial effects of metformin may be due to decreased intestinal absorption of glucose and hence a decreased insulinaemic response [148]. Pioglitazone is an insulin-sensitizing drug used in humans. It has been shown that a dose of 1 mg/kg po SID achieves a plasma concentration in horses similar to the human therapeutic concentration [149]. However, when healthy horses were pre-treated with pioglitazone for 14 days prior to an infusion of endotoxin to induce IR, there was no significant effect on insulin sensitivity values [150].

Prognosis

A variety of studies have attempted to determine prognostic indicators for laminitis. Colles and Jeffcott [151] related the prognosis for a case of laminitis to the severity of onset, number of feet affected and speed of recovery although no clinical data were published [151]. A retrospective survey of referred laminitis cases indicated that horses with distal phalangeal rotation (relative to the dorsal hoof wall) of more than 11.5′ tended to remain lame [152]. It was suggested that only in exceptional circumstances should treatment be considered for cases in which the distal phalanx has penetrated the sole [153]. Eustace and Caldwell [154] demonstrated that animals with degrees of distal phalangeal rotation greater than the suggested threshold of 11.5′ can return to full athletic function if they are treated using a heart bar shoe and the technique of dorsal wall resection [154].

The significance of clinical and radiological parameters as prognostic indicators for laminitis was studied by Cripps and Eustace [155]. Of the animals included in the study, 77% returned to their original athletic function and 16% died or were killed within 6 months of diagnosis. The cases were assigned to four groups on the basis of initial clinical examination alone, namely laminitis, acute founder, sinker and chronic founder. This grouping was found in itself to be the most important prognostic parameter which was studied. Stepwise regression analysis of data from animals by group indicated that the radiological measurement of founder distance was the most significant radiological prognostic measurement for acute founder cases; a founder distance >14 mm had a probability of success of <50%. Less significant prognostic parameters were the severity of lameness, rotation angles, the presence of solar prolapse and the number of feet affected.

More recently, in study of cases of pasture-associated laminitis treated in first opinion practice, 8 weeks after the onset of the episode of laminitis, 95% of animals were still alive, 72% were considered to have a good clinical outcome and 59% of those that had previously been ridden were being ridden again [133]. Lower bodyweight, optimal body condition, mild laminitis and category of acute/chronic founder as defined by Cripps and Eustace [155] were significantly associated with survival; being ridden again was 2.6 times more likely in animals without previous laminitis; and a good clinical outcome was significantly associated with horse type: the outcome was 'bad' in none of the small horses, compared with 15 of 45 large ponies/cobs, 11 of 33 small ponies and three of 12 large horses [133]. Another study retrospectively evaluated 23 cases of laminitis that developed coronary band separation >8 cm long secondary to displacement of the distal phalanx relative to the hoof revealed that using a cut-off of ≥11.8 mm, founder distance correctly predicted death for 100% (13/13) of the cases and correctly predicted survival for 100% (7/7) [156]. Finally, data from 48 cases of chronic laminitis were reviewed. There was no relationship between angle of rotation of P3, clinical recovery and treatment time. However, a significant percentage of animals treated with a foam pad plus shoeing recovered compared with those treated with foam pads alone [157].

In PPID, horses with very high serum insulin concentration at presentation (>188 μIU/ml) were much more likely to develop laminitis and survive <2 years after diagnosis than those with normal or only moderately increased serum insulin concentrations (<62 μIU/ml) [158].

Prevention

Prevention in animals prone to recurrent pasture-associated laminitis focuses on two areas, namely mitigation of metabolic predispositions (i.e. IR and obesity) through management changes to promote weight loss and increase exercise which in turn will improve insulin sensitivity [101, 139, 140] and suppress inflammation [141] and limiting the ingestion of NSC from pasture and other feedstuffs [159].

The type of carbohydrate in pasture is a key factor in the development of laminitis. Fructans are part of the group of NSCs in pasture which includes starch, sugars and fructans that are storage compounds in grass. At certain times of the year, the quantity of pasture NSCs ingested by grazing equids may approach or exceed the amount of starch or fructan known to induce laminitis when administered as a single dose [8]. Many factors influence the NSC content of plants including plant species and part, stage of development and environmental conditions such as growth temperatures, light intensity and nutrient and water availability [160]. Fructans accumulate in grass under conditions which favour photosynthesis over growth, including days with bright sunshine and cool nights [161]. NSC concentrations vary over time, being low in the morning and high in the afternoon [162]. Fructan content also varies seasonally and is highest in spring, decreases during the summer and then increases again in autumn [163]. Ideally, the pasture should be analysed and access to pasture should be limited to 1–2 h a day when the environmental conditions are conducive to NSC accumulation [159]. It is best to turn animals out late at night or early in the morning, removing them from pasture by mid-morning at the latest because NSC levels are likely to be at their lowest late at night through early morning [159]. Pastures that have not been properly managed by regular grazing or cutting should be avoided because mature stemmy grasses may contain more fructan (it is stored in the stem) [159]. Horses should not be turned out onto pasture that has been exposed to low temperatures in conjunction with bright sunlight, such as in the autumn after a flush of growth or on bright cool winter days, because cold temperatures reduce grass growth, resulting in the accumulation of fructan [159]. Animals should not be allowed to graze on recently cut stubble, because fructan is stored predominantly in the stem [159].

Animals denied access to pasture for most or all of the day require provision of alternative feedstuffs. Horses at maintenance require ~2.0% of their body weight as forage or forage plus supplement to meet daily nutrient requirements. Forage (as hay or hay substitute, such as chop, chaff or haylage) should be the primary, if not sole, energy-providing component of the ration. Mature grass hay has higher fibre and lower NSCs when compared with immature hay and is suitable forage for the obese horse or pony. Alfalfa hay or other legumes, such as clover, are less preferred because, on average, these forages have higher energy and NSC content when compared with grass hay. An NSC content of less than 10% has been recommended [164]. Ideally, the hay should be analysed as soaking will not reliably reduce NSC content to below this value [165]. Forage-only diets do not provide adequate protein, minerals or vitamins. Therefore, the forage diet should be supplemented with a low-calorie commercial ration balancer product that contains sources of high-quality protein and a mixture of vitamins and minerals to balance the low vitamin E, copper, zinc, selenium and other minerals typically found in mature grass hays [159].

The ability to identify animals at increased risk of laminitis would be priceless, as it would enable the appropriate management countermeasures to be focused on these individuals and implemented prior to the onset of the disease. Previously it has been shown that meeting three or more criteria out of body condition score >6/9, plasma triglyceride concentration >57.0 mg/dl, proxy measurements of insulin sensitivity reciprocal of the square root of insulin (RISQI) <0.32 $[\text{mU/l}]^{-0.5}$ and insulin secretory response modified insulin-to-glucose (MIRG) ratio >5.6 $\text{mUinsulin}^2/[10.\text{L}.\text{mg}_{\text{glucose}}]$ differentiated a group of previously laminitic ponies from a group of normal ponies with a total predictive power of 78% [98]. There have been several studies more recently attempting to devise a test or panel of tests capable of identifying individual animals at risk of pasture-associated laminitis rather than groups of animals and of identifying the imminent onset of an episode of disease. Carter *et al.* [111] sought to identify variables and clinically useful cut-off values with reproducible diagnostic accuracy for the prediction of the future occurrence of laminitis following exposure to nutrient dense pasture within a cohort of ponies from a single inbred herd. Variables with diagnostic accuracy for the prediction of an episode of clinical laminitis in individual ponies within the next 3 months included serum insulin concentration >32 mu/l, plasma leptin concentration >73 ng/ml, BCS≥7/9, cresty neck score (CNS)≥4 and neck circumference (NC):height ratio >0.71 [111]. Combining tests did not result in higher diagnostic accuracy than individual tests of insulin or leptin. Borer *et al.* [166] sought to determine whether proxy measurements of insulin sensitivity (RISQI and quantitative insulin sensitivity check index: QUICKI) and insulin secretory response (MIRG and MIGRP (MIRG ratio for ponies)) based on basal glucose and insulin concentrations could distinguish between outbred normal and previously laminitic ponies kept under identical management regimes. While RISQI, QUICKI and MIGRP, but not MIRG, differentiated between normal and previously laminitic as a group, none accurately identified individual laminitis prone animals [166], thus limiting their usefulness as a clinical test. In a second study, Borer *et al.* [167] evaluated the glycaemic and insulinaemic responses of normal and previously laminitic ponies to glucose, fructose and inulin in spring and autumn after a 7 days adaptation to a pasture or hay diet. Measurement of peak serum insulin concentrations 2 h after feeding of a single dose of glucose (1 g/kg) discriminated between individual normal and laminitis prone animals, particularly when ponies are adapted to eating autumn pasture [167]. All normal ponies had peak insulin concentrations <350 μIU/ml, whereas all previously laminitic ponies had peak insulin concentrations >500 μIU/ml in autumn when adapted to eating pasture. Thus, this may be a clinically useful test of laminitis predisposition when an animal's history is unknown.

In animals with diseases associated with laminitis such as colic, colitis and retained foetal membranes, prevention is also important. Preventive measures include the use of foot supports, anti-endotoxic therapy and cryotherapy. Frog or frog and sole supports identical to those used in the treatment of acute laminitis can be applied to the feet prophylactically. Anti-endotoxic therapy is aimed at removal of the cause, neutralization of circulating endotoxin, inhibition of endotoxin-induced inflammation and modulation of the coagulation cascade. The cause can be removed by appropriate surgical correction of intestinal lesions, drainage of pleural or peritoneal effusions and appropriate antimicrobial therapy. Circulating endotoxin can be neutralized using polymyxin B [168]. Endotoxin-induced inflammation can be reduced using NSAIDs [169–171] and lignocaine infusion [172]. Modulation of the endotoxin-induced activation of the coagulation system can be achieved through the use of heparin [173] and in one study the administration of low molecular weight heparin appeared to be effective in the prophylaxis of laminitis following colic surgery [174]. Cryotherapy has been shown to prevent experimental oligofructose-induced laminitis [175]. Further studies have shown that the preventative effect of cryotherapy may be the result of blockade of the early lamellar inflammatory events likely to play an important role in lamellar injury, including the expression of chemokines, pro-inflammatory cytokines, COX-2 and endothelial adhesion molecules [176]. However, continuous adequate and controlled chilling of the digits over a prolonged period of time may be difficult to achieve in the clinical situation. A study evaluated the cooling effect of an ice water bag system applied for 90 min between the carpus and the hooves on the hoof surface temperature in six healthy horses. At 60 and 90 min it had a very small cooling effect of only 1.4°C at the coronary band, while at mid wall no significant temperature change was found [177]. A second study showed that cryotherapy with a wader boot with ice water or a 5-litre intravenous fluid bag filled with ice water to the level of the fetlock seemed to be effective methods of providing lamellar and venous cooling reaching temperatures similar to those previously shown to prevent laminitis [178].

Conclusion/Summary

Although there is abundant evidence to support a role for inflammation, ECM degradation, metabolic disturbances and vascular and endothelial dysfunction, the exact pathogenesis of laminitis and the link between the gastrointestinal system and the foot remain unclear. Despite the considerable efforts of a large number of eminent research workers over several decades, further research into all areas of the disease including epidemiology, risk factors, pathogenesis, prognosis and identification of at risk animals is required.

References

1. Pollitt CC. Basement membrane pathology: a feature of acute equine laminitis. Equine Veterinary Journal 1996;28(1):38–46.
2. Hood DM, Grosenbaugh DA, Mostafa MB, Morgan SJ, Thomas BC. The role of vascular mechanisms in the development of acute equine laminitis. Journal of Veterinary Internal Medicine 1993;7(4):228–34.
3. Johnson PJ. The equine metabolic syndrome peripheral Cushing's syndrome. Veterinary Clinics of North America, Equine Practice 2002;18:271–93.
4. Hinckley KA, Henderson IW. The epidemiology of equine laminitis in the UK. In: 35th Congress of the British Equine Veterinary Association 1996. p. 62.
5. Kane AJ, Traub-Dargatz JL, Garber LP, Losinger WP, Wagner BA. Lameness and Laminitis in US Horses. USDA APHIS VS CEAH, National Animal Health Monitoring System, Fort Collins, CO; 2000.
6. Wylie CE, Collins SN, Verheyen KL, Richard Newton J. Frequency of equine laminitis: a systematic review with quality appraisal of published evidence. Veterinary Journal 2011;189(3):248–56.
7. Garner HE, Coffman JR, Hahn AW, Hutcheson DP, Tumbleson ME. Equine laminitis of alimentary origin: an experimental model. American Journal of Veterinary Research 1975;36(4 Pt.1):441–4.
8. van Eps AW, Pollitt CC. Equine laminitis induced with oligofructose. Equine Veterinary Journal 2006;38(3):203–8.

9. Galey FD, Twardock AR, Goetz TE, Schaeffer DJ, Hall JO, Beasley VR. Gamma scintigraphic analysis of the distribution of perfusion of blood in the equine foot during black walnut (*Juglans nigra*)-induced laminitis. American Journal of Veterinary Research 1990;51(4):688–95.
10. de Laat MA, McGowan CM, Sillence MN, Pollitt CC. Equine laminitis: induced by 48 h hyperinsulinaemia in standardbred horses. Equine Veterinary Journal 2010;42(2):129–35.
11. Loftus JP, Black SJ, Pettigrew A, Abrahamsen EJ, Belknap JK. Early laminar events involving endothelial activation in horses with black walnut-induced laminitis. American Journal of Veterinary Research 2007;68(11):1205–1.
12. Black SJ, Lunn DP, Yin C, Hwang M, Lenz SD, Belknap JK. Leukocyte emigration in the early stages of laminitis. Veterinary Immunology and Immunopathology 2006;109(1–2):161–6.
13. Belknap JK, Giguere S, Pettigrew A, Cochran AM, Van Eps AW, Pollitt CC. Lamellar pro-inflammatory cytokine expression patterns in laminitis at the developmental stage and at the onset of lameness: innate vs. adaptive immune response. Equine Veterinary Journal 2007;39(1):42–7.
14. Leise BS, Faleiros RR, Watts M, Johnson PJ, Black SJ, Belknap JK. Laminar inflammatory gene expression in the carbohydrate overload model of equine laminitis. Equine Veterinary Journal 2011;43(1):54–61.
15. Belknap JK, Moore JN, Crouser EC. Sepsis – from human organ failure to laminar failure. Veterinary Immunology and Immunopathology 2009;129(3–4):155–7.
16. Riggs LM, Krunkosky TM, Noschka E, Boozer LA, Moore JN, Robertson TP. Comparison of characteristics and enzymatic products of leukocytes in the skin and laminar tissues of horses administered black walnut heartwood extract or lipopolysaccharide. American Journal of Veterinary Research 2009;70(11):1383–90.
17. Stewart AJ, Pettigrew A, Cochran AM, Belknap JK. Indices of inflammation in the lung and liver in the early stages of the black walnut extract model of equine laminitis. Veterinary Immunology and Immunopathology 2009;129(3/4):254–60.
18. Faleiros RR, Nuovo GJ, Belknap JK. Calprotectin in myeloid and epithelial cells of laminae from horses with black walnut extract-induced laminitis. Journal of Veterinary Internal Medicine 2009;23(1):174–81.
19. Waguespack RW, Kemppainen RJ, Cochran A, Lin HC, Belknap JK. Increased expression of MAIL, a cytokine-associated nuclear protein, in the prodromal stage of black walnut-induced laminitis. Equine Veterinary Journal 2004;36(3):285–91.
20. Gustot T. Multiple organ failure in sepsis: prognosis and role of systemic inflammatory response. Current Opinion in Critical Care 2011;17(2):153–9.
21. Visser MB, Pollitt CC. Lamellar leukocyte infiltration and involvement of IL-6 during oligofructose-induced equine laminitis development. Veterinary Immunology and Immunopathology 2011;144(1/2):120–8.
22. Bailey SR, Adair HS, Reinemeyer CR, Morgan SJ, Brooks AC, Longhofer SL, *et al.* Plasma concentrations of endotoxin and platelet activation in the developmental stage of oligofructose-induced laminitis. Veterinary Immunology and Immunopathology 2009;129(3–4):167–73.
23. de Laat MA, van Eps AW, McGowan CM, Sillence MN, Pollitt CC. Equine Laminitis: comparative histopathology 48 hours after experimental induction with insulin or alimentary oligofructose in standardbred horses. Journal of Comparative Pathology 2011;145(4):399–409.
24. Asplin KE, Patterson-Kane JC, Sillence MN, Pollitt CC, Mc Gowan CM. Histopathology of insulin-induced laminitis in ponies. Equine Veterinary Journal 2010;42(8):700–6.
25. Budak MT, Orsini JA, Pollitt CC, Rubinstein NA. Gene expression in the lamellar dermis-epidermis during the developmental phase of carbohydrate overload-induced laminitis in the horse. Veterinary Immunology and Immunopathology 2009;131(1/2):86–96.
26. Noschka E, Vandenplas ML, Hurley DJ, Moore JN. Temporal aspects of laminar gene expression during the developmental stages of equine laminitis. Veterinary Immunology and Immunopathology 2009;129(3/4):242–53.
27. Suagee JK, Corl BA, Crisman MV, Hulver MW, McCutcheon LJ, Geor RJ. Effects of acute hyperinsulinemia on inflammatory proteins in horses. Veterinary Immunology and Immunopathology 2011;142(3/4):141–6.
28. Holbrook TC, Tipton T, McFarlane D. Neutrophil and cytokine dysregulation in hyperinsulinemic obese horses. Veterinary Immunology and Immunopathology 2012;145(1–2):283–9.
29. Faleiros RR, Leise BS, Watts M, Johnson PJ, Black SJ, Belknap JK. Laminar chemokine mRNA concentrations in horses with carbohydrate overload-induced laminitis. Veterinary Immunology and Immunopathology 2011;144(1/2):45–51.
30. Faleiros RR, Leise BB, Westerman T, Yin C, Nuovo GJ, Belknap JK. *In vivo* and *in vitro* evidence of the involvement of CXCL1, a keratinocyte-derived chemokine, in equine laminitis. Journal of Veterinary Internal Medicine 2009;23(5):1086–96.
31. Heyman SN, Rosen S, Rosenberger C. A role for oxidative stress. Contributions to Nephrology 2011;174:138–48.
32. Burns TA, Westerman T, Nuovo GJ, Watts MR, Pettigrew A, Yin C, *et al.* Role of oxidative tissue injury in the pathophysiology of experimentally induced equine laminitis: a comparison of 2 models. Journal of Veterinary Internal Medicine 2011;25(3):540–8.
33. de Laat MA, Kyaw-Tanner MT, Sillence MN, McGowan CM, Pollitt CC. Advanced glycation endproducts in horses with insulin-induced laminitis. Veterinary Immunology and Immunopathology 2012;145(1–2):395–401.
34. Keen JA, McLaren M, Chandler KJ, McGorum BC. Biochemical indices of vascular function, glucose metabolism and oxidative stress in horses with equine Cushing's disease. Equine Veterinary Journal 2004;36(3):226–9.
35. Alberts B, Johnson AJL. The extracellular matrix. In: Alberts B, Johnson AJL, editors. Molecular Biology of the Cell, 5th edn. Garland Science, New York; 2007. p. 1065–126.
36. French KR, Pollitt CC. Equine laminitis: congenital, hemidesmosomal plectin deficiency in a quarter horse foal. Equine Veterinary Journal 2004;36(3):299–303.
37. French KR, Pollitt CC. Equine laminitis: loss of hemidesmosomes in hoof secondary epidermal lamellae correlates to dose in an oligofructose induction model: an ultrastructural study. Equine Veterinary Journal 2004;36(3):230–5.

38. Johnson PJ, Tyagi SC, Katwa LC, Ganjam VK, Moore LA, Kreeger JM, *et al.* Activation of extracellular matrix metalloproteinases in equine laminitis. Veterinary Record 1998;142(15):392–6.

39. Pollitt CC, Pass MA, Pollitt S. Batimastat (BB-94) inhibits matrix metalloproteinases of equine laminitis. Equine Veterinary Journal Supplement 1998;26(26):119–24.

40. Kyaw-Tanner M, Pollitt CC. Equine laminitis: increased transcription of matrix metalloproteinase-2 (MMP-2) occurs during the developmental phase. Equine Veterinary Journal 2004;36(3):221–5.

41. Visser MB, Pollitt CC. The timeline of metalloprotease events during oligofructose induced equine laminitis development. Equine Veterinary Journal 2012;44(1):88–93.

42. Visser MB, Pollitt CC. Immunohistochemical distribution of laminin-332 and collagen type IV in the basement membrane of normal horses and horses with induced laminitis. Journal of Comparative Pathology 2011;145(1):80–7.

43. Visser MB, Pollitt CC. The timeline of lamellar basement membrane changes during equine laminitis development. Equine Veterinary Journal 2011;43(4):471–7.

44. Black RA, Rauch CT, Kozlosky CJ, Peschon JJ, Slack JL, Wolfson MF, *et al.* A metalloproteinase disintegrin that releases tumour-necrosis factor-alpha from cells. Nature 1997;385(6618):729–33.

45. Malfait AM, Liu RQ, Ijiri K, Komiya S, Tortorella MD. Inhibition of ADAM-TS4 and ADAM-TS5 prevents aggrecan degradation in osteoarthritic cartilage. Journal of Biological Chemistry 2002;277(25):22201–8.

46. Coyne MJ, Cousin H, Loftus JP, Johnson PJ, Belknap JK, Gradil CM, *et al.* Cloning and expression of ADAM-related metalloproteases in equine laminitis. Veterinary Immunology and Immunopathology 2009;129(3–4):231–41.

47. Wang L, Pawlak E, Johnson PJ, Belknap JK, Alfandari D, Black SJ. Effects of cleavage by a disintegrin and metalloproteinase with thrombospondin motifs-4 on gene expression and protein content of versican and aggrecan in the digital laminae of horses with starch gruel-induced laminitis. American Journal of Veterinary Research 2012;73(7):1047–56.

48. Handley CJ, Samiric T, Ilic MZ. Structure, metabolism, and tissue roles of chondroitin sulfate proteoglycans. Advances in Pharmacology 2006;53:219–32.

49. Black SJ. Extracellular matrix, leukocyte migration and laminitis. Veterinary Immunology and Immunopathology 2009;129(3–4):161–3.

50. de Laat MA, Kyaw-Tanner MT, Nourian AR, McGowan CM, Sillence MN, Pollitt CC. The developmental and acute phases of insulin-induced laminitis involve minimal metalloproteinase activity. Veterinary Immunology and Immunopathology 2011;140(3–4):275–81.

51. Johnson PJ, Messer NT, Slight SH, Wiedmeyer C, Buff P, Ganjam VK. Endocrinopathic laminitis in horses. Clinical Techniques in Equine Practice 2004;3:45–56.

52. Johnson PJ, Messer NT, Ganjam VK. Cushing's syndromes, insulin resistance and endocrinopathic laminitis. Equine Veterinary Journal 2004;36(3):194–8.

53. Johnson PJ, Slight SH, Ganjam VK, Kreeger JM. Glucocorticoids and laminitis in the horse. Veterinary Clinics of North America, Equine Practice 2002;18(2):219–36.

54. Karikoski NP, Horn I, McGowan TW, McGowan CM. The prevalence of endocrinopathic laminitis among horses presented for laminitis at a first-opinion/referral equine hospital. Domestic Animal Endocrinology 2011;41(3):111–7.

55. Donaldson MT, Jorgensen AJ, Beech J. Evaluation of suspected pituitary pars intermedia dysfunction in horses with laminitis. Journal of American Veterinary Medical Association 2004;224(7):1123–7.

56. Bailey SR, Habershon-Butcher J, Ransom K, Elliott J, Menzies-Gow NJ. Hypertension and insulin resistance in a mixed-breed population of ponies predisposed to laminitis: further characterizing pre-laminitic metabolic syndrome. American Journal of Veterinary Research 2008;69(1):122–9.

57. Parsons CS, Orsini JA, Krafty R, Capewell L, Boston R. Risk factors for development of acute laminitis in horses during hospitalization: 73 cases (1997–2004). Journal of American Veterinary Medical Association 2007;230(6):885–9.

58. Slater MR, Hood DM, Carter GK. Descriptive epidemiological study of equine laminitis. Equine Veterinary Journal 1995;27:364–7.

59. Wattle O, Pollitt CC. Lamellar metabolism. Clinical Techniques in Equine Practice 2004;3(1):22–33.

60. Pass MA, Pollitt S, Pollitt CC. Decreased glucose metabolism causes separation of hoof lamellae *in vitro*: a trigger for laminitis?Equine Veterinary Journal 1998;26(Suppl.):133–8.

61. French KR, Pollitt CC. Equine laminitis: Glucose deprivation and MMP activation induce dermo-epidermal separation *in vitro*. Equine Veterinary Journal 2004;36(3):261–6.

62. Mobasheri A, Critchlow K, Clegg PD, Carter SD, Canessa CM. Chronic equine laminitis is characterised by loss of GLUT1, GLUT4 and ENaC positive laminar keratinocytes. Equine Veterinary Journal 2004;36(3):248–54.

63. Asplin KE, McGowan CM, Pollitt CC, Sillence MN. Role of insulin in gluocse uptake in the equine hoof. Journal of Veterinary Internal Medicine 2007;21:668.

64. Asplin KE, Curlewis JD, McGowan CM, Pollitt CC, Sillence MN. Glucose transport in the equine hoof. Equine Veterinary Journal 2011;43(2):196–201.

65. Asplin KE, Sillence MN, Pollitt CC, McGowan CM. Induction of laminitis by prolonged hyperinsulinaemia in clinically normal ponies. Veterinary Journal 2007;174(3):530–5.

66. Laat MAd, Sillence MN, McGowan CM, Pollitt CC. Continuous intravenous infusion of glucose induces endogenous hyperinsulinaemia and lamellar histopathology in standard bred horses. Veterinary Journal 2012;191(3):317–22.

67. Frank N, Elliott SB, Chameroy KA, Toth F, Chumbler NS, McClamroch R. Association of season and pasture grazing with blood hormone and metabolite concentrations in horses with presumed pituitary pars intermedia dysfunction. Journal of Veterinary Internal Medicine 2010;24(5):1167–75.

68. Treiber KH, Kronfeld DS, Hess TM, Byrd BM, Splan RK. Pre-laminitic metabolic syndrome in genetically predisposed ponies involves compensated insulin resistance. Journal of Animal Physiology and Animal Nutrition 2005;89(11–12):430–1.

69. de Laat MA, Pollitt CC, Walsh DM, McGowan CM, Sillence MN. Persistent digital hyperthermia over a 48 h

period does not induce laminitis in horses. Veterinary Journal 2012;192(3):435–40.

70. Kim JA, Montagnani M, Koh KK, Quon MJ. Reciprocal relationships between insulin resistance and endothelial dysfunction: molecular and pathophysiological mechanisms. Circulation 2006;113(15):1888–904.

71. Venugopal CS, Eades S, Holmes EP, Beadle RE. Insulin resistance in equine digital vessel rings: an *in vitro* model to study vascular dysfunction in equine laminitis. Equine Veterinary Journal 2011;43(6):744–9.

72. Hood DM, Wagner IP, Brumbaugh GW. Evaluation of hoof wall surface temperature as an index of digital vascular perfusion during the prodromal and acute phases of carbohydrate-induced laminitis in horses. American Journal of Veterinary Research 2001;62(7):1167–72.

73. Adair HS, 3rd, Goble DO, Schmidhammer JL, Shires GM. Laminar microvascular flow, measured by means of laser Doppler flowmetry, during the prodromal stages of black walnut-induced laminitis in horses. American Journal of Veterinary Research 2000;61(8):862–8.

74. Bailey SR, Elliott J. Evidence for different 5-HT1B/1D receptors mediating vasoconstriction of equine digital arteries and veins. European Journal of Pharmacology 1998;355(2–3):175–87.

75. Bailey SR, Elliott J. Plasma 5-hydroxytryptamine constricts equine digital blood vessels *in vitro*: implications for pathogenesis of acute laminitis. Equine Veterinary Journal 1998;30(2):124–30.

76. Bailey SR, Marr CM, Elliott J. Current research and theories on the pathogenesis of acute laminitis in the horse. Veterinary Journal 2004;167(2):129–42.

77. Elliott J, Berhane Y, Bailey SR. Effects of monoamines formed in the cecum of horses on equine digital blood vessels and platelets. American Journal of Veterinary Research 2003;64(9):1124–31.

78. Bailey SR, Wheeler-Jones C, Elliott J. Uptake of 5-hydroxytryptamine by equine digital vein endothelial cells: inhibition by amines found in the equine caecum. Equine Veterinary Journal 2003;35(2):164–9.

79. Robinson NE. Step by painful step: increasing knowledge about laminitis. Equine Veterinary Journal 1998;30(2):89–90.

80. Weiss DJ, Geor RJ, Johnston G, Trent AM. Microvascular thrombosis associated with onset of acute laminitis in ponies. American Journal of Veterinary Research 1994;55(5):606–12.

81. Weiss DJ, Trent AM, Johnston G. Prothrombotic events in the prodromal stages of acute laminitis in horses. American Journal of Veterinary Research. 1995;56(8):986–91.

82. Allen D, Jr., Clark ES, Moore JN, Prasse KW. Evaluation of equine digital Starling forces and hemodynamics during early laminitis. American Journal of Veterinary Research 1990;51(12):1930–4.

83. Moore RM, Eades SC, Stokes AM. Evidence for vascular and enzymatic events in the pathophysiology of acute laminitis: which pathway is responsible for initiation of this process in horses?Equine Veterinary Journal 2004;36(3):204–9.

84. Dabareiner RM, Snyder JR, White NA, Pascoe JR, Harmon FA, Gardner I, *et al.* Microvascular permeability and endothelial cell morphology associated with low-flow ischemia/reperfusion injury in the equine jejunum. American Journal of Veterinary Research 1995;56(5):639–48.

85. Henninger DD, Snyder JR, Pascoe JR, Dilling GW. Microvascular permeability changes in ischemia/reperfusion injury in the ascending colon of horses. Journal of the American Veterinary Medical Association 1992;201(8):1191–6.

86. Loftus JP, Belknap JK, Stankiewicz KM, Black SJ. Laminar xanthine oxidase, superoxide dismutase and catalase activities in the prodromal stage of black-walnut induced equine laminitis. Equine Veterinary Journal 2007;39(1):48–53.

87. Noschka E, Moore JN, Peroni JF, Lewis SJ, Morrow JD, Robertson TP. Thromboxane and isoprostanes as inflammatory and vasoactive mediators in black walnut heartwood extract induced equine laminitis. Veterinary Immunology and Immunopathology 2009;129(3–4):200–10.

88. Galley HF, Webster NR. Physiology of the endothelium. British Journal of Anaesthesia 2004;93(1):105–13.

89. Elliott J, Bryant CE, Soydan J. The role of nitric oxide in the responses of equine digital veins to vasodilator and vasoconstrictor agents. Equine Veterinary Journal 1994;26(5):378–84.

90. Potenza MA, Addabbo F, Montagnani M. Vascular actions of insulin with implications for endothelial dysfunction. American Journal of Physiology Endocrinology and Metabolism 2009;297(3):E568–77.

91. Verma S, Buchanan MR, Anderson TJ. Endothelial function testing as a biomarker of vascular disease. Circulation 2003;108(17):2054–9.

92. Bailey SR, Marr CM, Elliott J. Current research and theories on the pathogenesis of acute laminitis in the horse. Veterinary Journal 2004;36(3):255–60.

93. Milinovich GJ, Klieve AV, Pollitt CC, Trott DJ. Microbial events in the hindgut during carbohydrate-induced equine laminitis. Veterinary Clinics of North America, Equine Practice 2010;26(1):79–94.

94. Keller MD, Pollitt CC, Marx UC. Nuclear magnetic resonance-based metabonomic study of early time point laminitis in an oligofructose-overload model. Equine Veterinary Journal 2011;43(6):737–43.

95. Bailey SR, Rycroft A, Elliott J. Production of amines in equine cecal contents in an *in vitro* model of carbohydrate overload. Journal of Animal Science 2002;80(10):2656–62.

96. Alford P, Geller S, Richardson B, Slater MR, Honnas C, Foreman JH, *et al.* A multicenter, matched case-control study of risk factors for equine laminitis. Preventative Veterinary Medicine 2001;49:209–22.

97. Treiber KH, Hess TM, Kronfeld DS, Boston RC, Geor RJ, Harris PA, editors. Insulin resistance and compensation in laminitis-predisposed ponies characterized by the Minimal Model. Equine Nutrition Symposium. Pferdeheilkunde, Hannover; 2005.

98. Treiber KH, Kronfeld DS, Hess TM, Byrd BM, Splan RK,Staniar WB. Evaluation of genetic and metabolic predispositions and nutritional risk factors for pasture-associated laminitis in ponies. Journal of American Veterinary Medical Association 2006;228:1538–45.

99. Eades SC, Stokes AM, Johnson PJ, LeBlanc CJ, Ganjam VK, Buff PR, *et al.* Serial alterations in digital hemodynamics

and endothelin-1 immunoreactivity, platelet-neutrophil aggregation, and concentrations of nitric oxide, insulin, and glucose in blood obtained from horses following carbohydrate overload. American Journal of Veterinary Research 2007;68(1):87–94.

100. Treiber KH, Boston RC, Kronfeld DS, Staniar WB, Harris PA. Insulin resistance and compensation in Thoroughbred weanlings adapted to high-glycaemic meals. Journal of Animal Science 2005;83:2357–64.

101. Hoffman RM, Boston RC, Stefanovski D, Kronfeld DS, Harris PA. Obesity and diet affect glucose dynamics and insulin sensitivity in Thoroughbred geldings. Journal of Animal Science 2003;81:2333–42.

102. Bailey SR, Harris PA. Effects of dietary fructan carbohydrates on plasma insulin levels in laminitis-prone ponies. Journal of Veterinary Internal Medicine 2006;20:799–800.

103. Buff PR, Johnson PJ, Wiedmeyer CE, Ganjam VK, Messer INT, Keisler DH. Modulation of leptin, insulin and growth hormone in obese pony mares under chronic nutritional restriction and supplementation with ractopamine hydrochloride. Veterinary Therapeutics 2006;7:64–72.

104. Bujalska IJ, Kumar S, Stewart PM. Does central obesity reflect 'Cushing's disease of the omentum'? Lancet 1997;349(9060):1210–3.

105. Picchi A, Gao X, Belmadani S, Potter BJ, Focardi M, Chilian WM, *et al.* Tumor necrosis factor-alpha induces endothelial dysfunction in the prediabetic metabolic syndrome. Circulation Research 2006;99(1):69–77.

106. Houstis N, Rosen ED, Lander ES. Reactive oxygen species have a causal role in multiple forms of insulin resistance. Nature 2006;440(7086):944–8.

107. Nishikawa T, Kukidome D, Sonoda K, Fujisawa K, Matsuhisa T, Motoshima H, *et al.* Impact of mitochondrial ROS production in the pathogenesis of insulin resistance. Diabetes Research in Clinical Practice 2007;77(Suppl. 1):S161–4.

108. Treiber K, Carter R, Gay L, Williams C, Geor R. Inflammatory and redox status of ponies with a history of pasture-associated laminitis. Veterinary Immunology and Immunopathology 2009;129:216–20.

109. Vick MM, Adams AA, Murphy BA, Sessions DR, Horohov DW, Cook RF, *et al.* Relationships among inflammatory cytokines, obesity, and insulin sensitivity in the horse. Journal of Animal Science 2007;85(5):1144–55.

110. Suagee JK, Corl BA, Crisman MV, Hulver MW, McCutcheon LJ, Geor RJ. Effects of acute hyperinsulinemia on inflammatory proteins in horses. Veterinary Immunology and Immunopathology 2011;142(3–4):141–6.

111. Carter RA, Treiber KH, Geor RJ, Douglass L, Harris PA. Prediction of incipient pasture-associated laminitis from hyperinsulinaemia, hyperleptinaemia and generalised and localised obesity in a cohort of ponies. Equine Veterinary Journal 2009;41(2):171–8.

112. Burns TA, Geor RJ, Mudge MC, McCutcheon LJ, Belknap JK. Pro-inflammatory cytokine and chemokine expression profiles of various adipose tissue depots of insulin-resistant and insulin-sensitive light breed horses. Journal of Veterinary Internal Medicine 2009;23(3):722.

113. Burns TA, Geor RJ, Mudge MC, McCutcheon LJ, Hinchcliff KW, Belknap JK. Proinflammatory cytokine and chemokine gene expression profiles in subcutaneous and visceral adipose tissue depots of insulin-resistant and insulin-sensitive light breed horses. Journal of Veterinary Internal Medicine 2010;24(4):932–9.

114. Suagee JK, Burk AO, Quinn RW, Hartsock TG, Douglass LW. Effects of diet and weight gain on circulating tumour necrosis factor-alpha concentrations in Thoroughbred geldings. Journal of Animal Physiology and Animal Nutrition (Berlin) 2011;95(2):161–70.

115. Elliott J, Bailey SR. Gastrointestinal derived factors are potential triggers for the development of acute equine laminitis. Journal of Nutrition 2006;136 7 Suppl.:2103S–7S.

116. Bailey SR, Baillon ML, Rycroft AN, Harris PA, Elliott J. Identification of equine cecal bacteria producing amines in an *in vitro* model of carbohydrate overload. Applied Environmental Microbiology 2003;69(4):2087–93.

117. Bailey SR, Katz LM, Berhane Y, Samuels T, De Brauvere N, Marr CM, *et al.* Seasonal changes in plasma concentrations of caecum-derived amines in clinically normal ponies and ponies predisposed to laminitis. American Journal of Veterinary Research 2003;64(9):1132–8.

118. Crawford C, Sepulveda MF, Elliott J, Harris PA, Bailey SR. Dietary fructan carbohydrate increases amine production in the equine large intestine: implications for pasture-associated laminitis. Journal of Animal Science 2007;85(11):2949–58.

119. Virgin JE, Goodrich LR, Baxter GM, Rao S. Incidence of support limb laminitis in horses treated with half limb, full limb or transfixation pin casts: a retrospective study of 113 horses (2000–2009). Equine Veterinary Journal Supplement 2011;43(s40):7–11.

120. van Eps A, Collins SN, Pollitt CC. Supporting limb laminitis. Veterinary Clinics of North America, Equine Practice 2010;26(2):287–302.

121. Belknap JK. Pathogenesis of laminitis. In: Robinson NE, Sprayberry KA, editors. Current Therapy in Equine Medicine. 6th ed. Saunders, Philadelphia; 2008. p. 541–3.

122. Baxter GM, Morrison S. Complications of unilateral weight bearing. Veterinary Clinics of North America, Equine Practice 2008;24(3):621–42.

123. Obel N. Studies on the Histopathology of Acute Laminitis. Almquisst and Wiksells Boktryckeri, Uppsala, Sweden; 1948.

124. Arble JB, Mattoon JS, Drost WT, Weisbrode SE, Wassenaar PA, Pan X, *et al.* Magnetic resonance imaging of the initial active stage of equine laminitis at 4.7 T. Veterinary Radiology and Ultrasound 2009;50(1):3–12.

125. Griffin RL, McKnight AL, Rucker A, Bennett SD, Fiser DM. Initial experience using contrast magnetic resonance imaging in laminitic horses: 18 studies. Journal of Equine Veterinary Science 2011;31(12):685–92.

126. Kruger EF, Puchalski SM, Pollard RE, Galuppo LD, Hornof WJ, Wisner ER. Measurement of equine laminar blood flow and vascular permeability by use of dynamic contrast-enhanced computed tomography. American Journal of Veterinary Research 2008;69(3):371–7.

127. McFarlane D. Equine pituitary pars intermedia dysfunction. Veterinary Clinics of North America, Equine Practice 2011;27(1):93–113.

128. Dybdal NO, Hargreaves KM, Madigan JE, Gribble DH, Kennedy PC, Stabenfeldt GH. Diagnostic testing for pituitary pars intermedia dysfunction in horses. Journal of American Veterinary Medical Association 1994;204(4):627–32.

129. Meissner TJ, Beard LA, Schmall SM, Reed SM. Results of overnight dexamethasone suppression test repeated over time in horses suspected of having equine cushing's disease. Journal of Veterinary Internal Medicine 2003;17(3):420.

130. Donaldson MT, McDonnell SM, Schanbacher BJ, Lamb SV, McFarlane D, Beech J. Variation in plasma adrenocorticotropic hormone concentration and dexamethasone suppression test results with season, age, and sex in healthy ponies and horses. Journal of Veterinary Internal Medicine 2005;19(2):217–22.

131. Beech J, Boston R, Lindborg S, Russell GE. Adrenocorticotropin concentration following administration of thyrotropin-releasing hormone in healthy horses and those with pituitary pars intermedia dysfunction and pituitary gland hyperplasia. Journal of American Veterinary Medical Association 2007;231(3):417–26.

132. Frank N. Equine metabolic syndrome. Veterinary Clinics of North America: Equine Practice 2011;27(1):73–92.

133. Menzies-Gow NJ, Stevens K, Barr A, Camm I, Pfeiffer D, Marr CM. Severity and outcome of equine pasture-associated laminitis managed in first opinion practice in the UK. Veterinary Record 2010;167(10):364–9.

134. West E, Bardell D, Morgan R, Senior M. Use of acetaminophen (paracetamol) as a short-term adjunctive analgesic in a laminitic pony. Veterinary Anaesthesia and Analgesia 2011;38(5):521–2.

135. Leise BS, Fugler LA, Stokes AM, Eades SC, Moore RM. Effects of intramuscular administration of acepromazine on palmar digital blood flow, palmar digital arterial pressure, transverse facial arterial pressure, and packed cell volume in clinically healthy, conscious horses. Veterinary Surgery 2007;36(8):717–23.

136. Donaldson MT, LaMonte BH, Morresey P, Smith G, Beech J. Treatment with pergolide or cyproheptadine of pituitary pars intermedia dysfunction (equine Cushing's disease). Journal of Veterinary Internal Medicine 2002;16(6):742–6.

137. Perkins GA, Lamb S, Erb HN, Schanbacher B, Nydam DV, Divers TJ. Plasma adrenocorticotropin (ACTH) concentrations and clinical response in horses treated for equine Cushing's disease with cyproheptadine or pergolide. Equine Veterinary Journal 2002;34(7):679–85.

138. McGowan CM, Neiger R. Efficacy of trilostane for the treatment of equine Cushing's syndrome. Equine Veterinary Journal 2003;35(4):414–8.

139. Powell DM, Reedy SE, Sessions DR, Fitzgerald BP. Effect of short-term exercise training on insulin sensitivity in obese and lean mares. Equine Veterinary Journal Supplement 2002;34:81–4.

140. Treiber KH, Hess TM, Kronfeld DS, Boston RC, Geor RJ, Friere M, *et al.* Glucose dynamics during exercise: dietary energy sources affect minimal model parameters in trained Arabian geldings during endurance exercise. Equine Veterinary Journal Supplement 2006;36:631–6.

141. Bruunsgaard H. Physical activity and modulation of systemic low-level inflammation. Journal of Leukocyte Biology 2005;78(4):819–35.

142. Frank N, Elliott SB, Boston RC. Effects of long-term oral administration of levothyroxine sodium on glucose dynamics in healthy adult horses. American Journal of Veterinary Research 2008;69(1):76–81.

143. Inzucchi SE. Oral antihyperglycemic therapy for type 2 diabetes: scientific review. Journal of American Medical Association 2002;287(3):360–72.

144. Durham AE, Rendle DI, Newton JE. The effect of metformin on measurements of insulin sensitivity and beta cell response in 18 horses and ponies with insulin resistance. Equine Veterinary Journal 2008;40(5):493–500.

145. Hustace JL, Firshman AM, Mata JE. Pharmacokinetics and bioavailability of metformin in horses. American Journal of Veterinary Research 2009;70(5):665–8.

146. Tinworth KD, Edwards S, Noble GK, Harris PA, Sillence MN, Hackett LP. Pharmacokinetics of metformin after enteral administration in insulin-resistant ponies. American Journal of Veterinary Research 2010;71(10):1201–6.

147. Tinworth KD, Boston RC, Harris PA, Sillence MN, Raidal SL, Noble GK. The effect of oral metformin on insulin sensitivity in insulin-resistant ponies. Veterinary Journal 2012;191(1):79–84.

148. Sakar Y, Meddah B, Faouzi MA, Cherrah Y, Bado A, Ducroc R. Metformin-induced regulation of the intestinal D-glucose transporters. Journal of Physiology and Pharmacology 2010;61(3):301–7.

149. Wearn JM, Crisman MV, Davis JL, Geor RJ, Hodgson DR, Suagee JK, *et al.* Pharmacokinetics of pioglitazone after multiple oral dose administration in horses. Journal of Veterinary Pharmacology and Therapeutics 2011;34(3):252–8.

150. Suagee JK, Corl BA, Wearn JG, Crisman MV, Hulver MW, Geor RJ, *et al.* Effects of the insulin-sensitizing drug pioglitazone and lipopolysaccharide administration on insulin sensitivity in horses. Journal of Veterinary Internal Medicine 2011;25(2):356–64.

151. Colles CM, Jeffcott LB. Laminitis in the horse. Veterinary Record 1977;100(13):262–4.

152. Stick JA, Jann HW, Scott EA, Robinson NE. Pedal bone rotation as a prognostic sign in laminitis of horses. Journal of American Veterinary Medical Association 1982;180(3):251–3.

153. Stashak TS. Lameness. In: Stashak TS, editor. Adams' Lameness in Horses. 4th ed. Lea and Febiger, Philadelphia; 1987. p. 486–99.

154. Eustace RA, Caldwell MN. Treatment of solar prolapse using the heart bar shoe and dorsal hoof wall resection technique. Equine Veterinary Journal 1989;21(5):370–2.

155. Cripps PJ, Eustace RA. Factors involved in the prognosis of equine laminitis in the UK. Equine Veterinary Journal 1999;31:433–42.

156. Eustace RA, Emery SL, Cripps PJ. A retrospective study of 23 cases of coronary band separation longer than 8 cm as a sequel to severe laminitis. Journal of Equine Veterinary Science 2012;32(4):235–44.

157. Oliveira TM, Pereira MMF, Silva LCLCd, Fernandes WR, Baccarin RYA. Relationship between the use of therapeutic shoeing, period of treatment and rehabilitation of horses with chronic laminitis. Brazilian Journal of Veterinary Research and Animal Science 2011;48(5):355–60.

158. McGowan CM, Frost R, Pfeiffer DU, Neiger R. Serum insulin concentrations in horses with equine Cushing's syndrome: response to a cortisol inhibitor and prognostic value. Equine Veterinary Journal 2004;36(3):295–8.

159. Geor RJ. Pasture-associated laminitis. Veterinary Clinics of North America, Equine Practice 2009;25(1):39–50.

160. Chatterton NJ, Watts KA, Jensen KB, Harrison PA, Horton WH. Nonstructural carbohydrates in oat forage. Journal of Nutrition 2006;136 7 Suppl.:2111S–3S.

161. Longland AC, Byrd BM. Pasture nonstructural carbohydrates and equine laminitis. Journal of Nutrition 2006;136:2099S–102S.

162. Fisher DS, Mayland HF, Burns JC. Variation in ruminants' preference for tall fescue hays cut either at sundown or at sunup. Journal of Animal Science 1999;77(3):762–8.

163. Byrd BM, Treiber KH, Staniar WB. Circadian and seasonal variation on pasture NSC and circulating insulin concentrations in grazing horses. Journal of Animal Science 2006;84(Suppl. 1):330–1.

164. Frank N. Endocrinopathic laminitis, obesity-associated laminitis and pasture-associated laminitis. In 54th American Association of Equine Practitioners Meeting. San Diego, CA; 2008. p. 341–6.

165. Longland AC, Barfoot C, Harris PA. Effects of soaking on the water-soluble carbohydrate and crude protein content of hay. Veterinary Record 2011;168(23):618.

166. Borer KE, Bailey SR, Menzies-Gow NJ, Harris PA, Elliott J. Use of proxy measurements of insulin sensitivity and insulin secretory response to distinguish between normal and previously laminitic ponies. Equine Veterinary Journal 2012; doi:10.1111/j.2042-3306.2011.00475.x.

167. Borer KE, Bailey SR, Menzies-Gow NJ, Harris PA, Elliott J. Effect of feeding glucose, fructose, and inulin on blood glucose and insulin concentrations in normal ponies and those predisposed to laminitis. Journal of Animal Science 2012;90(9):3003–11.

168. Barton MH, Parviainen A, Norton N. Polymyxin B protects horses against induced endotoxaemia *in vivo*. Equine Veterinary Journal 2004;36(5):397–401.

169. Semrad SD, Moore JN. Effects of a specific thromboxane synthetase inhibitor in equine endotoxaemia. Research in Veterinary Science 1987;43(2):137–42.

170. Burrows GE. Therapeutic effect of phenylbutazone on experimental acute *Escherichia coli* endotoxemia in ponies. American Journal of Veterinary Research 1981;42(1):94–9.

171. Jackman BR, Moore JN, Barton MH, Morris DD. Comparison of the effects of ketoprofen and flunixin meglumine on the *in vitro* response of equine peripheral blood monocytes to bacterial endotoxin. Canadian Journal of Veterinary Research 1994;58(2):138–43.

172. Peiro JR, Barnabe PA, Cadioli FA, Cunha FQ, Lima VM, Mendonca VH, *et al.* Effects of lidocaine infusion during experimental endotoxemia in horses. Journal of Veterinary Internal Medicine 2010;24(4):940–8.

173. Sykes BW, Furr MO. Equine endotoxaemia–a state-of-the-art review of therapy. Australian Veterinary Journal 2005;83(1–2):45–50.

174. de la Rebiere de Pouyade G, Grulke S, Detilleux J, Salciccia A, Verwilghen DR, Caudron I, *et al.* Evaluation of low-molecular-weight heparin for the prevention of equine laminitis after colic surgery. Journal of Veterinary Emergency and Critical Care (San Antonio) 2009;19(1):113–9.

175. van Eps AW, Pollitt CC. Equine laminitis: cryotherapy reduces the severity of the acute lesion. Equine Veterinary Journal 2004;36(3):255–60.

176. van Eps AW, Leise BS, Watts M, Pollitt CC, Belknap JK. Digital hypothermia inhibits early lamellar inflammatory signalling in the oligofructose laminitis model. Equine Veterinary Journal 2012;44(2):230–7.

177. Buchner HHF, Wiesenhofer G. Effects of distal limb ice bag cryotherapy in horses – a thermographic evaluation. Pferdeheilkunde 2011;27(6):670–3.

178. Reesink HL, Divers TJ, Bookbinder LC, van Eps AW, Soderholm LV, Wildenstein MJ, *et al.* Cryotherapy of the Equine Digit for the Treatment of Laminitis: Assessment of a Novel Method by Measuring Digital Venous Temperature. American Association of Equine Practitioners (AAEP), Lexington; 2010. p. 149.

179. Aumailley M, Bruckner-Tuderman L, Carter WG, Deutzmann R, Edgar D, Ekblom P, *et al.* A simplified laminin nomenclature. Matrix Biology 2005;24:326–32.

A review of innate immune functions in crocodilians

John W. Finger Jr[1,2,*] and Sally R. Isberg[2]

Address: [1] Department of Environmental Health Science, College of Public Health, University of Georgia, Athens, GA, USA.
[2] Porosus Pty Ltd, PO Box 86, Palmerston, NT 0831, Australia

***Correspondence**: John W. Finger Jr. Department of Environmental Health Science, College of Public Health, University of Georgia, Athens, GA, USA. Email: finger20@uga.edu

Received: 4 September 2012
Accepted: 4 October 2012

Abstract

The immune function of crocodilians is understudied but is of interest for medical, ecological and evolutionary purposes. Crocodiles share a common ancestor with birds, comprising the archosaurian lineage, so they are an important link in our understanding of immune system evolution. As top trophic carnivores, they inhabit temperate and tropical climates in their semi-aquatic environment. However, they are also ectothermic, whereby environmental temperatures affect their physiological processes, including immune function, adding to the complexity of research in this area. Anecdotal observations and recent research have augmented enthusiasm in the realm of crocodilian immunology. Despite comprising both adaptive and innate systems, most research has investigated the innate system, which comprises peptides, proteins and leucocytes functioning in defence. Herein, we provide an overview into the innate immune system of the crocodile and areas for further research.

Keywords: Crocodilian, Innate, Immune, Leucocyte, Alligator

Review Methodology: We researched the following databases for this review: Web of Science, JSTOR, and ScienceDirect (keyword examples: alligator, immune, reptile). In addition, we performed searches via Google Scholar to broaden our hits. Lastly, we used references from papers found in the aforementioned databases to further increase material.

Introduction

The study of crocodilian immunology is in its infancy. However, recent research and a multitude of anecdotal evidence has suggested potent immune capabilities, and necessitated a greater understanding of the immunology in this ancient group of archosaurian reptiles [1–3]. Crocodilians comprised 23 different species divided into three different families: Crocodylidae, Alligatoridae and Gavialidae, with true crocodiles comprising Crocodylidae, alligators and caimans encompassing Alligatoridae, and gharials comprising Gavialidae [4]. Crocodilians are opportunistic, top trophic carnivores inhabiting a semi-aquatic environment. Similar among all non-avian reptiles, crocodilians are ectothermic, and thus lack the necessary high metabolic rate and insulation for endothermic [5, 6]. As such, all physiological processes are directly entwined with temperature, including immune function, feeding and assimilation, reproduction, and growth and development [7, 8]. Thus, for optimal physiology, crocodilians must regulate their body temperature through behavioural means (i.e. basking and submergence) [9]. As in all jawed vertebrates, the crocodilian immune system is dichotomous, composed of both innate and adaptive (acquired) portions [10]. Innate functions are present from birth (germline encoded), acute in their onset, require no previous stimulation and are non-specific, whereas adaptive functions are specific in nature, non-germline encoded (arise via somatic recombination) and require previous stimulation with subsequent infections/exposures increasing efficacy.

The purpose of this review is to provide an insight into the overall immunobiology of crocodilians, with the hope of revealing the gaps for continued research. Although the presence of the adaptive system has been documented in crocodilians able to produce both T and B lymphocytes

Table 1 Immune tissues and cell types of amniotic vertebrates

Tissue/cell	Non-archosaurian reptiles [13–16]	Crocodilians [11, 12, 16–19]	Birds [16, 20, 21]	Mammals [16, 22]
GALT	+	+	+	+
Thymus	+	+	+	+
Spleen	+	+	+	+
Lymph nodes	–	–	+	+
Bursa of Fabricius	–	–	+	–
Bone marrow	+	+	+	+
Heterophils	+	+	+	–
Azurophils	+	?	–	–
Eosinophils	+	+	+	+
Basophils	+	+	+	+
Neutrophils	–	–	–	+
Monocytes	+	+	+	+
Lymphocytes	+	+	+	+

and exhibiting immunological memory, most of research has investigated the innate immune system [11, 12]. Thus, discussion will mainly be restricted to innate aspects of the crocodilian immune system, although we realize the necessity of continued research into both components to fully elucidate the mechanisms of these two interdependent systems.

Lymphoid Tissues in Crocodilians: Structure and Hormonal Regulation

Crocodilian, as with other reptilian, lymphoid tissues consist of the spleen, thymus, bone marrow and gut-associated lymphoid tissue (GALT), but are devoid of germinal centres and lymph nodes (Table 1) [13, 14]. Analogous to hibernating mammals, reptilian lymphoid tissues exhibit involution during certain times of the year with functionality varying between changing seasons and temperatures [23, 24]. Given this variability, ascertaining lymphoid organ functionality and overall immunocompetence adds additional complexity. For example, temperature has been shown to affect haematology values of captive saltwater crocodiles, *Crocodylus porosus*, although interestingly there was no effect on plasma immunoglobulin levels in the same animals [15]. Similarly, innate and adaptive immune components exhibit seasonal variation in slider, *Trachemys scripta*, and painted turtles, *Chrysemys picta* [25, 26]. Thus, depending on season and temperature (and a host of other abiotic factors), immune functionality may display differing parameters.

Much of this seasonality in immune parameters may be the result of endogenous hormones, as some studies have shown quantitative changes in leucocyte numbers and lymphoid involution to be associated with changes in circulating levels of glucocorticoids and sex steroids [15, 24, 27–30]. In addition, exogenous exposure to hormonal mimics, such as endocrine disrupting chemicals (EDCs), may also affect immunity. The susceptibility of crocodilians to EDCs has previously been elucidated as they are top trophic carnivores, so EDCs may bioaccumulate and biomagnify in tissues, vastly affecting normal physiology [17, 31–34]. Crocodilian exposure to EDCs has also been associated with a number of immune abnormalities including histological alterations in immune organs [35].

By convention, lymphoid tissues are classified into primary and peripheral lymphoid tissues. Primary lymphoid tissues are sites of lymphocyte production, whereas peripheral organs facilitate innate and adaptive interaction and promote clearance of pathogens [36]. As in most vertebrates excluding birds, the thymus and bone marrow function as primary tissues in crocodilians from which mature cells disperse and interact with pathogens in peripheral tissues [16]. During reptilian embryonic development, haematopoiesis mainly occurs in the yolk sac, whereas upon hatching, the bone marrow serves as the primary site of haematopoiesis [20]. However, other organs, such as the thymus, spleen and liver, may also play an important role in haematopoiesis [37]. Immature T-cell progenitors migrate out of the bone marrow and finish maturation in the thymus, while B-cell maturation continues in the bone marrow. It is from these primary tissues that naive (never encountered antigen) lymphocytes disperse and engage pathogens in peripheral lymphoid tissues. Contrastingly in birds, the closest living relative of crocodilians, maturation of B-cells does not occur in the bone marrow but in an avian-specific organ located near the cloaca, the bursa of Fabricius [18].

Similarly, among vertebrates, the crocodilian thymus is composed of an outer cortex and inner medulla, with connective tissue septa partitioning the thymus into discrete lobules [35]. Histologically, the outer cortex shows a higher concentration of T-lymphocytes, mediators of cell-mediated adaptive immunity, than the inner medulla [38]. In mammals, T-cells mature as they traverse through the cortex and into the medulla, although this is yet to be investigated in crocodilians [35].

The spleen, a peripheral lymphoid organ, aids in removal of blood-borne pathogens and senescent erythrocytes. Histological staining reveals distinctive splenic compartments, including the red and white pulp and the splenic capsule. The splenic capsule encapsulates the organ itself, while the red pulp comprised reticulated splenic cords and venous vessels [39]. In mammals, the white pulp comprised splenic follicles, the periarteriolar lymphoid sheath and the marginal zone, which serves as a type of boundary between the red and white pulp [40]. In the American alligator, *Alligator mississippiensis*, the white pulp consists of nodular lymphoid tissue surrounding terminal arterioles and sheathed capillaries [39]. Although germinal centres arise in mammals, they are not present in the crocodilian spleen [39]. In mammals, antigenic stimulation activates B-cells inducing germinal centre formation. These serve as sites for somatic hypermutation and affinity maturation of antibodies, with subsequent differentiation into B-lymphocyte effectors, memory and plasma cells [41–43]. Somatic mutation, affinity maturation and lymphocyte differentiation enhance the adaptive immune response, helping rid the body of the invader. Memory cells provide enhanced responsiveness in subsequent infections, while plasma cells produce copious amounts of antibodies. Thus, the lack of germinal centres, along with lymph nodes, in crocodilians has led some to suggest that the spleen may play a more important role in immune defence, although further research is obviously needed to elucidate and understand antigen presentation and subsequent B-cell activation in crocodilians [35].

Proteins Functioning in Defence

Innate immunity is the first line of defence against microbial invaders, consisting of barriers (i.e. skin, gut mucosa, etc.), innate leucocytes, as well as antimicrobial peptides (AMPs), enzymes and proteins. Of particular notoriety among antimicrobial enzymes are lysozymes, which in mammals are components of saliva, tear and nasal secretions exhibiting bactericidal activities against both Gram-negative and Gram-positive bacteria [21, 44]. These enzymes function to hydrolyse the outer components of microbial membranes, namely the glycosidic bond between *N*-acetylmuramic acids and *N*-acetylglucosamines of bacterial peptidoglycans [45]. This enzymatic-mediated hydrolysis allows for the destruction of cellular membrane, subsequent microbial osmotic lysis and eventual clearance of the pathogen. Similar to the documentation of these enzymes in birds, lysozyme-like proteins have also been putatively identified in the egg white and leucocytic extracts of the Siamese crocodile, *Crocodylus siamensis* [46–48].

Another antimicrobial enzyme so far identified in crocodilians includes phospholipase A_2 (PLA2), which is an antibacterial enzyme catalysing the hydrolysis of phosphoglycerides via the *sn*-2 acyl bond, and is expressed in both cytosolic and secreted forms [49]. The cationic character of PLA2 permits affinity for anionic bacterial membranes, consequently enabling membrane destruction via increased permeabilization [50]. Serum extracts have revealed PLA2 in a host of crocodilians including *C. porosus*, *A. mississippiensis*, *C. siamensis*, *Crocodylus niloticus*, *Mesticops cataphractus* and *Osteolameous tetraspis* [51–53]. For example, to determine the activity of PLA2 in *A. mississippiensis*, Merchant *et al.* [51] treated fluorescently labelled (at sn-2 acyl bond) *Escherichia coli* with alligator serum. Fluorescence signified phosphoglyceride cleavage and consequently denoted PLA2 activity, as this particular enzyme hydrolyses this bond. The use of a PLA2 inhibitor decreased fluorescence, further illustrating the effects of cleavage are mediated by PLA2. Activity, which was both temperature and concentration dependent, was observed within 5 min of incubation, with peak activity observed after 20 min and at 40°C [51].

The antimicrobial enzyme dipeptidyl peptidase IV (DPP4), also known as CD26, has also been demonstrated in crocodilians [54]. As a serine ectoprotease, this enzyme has been shown to be important in a number of different physiological processes in mammals, particularly immunity [55]. DPP4/CD26 is expressed in both membrane and secreted forms, as well as on the surface of T-cells suggesting that it may play an important role in T-cell activation [56, 57]. Other data suggest that DPP4/CD26 activates and regulates the complement system [58]. In addition, studies with CD26 knockout mice illustrate that DPP4/CD26 is an important mediator of cellular maturation and migration, production of cytokines and various aspects of humoral immunity [59]. To determine the activity of DPP4 within a crocodilian, Merchant *et al.* [54] incubated both whole blood and plasma from *A. mississippiensis* with the DPP4 fluorogenic substrate Ala-Pro-AFC. Subsequent proteolytic product formation, and thus DPP4 activity, was measured through the amount of fluorescence emitted. DPP4 activity was higher in whole blood when compared with plasma, possibly due to increased production, catalytic efficiency or presence of membrane forms on T-lymphocytes [54]. Both whole blood and plasma displayed temperature dependence, with peak production occurring in the whole blood and plasma at 30 and 40°C, respectively [54]. So, while DPP4 has been identified within the blood and serum of crocodilians, much remains to be determined about the activity and function of this enzyme in terms of crocodilian immunology.

AMPs

AMPs, comprised up to 45 amino acid residues, are an ancient component of innate immunity found across divergent taxa, with most functioning to increase microbial membrane permeability [60–62]. AMPs are synthesized as precursor proteins, requiring post-translational modification for activation [63]. Many are cationic and amphipathic, allowing for increased affinity with microbial membranes [61]. Consequently, increased membrane attraction results in augmented membrane dysregulation, inducing osmotic lysis. Structurally, AMPs are assembled into distinct groups based on their conformation, including linear peptides lacking cysteine residues, peptides with disulfide bridges enabling β-sheet conformations, and linear peptides with an extraordinary amount of tryptophan, proline or histidine residues [60, 61]. Although AMPs are well known for their microbial membrane permeabilization abilities, they may also serve as antagonists for pathogen-specific enzymes or trigger production of antimicrobial substances, further assisting in clearance of pathogens [62].

Defensins and cathlecidins are two groups of AMPs produced in leucocytes that aid in pathogenic defence. Defensins are cationic peptides characterized by multiple β-sheet folds and three disulfide linkages. Functionally, defensins act to increase microbial membrane permeability, whereas cathlecidins are a group of cationic, mostly linear AMPs inducing membrane lysis in microbes [64, 65]. In addition to membrane permeabilization, cathelicidins may also serve as immunmodulators and function in leucocytic chemotaxis, directing immune responders to sites of infection [65, 66]. As of yet, neither defensins nor cathlecidins have been described in crocodilians, though defensin-like proteins and cathlecidins have been observed in the European pond turtle, *Emys orbicularis*, and in the venom of elapid snakes, respectively [67, 68]. Evidence suggests that defensins and cathlecidins are phylogenetically conserved, with analogues identified in fish, reptiles, birds, mammals and even plants [68]. Thus, one may posit their existence in crocodilians, although further research is needed for verification.

Although cathelicidins and defensins are yet to be found within crocodilians, other AMPs have been identified. Merchant *et al.* [69] demonstrated that the leucocytic extracts of *A. mississippiensis* exhibit anti-bacterial, anti-viral and anti-fungal activity. Results from zones of inhibition assays revealed that extracts were effective at inhibiting the growth of six out of the eight *Candida* species tested, all Gram-negative species tested (n=7), with greatest activity observed against *Shigella flexneri* and *Citrobacter freundii*, a number of Gram-positive *Streptococcus* bacteria (2 out of 4), and viruses, herpes simplex virus-1 and human immunodeficiency virus-1. Antimicrobial activity exhibited heat stability, was not affected by EDTA concentration, and sensitive to protease treatment [69]. As proteins are sensitive to EDTA concentration and heat, this suggested that the observed antimicrobial processes were mediated via cationic AMPs.

Pata *et al.* [70] recently identified four previously uncharacterized AMPs, termed leucrocins, in leucocytic extracts from *C. siamensis*. Leucrocin I and II demonstrated strong antibacterial activity against *Staphylococcus epidermis*, while Leucrocin III and IV were most effective at inhibiting the growth of *Vibrio cholerae*. In addition, Leucrocin I and II were shown to be effective in the lysis of human erythrocytes and scanning electron microscopy revealed the antimicrobial properties were mediated via permeabilization of microbial membranes [70].

Crocosin, another crocodilian-specific antibacterial compound identified in the plasma of *C. siamensis*, has also demonstrated antibacterial activity against both Gram-positive and Gram-negative bacteria [71]. Characteristic of all AMPs, crocosin demonstrated thermostability; however, the compound was impervious to protease digestion, leading the authors to suggest that it may be a possible AMP derivative or composed of other substituents instead of amino acids [71]. Further study is needed to determine the basis of this antimicrobial activity.

The Complement System of Crocodilians

The complement system consists of a group of about 30 soluble plasma proteins that function in host defence. Initially named for their complementary role with antibodies in ridding the body of invaders, these phylogenetically conserved proteins regulate various aspects of adaptive immunity, aid in microbial opsonization, and promote destruction of invaders through the formation of the membrane attack complex (MAC) [72, 73]. Opsonization increases phagocytic efficacy, while MAC directly destroys microbes through the formation of pores, triggering subsequent osmotic lysis. Complement proteins are synthesized as inactive precursors (zymogens), with activation occurring through proteolytic cleavage. These cleaved protein products may collectively associate to form convertases, act as opsonins for phagocytic cells or function as anaphylatoxins, the latter of which function in promotion of inflammation [74]. Complement activation proceeds through three distinct pathways including the classical (CP), alternative (AP) and lectin pathway (LP) (Figure 1).

Although the mechanism of induction differs among these three pathways, they similarly functionally converge in the formation of complement protein C3 products, C3a and C3b [74]. Important as an inducer of inflammation, C3a acts as an anaphylatoxin, while C3b functions as a component of C5 convertase, the next enzyme in the cascade [74]. Complement protein C5 serves as substrate for C5 convertase, producing C5a and C5b. While C5a promotes inflammation, interaction of C5b with proteins C6, C7 and C8 on the surfaces of microbial membranes induces polymerization and subsequent indention of multiple C9 molecules into microbial membranes (the MAC), triggering osmotic lysis [72].

Activation of the CP consists of complement protein C1q binding the Fc receptor of antibody–antigen complex or C-reactive proteins [21]. C1q then interacts with other proteins of the C1 complex, C1s and C1r, which possess serine protease activity. Subsequently, these proteins sequentially cleave C4 into C4a and C4b and C2 into C2a

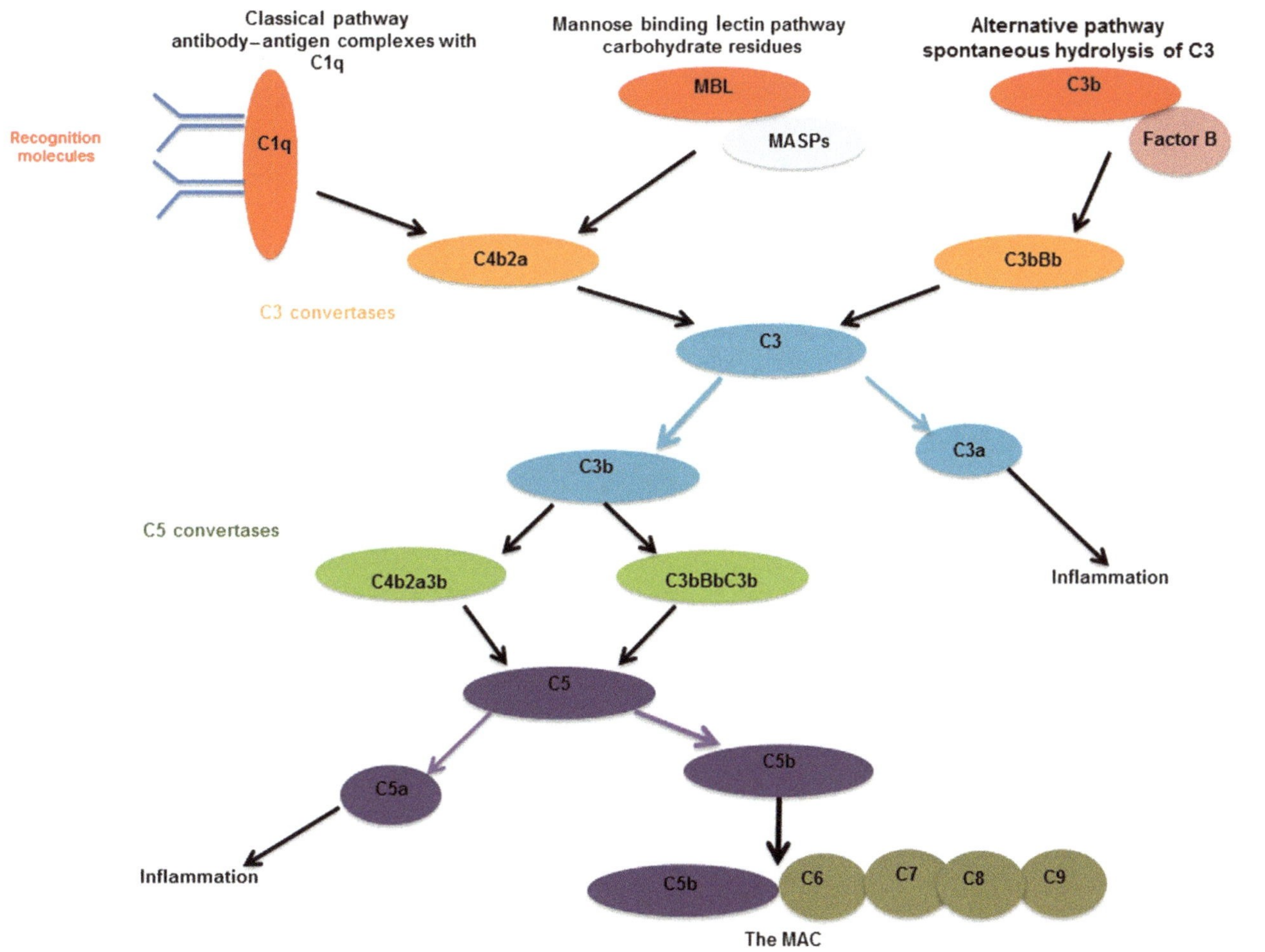

Figure 1 Pathways of complement activation. The complement system may be activated via three distinct pathways including the CP, LP and AP. CP activation proceeds through C1q binding of antigen–antibody complexes, binding of MBL to terminal mannose residues present on microbes activates the LP, whereas the AP is activated via spontaneous hydrolysis of C3. Only the AP has been demonstrated in crocodilians [3, 75–77]

and C2b [72]. Mannose residues present on microbes function as ligands, recognized by mannose-binding lectin (MBL), for the LP. Ligand-MBL binding activates MBL-associated serine proteases (MASPs), which sequentially cleave C4 and then C2, with the protein products summing to generate C3 convertase, further activating downstream mediators of the complement cascade [78]. Activation of the AP is achieved via the spontaneous cleavage of an internal thioester linkage in the C3 molecule; production and interaction of C3b with Factor B, a homologue of C2, and subsequent cleavage of Factor B forms C3 covertase C3bBb [21]. Amplified production of cleavage product C3b allows for association with C3bBb, forming C5 convertase C3bBbC3b [73, 79].

The presence of the AP and complement protein C3 has previously been demonstrated in the serum of the American alligator, *A. mississippiensis* [3]. Lysis of exogenous erythrocytes was observed in the presence of CP inactivators (ammonium hydroxide and methylamine), with activity being heat-labile (AMPs are heat-stable), as the prevention of lysis occurred when serum was heated at 56°C for 30 min, thus suggesting that lysis was induced by complement proteins activated via the AP [3]. Similar findings were observed in the American crocodile (*Crocodylus acutus*), broad snouted caiman (*Caiman latirostris*), freshwater crocodile (*Crocodylus johnstoni*) and *C. porosus* [75–77]. Expectedly, temperature influenced haemolytic activity in each of the species studied. As shown in Figure 2, both higher and lower temperatures reduced efficacy of haemolysis, with highest activity observed between 20 and 35°C. Although these studies demonstrate that crocodilians possess the AP for complement activation, to our knowledge, the other avenues of activation have not yet been elucidated, providing a gap in our understanding of this system in crocodilians.

As in mammals, crocodilian complement proteins may act as opsonins, enhancing phagocytic efficacy. Aree *et al.* [80] demonstrated the opsonizing effects of crocodilian serum via incubation of murine macrophages and two types of bacteria, *Staphylococcus aureus* and *E. coli*, in the presence of *C. siamensis* serum. Incubation with serum induced 2- and 4-fold increases in phagocytic uptake of

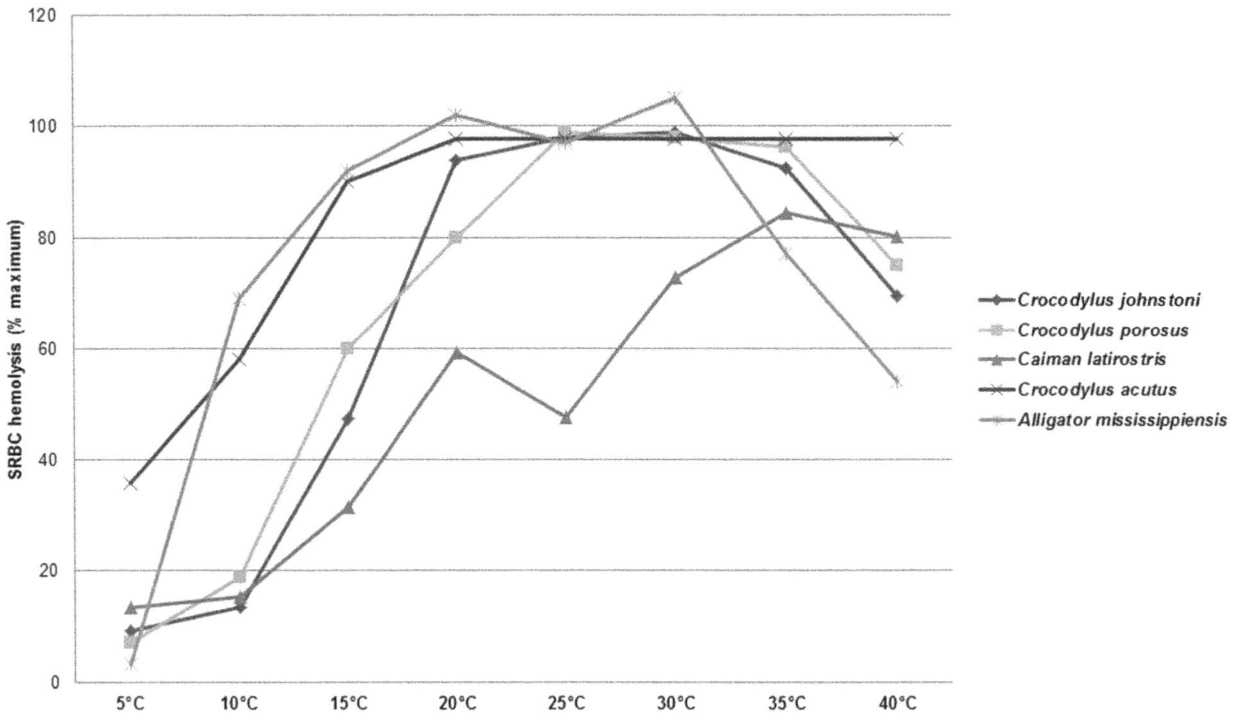

Figure 2 The effects of temperature on complement mediated sheep red blood cell haemolysis in five species of crocodilian [3, 75, 78, 79]

S. aureus and *E. coli*, respectively, while pre-treatment heating of serum reduced this phagocytic ability [80].

Innate Leucocytes

Leucocytes or white blood cells (WBCs) are important mediators of the immune system, carrying out the tasks of sensing and ridding the body of pathogenic microbes. WBCs arise from pluripotent haematopoetic stem cells located within the bone marrow, which further differentiate into common lymphoid and common myeloid progenitors. Myeloid cells include erythrocytes, thrombocytes, granulocytes and monocytes, while natural killer cells and the lymphocytes, B- and T-cells (adaptive cells) are members of the lymphoid lineage. Lymphocytes possess specific receptors owing to the events of somatic recombination, while innate leucocytes express germline encoded, non-specific, phylogentically conserved receptors termed pattern recognition receptors (PRRs), which recognize conserved molecules or molecular patterns on pathogenic microbes, subsequently inducing production of inflammatory mediators, antiviral molecules (interferons), chemotactic factors and phagocytosis, depending on cell type [81–83].

The innate leucocytes of crocodilians include heterophils, eosinophils, basophils and monocytes [19, 84–86]. Although the phagocytic azurophil has been identified in squamates (lizards and snakes), its presence remains uncertain within the crocodilian vasculature [87]. However, some haematological studies have observed staining representative of azurophils in the Nile crocodile, *C. niloticus*, and Morelet's crocodile, *Crocodylus moreletti* [88, 89]. Given the presence of numerous cytoplasmic granules (which influence staining characteristics), heterophils, eosinophils and basophils are collectively termed granulocytes. In mammals, granulocytes are also referred to as polymorphonuclear leucocytes because of the multilobed appearance of the nuclei, whereas the nuclei of reptilian granulocytes are more heterogeneous in appearance and may appear round (non-segmented), multilobed or bi-lobed [87].

Heterophils are the functional equivalent of mammalian neutrophils, and as such, function as phagocytes. However, unlike mammalian neutrophils, crocodilian heterophils stain eosinophilic, possessing fusiform-shaped eosinophilic cytoplasmic granules [37, 46]. Heterophils comprise more than 50% of the circulating leucocytes in *A. mississippiensis*, although quantitative differences may arise because of other factors, such as infection, seasonality or age [84, 86]. The nuclei of these cells are oval in appearance, lacking the multi-lobed appearance characteristic of mammalian neutrophils [84, 86]. Similar to mammalian neutrophils, heterophils are the first cells to arrive at the site of infection, followed by mononuclear phagocytes [37]. For example, Mateo *et al.* [90] found that alligator heterophils infiltrate damaged tissue within 4 h following administration of turpentine.

Eosinophils and basophils make up about 10 and 12% of circulating leucocytes in crocodilians, respectively [86].

Although both cell types are imperative in regulating the inflammatory milieu, eosinophils may also assist in parasitic infections, with eosinophilia (an increase in eosinophil numbers) commonly observed during helminth-associated infections in mammals [91]. Similarly, in *A. mississippiensis* eosinophilia has also been observed in response to leech infections [92]. Basophils are innate cells commonly implicated in allergic reactions, releasing pro-inflammatory mediators, such as histamine, from cytoplasmic granules following antibody-Fc receptor binding [93, 94].

Important innate phagocytic cells include the mononuclear phagocytes, comprising monocytes, macrophages and dendritic cells (DCs) [44]. Aptly termed antigen-presenting cells, these cell types function in linking innate and adaptive immunity through antigen presentation and subsequent activation of adaptive responses. In *A. mississippiensis*, monocytes comprise about 1% of peripheral leucocytes (1.5 $\pm$ SD 0.3), whereas in *C. porosus* yearlings, the range (4.5–21.6) was much greater [86, 95]. In mammals, these cells give rise to tissue-associated macrophages and some types of DCs [96, 97]. Macrophages are important in a number of physiological processes including ingestion of senescent erythrocytes, tissue repair, disposing of apoptotic cells, and notably, 'eating' invading microbes [98]. In mammals, many subsets of DCs have been identified including plasmacystoid DCs, producers of copious amounts of interferon aiding in defence against viruses, and Langerhans cells, epidermal cells important in the transport of ingested microbes to lymph nodes for antigen presentation [99, 100]. However, in reptiles, notably crocodilians, characterization and understanding of DCs remain scanty, although the presence of such cells has been noted in snakes [101]. Understanding of these cell types could provide further insights into how the adaptive immune response is regulated.

Phagocytosis of pathogenic microbes requires participation of PRR–ligand interaction, stimulating ingestion and formation of a membrane-enclosed vesicle containing ingested microbe, the phagosome [102]. Following ingestion, the phagocytic vesicle is acidified, subsequently fuses with lysosomes, establishing the phagolysosome, composed of proteases and a host of antimicrobial substances [21]. Further, ingestion stimulates the production of reactive oxygen species (ROS), the so-called oxidative burst, through the activation of NADPH oxidase, an enzyme complex assembled on the phagosome membrane [103, 104]. NADPH oxidase initially acts to reduce molecular oxygen (O_2) producing superoxide anion (O_2^-), the basis for all phagosome-derived ROS [105]. Subsequent dismutation of O_2^- produces hydrogen peroxide (H_2O_2), which may further react with iron, via the Fenton reaction, to produce hydroxyl radical ($OH^·$), a potent oxidizing agent capable of inducing alterations in biomolecules [103–105]. Although the exact mechanisms underlying the bactericidal activities of the oxidative burst are unknown, production of ROS is thought to be of paramount importance [103].

Leucocytic production of ROS has been documented within crocodilian phagocytes [106]. To demonstrate ROS production within crocodilian leucocytes, Merchant *et al.* [106] incubated whole blood from *A. mississippiensis* with water-soluble tetrazolium salts to determine spectrophotometrically the amount of O_2^- produced through the formation of reduced formazan compounds. Leucocytic production of O_2^- demonstrated temperature, concentration and time dependence, respectively. Peak production of O_2^- was observed at 30 and 35°C. Production increased up until 48 h post initial incubation, with no observed statistical differences in product observed after 48 h. Staining of *A. mississippiensis* leucocytes indicated the principal producers of O_2^- to be heterophils and monocytes [106]. Further investigation may provide insight into how production is regulated and to determine the effect of ROS generation on overall crocodilian immunocompetence.

Fever in Crocodilians

Fever is characterized as a hypothalamic-mediated elevation of body temperature above normal levels, with induction triggered through the action of pyrogenic molecules [107–109]. Pyrogenic molecules activate cyclooxygenases, consequently increasing prostaglandin E2 synthesis and raising the hypothalamic thermoregulatory set point [109, 110]. Fever is phylogenetically conserved, being observed across many distantly related species [111, 112]. In particular, studies with squamates have demonstrated that bacterial-exposed reptiles are able to mount a febrile response when provided with thermal gradients [113, 114].

Evidence suggests that crocodilians may also demonstrate the ability to induce a febrile response, though as in other reptiles, the fevers of crocodilians are behavioural in nature, relying on ambient temperatures for induction of said response [115]. This is in contrast to mammalian fevers, which are derived of increased metabolic heat production [112]. To determine the febrile response of crocodilians, the body temperatures of *A. mississippiensis* injected with lipopolysaccharide (LPS; membrane component of Gram negative bacteria) were monitored to determine internal temperature fluctuations. LPS-injected alligators kept at a constant temperature exhibited no change in internal body temperature, while those exposed to thermal gradients demonstrated increased internal body temperatures (4.2 and 3.5°C higher at 1 and 2 days post injection, respectively) [115]. Thus, immune-challenged alligators, when presented with thermal gradients, were able to select sites for optimal heat gain and induce up-regulation of body temperature, that is, producing a febrile response. Interestingly, intra-peritoneal injections with the Gram-positive bacterium, *S. aureus*, did not induce fever, whereas injections with Gram-negative bacteria did in fact provoke a febrile response [115].

Conclusions

The immune system of crocodilians is currently understudied. Anecdotal observations of conspecific combat resulting in limb amputation or mighty wounds, with combatants rarely seeming to succumb to infection or disease, have stimulated postulation of potent immune capabilities and encouraged exploration of crocodilian immune functions. Although such evidence suggests a heightened immune response, this is unsubstantiated, as our understanding of such observations is not known, requiring further study for elucidation. The most recent exploration of crocodilian immunity has been restricted to innate immune aspects, an ancient arm of immunity conserved in all metazoans, functioning as a first line of defence against invading microbes and subsequently activating the adaptive immune system via antigen presentation and release of chemical mediators [1–3, 10].

In conclusion, we have attempted to provide insight into the innate mechanisms of crocodilian immunity. As we have illustrated, understanding of the innate immune system in crocodilians has increased over the past 30 years, with important cell types, proteins and peptides identified. However, there are still many avenues awaiting exploration. Some have suggested that crocodilians may serve as important indicators of environmental health, as they are top trophic carnivores, with optimal physiology and development perturbed following exposure to exogenous toxicants [34]. Others have suggested that the potent blood components of crocodilians may prove useful in the realm of human medicine in combating sickness and disease [70]. In addition, further exploration may enable further understanding of the evolution of immune mechanisms. Thus, increased insight into crocodilian immunity is necessary not only to further knowledge of crocodilian biology, but it may also improve our understanding in toxicology, medicine, ecology and evolution.

References

1. Merchant ME, Roche C, Elsey RM, Prudhomme J. Antibacterial properties of serum from the American alligator (*Alligator mississippiensis*). Comparative Biochemistry and Physiology Part B 2003;136:505–13.
2. Merchant M, Thibodeaux D, Loubser K, Elsey RM. Amoebacidal effects of serum from the American alligator (*Alligator mississippiensis*). Journal of Parasitology 2004;90:1480–3.
3. Merchant ME, Roche CM, Thibodeaux D, Elsey RM. Identification of alternative pathway serum complement activity in the blood of the American alligator (*Alligator mississippiensis*). Comparative Biochemistry and Physiology Part B 2005;141:281–8.
4. Brochu CA. Phylogenetic approaches toward Crocodylian history. Annual Review of Earth and Planetary Sciences 2003;31:357–97.
5. Coulson RA, Herbert JD. Relationship between metabolic rate and various physiological and biochemical parameters. A comparison of alligator, man and shrew. Comparative Biochemistry and Physiology Part A: Physiology 1981;69:1–13.
6. Seebacher F, Franklin CE. Physiological mechanisms of thermoregulation in reptiles: a review. Journal of Comparative Physiology B: Biochemical, Systemic, and Environmental Physiology 2005;175:533–41.
7. Lance VA. Alligator physiology and life history: the importance of temperature. Experimental Gerontology 2003;38:801–5.
8. Lang JW, Andrews HV. Temperature-dependent sex determination in crocodilians. Journal of Experimental Zoology 1994;270:28–44.
9. Smith NE. Thermoregulation of the American alligator, *Alligator mississippiensis*. Physiological Zoology 1975;48:177–94.
10. Dzik JM. The ancestry and cumulative evolution of immune reactions. Acta Biochimica Polonica 2010;57:443–66.
11. Lerch EG, Huggins SE, Bartel AH. Comparative immunology. Active immunization of young alligators with hemocyanin. Proceedings of the Society for Experimental Biology and Medicine 1967;124:448–51.
12. Cuchens MA, Clem LW. Phylogeny of lymphocyte heterogeneity IV. Evidence for T-like and B-like cells in reptiles. Developmental and Comparative Biology 1979;3:465–75.
13. Solas MT, Zapata A. Gut-associated lymphoid tissue (GALT) in reptiles: intraepithelial cells. Developmental and Comparative Immunology 1980;4:87–97.
14. Zimmerman LM, Vogel LA, Bowden RM. Understanding the vertebrate immune system: insights from the reptilian perspective. Journal of Experimental Biology 2010;213:661–71.
15. Turton JA, Ladds PW, Manolis SC, Webb GJ. Relationship of blood corticosterone, immunoglobulin and haematological values in young crocodiles (*Crocodylus porosus*) to water temperature, clutch of origin and body weight. Australian Veterinary Journal 1997;75:114–9.
16. Scollay RG, Butcher EC, Weissman IL. Thymus cell migration: quantitative aspects of cellular traffic from the thymus to the periphery in mice. European Journal of Immunology 1980;10:210–18.
17. Stoker C, Rey F, Rodriguez H, Ramos JG, Sirosky P, Larriera A, *et al.* Sex reversal effects on *Caiman latirostris* exposed to environmentally relevant doses of the xenoestrogen bisphenol A. General and Comparative Endocrinology 2003;13:287–96.
18. Warner NL, Szenberg A. The immunological function of the bursa of Fabricius in the chicken. Annual Review of Microbiology 1964;18:253–66.
19. Zayas MA, Rodriguez HA, Galoppo GH, Stoker C, Durando M, Luque EH, *et al.* Hematology and blood biochemistry of young healthy broad-snouted caimans (*Caiman latirostris*). Journal of Herpetology 2011;45:516–24.
20. El Deeb S, Zada S, El Ridi R. Ontogeny of hemopoietic and lymphopoietic tissues in the lizard *Chalcides ocellatus* (Reptilia, Sauna, Scincidae). Journal of Morphology 1985;185:241–53.

21. Murphy K, Travers P, Walport M. Janeway's Immunobiology. 7th ed. Garland Science, New York; 2007.

22. McHeyzer-Williams MG. B cells of effectors. Current Opinion in Immunology 2003;15:354–61.

23. El Ridi R, Zada S, Afifi A, El Deeb S, El Rouby S, Farag M, *et al.* Cyclic changes in the differentiation of lymphoid cells in reptiles. Cell Differentiation 1988;24:1–8.

24. Zapata AG, Varas A, Torroba M. Seasonal variations in the immune system of lower vertebrates. Immunology Today 1992;13:142–7.

25. Schwanz L, Warner DA, McGaugh S, Di Terlizzi R, Bronikowski A. State-dependent physiological maintenance in a long-lived ectotherm, the painted turtle (*Chrysemys picta*). Journal of Experimental Biology 2011;214:88–97.

26. Zimmerman LM, Paitz RT, Vogel LA, Bowden RM. Variation in the seasonal patterns of innate and adaptive immunity in the red-eared slider (*Trachemys scripta*). Journal of Experimental Biology 2010;213:1477–83.

27. Davis MC, Zautra AJ, Younger J, Motivala SJ, Attrep J, Irwin MR. Chronic stress and regulation of cellular markers of inflammation in rheumatoid arthritis: implications for fatigue. Brain, Behavior, and Immunity 2008;22:24–32.

28. Dai R, Phillips RA, Karpuzoglu E, Khan D, Ahmed SA. Estrogen regulates transcription factors STAT-1 and NF-kappaB to promote inducible nitric oxide synthase and inflammatory responses. Journal of Immunology 2009;183:6998–7005.

29. Ruiz M, French SS, Deman GE, Martins EP. Food supplementation and testosterone interact to influence reproductive behavior and immune function in *Sceloporus graciosus*. Hormones and Behavior 2010;57:134–9.

30. Morici LA, Elsey RM, Lance VA. Effects of long-term corticosterone implants on growth and immune function in juvenile alligators, *Alligator mississippiensis*. Journal of Experimental Zoology 1997;279:156–162.

31. Guillette LJ, Gross TS, Masson GR, Matter JM, Percival HF, Woodward AR. Developmental abnormalities of the gonad and abnormal sex hormone concentrations in juvenile alligators from contaminated and control lakes in Florida. Environmental Health Perspectives 1994;102:680–8.

32. Guillette LJ, Pickford DB, Crain DA, Rooney AA, Percival HF. Reduction in penis size and plasma testosterone concentrations in juvenile alligators living in a contaminated environment. General and Comparative Endocrinology 1996;101:32–42.

33. Milnes MR, Guillette LJ. Alligator tales: new lessons about environmental contaminants from a sentinel species. BioScience 2008;58:1027–36.

34. Rey F, Gonzalez M, Zayas MA, Stoker C, Durando M, Luque EH, *et al.* Prenatal exposure to pesticides disrupts testicular histoarchitecture and alters testosterone levels in male *Caiman latirostris*. General and Comparative Endocrinology 2009;162:286–92.

35. Rooney AA, Bermudez DS, Guillette LJ. Altered histology of the thymus and spleen in contaminant-exposed juvenile American alligators. Journal of Morphology 2003;256: 349–59.

36. Pabst R. Plasticity and heterogeneity of lymphoid organs. What are the criteria to call a lymphoid organ primary, secondary, or tertiary. Immunology Letters 2007;112:1–8.

37. Jacobson ER, editor. Infectious Diseases and Pathology of Reptiles. CRC Press, Boca Raton, FL; 2007.

38. Saad AHM, Zapata AG. Reptilian thymus gland: an ultrastructural overview. Thymus 1992;20:135–52.

39. Tanaka Y, Elsey RM. Light microscopic study of the alligator (*Alligator mississippiensis*) spleen with special reference to vascular architecture. Journal of Morphology 1997;233:43–52.

40. Cesta MF. Normal structure, function, and histology of the spleen. Toxicologic Pathology 2006;34:455–65.

41. Crotty S, Ahmed R. Immunological memory in humans. Seminars in Immunology 2004;16:197–203.

42. Klein U, Dalla-Favera R. Germinal centres: role in B-cell physiology and malignancy. Nature Reviews Immunology 2008;8:22–33.

43. Schlomick MJ, Weisel F. Germinal center selection and the development of memory B and plasma cells. Immunological Reviews 2012;247:52–63.

44. Beutler B. Innate immunity: an overview. Molecular Immunology 2004;40:845–59.

45. Callewaert L, Michiels CW. Lysozymes in the animal kingdom. Journal of Biosciences 2010;35:127–60.

46. Maxwell MH, Robertson GW. The avian heterophil leukocyte: a review. World's Poultry Science Journal 1998;54:155–78.

47. Pata S, Daduang S, Svasti J, Thammasirirak S. Isolation of lysozyme like protein from crocodile leukocyte extract (*Crocodylus siamensis*). KMITL Science and Technology Journal 2007;7:70–85.

48. Prajanban BO, Shawsuan L, Daduang S, Kommanee J, Roytrakul S, Dhiravisit A *et al.* Identification of five reptile egg whites protein using MALDI-TOF mass spectrometry and LC/MS-MS analysis. Journal of Proteomics 2012;75:1940–59.

49. Schaloske RH, Dennis EA. The phospholipase A2 superfamily and its group numbering system. Biochimica et Biophysica Acta 2006;1761:1246–59.

50. Buckland AG, Wilton DC. The antibacterial properties of secreted phospholipases A(2). Biochimica et Biophysica Acta 2000;1488:71–82.

51. Merchant M, Heard R, Monroe C. Characterization of Phospholipase A2 activity in serum of the American alligator (*Alligator mississippiensis*). Journal of Experimental Zoology 2009;311A:662–6.

52. Nevalainen TJ, Kanchanapangka S, Youngprapakorn P, Webb GJW, Manolis SC, Scott KF. Phospholipase A2 activity of crocodile serum. Amphibia-Reptilia 2009;30:119–25.

53. Merchant M, Juneau K, Gemillion J, Falconi R, Doucet A, Shirley MH. Characterization of serum Phospholipase A2 activity in three diverse species of West African crocodiles. Biochemistry Research International 2011;2011:1–7.

54. Merchant M, Monroe C, Falconi R. Dipeptidyl peptidase IV activity in the blood of the American alligator (*Alligator mississippiensis*). Comparative Biochemistry and Physiology, Part B, Biochemistry and Molecular Biology 2009;154:341–45.

55. Kubota T, Flentke GR, Bachovchin WW, Stollar BD. Involvement of dipeptidyl peptidase IV in an *in vivo* immune response. Clinical and Experimental Immunology 1992;89:192–97.

56. Fleischer B. CD26: a surface protease involved in T-cell activation. Immunology Today 1994;15:180–4.

57. Morimoto C, Schlossman SF. The structure and function of CD26 in the T-cell immune response. Immunological Reviews 1998;161:55–70.

58. Shinosaki T, Kobayashi T, Kimura K, Kurihara H. Involvement of dipeptidyl peptidase IV in immune complex-mediated glomerulonephritis. Laboratory Investigation 2002;82:505–13.

59. Yan S, Marguet D, Dobers J, Reutter W, Fan H. Deficiency of CD26 results in a change of cytokine and immunoglobulin secretion after stimulation by pokeweed mitogen. European Journal of Immunology 2003;33:1519–27.

60. Boman HG. Antibacterial peptides: basic facts and emerging concepts. Journal of Internal Medicine 2003;254:197–215.

61. Hancock RE, Lehrer R. Cationic peptides: a new source of antibiotics. Trends in Biotechnology 1998;16:82–8.

62. Andreu D, Rivas L. Animal antimicrobial peptides: an overview. Biopolymers 1998;47:415–33.

63. Zasloff M. Antimicrobial peptides of multicellular organisms. Nature 2002;415:389–95.

64. Ganz T. Defensins: antimicrobial peptides of innate immunity. Nature Reviews Immunology 2003;3:710–20.

65. Zanetti M. Cathelicidins, multifunctional peptides of the innate immunity. Journal of Leukocyte Biology 2004;75:39–48.

66. da Silva FP, Gallo RL, Nizet V. Differing effects of exogenous or endogenous cathelicidin on macrophage toll-like receptor signaling. Immunology and Cell Biology 2009;87:496–500.

67. Stegemann C, Kolobov A, Leonova YF, Knappe D, Shamova O, Ovchinnikova TV, *et al.* Isolation, purification and *de novo* sequencing of TBD-1, the first beta-defensin from leukocytes of reptiles. Proteomics 2009;9:1364–73.

68. Wang Y, Hong J, Liu X, Yang H, Liu R, Wu J, *et al.* Snake cathelicidin from *Bungarus fasciatus* is a potent peptide antibiotic. PLoS One 2008;3:1–9.

69. Merchant ME, Leger N, Jenkins E, Mills K, Pallansch MB, Paulman RL, *et al.* Broad spectrum antimicrobial activity of leukocyte extracts from the American alligator (*Alligator mississippiensis*). Veterinary Immunology and Immunopathology 2006;110:221–8.

70. Pata S, Yaraksa N, Daduang S, Temsiripong Y, Svasti J, Araki T, *et al.* Characterization of the novel antibacterial peptide Leucrocin from crocodile (*Crocodylus siamensis*) white blood cell extracts. Developmental and Comparative Immunology 2011;35:545–53.

71. Preecharram S, Jearranaiprepame P, Daduang S, Temsiripong Y, Somdee T, Fukamizo T, *et al.* Isolation and characterisation of crocosin, an antibacterial compound from crocodile (*Crocodylus siamensis*) plasma. Animal Science Journal 2010;81:393–401.

72. Bohana-Kastan O, Ziporen L, Donin N, Kraus S, Fishelson Z. Cell signals transduced by complement. Molecular Immunology 2004;41:583–97.

73. Holland MCH, Lambris JD. The complement system of teleosts. Fish and Shellfish Immunology 2002;12:399–420.

74. Trouw LA, Daha MR. Role of complement in innate immunity and host defense. Immunology Letters 2011;138:35–7.

75. Merchant M, Britton A. Characterization of serum complement activity of saltwater (*Crocodylus porosus*) and freshwater (*Crocodylus johnstoni*) crocodiles. Comparative Biochemistry and Physiology, Part A 2006;143:488–493.

76. Merchant M, McFatter J, Mead S, McAdon C, Wasilewski J. Identification and characterization of serum complement activity in the American crocodile (*Crocodylus acutus*). Veterinary Immunology and Immunopathology 2010;133:165–9.

77. Siroski PA, Merchant M, Marco MVP, Pina CI, Ortega HH. Characterization of the serum complement activity of the broad-snouted caiman *Caiman latirostris* (Crocodilia: Alligatoridae). Zoological Studies 2010;49:64–70.

78. Degn SE, Theil S, Jensenius JC. New perspectives on mannan-binding lectin-mediated complement activation. Immunobiology 2007;212:301–11.

79. Harboe M, Mollnes TE. The alternative complement pathway revisited. Journal of Cellular and Molecular Medicine 2008;12:1074–84.

80. Aree K, Siruntawineti J, Chaeychomsri W. *Crocodylus siamensis* serum and macrophage phagocytic activity. Journal of the Medical Association of Thailand 2011;94(Suppl. 7):S131–8.

81. Hoffmann JA, Kafatos FC, Janeway CA, Ezekowitz RAB. Phylogenetic perspectives in innate immunity. Science 1999;284:1313–18.

82. Kimbrell DA, Beutler B. The evolution and genetics of innate immunity. Nature Reviews Genetics 2001;2:256–67.

83. Kawai T, Akira S. TLR signaling. Cell Death and Differentiation 2006;13:816–25.

84. Glassman AB, Bennett CE, Hazen TC. Peripheral blood components in *Alligator mississippiensis*. Transactions of the American Microscopical Society 1981;100:210–5.

85. Canfield PJ. Characterization of the blood cells of Australian crocodiles (*Crocodylus porosus* [Schneider] and *C. johnstoni* [Krefft]). Anatomia, Histologia, Embryolgia 1985;14:269–88.

86. Mateo MR, Roberts ED, Enright FM. Morphologic, cytochemical, and functional studies of peripheral blood cells of young healthy American alligators (*Alligator mississippiensis*). American Journal of Veterinary Research 1984;45:1046–53.

87. Montali RJ. Comparative pathology of inflammation in the higher vertebrates (reptiles, birds and mammals). Journal of Comparative Pathology 1988;99:1–26.

88. Lovely CJ, Pittman JM, Leslie AJ. Normal haematology and blood biochemistry of wild Nile crocodiles (*Crocodylus niloticus*) in the Okavango Delta, Botswana. Journal of the South African Veterinary Association 2007;78:137–44.

89. Padilla SE, Weber M, Jacobson ER. Hematologic and plasma biochemical reference intervals for Morelet's crocodiles (*Crocodylus moreletii*) in the northern wetlands of Campeche, Mexico. Journal of Wildlife Diseases 2011;47:511–22.

90. Mateo MR, Roberts ED, Enright FM. Inflammation induced by subcutaneous turpentine inoculation of young American alligators (*Alligator mississippiensis*). American Journal of Veterinary Research 1984;45:1870–5.

91. Behm CA, Ovington KS. The role of eosinophils in parasitic helminth infections: insights from genetically modified mice. Parasitology Today 2000;16:202–9.

92. Glassman AB, Holbrook TW, Bennett CE. Correlation of leech infestation and eosinophilia in alligators. Journal of Parasitology 1979;65:323–24.

93. Gibbs BF. Human basophils as effectors and immunomodulators of allergic inflammation and innate immunity. Clinical and Experimental Medicine 2005;5:43–9.

94. Sullivan BM, Locksley RM. Basophils: a nonredundant contributor to host immunity. Immunity 2009;30:12–20.

95. Millan JM, Janmaat A, Richardson KC, Chambers LK, Fomiatti KR. Reference ranges for biochemical and haematological values in farmed saltwater crocodile (*Crocodylus porosus*) yearlings. Australian Veterinary Journal 1997;75:814–817.

96. Gordon S, Taylor PR. Monocyte and macrophage heterogeneity. Nature Reviews Immunology 2005;5:953–64.

97. Geissmann F, Manz MG, Jung S, Sieweke MH, Merad M, Ley K. Development of monocytes, macrophages, and dendritic cells. Science 2010;327:656–61.

98. Stuart LM, Ezekowitz RA. Phagocytosis: elegant complexity. Immunity 2005;22:539–50.

99. Naik SH. Demystifying the development of dendritic cell subtypes, a little. Immunology and Cell Biology 2008;86:439–52.

100. Romani N, Holzmann S, Tripp CH, Koch F, Stoitzner P. Langerhans cells – dendritic cells of the epidermis. AMPIS: Acta Pathologica, Microbiologica et Immunologica Scandinavica 2003;111:725–40.

101. Kroese FG, Leceta J, Döpp EA, Herraez MP, Nieuwenhuis P, Zapata A. Dendritic immune complex trapping cells in the spleen of the snake, *Python reticulatus*. Developmental and Comparative Immunology 1985;9:641–52.

102. Lee MS, Kim YJ. Signaling pathways downstream of pattern-recognition receptors and their cross talk. Annual Review of Biochemistry 2007;76:447–80.

103. Slauch JM. How does the oxidative burst of macrophages kill bacteria? Still an open question. Molecular Microbiology 2011;80:580–3.

104. Babior BM. Oxidants from phagocytes: agents of defense and destruction. Blood 1984;64:959–66.

105. El-Benna J, Dang PM, Gougerot-Pocidalo MA, Elbim C. Phagocyte NADPH oxidase: a multicomponent enzyme essential for host defenses. Archivum Immunologiae et Therapiae Experimentalis 2005;53:199–206.

106. Merchant M, Williams S, Hardy R. Production of superoxide ions by leukocytes of the American alligator (*Alligator mississippiensis*). Comparative Biochemistry and Physiology Part B: Biochemistry and Molecular Biology 2009;152:67–71.

107. Atkins E. Fever: its history, cause, and function Yale. Journal of Biology and Medicine 1982;55:283–9.

108. Dascombe MJ. The pharmacology of fever. Progress in Neurobiology 1985;25:327–73.

109. Moltz H. Fever: causes and consequences. Neuroscience and Biobehavioral Reviews 1993;17:237–69.

110. Dinarello CA. Infection, fever, and exogenous and endogenous pyrogens: some concepts have changed. Journal of Endotoxin Research 2004;10:201–22.

111. Kluger MJ. The evolution and adaptive value of fever. American Scientist 1978;66:38–43.

112. Hasday JD, Fairchild KD, Shanholtz C. The role of fever in the infected host. Microbes and Infection 2000;2:1891–904.

113. Bernheim HA, Kluger MJ. Fever and antipyresis in the lizard *Dipsosaurus dorsalis*. American Journal of Physiology 1976;231:198–203.

114. Burns G, Romas A, Muchlinski A. Fever response in North American snakes. Journal of Herpetology 1996;30:133–9.

115. Merchant M, Williams S, Trosclair PL 3rd, Elsey RM, Mills K. Febrile response to infection in the American alligator (*Alligator mississippiensis*). Comparative Biochemistry and Physiology Part A: Molecular and Integrative Physiology 2007;148:921–5.

Novel techniques for preserving genetic diversity in poultry germplasm

Frederick G. Silversides* **and Jianan Liu**

Address: Agassiz Research Centre, P.O. Box 1000, Agassiz, British Columbia, V0M 1A0, Canada.

***Correspondence**: Email: Fred.Silversides@hotmail.com

Received: 15 August 2012
Accepted: 12 October 2012

Abstract

There have been massive losses of avian genetic material over the past several decades, partly because of a belief that cryoconservation techniques used for mammals do not work for birds. For several avian species, cryopreservation of semen may not be sufficiently efficient to use in commercial practice but the fertility it produces is fully adequate for conservation purposes. Technology around embryonic primordial germ cells, which will produce the gametes, has been used to produce transgenic birds. These cells can be isolated, cryopreserved and recovered in chimaeras but the efficiency of recovery is far too low for conservation purposes because several dozen individuals must be produced to recover the variation from the original population. Cryopreservation of avian gonads with subsequent recovery of the germline through transplantation has been demonstrated recently and should provide the ability to both cryoconserve and recover avian populations.

Keywords: Cryoconservation, Avian, Genetic conservation, Gonadal transplantation, Primordial germ cells

Introduction

Poultry species, in particular chickens and turkeys, have played an important role in industry and research. The rapid development of the chicken and turkey industries has been accomplished in part because of the advantages that these species provide for genetic improvement, including short generation times and large families that can be kept in homogenous environments, and the advances in genetics can be distributed anywhere in the world that has access to an airport. In the past, poultry breeding was widely distributed, and in the 1940s Canada alone had over 300 active commercial poultry breeders [1]. Currently, producers of industrial poultry worldwide are principally served by two primary breeders; EW Group based in Germany produces layers, broilers and turkeys, and Hendrix Poultry based in the Netherlands produces layers and turkeys and has a marketing agreement with Cobb-Vantress, a broiler breeder. Hubbard, owned by Groupe Grimaud, comes a distant third in industrial broiler breeding. Industrial breeders produce their products with a small number of highly selected pedigreed lines. The breeding structure is based on a pedigree generation, which is multiplied to produce a grandparent generation. The grandparents are mated to produce a parent generation, which are mated in turn to produce the commercial generation. Combined with the reproductive efficiency of chickens this structure allows a population containing as few as 100 males in the pedigree generation to contribute to as many as 15 billion commercial offspring. Chickens have been used extensively for research because the species contains a vast array of genetic variation, the developing embryo is accessible to investigators and the management and reproduction is well understood. Unfortunately, as with the lines used for industry, those used for research have also been decimated [2] with hundreds of lines being lost in North America alone.

The decline in avian genetic variation is outlined in The State of the World [3]. A significant complicating factor in the decline in avian genetic resources has been slow progress in the development of cryoconservation and reproductive technologies for birds. Semen of dairy cattle has been frozen for more than 50 years and conserving genetic variation for the most popular breeds is a problem of sampling rather than availability. Industrial poultry breeders and researchers who wish to maintain variation have incurred the cost of keeping live populations, resulting in populations being discarded when companies merge, researchers retire or budgets are cut.

Mazur *et al.* [4] defined 'germplasm' as cells that lead to the development of offspring, singly or in combination, although in the context of their review, these were mainly sperm and preimplantation embryos. Here we continue to use the definition of Mazur *et al.* [4], but take a broader view because genetic variation can be accessed at various points in the germline cycle. Germplasm can be kept as living birds, but it can also be isolated from the host in which it resides and resume the germline cycle in another host, making it possible to use cryoconservation as an *ex situ* strategy for genetic conservation. Effective cryopreservation conserves the germplasm with minimal damage to its reproductive potential, stores it at sub-zero temperatures, and recovers it in a way that produces sufficient progeny to regenerate the variation represented in the original population. Avian semen can be frozen, but ova and embryos cannot be because of the structure that they are attached to. Primordial germ cells (PGC), the embryonic cells that will produce sperm and eggs, have been used to provide access to the genome for manipulation and they can be cryopreserved and recovered. Finally, the gonads, where the PGC accumulate, can be used to cryogenically store genetic variation and can be recovered through transplantation.

Cryoconservation of Avian Semen

Semen preservation is not a novel technique for many mammalian species and it should not be considered to be novel for several species of poultry. However, there is a belief or ideology among some that it either cannot be done or that efficiency is too low to be useful, which is clearly not true. The first descriptions of using glycerol as a cryoprotectant were with poultry [5], but freezing technology subsequently developed more rapidly for mammalian species such as cattle and humans than it did for poultry species. Avian sperm and the requirements for fertilization are very different from those of mammals.

Massip *et al.* [6] summarized progress in avian semen cryopreservation beginning with Shaffner *et al.* [7] who produced a chick from frozen chicken semen. Much of the subsequent work with avian semen has used low cooling rates and either glycerol [8] or dimethylsulfoxide (DMSO, [9]) as cryoprotectants, although Tselutin *et al.* [10] used rapid cooling and warming and dimethylacetamide (DMA) as a cryoprotectant. Both slow and rapid cooling have since been optimized and used in various laboratories [11]. Massip *et al.* [6] reported that cryoconserved chicken semen produces fertility of between 10 and 90% with an average of 60%, but that some lines have poor fertility. Fertility of geese inseminated with cryopreserved semen is similar to that of chickens, and that of turkeys is somewhat lower.

The low level of fertility needed for conservation and recovery of avian lines [12, 13] means that fertility is not the most important barrier to cryoconservation using poultry semen. However, there is variation between lines in the ability to survive the freeze/thaw cycle [14], and it may not be possible to conserve some lines using cryconserved semen. In addition, the female is the heterogametic sex in birds and storing semen does not preserve the w chromosome. Finally, recovering a line by backcrossing requires several generations, five generations is most often suggested, and incurs the cost of keeping birds during this time [15]. Several significant cryoconservation programmes have been established with semen [12, 16, 17].

Cryoconservation and Recovery of Embryonic Cells

There has been extensive work with embryonic cells of poultry and the technology developed could in principle be used for genetic conservation [18–20]. Storing genetic material using embryonic cells at first appears attractive because the cells can be isolated, grown in culture, cryoconserved and used to produce germline chimaeras. However, the technology around isolation of embryonic cells and their use to form chimaeras was developed primarily to provide access to the genome to allow genetic manipulation of poultry [21–25], which requires much lower efficiencies than those needed for genetic conservation. Genetic manipulation requires a minimum of one manipulated individual, whereas regeneration of a population requires several dozen individuals (FAO [26] suggests an effective population size of 50) to capture the genetic information in a population. In addition, the cells used as a starting point for genetic manipulation are likely to be readily available and are unlikely to be a limiting factor, whereas the cells obtained for genetic conservation will be derived from rare breeds or species.

The literature on isolation of the cells, injection and production of germline chimaeras is extensive and it is not the aim of this review to present it all because their goal is largely to manipulate the genome rather than conserve it. Cells can be isolated from the blastoderm [27], the blood [28] or the gonads [29] of embryos. Determining the gender [30, 31] allows the donors and recipients to be matched. The cells can be concentrated [32] or grown in culture [33, 34]. The PGC can be cryopreserved and warmed [35–37] before injection into the recipient [38, 39, and many others]. Manipulations can be performed through a window in the shell [40, 41] or the embryos can develop *in vitro* using a culture system [42]. The host PGC may be compromised [43, 44] to prevent them from competing with the donor cells. Somatic chimaerism can be determined [45, 23] but it may not be associated with germline chimaerism so many investigators do not report on it. Germline chimaerism is determined by mating tests [46, 47] and a few researchers have crossed germline chimaeras to produce pure-line offspring from transferred PGC, which is discussed below.

Many authors have described cryoconservation of embryonic cells, but few have described using cryoconserved cells to produce donor-derived offspring. Naito *et al.* [35] appears to be the first to produce donor-derived offspring from cryopreserved PGC. Of the 100 embryos they manipulated, they obtained 10 birds that produced from 0.9 to 26% donor-derived offspring, and a mating between chimaeras produced two donor-derived chicks. Kino *et al.* [27] produced five germline chimaeras from 234 embryos injected with frozen–thawed blastodermal cells and mating the individuals with the highest germline chimaerism (6 and 17%) produced four donor-derived offspring. Tajima *et al.* [29] produced three germline chimaeras from cyroconserved PGC that produced up to 18% donor-derived offspring. Tajima *et al.* [48] injected 21 host embryos, and obtained two male germline chimaeras, which produced 1.7 and 2.9% donor-derived offspring. All of these reports used DMSO as a cryoprotectant and controlled rate freezing of 1°C/min through the freezing point.

Kuwana *et al.* [28] proposed using PGC to recover a line of Japanese fowl called the Kureko Dori. They picked PGC individually from pooled blood of 49 Stage 13 to 16 [49] embryos, cryopreserved them using DMSO and injected 2024 thawed cells into 13 embryos. This produced seven chicks (one male, six females) and four females were germline chimaeras, which produced donor-derived chicks at a rate of 0.5–5.3%. Kuwana *et al.* [28] produced no pure Kureko Dori chicks and they used 49 donor chicks to produce seven germline chimaeras, but they used mixed donor cells, did not know the sex of the donors or recipients and did not compromise the recipient embryos' PGC.

Nakamura *et al.* [47] injected cryopreserved PGC from 88 Gifujidori embryos into 24 recipient embryos. In total, 18 chicks hatched, 15 chicks reached sexual maturity and produced from 0.6 to 41.7% donor-derived offspring. Matings between germline chimaeras produced six pure Gifujidori individuals. Despite sexing both donor and recipient, concentrating the PGC and compromising the host PGC, the result was the production of six pureline Gifujidori chickens from 88 donors, and although the procedures were designed to obtain PGC without sacrificing the embryo, hatchability of the manipulated embryos was only 28%. Nakamura *et al.* [47] suggested that sterilization of the hosts could be improved [50] which should increase the frequency of donor-derived chicks.

Two additional reports describe attempts to use embryonic cells to recover a rare breed but without cryoconservation. Chang *et al.* [46] produced six chimaeric chickens by injecting 372 White Leghorn embryos with cultured PGC from Korean Native Ogol Chickens, which generated 1.3–3.1% donor-derived offspring. Bednarczyk *et al.* [45] used cells from 20 blastoderms of a Polish breed, the Green-Legged Partridge like, and cells from 'several' embryos were pooled to treat 41 embryos, producing 23 chicks [51]. Two males and four females were germline chimaeras, and mating one of the males with the four females produced 477 chicks, of which 10 were donor-derived.

Some effort has been made to use the PGC of one species to generate offspring in another, which could be used to recover a rare species of bird. Li *et al.* [52] injected PGC from ducks into White Leghorn chickens and produced five fertile chimaeras. Mating male and female chimaeras produced only chickens but mating one of the fertile males to ducks produced six ducks from 622 eggs. Liu *et al.* [53] transferred chicken PGC to ducks and produced five chickens from 1057 eggs from a cross of chimaeric males to chickens, but no fertile eggs (of 2076) when chimaeric ducks were inseminated with chicken semen. Kang *et al.* [41] injected male pheasant PGC into 232 chicken embryos and obtained 39 fertile chimaeric offspring, 30 of which were male. Five of these were test mated with pheasant females and two produced pheasant offspring. Crosses of chimaera×chimaera and chimaera×White Leghorn produced 1846 and 2591 chicken offspring with no pheasants or hybrids but a pheasant male crossed to a chimaeric female produced one hybrid from 161 eggs. Chimaeric semen was produced by 22 of 25 male chickens receiving male pheasant PGC, but only 11 of 3206 progeny were donor-derived, and female PGC in male embryos did not produce donor-derived progeny. Wernery *et al.* [54] claimed to have produced a Houbara bustard by injecting dispersed Houbara gonadal cells into chicken embryos. Eight chimaeric roosters that resulted were mated to three Houbara females, one of which produced two fertile eggs, one of which produced a Houbara chick. Finally, van de Lavoir *et al.* [34] cultured the PGC from two chicken cell lines and injected 3000 cells into each of 141 guinea fowl embryos at 60 h of incubation. In total, 51 embryos hatched and 14 males were reared to sexual maturity. Semen from five of these males showed evidence of chicken sperm and were mated to chicken females, producing significant numbers of chicken offspring. Culturing the PGC allowed a greater production of offspring from the donor PGC, and injecting large numbers appeared to have reduced or eliminated the need to compromise the host. Efficient culturing of PGC provides a significant advance but the approach is limited by the technical effort required and the limited variation that it can regenerate. The authors suggest that thousands of birds could be produced, but the variation that can be conserved or recovered is limited because each donor cell line represents only one individual.

Several steps of the process of cryoconservation and reconstitution of PGC have been demonstrated to work quite well, but overall the efficiency of reconstitution remains low. Perhaps as important, the procedures involved are complex and require significant resources and training to carry them out. The development of the techniques has taken over 20 years of intensive work by laboratories around the world and the efficiencies are still far below what is needed for cryoconservation of avian genetic resources.

Cryoconservation and Recovery of Avian Gonads

Avian folliculogenesis features vitellogenesis, in which precursors of the yolk are synthesized in the liver and transported to the developing follicles [55], resulting in a very large ovum with a yolky ooplasm. At the time of ovulation, the ovum has a very low permeability to water and it has a very low surface-to-volume ratio. After ovulation, the albumin and shell are deposited during the passage of the developing egg down the oviduct and prevent the application of cryopreservation procedures to the whole egg. Although the egg cannot be cryopreserved, ovarian tissue can be which allows the preservation of a pool of developing follicles and their potential to produce fertile gametes in their natural microenvironment. Testicular tissue can also be cryopreserved and recovered, although the possibility of freezing semen makes it less essential.

The objective of gonadal cryoconservation is to produce donor-derived offspring, which requires the successful resumption of germline development of the gonadal tissue that has been cryopreserved and warmed along with successful fertilization. Techniques such as *in vitro* maturation of follicles or intracytoplasmic sperm injection are available for mammals, but these are not practical for avian species. The transplantation can be used for functional recovery of cryopreserved avian gonadal tissue.

Gonadal transplantation has been successfully used in a number of mammalian species, including humans [56]. However, until recently, ovarian transplantation has not resulted in the production of donor-derived progeny in birds [57, 58] because of the challenges of the surgical procedures and immunological rejection. More recently, Song and Silversides [59] described a technique of chicken ovarian transplantation that enables the transplanted tissues from newly hatched donors to survive and resume development in recipients of the same age. This technique includes surgical ovariectomy and subsequent implantation of the donor ovarian tissue into the normal anatomical location (orthotopic transplantation). This technique has since been used to produce donor-derived offspring from the transplantation of fresh ovarian chicken [60] and duck [61] tissue and from fresh [62, 63] and cryopreserved Japanese quail tissue [63].

The transplanted ovarian tissue used for most species is usually small fragments up to several mm thick. Vascular anastomosis cannot be achieved surgically after transplantation of these tissue pieces, so they may be exposed to ischaemic stress before vascularization can be restored by the growth of new blood vessels. In fact, Liu *et al.* [64] found that the loss of PGC caused by ischaemia is more severe than that caused by cryopreservation. This can be reduced [56, 65]. Song and Silversides [59, 60, 62] and Liu et al. [63] placed the ovarian fragments on the wounded surface created by ovariectormy of the donors in order to facilitate revascularization of the transplants through the wound healing process.

In some species of mammals, notably rodents, immunodeficient lines are available and can be used as recipients for allogeneic and xenogeneic gonadal transplantation [66, 67]. Immunodeficient strains of birds are not available and an immunosuppressant, mycophenolate mofetil, can be used to prevent tissue rejection. This has produced success in allogeneic ovarian transplantation in chickens and Japanese quail [60, 62, 63], xenogeneic ovarian transplantation in ducks [61], and allogeneic testicular transplantation in chickens [68, 69] and Japanese quail (Liu *et al.*, unpublished data). Song and Silversides [70] showed that the short-term immunosuppressive treatment used for chicken ovarian transplantation produces long-term allogeneic tolerance.

Avian testicular tissue can be recovered by heterotopic transplantation, in which the transplantation site is not the normal anatomical location of the testes. Heterotopic transplantation may be necessary because it is difficult to re-establish the connection between the transplants and the vasa deferentia but it gives the advantage of knowing the origin of the germinal cells with certainty. Song and Silversides [69] demonstrated that transplantation of fresh chicken testicular tissue under the dorsal skin and into the abdominal cavity was equally efficient in producing donor progeny. In chickens, cell suspensions from the testicular transplants can be used to produce donor-derived offspring through intramagnal insemination [68, 69]. Complete castration of the recipients appears to be important for the recovery of the testicular transplants, and in chickens, Song and Silversides [68, 69] obtained fertile transplants only from recipients that were completely castrated. Similar observations have been reported for mammalian testicular allogeneic transplantation but the requirement may not be as strict for xenogeneic transplantation [71–73], which may suggest species-specific mechanisms in the functional recovery of testicular transplants. Intracytoplasmic sperm injection can be used in mammals for fertilization using spermatozoa retrieved from testicular transplants [74]. Similar techniques may be available for birds [75, 76], but they require collection of the preovulatory follicle and development of the embryo in host egg shells, which may be impractical for conservation programmes where several dozen offspring are needed to recreate a population.

Slow-freezing procedures that have been used in fundamental cryobiological research and previous studies have most commonly cryopreserved cell suspensions. The objective of cryopreservation is to prevent detrimental intracellular ice formation and solution effects during cooling and warming procedures [77]. To prevent intracellular ice formation, extracellular ice nucleation (seeding) is induced at a seeding temperature and the cooling rate is slow enough for sufficient osmotic dehydration to keep the chemical potential of intracellular water and the partly frozen extracellular medium near to equilibrium [4]. Solution effects that are associated with long exposure of cells to concentrated solutions are diminished by cryoprotective agents [78]. Silversides *et al.* [79] used a slow-freezing protocol for cryobanking chicken gonadal tissue and its effectiveness was evaluated for ovarian tissue from Japanese quail [63].

From a cryobiological perspective, slow-freezing procedures provide limited protection to multicellular structures such as tissues and organs. The different cell types that make up a tissue can be very different in the inherent characteristics that affect the outcome of cryopreservation [77]. As a result, a protocol that leads to high survival of one type of cell in a tissue may lead to death of others. In addition, formation of ice in extracellular space is harmful to the extracellular matrix, which can affect the functional recovery of tissue and organs. These problems can be solved by vitrification procedures.

The definition of vitrification is the process of solidification of a liquid through formation of an amorphous solid (glass) instead of crystallization [80]. This can be achieved by ultra-rapid cooling of biological materials subjected to high concentrations of glass-promoting cryoprotectants, so that both intracellular and extracellular components convert to glass in a very short time. A vitrification protocol that has been successfully used to preserve mammalian ovarian tissue [81, 82] was adopted to preserving ovarian tissue of Japanese quail [63] and resulted in higher efficiency in preserving cell viability, tissue morphology and functional recovery than a slow-freezing protocol.

The prerequisite for semen cryopreservation is the availability of fertile semen, and the application is limited to preservation of the male germplasm. Cryopreservation of testicular tissue provides an alternative, targeting immature male germ cells that have the potential to generate fertile spermatozoa. Slow freezing has been used by most previous studies with limited success [83], but vitrification procedures have been investigated for preserving testicular tissue and results have been promising for humans [84] and various other animals [85, 86, Liu *et al.*, unpublished data). Live offspring have been obtained from sexually immature and mature testicular donors in murine species [67]. In chickens, immature testicular tissue can be cryopreserved by a slow-freezing protocol and recovered by allogeneic transplantation, and cell suspensions retrieved from the cryopreserved transplants after their maturation in a host can be used to produce donor-derived offspring [68, 69]. Validation of a practical vitrification procedure that can be used in other avian species is needed.

Summary

Those in industry and research who work with poultry would like to keep living birds, but the high cost of maintaining living populations means that they are inevitably discarded if they are not used. Criticism of cryoconservation and the inertia that this has created have resulted in massive losses of avian genetic variation, but cryoconservation must be considered for the genetic variation that remains. Fertility of frozen–thawed semen may be too low for commercial application, but it is fully adequate for conserving avian genetic resources. In principle, embryonic cells, in particular the PGC, can be stored and recovered, but the number and complexity of procedures involved and the low efficiency of isolation and recovery create difficulties for maintaining genetic resources. Questions remain for the storage and recovery of avian gonads, but the technology is promising and would allow rapid recovery of stored material.

References

1. Silversides FG, Shaver DMcQ, Song Y. Pure line laying chickens at the Agassiz Research Centre. Animal Genetic Resources Information 2007;40:79–85.
2. Fulton JE, Delany ME. Poultry genetic resources – operation rescue needed. Science Magazine 2003;300:1667–8.
3. FAO. The State of the World's Animal Genetic Resources for Food and Agriculture. FAO, Rome, Italy; 2007.
4. Mazur P, Leibo SP, Seidel GE Jr. Cryopreservation of the germplasm of animals used in biological and medical research: importance, impact, status, and future directions. Biology of Reproduction 2008;78:2–12.
5. Polge C. Functional survival of fowl spermatozoa after freezing at −79°C. Nature 1951;167:949–50.
6. Massip A, Leibo SP, Blesbois E. Cryobiolgoy of gametes and the breeding of domestic animals. In: Fuller, BJ, Lane N, Benson EE, editors. Life in the Frozen State. CRC Press, New York, USA; 2004. p. 371–92.
7. Shaffner CS, Henderson EW, Card CG. Viability of spermatozoa of the chicken under various environmental conditions. Poultry Science 1941;20:259–65.
8. Lake PE, Stewart JM. Preservation of fowl semen in liquid nitrogen-an improved method. British Poultry Science 1978;19:187–94.
9. Sexton TJ. Optimal rates for cooling chicken semen from +5 to −196°C. Poultry Science 1980;59:2765–70.
10. Tselutin K, Narubina L, Mavrodina T, Tur B. Cryopreservation of poultry semen. British Poultry Science 1995;36:805–11.
11. Blesbois E. Current status in avian semen cryopreservation. World's Poultry Science Journal 2007;63:213–22.
12. Blesbois E, Seigneurin F, Grasseau I, Limouzin C, Besnard J, Gourichon D, *et al.* Semen cryopreservation for *ex situ* management of genetic diversity in chicken; creation of the French avian cryobank. Poultry Science 2007;86:555–64.

13. Blackburn HD, Silversides FG, Purdy PH. Inseminating fresh or cryopreserved semen from maximum efficiency: implications for gene banks and industry. Poultry Science 2009;88:2184–91.

14. Ansah G, Buckland R. Genetic variation in fowl semen cholesterol and phospholipid levels and the relationship of these lipids with fertility of frozen-thawed and fresh semen. Poultry Science 1982;61:623–37.

15. Silversides FG, Purdy PH, Blackburn HD. Comparative costs of programs to conserve chicken genetic variation based on maintaining living populations or storing cryopreserved material. British Poultry Science 2012; doi:10.1080/00071668.2012.727383.

16. Blackburn HD. The National Animal Germplasm Program: challenges and opportunities for poultry genetic resources. Poultry Science 2006;85:210–5.

17. Woelders H, Zuidberg CA, Hiemstra SJ. Animal genetic resources conservation in The Netherlands and Europe: poultry perspective. Poultry Science 2006;85:216–22.

18. Tajima A. Production of germline chimeras and their application in domestic chicken. Avian and Poultry Biology Reviews 2002;13:15–30.

19. Naito M. Cryopreservation of avian germline cells and subsequent production of viable offspring. Journal of Poultry Science 2003;40:1–12.

20. Petitte JN. Avian germplasm preservation: embryonic stem cells or primordial germ cells. Poultry Science 2006;85: 237–42.

21. Pettite JN, Brazolot CL, Clark ME, Liu G, Verrider Gibbins AM, Etches RJ. Accessing the genome of the chicken using germline chimeras. In Etches RJ, Verrinder Gibbins AM, editors. Manipulation of the Avian Genome. CRC Press, New York, USA; 2000. p. 81–101.

22. Park TS, Jeong DK, Kim JN, Song GH, Hong JH, Lim JM, *et al.* Improved germline transmission in chicken chimeras produced by transplantation of gonadal primordial germ cells into recipient embryos. Biology of Reproduction 2003;68:1657–62.

23. van de Lavoir MC, Mather-Love C, Leighton P, Diamond JH, Heyer BS, Roberts R, *et al.* High-grade transgenic somatic chimeras from chicken embryonic stem cells. Mechanisms of Development 2006;123:31–41.

24. Kim JN, Park TS, Park SH, Park KJ, Kim TM, Lee SK, *et al.* Migration and proliferation of intact and genetically modified primordial germ cells and the generation of a transgenic chicken. Biology of Reproduction 2010;82:257–62.

25. Li JJ, Lu LZ. Recent progress on technologies and applications of transgenic poultry. African Journal of Biotechnology 2010;9:3481–8.

26. FAO. Draft Guidelines for the Cryoconservation of Animal Genetic Resources. Available from [Accessed August 2, 2012]: URL: http://www.fao.org/docrep/meeting/022/mb553e.pdf. FAO, Rome, Italy; 2011.

27. Kino K, Pain B, Lebo SP, Cochran M, Clark ME, Etches RJ. Production of chicken chimeras from injection of frozen-thawed blastodermal cells. Poultry Science 1997;76:753–60.

28. Kuwana T, Kawashima T, Naito M, Yamashita H, Matsuzaki M, Takano T. Conservation of a threatened indigenous fowl (Kureko Dori) using the germline chimeras transplanted from primordial germ cells. Journal of Poultry Science 2006;43:60–6.

29. Tajima A, Naito M, Yasuda Y, Kuwena T. Production of germline chimeras by transfer of cryopreserved gonadal primordial germ cells (gPGCs) in chicken. Journal of Experimental Zoology 1998;280:265–7.

30. Clinton M, Haines L, Belloir B, McBride D. Sexing chick embryos: a rapid and simple protocol. British Poultry Science 2001;42:134–8.

31. Shaw DL, Carscience RS, Etches RJ, Verrinder Gibbins AM. The fate of female donor blastodermal cells in male chimeric chickens. Biochemistry and Cell Biology 1992;70:1218–29.

32. Kim JN, Kim MA, Pak TS, Kim DK, Park HJT, Ono T, *et al.* Enriched gonadal migration of donor-derived gonadal primordial germ cells by immunomagnetic cell sorting in birds. Molecular Reproduction and Development 2004;68:81–87.

33. Pain B, Clark ME, Shen M, Nakazawa H, Sakurai M, Samarut J, *et al.* Long-term *in vitro* culture and characterisation of avian embryonic stem cells with multiple morphogenetic potentialities. Development 1996;122:2338–48.

34. van de Lavoir MC, Collarini EJ, Leighton PA, Fesler J, Lu DR, Harriman WD, *et al.* Interspecific germline transmission of cultured primordial germ cells. PLoS ONE 2012;7(5):e0035664.

35. Naito M, Tajima A, Tagami T, Yasuda Y, Kuwana T. Preservation of chick primordial germ cells in liquid nitrogen and subsequent production of viable offspring. Journal of Reproduction and Fertility 1994;102:321–5.

36. Setioko AR, Tagami T, Tase H, Nakamura Y, Takeda K, Nirasawa K. Cryopreservation of primordial germ cells (PGCs) from White Leghorn embryos using commercial cryoprotectants. Journal of Poultry Science 2007;44:73–7.

37. Kohara Y, Kanai Y, Tajima A. Cryopreservation of gonadal germ cells (GGCs) from the domestic chicken using vitrification. Journal of Poultry Science 2008;45:57–61.

38. Reynaud G. Transfert de cellules germinales primordiales de dindon à l'embryon de poulet par injection intravasculaire. Journal of Embryology and Experimental Morphology 1969;21:485–507.

39. Shiue YL, Tailiu JJ, Liou JF, Lu HT, Tai C, Shiau JW, *et al.* Establishment of the long-term *in vitro* culture system for chicken primordial germ cells. Reproduction in Domestic Animals 2009;44:55–61.

40. Han JY, Park TS, Hong YH, Jeong DK, Kim JN, Kim KD, *et al.* Production of germline chimeras by transfer of chicken gonadal primordial germ cells maintained *in vitro* for an extended period. Theriogenology 2002;58:1531–9.

41. Kang SJ, Choi JW, Kim SY, Park KJ, Kim TM, Lee YM, *et al.* Reproduction of wild birds via interspecies germ cell transplantation. Biology of Reproduction 2008;79:931–7.

42. Perry MM. A complete culture system for the chick embryo. Nature 1988;331:70–2.

43. Carscience RS, Clark ME, Verrinder Gibbins AM, Etches RJ. Germline chimeric chickens from dispersed donor blastodermal cells and compromised recipient embryos. Development 1993;117:669–75.

44. Song YH, D'Costa S, Pardue SL, Petitte JN. Production of germline chimeric chickens following administration of a busulfan emulsion. Molecular Reproduction and Development 2005;70:438–44.

45. Bednarcyzk M, Lakota P, Slomski R, Plawski P, Lipinski D, Siemieniako B, *et al.* Reconstitution of a chicken breed by inter se mating of germline chimeric birds. Poultry Science 2002;81:1347–53.

46. Chang IK, Jeong DK, Hong YH, Park TS, Moon YK, Ohno T, *et al.* Production of germline chimeric chickens by transfer of cultured primordial germ cells. Cell Biology International 1997;21:495–9.

47. Nakamura Y, Usai F, Ono T, Takeda K, Nirasawa K, Kagami H, *et al.* Germline replacement by transfer of primordial germ cells into partially sterilized embryos in the chicken. Biology of Reproduction 2010;83:130–7.

48. Tajima A, Minematsu T, Ohara M. Production of germ-line chimeras by the transfer of cryopreserved gonadal germ cells collected from 7- and 9-day-old chick embryos. Animal Science Journal 2004;75:85–8.

49. Hamburger V, Hamilton HL. A series of normal stages in the development of the chick embryo. Journal of Morphology 1951;88:49–92.

50. Nakamura Y, Usui F, Miyahara D, Mori T, Ono T, Takeda K, *et al.* Efficient system for preservation and regeneration of genetic resources in chicken: concurrent storage of primordial germ cells and live animals from early embryos of a rare indigenous fowl (Gifujidori). Reproduction, Fertility and Development 2010;22:1237–46.

51. Bednarcyzk M, Lakota P, Siwek M. Improvement of hatchability of chicken eggs injected by blastoderm cells. Poultry Science 2000;79:1823–8.

52. Li ZD, Deng H, Liu CH, Song YH, Sha J, Wang N, *et al.* Production of duck-chicken chimeras by transferring early blastodermal cells. Poultry Science 2002;81:1360–4.

53. Liu C, Khazanehdari KA, Baskar V, Saleem S, Kinne J, Wernery U, *et al.* Production of chicken progeny (*Gallus gallus domesticus*) from interspecies germline chimeric duck (*Anas domesticus*) by primordial germ cell transfer. Biology of Reproduction 2012;86:1–8.

54. Wernery U, Liu C, Baskar V, Guerineche Z, Khazanehdari KA, Saleem S, *et al.* Primordial germ cell-mediated chimera technology produces viable pure-line Houbara Bustard offspring: potential for repopulating an endangered species. PLoS ONE 2010;5(12):e15824.

55. Johnson AL. Reproduction in the female. In: Sturkie PDeditor. Avian Physiology. 4th ed. Springer-Verlag, New York, USA; 1986. p. 403–31.

56. Silber SJ. Ovary cryopreservation and transplantation for fertility preservation. Molecular Human Reproduction 2012;18:59–67.

57. Davenport CB. The transplantation of ovaries in chickens. Journal of Morphology 1911;22:111–22.

58. Grossman M, Siegel PB. Orthotopic ovarian transplants in chickens. Poultry Science 1966;45:1434–6.

59. Song Y, Silversides FG. The technique of orthotopic ovarian transplantation in the chicken. Poultry Science 2006;85: 1104–6.

60. Song Y, Silversides FG. Offspring produced from orthotopic transplantation of chicken ovaries. Poultry Science 2007;86:107–11.

61. Song Y, Cheng KM, Robertson MC, Silversides FG. Production of donor-derived offspring after ovarian transplantation between Muscovy (*Cairina moschata*) and Pekin (*Anas platyrhynchos*) ducks. Poultry Science 2012;91:197–200.

62. Song Y, Silversides FG. Transplantation of ovaries in Japanese quail (*Coturnix japonica*). Animal Reproduction Science 2008;105:430–7.

63. Liu J, Song Y, Cheng KM, Silversides FG. Production of donor-derived offspring from cryopreserved ovarian tissue in Japanese quail (*Coturnix japonica*). Biology of Reproduction 2010;83:15–9.

64. Liu J, Van der Elst J, Van den Broecke R, Dhont M. Early massive follicle loss and apoptosis in heterotopically grafted newborn mouse ovaries. Human Reproduction 2002;17:605–11.

65. Israely T, Nevo N, Harmelin A, Neeman M, Tsafriri A. Reducing ischaemic damage in rodent ovarian xenografts transplanted into granulation tissue. Human Reproduction 2006;21:1368–79.

66. Bols PE, Aerts JM, Langbeen A, Goovaerts IG, Leroy JL. Xenotransplantation in immunodeficient mice to study ovarian follicular development in domestic animals. Theriogenology 2010;73:740–7.

67. Ehmcke J, Schlatt S. Animal models for fertility preservation in the male. Reproduction 2008;136:717–23.

68. Song Y, Silversides FG. Production of offspring from cryopreserved chicken testicular tissue. Poultry Science 2007;86:1390–6.

69. Song Y, Silversides FG. Heterotopic transplantation of testes in newly hatched chickens and subsequent production of offspring via intramagnal insemination. Biology of Reproduction 2007;76:598–603.

70. Song Y, Silversides FG. Long-term production of donor-derived offspring from chicken ovarian transplants. Poultry Science 2008;87:1818–22.

71. Schlatt S, Honaramooz A, Boiani M, Schöler HRI Dobrinski I. Progeny from sperm obtained after ectopic grafting of neonatal mouse testes. Biology of Reproduction 2003;68:2331–5.

72. Paris MCJ, Schlatt S. Ovarian and testicular tissue xenografting: its potential for germline preservation of companion animals, non-domestic and endangered species. Reproduction, Fertility and Development 2007;19:771–82.

73. Shinohara T, Inoue K, Ogonuki N, Kanatsu-Shinohara M, Miki H, Nakata K, *et al.* Birth of offspring following transplantation of cryopreserved immature testicular pieces and *in vitro* microinsemination. Human Reproduction 2002;17:3039–45.

74. Nakai M, Kaneko H, Somfai T, Maedomari N, Ozawa M, Noguchi J, *et al.* Production of viable piglets for the first time using sperm derived from ectopic testicular xenografts. Reproduction 2010;139:331–5.

75. Tanaka K, Wada T, Koga O, Nishio Y, Hertelendy F. Chick production by *in vitro* fertilization of the fowl ovum. Journal of Reproduction and Fertility 1994;100:447–9.

76. Hrabia A, Takagi S, Ono T, Shimada K. Fertilization and development of quail oocytes after intracytoplasmic sperm injection. Biology of Reproduction 2003;69:1651–7.

77. Mazur P. Cryobiology: the freezing of biological systems. Science 1970;168:939–49.

78. Muldrew K, Acker JP, Elliott JAW, McGann LE. 2004. The water to ice transition: implications for living cells.

In: Fuller BJ, Lane N, Benson EE, editors. Life in the Frozen State. CRC Press, New York, USA; 2004. p. 67–108.

79. Silversides FG, Song Y, Renema R, Rathgeber BR, Classen HL. Cryopreservation of germplasm from chickens kept in Canadian research institutions. Canadian Journal of Animal Science 2008;88:577–80.

80. Fahy GM, MacFarlane DR, Angell CA, Meryman HT. Vitrification as an approach to cryopreservation. Cryobiology 1984;21:407–26.

81. Chen SU, Chien CL, Wu MY, Chen TH, Lai SM, Lin CW, *et al.* Novel direct cover vitrification for cryopreservation of ovarian tissues increases follicle viability and pregnancy capability in mice. Human Reproduction 2006;21:2794–800.

82. Wang Y, Xiao Z, Li L, Fan W, Li SW. Novel needle immersed vitrification: a practical and convenient method with potential advantages in mouse and human ovarian tissue cryopreservation. Human Reproduction 2008;23:2256–65.

83. Woods EJ, Benson JD, Agca Y, Critser JK. Fundamental cryobiology of reproductive cells and tissues. Cryobiology 2004;48:146–56.

84. Curaba M, Poels J, van Langendonckt A, Donnez J, Wyns C. Can prepubertal human testicular tissue be cryopreserved by vitrification? Fertility and Sterility 2011;95:2123.e9–2123e12.

85. Abrishami M, Anzar M, Yang Y, Honaramooz A. Cryopreservation of immature porcine testis tissue to maintain its developmental potential after xenografting into recipient mice. Theriogenology 2010;73:86–96.

86. Curaba M, Verleysen M, Amorim AC, Dolmans MM, van Langendonckt A, Hovatta O, *et al.* Cryopreservation of prepubertal mouse testicular tissue by vitrification. Fertility and Sterility 2011;95:1229–34.

Recombinant vaccines and infectious bursal disease virus

M.S. Lucero[1,2]†, E. Gómez[1,2]†, J.M. Carballeda[1,2], M.J. Gravisaco[1], S.Chimeno Zoth[1,2] and A. Berinstein[1,2]*

Address: [1] Instituto de Biotecnología, CICVyA, INTA, Castelar, Cc 25, B1712WAA, Buenos Aires, Argentina.[2] Consejo Nacional de Investigaciones Científicas y Técnicas (CONICET), Rivadavia 1917, C1033AAJ, CABA, Argentina.

***Correspondence:** Analía Berinstein. Email: aberinstein@cnia.inta.gov.ar

Received: 23 August 2012
Accepted: 19 October 2012

Abstract

Infectious bursal disease virus (IBDV) is the aetiological agent of infectious bursal disease (IBD), an immunosuppressive and highly contagious disease that affects young birds causing important economic losses in the poultry industry worldwide. Currently, vaccination programmes with inactivated and live-attenuated viruses have been used to prevent IBD. However, these vaccines present a number of disadvantages, mainly because of their viral nature. Consequently, in the last two decades, many studies have been conducted in order to replace conventional virus-based vaccines by new, rationally designed vaccines that are safer as well as effective. In this review, we will present a background on the disease and its causative agent, and focus on the development of new generation vaccines against this significant poultry disease.

Keywords: Infectious bursal disease, Recombinant vaccines, Chicken immune response, VP2

Review Methodology: We searched PubMed database (Keyword search terms used: Infectious bursal disease, recombinant vaccines, VP2, chicken immune response). In addition, we used the references from the articles and reviews obtained by this method to check for additional relevant material.

Basic Concepts in Infectious Bursal Disease Virus (IBDV) and its Impact on the Chicken

Production of poultry products have increased significantly in the last few years as the demand for avian meat has gained ground over other meat types worldwide. The chicken is an economically relevant livestock animal, and poultry meat accounts for almost 40% of global meat consumption. In this context, sanitary conditions in commercial flocks are of utmost importance. Almost all chicken viral diseases are prevented by vaccination and infectious bursal disease (IBD) is not an exception.

IBDV belongs to the *Birnaviridae* family. It is a non-enveloped, icosahedral, double-stranded RNA virus and its genome is presented in two segments, A and B [1–3]. The segment A of the IBDV genome has two open reading frames (ORFs): one ORF encodes a polyprotein, pVP2-VP4-VP3, which undergoes cleavage by the viral protease VP4; and the other, encodes a small protein called VP5 of controversial function. The autoproteolytic processing of the polyprotein yields pVP2, VP3 and VP4 products [4]. Then, the precursor pVP2 (512 aa) matures to generate VP2 (441 aa) and four peptides that remain associated with VP2 within the virion [5]. The last step of maturation of VP2 also requires interaction with VP3 and it is supposed to occur during viral particle assembly [6].

IBDV is the aetiological agent of IBD, a highly contagious, worldwide-spread immunosuppressive chicken disease, also known as 'Gumboro disease'. The severity of the disease depends, in general, on the age of the bird infected and the virulence of the infecting virus. The clinical manifestation of IBD usually occurs in chickens between 3 and 6 weeks of age and has a sudden onset, with a rapid increase in the mortality rate of the affected flock. Clinical signs of the disease include dehydration, trembling, ruffled feathers, vent pecking and depression.

†These authors contributed equally to this work.

More importantly, the disease can cause severe, long-lasting suppression of the immune system. Chickens that are immunosuppressed by early IBDV infections do not respond well to vaccination [7, 8] and are more susceptible to other diseases, including those that do not normally affect healthy chickens. On necropsy, the principal lesions are found in the bursa of Fabricius (BF). This organ is found exclusively in birds and it is a specialized and essential organ for the amplification and differentiation of B-cells.

Two serotypes of IBDV (1 and 2) have been described, being serotype 1 viruses the only ones pathogenic for chickens [9]. Serotype 2 viruses are generally isolated from turkeys and are non-pathogenic [10–12]. In increasing order of virulence, different pathotypes of IBDV within type 1 viruses were characterized as mild, intermediate, classical virulent and very virulent strains. Antigenic variant strains are also described; typically, they do not cause clinical signs of disease but they can cause a marked immunosuppression and they are eventually able to infect vaccinated chickens, because these strains are generally serologically different from the classic pathotypes [13–16].

After oral infection or inhalation, IBDV replicates in gut-associated macrophages and B-cells, causing a primary and transient viraemia. Subsequently, virus travels to the bursa by the blood stream, and massive virus replication occurs in this target organ shortly after inoculation, leading to a secondary and pronounced viraemia and causing lesions in other organs [17]. As the predominant feature of the pathogenesis of IBDV, infection results in extensive lymphoid depletion in the medullar and the cortical regions of the follicles of the bursa. In a lesser degree, damage of caecal tonsils and spleen can be seen. The arrival of the virus to the bursa and its replication is accompanied by a high T-lymphocyte (CD4+ and CD8+) infiltration that begins at 4 days post infection (d.p.i.) and reaches a peak at 7 d.p.i. Before influx of T-cells into the bursa, a strong up-regulation of pro-inflammatory cytokines such as interleukin (IL)-1b, IL-6, nitric oxide synthase and the chemokine IL-8 occurs at 3 d.p.i.

Infection with IBDV frequently results in immunosuppression, even in a subclinical course of the disease. This immunosuppression often provokes secondary infection of the respiratory tract, and unresponsiveness to live-attenuated vaccines against respiratory diseases such as infectious bronchitis, Newcastle disease and avian influenza [18–20].

Currently, when maternal antibodies have disappeared, vaccination against IBD with the inactivated and live-attenuated intermediate virus (used for commercial birds such as broilers/layers) serves to create an immunological state in the flock against virulent strains. However, these vaccines present a number of disadvantages because of their viral nature. Live-attenuated vaccines can revert to virulence by the recombination of genomic segments [21]; they usually produce a state of immunosuppression in young chickens (although it is a short period from which birds can recover); and, they can be, eventually, inefficient in protecting animals from very virulent strains and variant strains [13–16]. On the other hand, inactivated vaccines are costly and lack efficient immunogenicity unless they are adjuvanted and administered in multiple inoculations, or delivered as a boost after priming with a replicating antigen [22]. Generally, their use is constrained to breeder birds just before laying in order to induce high levels of transferred maternal immunity in the progeny. Thus, there is a genuine need for replacing the conventional virus-based vaccines by new ones that could solve these concerns, having higher efficacy and fewer side-effects.

A new technology being used in the field is immune complex vaccines, which are obtained by mixing a certain amount of specific antibodies obtained from the serum of hyperimmunized chickens with live IBD vaccine virus [23, 24]. Since it has been shown that these vaccines were effective in the presence of maternally derived antibodies (MDA) [25, 26], they are suitable for *in ovo* vaccination of 18–day-old embryonated chicken eggs with commercial egg-injection machines or delivery by subcutaneous injection in one-day-old chickens [27]. Both methods of administration present some advantages over conventional live vaccines, which are usually given via drinking water: they allow earlier immunity and more automated, uniform and systematic administration process [22]. Additionally, *in ovo* vaccination offers reduced labour cost and reduced stress to birds compared with the subcutaneous route.

New generation vaccines act upon the immune system in different ways depending on the type of vaccine. The election of the production system of a recombinant vaccine will depend on the immune response that needs to be elicited. The development of recombinant vaccines includes the study of the immune response against the pathogen for which it is designed. In this way, it would be possible to design an immunogen that fulfils the requirements for rendering the best performance against IBDV. Given its importance in eliciting the antibody response, VP2 has been extensively used to develop recombinant vaccines against IBDV, as will be seen in the following section.

Recombinant Subunit Vaccines

Advances achieved in the fields of molecular biology, genetic engineering, crystallography and immunology, have allowed the genetic and structural characterization of several pathogens, identifying the epitopes or protein regions responsible for inducing a protective immune response. This knowledge promoted the development of new pathogen-free vaccines named recombinant subunit vaccines, where only the immunogenic element of a pathogen is delivered to the host for stimulating a specific immune response. In this context, the new generation vaccines to combat Gumboro disease are based on: protein immunogens, live vectors and nucleic acids [22, 28]. Table 1 recapitulates some of the most important properties of each strategy, as well as the major immunological response elicited by each system. As can be seen, each strategy presents particular features that have to be taken into consideration when selecting a method for a new development. However, a common feature is that all recombinant subunit vaccines have a defined chemical composition and low risk of pathogen escapes because of manipulation during the production stage.

Table 2 summarizes almost all of the studies found in the literature where the immunogenic properties of recombinant VP2 are assayed. Recombinant vaccines are organized as follows: DNA vaccines, live delivery system (viral and bacterial vectors) and protein immunogens, where VP2 is present in the soluble form or forming particles.

Naked plasmid DNA and live viral or bacterial vectors carrying either *vp2* or *vp2-vp4-vp3* genes have been assayed as vaccines against IBDV since they present a number of advantages as cost-effective production and are capable of eliciting both humoral and cellular immune responses (Table 1). However, the levels of protection produced by these vaccines have been variable, ranging from partial to complete protection against IBDV challenge. DNA vaccines are based on DNA encoding the target gene that is then injected into an animal. Animal cells capture these plasmids and incorporate them into the cell nucleus, allowing the expression of the foreign gene and the production of the desired protein [73]. The possibility of inducing an effective immune response by this strategy has been demonstrated, although with variable success [31–39]. The *in ovo* inoculation with DNA vaccine without a boost was insufficient to evoke the protective immunity [39].

Live vector vaccines are based on the use of a live micro-organism that acts as a vector for the expression of heterologous genes. An interesting application of vectored vaccines is the possibility of dealing with two diseases using the same sanitary tool. This is the case, for example, of the use of recombinant turkey herpes virus (HVT) as a vectored vaccine for viral avian diseases. HVT has the ability to confer protection against Marek's disease (MD). A recombinant HVT carrying the *VP2* gene of IBDV was constructed and successfully assayed as a live vaccine against both illness, MD and IBD. Vaxxitek HVT+IBD (Merial) [49, 50] and VECTORMUNE® HVT IBD (CEVA Santé Animale) are two commercially products based on HVT serotype 3 that are already being used in the field and they have proved to be safe and efficient. These vaccines do not show interference with MDA and in consequence, they are indicated to be applied in one-day-old chickens by the subcutaneous route or in 18-day-old embryonated chicken eggs. Remarkably, chickens first immunized with HVT vaccines should not be vaccinated again with these recombinant vaccines because the former vaccination has a major inhibitory effect on the efficacy of the later ones [49]. Nowadays, the number of eggs and chickens inoculated with these viral vaccines is increasing and the new technology for the *in ovo* vaccination supports the scale up.

Protein immunogens have also been used as subunit vaccines against IBDV where soluble or non-soluble VP2 is delivered to the host. The most commonly used expression systems to produce VP2 at large scale are *Escherichia coli*, yeast and baculoviruses. While *E. coli* has the advantage of being the fastest, cheapest and high-yielding expression system, post-translation modifications and correct folding of proteins can be problematic to obtain and this may affect the immunogenicity and protective efficacy of the product. Hence, other expression system, although more costly, have been employed to produce recombinant proteins. However, vaccination with the isolated antigen obtained by these expression systems supposes the use of adjuvants and more than one inoculation as the protein itself is a poor immunogen. Nevertheless, industry has taken profit of the technology. An example that arrived to the market is the commercial vaccine Gumbin® VP2 (Phibro Animal Health Corporation), a water in oil bivalent vaccine, that contains high concentrations of IBD VP2 produced in yeast [58]. This vaccine is intended for vaccination of chickens after priming with live Newcastle disease and live IBD vaccine, although it can be used in broiler, layer and breeder chickens any age starting from day one.

As can be observed in Table 2, delivery of non-soluble VP2, whether it is displayed in a vector or forming particles, have proved to be more efficient in eliciting an immune response than the soluble protein. VP2 is present in the capsid of the virion forming trimmers in which each subunit folds into three domains named base (B), shell (S) and projection (P) [74]. In the P domain, four loops harbouring the conformational epitopes responsible for eliciting the neutralizing response (loops P_{BC} and P_{HI}) and the regions involved in cell tropism and virulence (loops P_{DE} and P_{FG}), are identified [75–77]. It was reported that recombinant bamboo mosaic virus displaying loop P_{BC} region on the coat protein was successful in eliciting a protective response with a 100% survival rate of chickens challenged 28 days after the intramuscular immunization [68].

The expression of VP2 or the polyprotein (pVP2-VP4-VP3) in a heterologous system may lead to the generation of subviral particles (SVP) and virus-like particles (VLP), respectively, which basically differ in size but conserve the immunogenic properties. VLPs mimic the conformational structure of the infectious virus from which they derive but they are highly safe as they are non-infectious because they lack the viral genome. As these particles present a redundant antigenic structure they are highly immunogenic, being a promising candidate for the development of alternative vaccines. The baculovirus and

Table 1 Comparison of crucial properties of different vaccine types1

Vaccine type	Composition	Pathogenicity	Cost	Immune response elicited	Vaccination scheme	Other characteristics
Whole cells vaccines						
Live, attenuated vaccines	Undefined	Risk of reversion to virulent strain. Certain risk of transmission	Low	Both humoral and cellular responses	One or few doses normally required No need for adjuvants	Long-lasting immunity Controlled attenuation normally required
Killed, inactivated vaccines	Undefined	No risk of reversion to pathogenicity. No risk of transmission	Moderate	Mainly humoral responses	Multiple doses typically required. Adjuvants normally needed.	Less powerful than live vaccines
Recombinant subunit vaccines						
DNA vaccines	Defined	No risk for pathogenicity	Lower than conventional vaccines	Both humoral and cellular responses.	Multiple doses typically required.	*In vivo* amplification systems. Inefficient transfection Risk of integration into genome not completely excluded.
Live vectors vaccines	Defined	Risk of reversion to virulent forms when using attenuated pathogens as carriers	Moderate	Both humoral and cellular responses	One or few doses normally required No need for adjuvants	Possibility to develop oral vaccines with live bacterial vectors
Protein immunogens	Defined	No risk for pathogenicity	Generally expensive. Cost depends on production system	Mainly humoral responses	Multiple doses typically required. Adjuvants needed	Soluble proteins are poor immunogens. Immunogenicity can be enhanced by VLPs or antigen display delivery systems

[1]Adapted from [29, 30].

Table 2 Developments in recombinant IBDV vaccines

Recombinant vaccine	Expression system	IBDV antigen	Delivery route	Protective efficacy against challenge	Reference
DNA vaccines					
		vp2	Intramuscular	52% protection was defined as the percentage of chickens with no or mild BF lesions	[31]
		vp2-vp4-vp3	Intramuscular	90–100% when chickens were vaccinated three times with 7.5 or 10 mg of plasmid. Lower doses or fewer immunizations could not elicit protection against challenge. Protection was evaluated by bursal weight/body weight (B/B) ratio and gross and histopathological bursal lesions examination	[32]
		vp2 and *vp2-vp4-vp3*	Intramuscular	15 and 30% when chickens were vaccinated three times with VP2 or VP243 plasmids, respectively. Protection was evaluated by mortality, gross and histopathological bursal lesions and B/B ratio	[33]
		vp2	Intramuscular	64% protection was evidenced by absence of clinical signs, atrophy of BF and mortality	[34]
		vp2-vp4-vp3, vp2, vp2-vp4, vp4, vp3, vp3-vp4	Intramuscular	80% with VP243, VP2 and VP24 plasmids and 0% with VP4, VP3 and VP34 plasmids. Protection was evaluated by mortality, gross and histopathological bursal lesions and B/B ratio	[35]
		vp2-vp4-vp3 and *vp2*	Intramuscular	66.6 and 20% with 200 μg of VP243 and VP2 plasmids when homologous virus was used in challenge and 40 and 0%, respectively, when heterologous virus was used in challenge. Protection rate was initially calculated based on gross bursal lesions and further confirmed by bursa histopathological examination	[36]
		vp2 and *vp2-vp4-vp3*	Intramuscular	50 and 70% survival rate and less bursal damage compared with unvaccinated group, with VP2 and VP243 plasmids, respectively	[37]
		vp2	Intramuscular	80% protection was evaluated by morbidity and gross bursal lesions observation	[38]
		vp2-vp4-vp3	*In ovo*	100 and 20% survival rate when *in ovo* prime with DNA vaccine was followed or not with killed-vaccine boost, respectively	[39]
Live delivery system					
Viral vectors	Adenovirus	VP2	Different delivery routes were evaluated	100% protection when CELO-VP2 vector under the CMV promoter was used and following intradermal or subcutaneous double (or single) inoculation(s) of chickens. Other vectors, delivery routes and vaccination schemes were evaluated yielding lower protection rates	[40]
	Adeno-associated virus	VP2	Intramuscular	80% survival rate and absence of IBD clinical signs in surviving chickens	[41]
	Fowlpox virus	VP2	Wing-web puncture	78% protection was defined as the percentage of chickens with no or mild BF lesions	[42]
	Fowlpox virus	VP2	Wing-web puncture	60–100% protection against different challenge titres in various chicken lines. Protection was evaluated based on the degree of bursal damage	[43]
	Fowlpox virus and MD virus	VP2	Subcutaneous	0 and 13% protection against gross bursal lesion and 33 and 42% protection against histopathological BF lesion with recombinant Fowlpox virus (rFPV) and recombinant Marek's Disease virus (rMDV), respectively. Co-administration of both viral vectors yielded 14 and 53–73% protection against gross and histopathological bursal lesions respectively	[44]
	MD virus	VP2	Subcutaneous	100% survival rate and 55% protection defined as the percentage of chickens with no or mild BF lesions	[45]

Table 2 (Continued)

Recombinant vaccine	Expression system	IBDV antigen	Delivery route	Protective efficacy against challenge	Reference
	MD virus	VP2		87 and 100% survival rate, and 53 and 73% protection defined as the percentage of chickens with no or mild BF lesions after vaccination with 10^3 and 10^4 PFU, respectively	[46]
	MD virus	VP2	Subcutaneous	50, 60 and 80% when chickens were vaccinated with 1000, 2000 and 5000 PFU of rMDV, respectively. Protection ratio (%)=(mortality of unvaccinated rMDV-challenged chickens – mortality of vaccinated rMDV-challenged chickens)/mortality of unvaccinated rMDV-challenged chickens × 100%	[47]
	Newcastle disease virus	VP2	Eyedrop	80–100% protection was determined by the absence of viral antigen in bursa 3 days post challenge	[48]
	HVT	VP2	Subcutaneous	100% when chickens were raised in experimental conditions. Chickens raised in field conditions had lower protection. Protection was determined based on the absence of gross bursal lesions and B/B ratio	[49]
	HVT	VP2	Subcutaneous	93% protection was defined as the percentage of chickens with absence of bursal atrophy	[50]
	HVT	VP2	Subcutaneous	Protection based on the absence of bursal atrophy and mild histopathological bursal lesions was observed in specific pathogen free (SPF) chickens and commercial broilers	[51]
	HVT	VP2	Intramuscular	100% protection was evaluated by gross bursal lesions examination	[52]
	HVT	VP2	Subcutaneous	100 and 58% when chickens were vaccinated with rHVT-pecVP2 or rHVT-cmvVP2, respectively. Protection was evaluated by gross bursal lesions examination	[53]
Bacterial vectors	*E. coli*	VP2	Oral	73, 91 and 95.4% when chickens were vaccinated with 10^7, 10^8 and 10^9 CFU, respectively. Protection was assessed on the basis of gross bursal lesions and confirmed by bursa histopathological examination	[54]
	E. coli	VP2	Intramuscular	85.7% protection was defined by the percentage of chickens no or mild BF lesions	[55]
	S. typhimurium	VP2-VP4-VP3	Oral	73.3% protection was initially calculated based on gross bursal lesions and further confirmed by bursa histopathological examination	[56]
Protein immunogens					
Soluble protein	Fowlpox virus	VP2	Intramuscular	17–100% depending on vaccine concentration and vaccination scheme. Protection was defined as the percentage of chickens with no or mild BF lesions	[57]
	Pichia pastoris	VP2	Intramuscular	100%. Protection was determined by the absence of IBDV in bursa 3 days post-challenge and the degree of bursal atrophy	[58]
	E. coli	VP2	Intramuscular	90–100%. Protection was determined by the absence of IBDV in BF 4 days post-challenge	[55]
	E. coli	rVP252–417	Intramuscular	100%. Protection was defined as the percentage with no or mild BF lesions	[59]
	A. thaliana	VP2	Subcutaneous and oral	60 and 80% with subcutaneous and oral vaccine respectively. Protection was determined based on B/B ratio	[60]

	Saccharomyces cerevisiae	VP2/VP4/VP3 and VP2	Intramuscular	100% passive protection. All chickens injected i.p. with chicken antisera to native VP2a/2b were protected against challenge. Protection was assessed by the absence of viral antigen in the bursa 3 days after challenge.	[61]
	E. coli	VP2	Intramuscular	20 and 57% survival rate when chickens were vaccinated with 50 µg of purified VP2 protein or 150 µg of crude VP2 protein, respectively	[62]
	E. coli	Multi-mimotope peptide 5EPIS	Intramuscular	100 and 40% survival rate when 5EPIS was administered with or without adjuvant, respectively. Surviving chickens did not show clinical signs and pathological lesions of IBDV infection	[63, 64]
Non-soluble protein	Rice	VP2	Oral	16.6, 33.3 and 83.3% after vaccination with 1, 3 or 5 g of transgenic rice seeds, respectively. Protection was assessed based on the absence of gross bursal lesions and BF histopathological examination	[65]
	Baculovirus display	VP2	Intramuscular	100% survival rate and 90% protection defined as the percentage of chickens that remained healthy and had no or mild gross bursal lesions	[66]
	T4 bacteriophage display	VP2	Subcutaneous priming and intramuscular boost	100% survival rate and no clinical signs of IBD	[67]
	Bamboo mosaic virus	VP2 P domain loop P_{BC}	Intramuscular	100% survival rate	[68]
	Semliki forest virus	VP2/VP2-VP4-VP3	Intramuscular	No challenge	[69]
	Baculovirus	N5-452H SVP	Intramuscular	Neither mortality nor presence of virus in bursa after challenge with very virulent IBDV	[70]
	Baculovirus	VP2, VPX or VP234 VLPs	Subcutaneous	100% survival rate and no severe clinical signs after vaccination with 25 µg of the purified VLPs. At lower doses, VP2 VLPs had higher protective efficacy than VP234 VLPs	[71]
	E. coli	VP2-VP4-VP3 VLPs		91% survival rate although bursal atrophy similar to the unvaccinated group was observed	[72]
	E. coli	HBc-5EPIS, VLPs	Intramuscular	100% survival rate. Surviving chickens did not show clinical signs and pathological lesions of IBDV infection	[64]

E. coli expression systems have also been used to obtain these particles as vaccines. Baculoviruses are able to invade mammalian or avian cells; however, they are not known to initiate a replication cycle and producing infectious virus in these hosts [78, 79]. Baculoviruses are viruses that can infect insects and replicate in insect cells, resulting in a very safe system for producing proteins to be used in superior organisms. The presence of any trace of baculovirus in the formulation has no harmful effect on the host receiving the vaccine; on the contrary, residues of baculovirus can even work as immunomodulators or as antivirals in chickens [80, 81]. These newly discovered features of baculoviruses may have also contributed to the success of such recombinant vaccines.

Among the different strategies to generate vaccines against the disease, the construction of recombinant virus is the most widely-used approach. This may be because of the advantage of obtaining bivalent vaccines, the auto-replicative attribute of the viral vectors, or even because of the type of immune response (cellular response) that viruses are able to elicit. Nevertheless, they must be inoculated by the subcutaneous route, and they must be stored in liquid nitrogen. Conversely, the method exploited to a smaller extent is the plant expression system. Molecular farming, which refers to the use of transgenic plants for the expression of different antigens has been increasingly employed for the production of experimental immunogens. For different reasons, the design of efficient edible vaccines by means of transgenic plants represents a challenging alternative to the conventional ones [82]. The simplicity of their production, handling and administration makes them an attractive option for developing affordable vaccines. In addition, products from transgenic plants are unlikely to be contaminated by animal pathogens, microbial toxins or oncogenic sequences [83, 84]. Several authors have reported antibody response to parenteral or oral administration of plant-derived antigens [85–91]. The use of this expression platform was also described for IBD, where transgenic *Arabidopsis thaliana* [60] or rice seeds [65] expressing VP2 were assayed as oral vaccines. Results show a dose-dependent efficacy, although a 100% of survival rate was not achieved and several doses of the immunogen were needed. It is worthy to notice that VP2 was resistant to gut degradation.

Concluding Remarks

The main issues that the development of recombinant vaccines needs to tackle are basically: safety, cold chain and differentiation between infected and vaccinated animals. New generation vaccines act upon the immune system in different ways, depending on the type of vaccine but the most important advantage of all of them is that they lack the infectious agent. This feature not only eliminates the possibility of infection but also makes possible the differentiation between infected and vaccinated animals, since diagnostic strategies can be approached for the detection of antibodies against specific proteins that allow this discrimination, which is of utmost importance during disease eradication campaigns. Subunit and synthetic peptides vaccines do not require cold chain as conventional vaccines do. This advantage makes it possible to reduce costs at the time of vaccine storage and distribution.

In this context, IBDV recombinant vaccines that include VP2 antigen have been successful in providing protection against IBDV infection, depending on the delivery route and dose of the vaccine, without the risk of reverting to virulence. Nevertheless, the cost-effectiveness and commercial viability of these vaccines is still an issue. We hope to witness an increase in the number of commercially available IBDV recombinant vaccines and the actual adoption of this so useful technology by the poultry industry in the near future.

References

1. Dobos P, Hill BJ, Hallett R, Kells DT, Becht H, Teninges D. Biophysical and biochemical characterization of five animal viruses with bisegmented double-stranded RNA genomes. Journal of Virology 1979;32(2):593–605.
2. Kibenge FS, Dhillon AS, Russell RG. Biochemistry and immunology of infectious bursal disease virus. Journal of General Virology 1988;69(Pt 8):1757–75.
3. Sharma JM, Kim IJ, Rautenschlein S, Yeh HY. Infectious bursal disease virus of chickens: pathogenesis and immunosuppression. Developmental and Comparative Immunology 2000;24(2–3):223–35.
4. Lejal N, Da Costa B, Huet JC, Delmas B. Role of Ser-652 and Lys-692 in the protease activity of infectious bursal disease virus VP4 and identification of its substrate cleavage sites. Journal of General Virology 2000;81(Pt 4):983–92.
5. Da Costa B, Chevalier C, Henry C, Huet JC, Petit S, Lepault J, *et al.* The capsid of infectious bursal disease virus contains several small peptides arising from the maturation process of pVP2. Journal of Virology 2002;76(5):2393–402.
6. Chevalier C, Lepault J, Erk I, Da Costa B, Delmas B. The maturation process of pVP2 requires assembly of infectious bursal disease virus capsids. Journal of Virology 2002;76(5):2384–92.
7. Allan WH, Faragher JT, Cullen GA. Immunosuppression by the infectious bursal agent in chickens immunised against Newcastle disease. Veterinary Record 1972;90(18):511–2.
8. Faragher JT, Allan WH, Wyeth PJ. Immunosuppressive effect of infectious bursal agent on vaccination against Newcastle disease. Veterinary Record 1974;95(17):385–8.
9. Berg TP. Acute infectious bursal disease in poultry: a review. Avian Pathology 2000;29(3):175–94.
10. Jackwood DJ, Saif YM, Hughes JH. Characteristics and serologic studies of two serotypes of infectious bursal disease virus in turkeys. Avian Diseases 1982;26(4):871–82.
11. McFerran JB, McNulty MS, McKillop ER, Connor TJ, McCracken RM, Collins DS, *et al.* Isolation and serological

studies with infectious bursal disease viruses from fowl, turkeys and ducks: demonstration of a second serotype. Avian Pathology 1980;9(3):395–404.

12. McNulty MS, Saif YM. Antigenic relationship of non-serotype 1 turkey infectious bursal disease viruses from the United States and United Kingdom. Avian Diseases 1988;32(2):374–5.
13. Alkhalaf A. Detection of variant strains of infectious bursal disease virus in broiler flocks in Saudi Arabia using antigen capture enzyme-linked immunosorbent assay. Pakistan Veterinary Journal 2009;29(4):161–4.
14. Berg TP, Gonze M, Morales D, Meulemans G. Acute infectious bursal disease in poultry: Immunological and molecular basis of antigenicity of a highly virulent strain. Avian Pathology 1996;25(4):751–68.
15. Heine HG, Haritou M, Failla P, Fahey K, Azad A. Sequence analysis and expression of the host-protective immunogen VP2 of a variant strain of infectious bursal disease virus which can circumvent vaccination with standard type I strains. Journal of General Virology 1991;72(Pt 8):1835–43.
16. Singh ABM, Sharma R, Sarkhel BC, Singh S, Jain SK. Detection of very virulent infectious bursal disease virus from a field outbreak in Central India. Acta Veterinaria Hungarica 2012;60(1):165–74.
17. Muller R, Kaufer I, Reinacher M, Weiss E. Immunofluorescent studies of early virus propagation after oral infection with infectious bursal disease virus (IBDV). Zentralblatt für Veterinärmedizin Reihe B 1979;26(5):345–52.
18. Jackwood DJ, Cookson KC, Sommer-Wagner SE, Le Galludec H, de Wit JJ. Molecular characteristics of infectious bursal disease viruses from asymptomatic broiler flocks in Europe. Avian Diseases 2006;50(4):532–6.
19. Muller H, Islam MR, Raue R. Research on infectious bursal disease – the past, the present and the future. Veterinary Microbiology 2003;97(1–2):153–65.
20. Ramirez-Nieto G, Shivaprasad HL, Kim CH, Lillehoj HS, Song H, Osorio IG, *et al.* Adaptation of a mallard H5N2 low pathogenicity influenza virus in chickens with prior history of infection with infectious bursal disease virus. Avian Diseases 2010;54(Suppl. 1):513–21.
21. He CQ, Ma LY, Wang D, Li GR, Ding NZ. Homologous recombination is apparent in infectious bursal disease virus. Virology 2009;384(1):51–8.
22. Muller H, Mundt E, Eterradossi N, Islam MR. Current status of vaccines against infectious bursal disease. Avian Pathology 2012;41(2):133–9.
23. Jeurissen SH, Janse EM, Lehrbach PR, Haddad EE, Avakian A, Whitfill CE. The working mechanism of an immune complex vaccine that protects chickens against infectious bursal disease. Immunology 1998;95(3):494–500.
24. Whitfill CE, Haddad EE, Ricks CA, Skeeles JK, Newberry LA, Beasley JN, *et al.* Determination of optimum formulation of a novel infectious bursal disease virus (IBDV) vaccine constructed by mixing bursal disease antibody with IBDV. Avian Diseases 1995;39(4):687–99.
25. Giambrone JJ, Dormitorio T, Brown T. Safety and efficacy of *in ovo* administration of infectious bursal disease viral vaccines. Avian Diseases 2001;45(1):144–8.
26. Haddad EE, Whitfill CE, Avakian AP, Ricks CA, Andrews PD, Thoma JA, *et al.* Efficacy of a novel infectious bursal disease virus immune complex vaccine in broiler chickens. Avian Diseases 1997;41(4):882–9.
27. Ivan J, Velhner M, Ursu K, German P, Mato T, Dren CN, *et al.* Delayed vaccine virus replication in chickens vaccinated subcutaneously with an immune complex infectious bursal disease vaccine: quantification of vaccine virus by real-time polymerase chain reaction. Canadian Journal of Veterinary Research 2005;69(2):135–42.
28. Mahgoub HA. An overview of infectious bursal disease. Archives of Virology 2012;157:2047–57.
29. Hansson M, Nygren PA, Stahl S. Design and production of recombinant subunit vaccines. Biotechnology and Applied Biochemistry 2000;32(Pt 2):95–107.
30. Liljeqvist S, Stahl S. Production of recombinant subunit vaccines: protein immunogens, live delivery systems and nucleic acid vaccines. Journal of Biotechnology 1999;73(1):1–33.
31. Kumar S, Ahi YS, Salunkhe SS, Koul M, Tiwari AK, Gupta PK, *et al.* Effective protection by high efficiency bicistronic DNA vaccine against infectious bursal disease virus expressing VP2 protein and chicken IL-2. Vaccine 2009;27(6):864–9.
32. Hsieh MK, Wu CC, Lin TL. DNA-mediated vaccination conferring protection against infectious bursal disease in broiler chickens in the presence of maternal antibody. Vaccine 2010;28(23):3936–43.
33. Sun JH, Yan YX, Jiang J, Lu P. DNA immunization against very virulent infectious bursal disease virus with VP2-4-3 gene and chicken IL-6 gene. Journal of Veterinary Medicine. B, Infectious Diseases and Veterinary Public Health 2005;52(1):1–7.
34. Mahmood MS, Siddique M, Hussain I, Khan A, Mansoor MK. Protection capability of recombinant plasmid DNA vaccine containing VP2 gene of very virulent infectious bursal disease virus in chickens adjuvanted with CpG oligodeoxynucleotide. Vaccine 2006;24(22):4838–46.
35. Chang HC, Lin TL, Wu CC. DNA vaccination with plasmids containing various fragments of large segment genome of infectious bursal disease virus. Vaccine 2003;21(5–6):507–13.
36. Li J, Liang X, Huang Y, Meng S, Xie R, Deng R, *et al.* Enhancement of the immunogenicity of DNA vaccine against infectious bursal disease virus by co-delivery with plasmid encoding chicken interleukin 2. Virology 2004;329(1):89–100.
37. Kim SJ, Sung HW, Han JH, Jackwood D, Kwon HM. Protection against very virulent infectious bursal disease virus in chickens immunized with DNA vaccines. Veterinary Microbiology 2004;101(1):39–51.
38. Zhang HH, Yang XM, Xie QM, Ma JY, Luo YN, Cao YC, *et al.* The potent adjuvant effects of chicken beta-defensin-1 when genetically fused with infectious bursal disease virus VP2 gene. Veterinary Immunology and Immunopathology 2010;136(1–2):92–7.
39. Park JH, Sung HW, Yoon BI, Kwon HM. Protection of chicken against very virulent IBDV provided by *in ovo* priming with DNA vaccine and boosting with killed vaccine and the adjuvant effects of plasmid-encoded chicken interleukin-2 and interferon-gamma. Journal of Veterinary Science 2009;10(2):131–9.
40. Francois A, Chevalier C, Delmas B, Eterradossi N, Toquin D, Rivallan G, *et al.* Avian adenovirus CELO recombinants expressing VP2 of infectious bursal disease virus induce

protection against bursal disease in chickens. Vaccine 2004;22(17–18):2351–60.

41. Perozo F, Villegas P, Estevez C, Alvarado IR, Purvis LB, Williams S. Protection against infectious bursal disease virulent challenge conferred by a recombinant avian adeno-associated virus vaccine. Avian Diseases 2008;52(2):315–9.

42. Butter C, Sturman TD, Baaten BJ, Davison TF. Protection from infectious bursal disease virus (IBDV)-induced immunosuppression by immunization with a fowlpox recombinant containing IBDV-VP2. Avian Pathology 2003;32(6):597–604.

43. Shaw I, Davison TF. Protection from IBDV-induced bursal damage by a recombinant fowlpox vaccine, fpIBD1, is dependent on the titre of challenge virus and chicken genotype. Vaccine 2000;18(28):3230–41.

44. Tsukamoto K, Sato T, Saito S, Tanimura N, Hamazaki N, Mase M, *et al.* Dual-viral vector approach induced strong and long-lasting protective immunity against very virulent infectious bursal disease virus. Virology 2000;269(2):257–67.

45. Tsukamoto K, Kojima C, Komori Y, Tanimura N, Mase M, Yamaguchi S. Protection of chickens against very virulent infectious bursal disease virus (IBDV) and Marek's disease virus (MDV) with a recombinant MDV expressing IBDV VP2. Virology 1999;257(2):352–62.

46. Zhou X, Wang D, Xiong J, Zhang P, Li Y, She R. Protection of chickens, with or without maternal antibodies, against IBDV infection by a recombinant IBDV-VP2 protein. Vaccine 2010;28(23):3990–6.

47. Liu HM, Qin AJ, Liu YL, Jin WJ, Ye JQ, Chen HJ, *et al.* Construction and immunological characterization of recombinant Marek's disease virus expressing IBDV VP2 fusion protein. Chinese Journal of Biotechnology 2006;22(3):391–6.

48. Huang Z, Elankumaran S, Yunus AS, Samal SK. A recombinant newcastle disease virus (NDV) expressing VP2 protein of infectious bursal disease virus (IBDV) protects against NDV and IBDV. Journal of Virology 2004;78(18):10054–63.

49. Bublot M, Pritchard N, Le Gros FX, Goutebroze S. Use of a vectored vaccine against infectious bursal disease of chickens in the face of high-titred maternally derived antibody. Journal of Comparative Pathology 2007;137(Suppl. 1):S81–4.

50. Le Gros FX, Dancer A, Giacomini C, Pizzoni L, Bublot M, Graziani M, *et al.* Field efficacy trial of a novel HVT-IBD vector vaccine for 1-day-old broilers. Vaccine 2009;27(4):592–6.

51. Perozo F, Villegas AP, Fernandez R, Cruz J, Pritchard N. Efficacy of single dose recombinant herpesvirus of turkey infectious bursal disease virus (IBDV) vaccination against a variant IBDV strain. Avian Diseases 2009;53(4):624–8.

52. Darteil R, Bublot M, Laplace E, Bouquet JF, Audonnet JC, Riviere M. Herpesvirus of turkey recombinant viruses expressing infectious bursal disease virus (IBDV) VP2 immunogen induce protection against an IBDV virulent challenge in chickens. Virology 1995;211(2):481–90.

53. Tsukamoto K, Saito S, Saeki S, Sato T, Tanimura N, Isobe T, *et al.* Complete, long-lasting protection against lethal infectious bursal disease virus challenge by a single vaccination with an avian herpesvirus vector expressing VP2 antigens. Journal of Virology 2002;76(11):5637–45.

54. Mahmood MS, Hussain I, Siddique M, Akhtar M, Ali S. DNA vaccination with VP2 gene of very virulent infectious bursal disease virus (vvIBDV) delivered by transgenic *E. coli* DH5alpha given orally confers protective immune responses in chickens. Vaccine 2007;25(44):7629–35.

55. Rong J, Cheng T, Liu X, Jiang T, Gu H, Zou G. Development of recombinant VP2 vaccine for the prevention of infectious bursal disease of chickens. Vaccine 2005;23(40):4844–51.

56. Li L, Fang W, Li J, Huang Y, Yu L. Oral DNA vaccination with the polyprotein gene of infectious bursal disease virus (IBDV) delivered by the attenuated Salmonella elicits protective immune responses in chickens. Vaccine 2006;24(33–34): 5919–27.

57. Su BS, Chiu HH, Lin CC, Shien JH, Yin HS, Lee LH. Adjuvant activity of chicken interleukin-12 co-administered with infectious bursal disease virus recombinant VP2 antigen in chickens. Veterinary Immunology and Immunopathology 2011;139(2–4):167–75.

58. Pitcovski J, Gutter B, Gallili G, Goldway M, Perelman B, Gross G, *et al.* Development and large-scale use of recombinant VP2 vaccine for the prevention of infectious bursal disease of chickens. Vaccine 2003;21(32):4736–43.

59. Pradhan SN, Prince PR, Madhumathi J, Roy P, Narayanan RB, Antony U. Protective immune responses of recombinant VP2 subunit antigen of infectious bursal disease virus in chickens. Veterinary Immunology and Immunopathology 2012;148(3–4):293–301.

60. Wu H, Singh NK, Locy RD, Scissum-Gunn K, Giambrone JJ. Immunization of chickens with VP2 protein of infectious bursal disease virus expressed in Arabidopsis thaliana. Avian Diseases 2004;48(3):663–8.

61. Fahey KJ, Chapman AJ, Macreadie IG, Vaughan PR, McKern NM, Skicko JI, *et al.* A recombinant subunit vaccine that protects progeny chickens from infectious bursal disease. Avian Pathology 1991;20(3):447–60.

62. Omar AR, Kim CL, Bejo MH, Ideris A. Efficacy of VP2 protein expressed in *E. coli* for protection against highly virulent infectious bursal disease virus. Journal of Veterinary Science 2006;7(3):241–7.

63. Wang YS, Fan HJ, Li Y, Shi ZL, Pan Y, Lu CP. Development of a multi-mimotope peptide as a vaccine immunogen for infectious bursal disease virus. Vaccine 2007;25(22):4447–55.

64. Wang YS, Ouyang W, Liu XJ, He KW, Yu SQ, Zhang HB, *et al.* Virus-like particles of hepatitis B virus core protein containing five mimotopes of infectious bursal disease virus (IBDV) protect chickens against IBDV. Vaccine 2012;30(12):2125–30.

65. Wu J, Yu L, Li L, Hu J, Zhou J, Zhou X. Oral immunization with transgenic rice seeds expressing VP2 protein of infectious bursal disease virus induces protective immune responses in chickens. Plant Biotechnology Journal 2007;5(5):570–8.

66. Xu XG, Tong DW, Wang ZS, Zhang Q, Li ZC, Zhang K, *et al.* Baculovirus virions displaying infectious bursal disease virus VP2 protein protect chickens against infectious bursal disease virus infection. Avian Diseases 2011;55(2):223–9.

67. Cao YC, Shi QC, Ma JY, Xie QM, Bi YZ. Vaccination against very virulent infectious bursal disease virus using recombinant T4 bacteriophage displaying viral protein VP2. Acta Biochimica et Biophysica Sinica (Shanghai) 2005;37(10):657–64.

68. Chen TH, Hu CC, Liao JT, Lee CW, Liao JW, Lin MY, *et al.* Induction of protective immunity in chickens immunized with plant-made chimeric Bamboo mosaic virus particles

expressing very virulent Infectious bursal disease virus antigen. Virus Research 2012;166(1–2):109–15.

69. Phenix KV, Wark K, Luke CJ, Skinner MA, Smyth JA, Mawhinney KA, *et al.* Recombinant Semliki Forest virus vector exhibits potential for avian virus vaccine development. Vaccine 2001;19(23–24):3116–23.

70. Ho J, Lee L, Lin Y, Tai Y, Chang C, Chou Y, *et al.* vaccine development through terminal deletions of an infectious bursal disease virus protein 2 precursor variant. Process Biochemistry 2010;45:786–93.

71. Martinez-Torrecuadrada JL, Saubi N, Pages-Mante A, Caston JR, Espuna E, Casal JI. Structure-dependent efficacy of infectious bursal disease virus (IBDV) recombinant vaccines. Vaccine 2003;21(23):3342–50.

72. Rogel A, Benvenisti L, Sela I, Edelbaum O, Tanne E, Shachar Y, *et al.* Vaccination with *E. coli* recombinant empty viral particles of infectious bursal disease virus (IBDV) confer protection. Virus Genes 2003;27(2):169–75.

73. Liu MA. DNA vaccines: a review. Journal of Internal Medicine 2003;253(4):402–10.

74. Coulibaly F, Chevalier C, Gutsche I, Pous J, Navaza J, Bressanelli S, *et al.* The birnavirus crystal structure reveals structural relationships among icosahedral viruses. Cell 2005;120(6):761–72.

75. Azad AA, Jagadish MN, Brown MA, Hudson PJ. Deletion mapping and expression in *Escherichia coli* of the large genomic segment of a birnavirus. Virology 1987;161(1):145–52.

76. Letzel T, Coulibaly F, Rey FA, Delmas B, Jagt E, van Loon AA, *et al.* Molecular and structural bases for the antigenicity of VP2 of infectious bursal disease virus. Journal of Virology 2007;81(23):12827–35.

77. Qi X, Gao H, Gao Y, Qin L, Wang Y, Gao L, *et al.* Naturally occurring mutations at residues 253 and 284 in VP2 contribute to the cell tropism and virulence of very virulent infectious bursal disease virus. Antiviral Research 2009;84(3):225–33.

78. Huser A, Hofmann C. Baculovirus vectors: novel mammalian cell gene-delivery vehicles and their applications. American Journal of PharmacoGenomics 2003;3(1):53–63.

79. Gao H, Wang Y, Li N, Peng WP, Sun Y, Tong GZ, *et al.* Efficient gene delivery into mammalian cells mediated by a recombinant baculovirus containing a whispovirus ie1 promoter, a novel shuttle promoter between insect cells and mammalian cells. Journal of Biotechnology 2007;131(2):138–43.

80. Chimeno Zoth S, Carballeda JM, Gomez E, Gravisaco MJ, Carrillo E, Berinstein A. Modulation of innate immunity in chickens induced by *in vivo* administration of baculovirus. Veterinary Immunology and Immunopathology 2012;145(1–2): 241–7.

81. Mingshan Niu WL, Yan Han and Lijia An. Baculovirus up-regulates antiviral systems and induces protection against infectious bronchitis virus challenge in neonatal chicken. International Immunopharmacology 2008;8(12):1609–15.

82. Ryan EJ, Daly LM, Mills KH. Immunomodulators and delivery systems for vaccination by mucosal routes. Trends in Biotechnology 2001;19(8):293–304.

83. Fischer R, Emans N. Molecular farming of pharmaceutical proteins. Transgenic Research 2000;9(4–5):279–99.

84. Fischer R, Hoffmann K, Schillberg S, Emans N. Antibody production by molecular farming in plants. Journal of Biological Regulators and Homeostatic Agents 2000;14(2):83–92.

85. Arakawa T, Chong DK, Langridge WH. Efficacy of a food plant-based oral cholera toxin B subunit vaccine. Nature Biotechnology 1998;16(3):292–7.

86. Berinstein A, Vazquez-Rovere C, Asurmendi S, Gomez E, Zanetti F, Zabal O, *et al.* Mucosal and systemic immunization elicited by Newcastle disease virus (NDV) transgenic plants as antigens. Vaccine 2005;23(48–49):5583–9.

87. Gomez E, Chimeno Zoth S, Carrillo E, Estela Roux M, Berinstein A. Mucosal immunity induced by orally administered transgenic plants. Immunobiology 2008;213(8):671–5.

88. Haq TA, Mason HS, Clements JD, Arntzen CJ. Oral immunization with a recombinant bacterial antigen produced in transgenic plants. Science 1995;268(5211):714–6.

89. Mason HS, Ball JM, Shi JJ, Jiang X, Estes MK, Arntzen CJ. Expression of Norwalk virus capsid protein in transgenic tobacco and potato and its oral immunogenicity in mice. Proceedings of the National Academy of Sciences, USA 1996;93(11):5335–40.

90. Thanavala Y, Yang YF, Lyons P, Mason HS, Arntzen C. Immunogenicity of transgenic plant-derived hepatitis B surface antigen. Proceedings of the National Academy of Sciences, USA 1995;92(8):3358–61.

91. Wigdorovitz A, Pérez Filgueira DM, Robertson N, Carrillo C, Sadir AM, Morris TJ, *et al.* Protection of mice against challenge with foot and mouth disease virus (FMDV) by immunization with foliar extracts from plants infected with recombinant tobacco mosaic virus expressing the FMDV structural protein VP1. Virology 1999;264(1):85–91.

Fish frauds: the DNA challenge

Andrea Armani*, Lorenzo Castigliego and Alessandra Guidi

Address: Department of Veterinary Sciences, University of Pisa, Via delle Piagge 2, 56124, Pisa, Italy.

***Correspondence**: Andrea Armani. Email: aarmani@vet.unipi.it

Received: 31 March 2012
Accepted: 23 October 2012

Abstract

The rising demand for seafood and trade globalization has brought about a rapid increase in the number of fish species traded. Consequently, the occurrence of mislabelling is growing as well, reaching levels of concerns in USA, Canada and Europe. In this light, the evolving consciousness of consumers and the new exigencies of commerce call for greater safety and quality requirements. These factors have made urgent the need for efficient traceability systems, aimed to ensure transparency on the identity and origin of the traded products and the compliance with the regulations concerning illegal fisheries and labelling. Moreover, greater efforts are necessary to create a list of market names that can be recognized both locally and internationally, in order to overcome the confusion regarding fish denominations. In this context, molecular analysis represents the most promising challenge to verify and support traceability in the seafood chain. Nowadays, the three mitochondrial genes *cytb*, *COI* and *16srRNA* are the most targeted for this purpose and, among the available procedures, the DNA bar coding is the most commonly applied to verify the labelling compliances, also at the official level. In this review, the most important issues relating to these topics have been reported and discussed.

Keywords: Fish traceability, Mislabelling, DNA analysis

Review Methodology: The method chosen for writing this review was to select the most relevant material by searching and analysing official documentation, such as regulations and laws issued worldwide by food safety agencies, related to traceability of the fishery chain, publications dealing with molecular methods for fish species identification, and by consulting official websites and scientific databases (FishTrace, Barcode of Life Project and GenBank)

Introduction

The human consumption of fish products has doubled during the last 30 years [1], and some species that were at one time discarded, such as the halibut and the Atlantic bluefin tuna, have been subsequently caught and commercialized worldwide, leading to a drastic stock reduction [2–5]. Current data indicate that overfishing has exploited or depleted up to 75% of the global fish stocks and authorities are facing the need to revise rules on fishery [6–9]. The Asian countries, among the most active in fishing, have been reporting increasing capture statistics in the last 10 years and there are concerns regarding the resources in the western part of the South China Sea [10]. This uncontrolled global catching has a negative impact not only on the targeted species but also on many by-catches, as well as on habitat and piscivorous species, which are being replaced by invertebrates [11–13]. This change in the marine ecosystem has prompted the industry to develop new products, which have influenced consumer preferences, so that they are no longer limited to consuming local seafood. At the same time, advances in conservation and transportation technology have boosted the global seafood trade, making it easier to import from abroad [14–16]. As a consequence, while new exotic products are sold on the markets, verification of the traceability system along the whole chain becomes even more complicated. For example, the jellyfish, an invertebrate increasingly exploited for trade, which in the past was available only in certain Asian countries, is sold in Western countries with many label non-conformities [17].

Seafood is advertised mainly for its healthy characteristics, but several factors, such as price, fashion and ease of use, can affect the typology of demand. The consumers are attracted more to clean cuts and ready-to-cook fishery products than to raw whole fishes, which represent only a fraction of importations (i.e. one-fifth in the USA) [18]. Processing, such as beheading, skinning and cutting, often makes seafood indistinguishable: tails and fillets are one of the most easily disguised forms of fish [19, 20] and in some countries a considerable amount of products enters the market under the anonymous denomination of 'fish fillets' [21].

The possibility of replacing valuable species with cheaper species makes mislabelling a convenient practice. In the USA, the consumer price index for fishery products has risen more than 27% over the past 10 years [22], creating significant economic incentives for fraud: the value of frozen catfish, when sold as grouper, quadruples [13]. In 1997, the National Seafood Inspection Laboratory issued a press release indicating that 37% of fish and 13% of other seafood were labelled incorrectly [23]. A recent review reported false labels on more than one-third of fish [24], while other research found that one-quarter of fish tested in the USA and Canada were mislabelled [25]. Seafood mislabelling has also reached high levels in Europe [26, 27]. Recently, some authors reported that a number of cheaper species have been sold under the name of red snapper, one of the most substituted species in the world [20, 28]. A survey carried out in Ireland discovered that a large amount of fish sold as cod was in reality constituted by a range of different species [29, 30]. The Patagonian toothfish has been often marketed as Chilean sea bass [31]. In South Africa, the shortfin mako shark has been sold as 'ocean fillets' to increase its appeal [32]. Among seafood, bivalves are also subjected to fraud, and many species are fraudulently sold under the very popular name of grooved carpet shell clam [33]. Paradoxically, the South Pacific hake, a pelagic sea fish, has been sold as tilapia, a farm-raised freshwater fish [34].

Whether accidental or deliberate, fish mislabelling not only is a form of economic deception, but it can also threaten fisheries [35], owing to the fact that protected species or species subjected to fishery regulations may be illegally placed on the market. This happened, for instance, to the juvenile form of European pilchard, which, in Italy, has been replaced with cheaper Asiatic fishes [26].

Illicit substitutions may also involve the commercialization of hazardous species, such as the poisonous puffer fish, which are banned from the market in Europe [36] or which can be marketed only if coming from selected areas in the USA [37]. Furthermore, samples labelled as 'swordfish' and 'white steenbras' have been replaced with oilfish, toxic species, whose commercialization has already been regulated in the European Union and in some Asian countries [38]. The examples reported represent only a few of the countless types of frauds that have been uncovered in the seafood sector. One thing is certain: if many of these frauds are successfully set on the market and escape official controls, the common consumers have little chance of identifying a substitution or verifying that what they have bought actually corresponds to what they wanted to buy. All this has made urgent the need for traceability systems on local and international markets, aimed at ensuring transparency on the identity and quality of the traded products and compliance with the regulations concerning the Illegal Unreported Unregulated (IUU) fisheries and labelling. In this context, the molecular analysis represents the most promising challenge to verify traceability in the seafood chain, as discussed later on. Thanks to these techniques, we are discovering that seafood mislabelling is highly widespread worldwide.

What is the Right Name?

It is not so easy to assess how many different kinds of seafood are marketed worldwide and what their right name is. FishBase, a relational database that support the work of professionals operating in the fish sector, developed at the World Fish Center in collaboration with the Food and Agriculture Organization (FAO) of the United Nations and the European Commission, currently reports 32 400 species of fish [39]. In 2003, Rehbein estimated that more than 20 000 species of seafood could be used for human consumption, of which only 500 in Europe [40]. Globalization of the market brought to an increase of 'tradeable species', especially in western countries. For instance, in less than 10 years, the number of species sold on the Italian market, according to an official denomination provided by the Ministry of Agricultural, Food and Forestry Policies, has increased from about 200 to more than 800. About 420 species of fish are sold in the German market alone [41]. The US Seafood List, a compendium that provides the Acceptable FDA Market names, includes approximately 1700 species of finfish and shellfish [42].

The FAO Statistics and Information Service (FIPS), based on the Aquatic Sciences and Fisheries Information System (ASFIS), could be used to estimate the most probable number of seafood marketed at the global level. In 2011, 11 562 species, selected according to their interest or relation to fisheries and aquaculture, have been listed. Unfortunately, some names may not correspond to those commonly used by people [43]. The FAO denominations are not intended to replace local names, but are considered necessary to overcome the great confusion in this field, where a single name can be used for many different species or several names for one species. Furthermore, different vernacular denominations can be used in different regions. In this chaos, mislabelling may also be unintentional. On the other hand, some unscrupulous traders deliberately take profit from this scenario and introduce illegally caught fish into legitimate markets [44]. For this reason, as FAO deals with aquatic species at the global level, names selected should be recognizable both locally and internationally and only those species for which there are fishery statistics can be officially taken into consideration [45, 46].

Traceability Issues in the Fishery Chain

In the American food industry, the history of traceability dates backs to the early part of the 20th century, when the development of refrigerated railcars allowed the transport of products from distant areas to the urban centres [47]. In 2009, 80.4% of the world's fishery production was exported [48] and the developed countries, in particular, have become more and more disconnected from the original source [49, 50]. Europe is one of the world's largest importers of seafood. A recent report estimated that in 2011, Europeans had hypothetically eaten the total annual quantity of fish produced by EU fisheries prior to July 2, relying after this date on fish sourced from foreign waters [50–52].

The long-distance transactions have introduced many intermediaries into the market chain, bringing new problems with the quality assessment of perishable products. If at one time the stakeholders were only interested in establishing their legal rights in case of a disputed transaction, more recently, the impetus to further develop traceability systems in the food chain is mainly related to its implications in food safety [47].

When in 1987 traceability was for the first time defined by the international norm NF EN ISO 8402, it was recognized as 'an immediate issue with long-term implications' [53]. After almost 25 years and the implementation of many regulations, the accomplishment of seafood traceability is still not without challenges. Even though the first level of traceability could be easily achieved within one link of the chain, the ability to track a product along the whole chain is much more complicated, because it changes hands an average of five to seven times [54–56]. In this way, the information about the product could be sometimes lost and, by the time it arrives at the final consumer, its name may be completely changed [35].

While the traditional trade barriers, such as tariffs, are being reduced, the food safety standards, and the regulations related to traceability and product certification have assumed a pivotal role. The minimum requirement for traceability is that each traceable unit must be uniquely labelled to permit identification. For this reason, even though the labelling does not provide traceability, it is an important factor for physical tracking and an effective mean for product diversification. In many countries, government regulations require the full disclosure of food product content and claim that food labelling must not be misleading. At the international level, it is recommended that all seafood labels report species, country of origin, and catching or production method, and many countries have a list of 'acceptable market names' for seafood products [57–59]. Nowadays, the irregular updating and a lack of harmonization between trading partners is one of the main factors affecting the usefulness of such expedients [25, 29, 31, 60].

The EU seems to have the strictest regulation regarding traceability about fishery and aquaculture products. In fact, with the Reg. 178/2001 [61], the traceability for all kinds of food in all stages of production, processing and distribution has been made compulsory. Furthermore, a specific traceability system in the fishery has been issued. With the Reg. 104/2000, a framework was created aimed to facilitate the fair commerce and to provide consumers with a minimum amount of information regarding the products [62]. With the Reg. No. 2065/2001 and the recent Reg. No. 404/2011, the commercial and scientific denomination, the production method and the catch area must be available at each stage of marketing [63, 64]. Moreover, by issuing the Reg. No. 1005/2008 and No. 1010/2009 [7, 8], the EU jointed the FAO's international plan of action to prevent, deter and eliminate IUU fishing (June 2001) [9]. Finally, considering that it is nevertheless advisable to assist consumers who have allergies or intolerances by providing them with more comprehensive information on the composition of foodstuffs, the EU law has made mandatory to include all the allergens, such as crustaceans fish, molluscs and products thereof, in the list of ingredients [65, 66].

The Food and Drug Administration (FDA) has established that both domestically processed and imported seafood should be properly labelled according to the 'Guidance for Industry: The Seafood List – FDA's Guide to Acceptable Market Names for Seafood Sold in Interstate Commerce' and that the substitution and/or mislabelling of seafood are considered a violation of federal law [67]. For this reason, FDA facilitates programmes that include inspection, sampling, analysis, research and education on seafood issues [68].

Even in Canada, the current regulations are not sufficient to comply with today's traceability requirements of some export markets, such as the European Union. Despite a lack of strict regulatory requirements, both government-sponsored traceability programmes and non-governmental seafood sustainability programmes strongly recommend the sharing of additional information along the supply chain. For example, the Fisheries and Oceans Canada (DFO) will provide the necessary catching certificate document when requested by the European company to fulfil the new regulations established to prevent, deter and eliminate IUU fishing [7, 8, 69].

The strict food safety and traceability standards requested from the EU for the importation can have big impacts on the economic performance of many African developing countries. For this reason, many traceability initiatives for perishable food products, such as beef and fish, have been started in the last decade in these countries [70, 71]. Nowadays, the existing regulations in South Africa specify that product labelling should not be misleading to the consumer. In addition, the National Regulator for Compulsory Specifications (NRCS) expressly stipulates that a 'true description (of the) variety of fish' must be stated on the packaging of frozen fish products, but there are no specifications on acceptable market names and the scientific name is not compulsory [72, 73]. Even a great importer such as Japan currently lacks specific traceability regulations for seafood. Finally, a glance at the situation of one of the bigger trader in the world: China, which has recently received negative media attention because of many safety concerns about seafood processed in the country and then exported. However, because its economic engine depends on exportation, China has promised to eventually adopt the USA and European Union's regulatory standards [74].

The DNA Power

Fish species identification is usually based on the analysis of many morphological and meristic features, mainly represented by body shape, colours, fin or teeth shape, number and position [75]. In some cases, officials also rely on sensory evaluation [76]. This kind of approach can be tricky in case of species with limited diagnostic morphological features, cryptic species, juvenile forms or lack of descriptions [77, 78]. Some authors suggest that almost the 34% of the world's fisheries, from 1950 to 2002, have not been identified at the species level and, for the remaining 66%, there is no possibility of verifying if the identification has been performed rigorously [79]. Considering that errors may accumulate in fishery statistics, making the fish stock assessment less precise and, as a consequence, threatening sustainability, even the traditional taxonomists are moving towards modern approaches to get the right diagnosis [79–81].

At the beginning, many methods based on the analysis of proteins had been proposed and the isoelectric focusing methodology [82] was used by FDA as the only official method to provide species-specific biochemical patterns for some of the species reported in the Regulatory Fish Encyclopaedia [42]. Even though this technique has been successfully used in several cases, it cannot always be applied, especially for processed products. This is because some treatments may make proteins unsuitable for this kind of analysis [83, 84].

In 1992, an article titled 'The label said Snapper, the lab said baloney' was published [85]. From that time on, the opportunity to combine outputs from fishery genetic research with applied regulation mechanisms has been recognized and researchers have attempted to develop new methods, based on DNA markers, for identifying fish species without relying on morphological features [86, 87]. As previously reported, the key requirement of any identification system is to be able to answer some questions: 'What species is it? Where was it caught? Can it be traced along the food supply chain?' [44]. Nowadays, most of the methods for fish species identification are based on DNA analysis, even at the official level. In fact, their flexibility, rapidity, and cost-effectiveness permit the identification of a potentially huge number of different species and products [83, 84, 88]. In addition, DNA is thermostable and provides more information than proteins because of the degeneracy of the genetic code, thus enabling even closely-related species to be differentiated and identified [89]. Moreover, the genetic make-up does not undergo changes during the lifespan or in different tissues, permitting the collection and storage of samples in every stage of the fishery chain [90]. Finally, the DNA can potentially be recovered when starting from simple tissues up to complex food matrices [83, 84, 91]. In the past decade, many DNA-based identification methods have been developed and validated. Some of them have been routinely applied in many labs. The most used techniques for fish species identification are polymerase chain reaction (PCR) coupled with restriction enzyme length polymorphism (PCR-RFLP), PCR with sequence-specific primers (PCR-SSP) and PCR-sequencing, better known as FINS (forensically informative nucleotide sequencing) [83, 84, 88, 92]. Moreover, real time-PCR and microarrays represent emerging methods in this field [41, 83, 84, 93, 94]. Briefly, in PCR-RFLP, after PCR amplification, DNA is cut by specific endonucleases producing fragments of different sizes, which, once separated by agarose gel electrophoresis, form a species-specific pattern. This technique is relatively rapid, reproducible, and robust and does not require expensive equipment. It is suitable for routine analysis, such as those performed by food inspection agencies [95, 96]. The PCR-SSP requires a previous knowledge of the target sequences. Authentication simply relies on the presence/absence of a specific amplicon, which can be visualized by agarose gel electrophoresis or by melting analysis in real time-PCR. Multiplex PCR is a variant of PCR-SSP, which permits the simultaneous amplification of many targets using more than one pair of primers in the same reaction. Even though this technique increases the speed of analysis and decreases costs, authentication can be only performed for a small number of putative species [94, 97, 98].

Although in the past PCR-sequencing was time-consuming and technically demanding, it is nowadays relatively rapid and simpler to perform [84]. Given the widespread use of mtDNA in genetic research, many universal primers have been designed to anneal to DNA regions that are generally conserved across species, but encase a DNA fragment with interspecies variation [99]. To permit universal amplification, these primers are sometimes degenerate on certain nucleotide positions, which are known to vary within species. The sequences obtained can be considered a sort of genetic 'bar code' [100]. However, it must be taken into consideration that the 'DNA bar code' may show a certain variation among individuals of the same species. Therefore, for an effective identification, the differences between species should be bigger than the variation within species [101].

The Fish Barcode of Life campaign is an international project aimed to establish a reference library of DNA bar codes for all fish species that have been officially identified [102] and actually includes about 12 000 species of Actinopterygii and almost 1000 of Elasmobranchii [101]. The process of DNA bar coding entails two basic steps: (1) building the DNA bar code library of known species and (2) matching the bar code sequence of the unknown sample against the bar code library for identification [103]. Obtaining high-quality sequence records from the targeted regions is a key requirement for the FISH-BOL enterprise. Once completed, the FISH-BOL database will enable a fast, accurate, and cost-effective system for molecular identification of the world's icthyofauna [104].

The FDA, recently joining the project, represents the most important example of application of FISH-BOL to combat fish frauds. At present, even in some European countries, many laboratories belonging to official organs of control are validating bar coding methods.

Mitochondrial DNA (mtDNA) and markers for species identification

The potential use for genetic identification in fishery traceability was one of the earliest recognized applications of non-human DNA forensics and the ability to generate genetic data is continuously and exponentially increasing. Both nuclear and mtDNA are suitable for fish species identification, but the latter is the most commonly used, because of some advantages:

- it generally evolves much faster than nuclear DNA: its higher relative mutation rate results in the accumulation of enough sequence differences to enable discrimination of closely related species [95, 105];
- it is maternally inherited: individuals have only one allele and this avoids sequence ambiguities from heterozygous genotypes [93, 106];
- the number of mtDNA copies inside a cell is several-fold higher than the number of nuclear DNA copies [18, 107];
- the complete sequence of mtDNA is known for several aquatic organisms, because investigated for evolutionary studies [108, 109];
- it is smaller (from 16 000 to 19 000 nucleotides) than nuclear DNA and has a circular structure, with a higher heat resistance [110, 111];

introns and repetitive sequences are absent [84, 89]. However, pseudogenes have been recently discovered in many fish species [112, 113] and this may require a particular attention, as discussed later.

The *cytochrome b* (*cytb*), *cytochrome oxidase subunit I* (*COI*) and *16s ribosomal RNA* (*16S rRNA*) mitochondrial genes are the most targeted for fish-species identification and can be successfully used for designing specific PCR primers [44, 83, 84]. Two dedicated reference sequence databases, FishTrace, [114] and Barcode of Life Project [101] have been set up for *cytb* and *COI*, respectively. Given their more rigorous rules for depositing data, they represent a step forward with respect to previous databases, such as GenBank [115], some sequences of which may be incorrect [116, 117]. Nevertheless, when relying on databases, as many reference sequences as possible, resulting from different studies, should be evaluated, or an in-house database should be created [44, 83, 118]. To produce a reference sequence, the correct morphological identification of the specimens is fundamental and, in case of samples that cannot be analysed in their whole state, photographic documentation should be produced. Moreover, all the samples used for the production of the database should be clearly identified by a univocal voucher and stored for further analysis.

COI: this gene has been recognized by some authors as the marker of choice for species discrimination, because it has a greater potential for phylogenetic studies than any other mitochondrial gene [100]. Hajibabaei *et al.* [119] reported that even relatively short nucleotide sequences (100–200 bp) can provide accurate identifications. In general, this assumption can be considered true, even though, in some cases, 'aberrant' specimens, belonging to different species, but sharing the same partial haplotypes, have been reported [120]. For the amplification of this gene, a broad range of very robust universal primers applicable across a wide range of taxa do exist (Table 1) [68, 78, 120–124].

The main purpose of the researchers involved in bar coding projects is to develop primers able to amplify the same *COI* gene region from as high as possible a number of fish species. This task is very complex and nowadays, even though much work has still to be accomplished, some very promising protocols exist. Ward *et al.* obtained the amplification of the *COI* gene starting from the DNA of more than 200 different kinds of fishes. This procedure requires the utilization of two pairs of primers in all four pair wise combinations, thus noticeably increasing the number of the reactions [120]. Ivanova *et al.* tried to solve the problems by both projecting degenerate primers and assembling a cocktail of different primers targeting the same regions of the gene. Even though the primers were very effective in amplifying the target region from 94 fish species belonging to different families, the sequencing was successful in only 86% of the samples [123]. In the protocol developed by the FDA, the primers proposed by Baldwin *et al.* [68, 124] were slightly modified and linked to shorts tails, projected by Steffens *et al.* [125], to facilitate sequencing of products. At present, 217 reference nucleotide sequences are available in the Library for Seafood Identification [126].

The DNA bar coding is a molecular method that uses a short standardized DNA sequence, usually from the

Table 1 Universal primers used for the amplification of the *COI* gene from fish. The primers' position has been calculated on the sequence of *Pagrus auriga* (GenBank Accession Number AB124801)

Primer name	Sequence code	Position	Amp. length (bp)	Reference
LCO1490	GGTCAACAAATCATAAAGATATTGG	4417–4441	708	[125]
HC02198	TAAACTTCAGGGTGACCAAAAAATCA	5100–5125		
FishF1	TCAACCAACCACAAAGACATTGGCAC	4419–4444	703/706	[124]
FishF2	TCGACTAATCATAAAGATATCGGCAC	4419–4444		
FishR1	TAGACTTCTGGGTGGCCAAAGAATCA	5100–5125		
FishR2	ACTTCAGGGTGACCGAAGAATCAGAA	5097–5122		
COIF-ALT	ACAAATCAYAARGAYATYGG	4422–4441	698	[126]
COIR-ALT	TTCAGGRTGNCCRAARAAYCA	5100–5120		
FF2d	TTCTCCACCAACCACAARGAYATYGG	4416–4441	707	[127]
FR1d	CACCTCAGGGTGTCCGAARAAYCARAA	5097–5123		
FISH-BCL	TCAACYAATCAYAAAGATATYGGCAC	4419–4444	706	[128]
FISH-BCH	TAAACTTCAGGGTGACCAAAAAATCA	5100–5125		
COI-Fish-F	TTCTCAACTAACCAYAAAGAYATYGG	4416–4441	709	[80]
COI-Fish-R	TAGACTTCTGGGTGGCCRAARAAYCA	5100–5125		
FISHCOILBC_ts	**CACGACGTTGTAAAACGA** CTCAACYAATCAYAAAGATATYGGCAC	4418–4444	705	[70]
FISHCOIHBC_ts	**GGATAACAATTTCACACAGG** ACTTCYGGGTGRCCRAARAATCA	5100–5123		

COI gene, for species identification. For this reason, most of the works are addressed to improve the efficiency of amplification and sequencing of the same fragment belonging to this gene, the so-called Folmer's fragment, of about 706 bp (Table 1). This explains why the sequences available on GenBank and BOLD have a maximum length of 655 bp. With the aim of providing a better understanding on this issue, we analysed all the *COI* sequences available on GenBank and BOLD belonging to the species of *Sparidae* with a commercial interest on the international markets. This survey showed that no sequences are available for 15 species (26%) out of the 56 selected, while, for the remaining 41 species, 388 sequences have been deposited. Of these, the complete sequence has been deposited for three species, while only one portion of 1100 bp is present. The mean sequence length is about 600 bp for all the other *Sparidae*. Even though the progressive increasing of the number of these sequences brings advantages by boosting the discriminative power of the DNA bar coding, the existence of some disadvantages should be addressed. In particular, the exclusion of other portions of the gene may represent a limit, especially when primers have to be modified to improve the efficiency of amplification, since sequencing leaves out the primer annealing regions. Moreover, different parts of the gene, potentially rich in polymorphisms, remain unknown. The use of the gene *COI* may implies some other pitfalls. In fact, the existence of the so-called nuclear mitochondrial pseudogenes (numts), which are non-functional copies of mtDNA from the nucleus, may create some problems. Many studies have demonstrated that numts are extremely pervasive in nature and commonly occur when a portion of the mtDNA jumps into the nuclear genome by recombination or crossing-over mechanisms [127]. *COI* numts have been found in many different species of animals and they are particularly abundant in crustaceans [128–130]. The possibility of co-amplifying non-functional copies of mtDNA from the nucleus represents the main drawback related to the *COI* gene and, as reported by Song *et al.* [129], this introduces serious ambiguity into DNA bar coding by overestimating the number of species. In fact, the basic assumption of a comparison of homologous species is violated when numts are present, thus compromising the phylogenetic analysis [127, 129]. The accidental co-amplification of numts is not only influenced by their abundance but also influenced by the primers, the extraction protocol and the tissue used [128, 130, 131]. In particular, the utilization of universal primers could preferentially amplify numts because of relatively ancestral sequence similarity of numts or divergence within the primer regions of the orthologous mtDNA. Moreover, the need to overcome 3-end mismatching using degenerate inosine-containing primers can increase the probability of co-amplifying other gene regions [132]. In this case, the absence of aspecific amplification, such as multiple bands on the gel, could lead to production of paralogous sequences, which can be mistaken as orthologous mtDNA [129]. As a consequence, many questionable sequences, reported as 'COI-like', are deposited in GenBank. In other cases, when numts are present, it is highly probable that a forward primer amplifies a different gene portion than a reverse primer, producing no matching sequences or unreadable chromatograms [127]. For this reason, several methods have been suggested as means to avoid numt co-amplification, including RT-PCR, long PCR, cloning and mtDNA enrichment [128]. Song *et al.* [129] also suggested a quality control procedure, which is, however, laborious and time-consuming, thus noticeably complicating the analysis. Despite these obvious shortcomings, the proponents of *COI* DNA bar coding have

Table 2 Universal primers used for the amplification of the *cytb* gene from fish. The primers' positions have been calculated based on the sequence of *Pagrus auriga* (GenBank accession number AB124801). The primer sequences underlined refer to forward primers while the italics sequence to the reverse

Primer name	Sequence code	Position	Amp. length (bp)	Reference
cytb-L14841	AAAAAGCTTCCATCCAACATCTCAGCATGATGAAA	13360–13397	378	[111]
cytb-H15149	AAACTGCAGCCCCTCAGAATGATATTTGTCCTCA	13705–13738		
GluFish-F	AACCACCGTTGTTATTCAACTACAA	13270–13294	From 156 to 1229	[140]
FishcytB-F	ACCACCGTTGTTATTCAACTACAAGAAC	13271–13298		
CytBI-6F	TTCTCAGTAGACAACGCCACCCT	13804–13826		
CytBI-7F	CTAACCCGATTCTTTGCCTTCCACTTCCT	13825–13853		
CytBI-1F	CGATTCTTCGCATTCCACTTCCT	13831–13853		
CytBI-5R	*GGTCTTTGTAGGAGAAGTATGGGTGGAA*	13960–13987		
CytBI-3R	*GGGGTAAAGTTGTCTGGGTCTCC*	14050–14075		
CytBI-2R	*GCGGGGGTAAAGTTGTCTGGGTC*	14060–14078		
CytBI-4R	*AGGAAGTATCATTCGGGCTTAATATG*	14101–14126		
TruccytB-R	*CCGACTTCCGGATTACAAGACCG*	14470–14494		
THR-Fish2-R	*AACCTCCGACATCCGGCTTACAAGACCG*	14472–14499		
THR-Fish-R	*ACCTCCGATCTTCGGATTACAAGACC*	14473–14498		
CytbF	GGCTGATTCGGAATATGCAYGCNAAYGG	13532–13559	625	[80]
CytbR	GGGAATGGATCGTAGAATTGCRTANGCRAA	14128–14157		

argued that 'numts have proven a minor limitation to using mitochondrial bar code in groups studied so far' [100] even though typical DNA bar coding studies often produce unclear results, reporting the presence of cryptic species [129]. As a matter of fact, when the presence of numts is suspected in a region, it would be advisable to analyse an additional marker.

Other than with bar coding [120, 123] the gene *COI* has been targeted to develop other techniques such as RFLP and multiplex PCR [133, 134].

cytb: this gene has been used in more than half of the phylogenetic studies of the past decade. The lower within-species variation with respect to between-species variation makes *cytb* suitable for identification of different organisms [86]. Interestingly, this gene displays not only regions with a high level of variability, useful to perform evolutionary studies of even closely related fish species, but also conserved regions [26, 135]. These latter, together with the flanking regions belonging to the tRNA-Glu and tRNA-Thr, have been used for designing universal primers (Table 2) [78, 108, 140], which allow the amplification of fragments of different length. In particular, the primers designed by Sevilla *et al.* [136], produce amplicons from 156 to more than 1200 bp, often allowing one to obtain the entire coding sequence of the *cytb* gene for many fish species.

As for *COI*, all the available *cytb* sequences belonging to the 56 species of *Sparidae* have been collected from GenBank. Also in this case, no sequences are available for 16 species (28%). For the other 40 species, 275 sequences have been collected (113 less than for *COI*). Interestingly, *cytb* complete sequence is available for 39 species while the mean length of the deposited sequences is of about 957 bp.

Given the fact that the available sequences in the databases are usually longer for *cytb* than for *COI*, a better efficiency of identification has been hypothesized [136]. Finally, the analysis of the entire sequence, by increasing the probability of finding species-specific polymorphisms, can facilitate the development of rapid methods [97, 137–139].

16S rRNA: *16S rRNA* has alternate well-conserved and variable inter and intraspecific sequences and has been already reported as more specific than *cytb* for discrimination of fish eggs and snail species [75, 140, 141]. The high conservation degree of this gene facilitates the use of universal primers (often those designed by Palumbi *et al.* [109]), or the designing of new primers for the amplification of the same fragment of DNA from a higher number of species, which can also be phylogenetically distant. For this reason, the amplification of this gene is often used to provide a control for the DNA extraction process and for the DNA degradation grade [97, 123].

A recent and innovative application of the DNA technologies, beyond fraud identification purposes, is the detection of DNA molecules belonging to seafood allergens. In fact, if shrimp crustaceans and molluscs are not properly reported on the ingredient list or in cases of cross-contamination, they might provoke sensibilization or allergic reactions in consumers [142]. Some promising results in this field are represented by the work of Brzezinski [143] and Herrero *et al.* [144], who have developed two different methods, based on PCR-RFLP and real-time PCR, respectively, to detect a short fragment of the 16srRNA gene belonging to crustaceans in different kinds of food products.

Conclusions

In the light of a growing globalization of seafood trade and of an increased call for safety and quality requirements, margins for mislabelling and frauds for substitution, which have been demonstrated to plague commerce worldwide, should be significantly reduced by the harmonization of the international labelling policies and the implementation of reliable traceability systems. In this scenario, effective measures could be the inclusion of additional information on labels, such as the scientific denomination, and the use of molecular analysis to support traceability verification. Among these, the DNA-based techniques, and in particular the mitochondrial sequence DNA bar coding, represent the most promising tools for fish identification, which can rely on a continuous enrichment of the DNA databases, thus permitting the identification of an always-growing number of species.

References

1. Delgado CL, Wada N, Rosegrant MW, Meijer S, Ahmed M. Outlook for Fish to 2020: Meeting Global Demand. International Food Policy Research Institute and The World Fish Center; 2003. Available from: URL: http://www.ifpri.org/sites/default/files/pubs/pubs/fpr/pr15.pdf
2. Cushing DH. The Provident Sea. Cambridge University Press, Cambridge; 1988. p. 234.
3. Pauly D. Anecdotes and the shifting baseline syndrome. Trends in Ecology and Evolution 1995;10:430.
4. Magnusson J, Safina C, Sissenwine M. Whose fish are they anyway? Science 2001;293:1267–8.
5. Pauly D, Maclean J. In a Perfect Ocean: the State of Fisheries and Ecosystems in the North Atlantic Ocean. Island Press, Washington, DC; 2003. p. 178.
6. Food and Agriculture Organization of the United Nations (FAO). The State of World Fisheries and Aquaculture 2008. Available from: URL: http://www.fao.org/docrep/011/i0250e/i0250e00.htm
7. Council Regulation (EC) No. 1005/2008 of 29 September 2008 establishing a Community system to prevent, deter and eliminate illegal, unreported and unregulated fishing, amending Regulations (EEC) No. 2847/93, (EC) No. 1936/2001 and (EC) No. 601/2004 and repealing Regulations (EC) No. 1093/94 and (EC) No. 1447/1999. Official Journal of the European Union L 286/1 of 28th October 2008.
8. Commission Regulation (EC) No. 1010/2009 of 22 October 2009 laying down detailed rules for the implementation of Council Regulation (EC) No. 1005/2008 establishing a Community system to prevent, deter and eliminate illegal, unreported and unregulated fishing. Official Journal of the European Communities L280, 5:41.
9. FAO. ©2005–2012. International Plan of Action to Prevent, Deter, and Eliminate Illegal, Unreported and Unregulated Fishing – Web site. Illegal, Unreported and Unregulated (IUU) fishing. FI Institutional Websites. Text by David J. Doulman. In: *FAO Fisheries and Aquaculture Department*[online]. FAO, Rome, Updated 20 June 2011. Available from URL: http://www.fao.org/fishery/topic/3195/en
10. Food and Agriculture Organization of the United Nations (FAO). The State of World Fisheries and Aquaculture 2010. Available from: URL: http://www.fao.org/docrep/013/i1820e/i1820e.pdf
11. Pitcher T, Pauly D. Rebuilding ecosystems, not sustainability, as the proper goal of fishery management. In: Pitcher T, Hart P, Pauly D, editors. Reinventing Fisheries Management. Kluwer Academic Publishers, Dordrecht; 1998. p. 311–29.
12. Watling L, Norse EA. Disturbance of the seabed by mobile fishing gear: a comparison to clear cutting. Conservation Biology 1998;12:1180–98.
13. Kelleher K. Discards in the world's marine fisheries: an update. FAO Fisheries Technical Paper 2005;470:131.
14. Mansfield B. Spatializing globalization: 'A geography of quality' in the seafood industry. Journal of Economic Geography 2003;79:1–16.
15. Kirby MX. Fishing down the coast: historical expansion and collapse of oyster fisheries along continental margins. Proceedings of the National Academy of Sciences, USA 2004;101(35):13096–9.
16. Hajipieris P. Recent Development in the Branding and Marketing of Fish Products. Proceedings of the OECD/FAO Roundtable on EcoLabelling and Certification in Fisheries, 22–23 April 2009, The Hague, Netherlands. Available from: URL: http://www.oecd.org/dataoecd/38/57/42719002.pdf
17. Armani A, D'Amico P, Castigliego L, Sheng G, Gianfaldoni D, Guidi A. Mislabeling of an 'unlabelable' seafood sold on the European market: the jellyfish. Food Control 2012;26:247–51.
18. Rasmussen RS, Morrissey MT. Application of DNA-based methods to identify fish and seafood substitution on the commercial market. Food Science and Food Safety 2009;8:118–54.
19. Garcia Vazquez E, Perez J, Martinez JL, Pardinas AF, Lopez B, Karaiskou N, *et al.* High level of mislabeling in Spanish and Greek hake markets suggests the fraudulent introduction of African species. Journal of Agricultural and Food Chemistry 2011;59(2):475–80.
20. Stiles ML, Lahr H, Lahey W, Shaftel E, Bethel D, Falls J, *et al.* Bait and Switch: How Seafood Fraud Hurts our Oceans, our Wallets and our Health. Oceana; 2011. Available from: URL: http://oceana.org/en/news-media/publications/reports/bait-and-switch-how-seafood-fraud-hurts-our-oceans-our-wallets-and-our-health
21. National Environmental Trust (NET). Black Market for White Gold: The Illegal Trade in Chilean Sea Bass. 2004 Available from: URL: http://www.illegal-fishing.info/uploads/Black-mkt-for-white-gold-csb-report.pdf
22. Brown Jr, PB, Kenny J, Ashley C, Rubio A, Juszczak G, Conde D, *et al.* Annual Report on the United States Seafood Industry. 16th edn. Urner Barry, New Jersey; 2009.
23. Tennyson JM, Winters KS, Powell K. A fish by any other name: a report on species substitution. In Proceeding of the 22nd Annual Meeting of Seafood Science the Technology Society of the Americas, 6–7 October 1997, Biloxi, Mississippi. Available from: URL: http://sst.ifas.ufl.edu/22ndAnn/file08.pdfS
24. Jacquet JL, Pauly D. Trade secrets: renaming and mislabeling of seafood. Marine Policy 2008;32:309–18.
25. Wong EH, Hanner RH. DNA barcoding detects market substitution in North American seafood. Food Research International 2008;41:828–37.
26. Armani A, Castigliego L, Tinacci L, Gianfaldoni D, Guidi A. Molecular characterization of icefish, (*Salangidae* family),

using direct sequencing of mitochondrial cytochrome *b* gene. Food Control 2011;22:888–95.

27. Filonzi L, Chiesa S, Vaghi M, Nonnis, Marzano F. Molecular barcoding reveals mislabelling of commercial fish products in Italy. Food Research International 2010;43:1383–8.

28. Cawthorn DM, Steinman HA, Corli Witthuhn R. DNA barcoding reveals a high incidence of fish species misrepresentation and substitution on the South African market. Food Research International 2012;46:30–40.

29. Miller DM, Mariani S. Smoke, mirrors and mislabelled cod: poor transparency in the European seafood industry. Frontiers in Ecology and the Environment 2010;8:517–1.

30. Miller DM, Jessel A, Mariani S. Seafood mislabelling: comparisons of two western European case studies assist in defining influencing factors, mechanisms and motives. Fish and Fisheries 2011. Available from: URL: http://onlinelibrary.wiley.com/doi/10.1111/j.1467-2979.2011.00426.x/pdf

31. Marko PB, Lee SC, Rice AM, Gramling JM, Fitzhenry TM, McAlister JS. Fisheries: mislabeling of a depleted reef fish. Nature 2004;430:309–10.

32. Atkins H. You could be eating shark meat and not even know it. Cape Times 2010. Available from: URL: http://www.iol.co.za/news/south-africa/you-could-be-eating-shark-meat-1.483896

33. Mafra I, Ferreira MPLVO, Beatriz MBPP. Food authentication by PCR-based methods. European Food Research and Technology 2008;227:649–65.

34. Martinez Ortiz J. White Fish Handbook of Ecuador: 45 species of commercial interest. Quito: Asoexpebla; 2005.

35. Jacquet JL, Pauly D. The rise of seafood awareness campaigns in an era of collapsing fisheries. Marine Policy 2007;31:308–13.

36. REGULATION (EC) No. 854/2004 of the European Parliament and of the Council of 29 April 2004 laying down specific rules for the organisation of official controls on products of animal origin intended for human consumption. Official Journal of the European Union L 139 of 30th April 2004.

37. Food and Drug Administration (FDA). Advisory on Puffer Fish. Industry Advisory 2007. Available from: URL: http://www.fda.gov/Food/ResourcesForYou/FoodIndustry/ucm085458.htm

38. Ling KH, Nichols PD, But PPH. Fish-induced keriorrhea. Advances in Food and Nutrition Research 2009;57:1–52.

39. Froese R, Pauly D (editors). 2012. FishBase. World Wide Web electronic publication. Available from: URL: http://www.fishbase.org version (08/2012).

40. Rehbein H, Horstkotte B. Determination of the composition of multi-species fishery products by PCR-based techniques. In Proceedings of the First Trans-Atlantic Fisheries Technology Conference, 10–14 June 2003, Reykjavik, Iceland; 2003; 190–2.

41. Kochzius M, Nölte M, Weber H, Silkenbeumer N, Hjörleifsdottir S, Hreggvidsson GO. DNA microarrays for identifying fishes. Marine Biotechnology 2008;10:207–17.

42. US Food and Drug Administration (USFDA). Regulatory Fish Encyclopedia (RFE). 2012 Available from: URL: http://www.fda.gov/food/foodsafety/product-specificinformation/seafood/regulatoryfishencyclopediarfe/default.htm

43. Food and Agriculture Organization of the United Nations (FAO). Fishery Fact Sheets Collections. ASFIS List of Species for Fishery Statistics Purposes. 2012 Available from: URL: http://www.fao.org/fishery/collection/asfis/en

44. Ogden R. Fisheries forensics: the use of DNA tools for improving compliance, traceability and enforcement in the fishing industry. Fish and Fisheries 2008;9:462–72.

45. Coordinating Working Party on Atlantic Fishery Statistics (CWP). Handbook of Fishery Statistics. FAO; 1990. Available from: URL: http://www.fao.org/fishery/cwp/search/en

46. Garibaldi L, Busilacchi S. ASFIS List of Species for Fishery Statistics Purposes. Aquatic Sciences and Fisheries Information System 2002, Reference Series No. 15. Available from: URL: ftp://ftp.fao.org/docrep/fao/006/y7527t/y7527t00.pdf

47. Golan E, Krissoff B, Kuchler F, Calvin E, Nelson K, Price G. Traceability in the U.S. Food Supply: Economic Theory and Industry Studies. Agricultual Economic Report No. (AER-830) March 2004, pp. 56.

48. Food and Agriculture Organization of the United Nations (FAO). Estimated Total International Trade in Fishery Commodities; 2012. Available from: URL: ftp://ftp.fao.org/FI/STAT/summary/a2ybc.pdf

49. Taylor WW, Schechter MG, Wolfson LG. Globalization: Effects on Fisheries Resources. Cambridge University Press, Cambridge; 2007.

50. Swartz W, Sumaila UR, Watson R, Pauly D. Sourcing seafood for the three major markets: the EU, Japan and the USA. Marine Policy 2010;34:1366–73.

51. European Commission. Facts and Figures on the Common Fisheries Policy. Basic Statistical Data Publications Office of the European Union; 2010. Available from: URL: http://ec.europa.eu/fisheries/documentation/publications/pcp_en.pdf

52. Esteban A, Crilly R. Fish Dependence: 2011 Update: The Increasing Reliance of the EU on Fish from Elsewhere. Available from: URL: http://assets.ocean2012.eu/publication_documents/documents/104/original/2011_Fish_Dependence_UPDATE.pdf

53. Norme NF EN ISO 8402 'Management de la qualité et assurance de la qualité – vocabulaire'. 1994.

54. Moe T. Perspectives on traceability in food manufacture. Trends in Food Science and Technology 1998;9:211–4.

55. Food Standards Agency (FSA). Traceability in the Food Chain: A Preliminary Study. FSA, Food Chain Strategy Division, London, England; 2002. Available from: URL: http://www.food.gov.uk/multimedia/pdfs/traceabilityinthefoodchain.pdf

56. Lovejoy H. Making the seafood industry more sustainable: creating production chain transparency and accountability. Journal of Cleaner Production 2007;15:577–89.

57. Canadian Food Inspection Agency (CFIA). CFIA Fish List. 2010. Available from: URL: http://www.inspection.gc.ca/english/fssa/fispoi/product/comnome.shtml

58. Food and Drug Administration (FDA). The Seafood List. FDA's Guide to Acceptable Market Names for Seafood Sold in Interstate Commerce. 2012. Available from: URL: http://www.fda.gov/Food/GuidanceComplianceRegulatoryInformation/GuidanceDocuments/Seafood/ucm113260.htm

59. Food Standards Agency (FSA). Table of Commercial Designations for Specific Fish Species Derived from the Fish Labelling. Regulations 2010. Available from: URL: http://www.food.gov.uk/multimedia/pdfs/fishcomdesig.pdf

60. Logan CA, Alter SE, Haupt AJ, Tomalty K, Palumbi SR. An impediment to consumer choice: overfished species are sold as Pacific red snapper. Biological Conservation 2008;141:1591–9.

61. Regulation (EC) No. 178/2002 of the European Parliament and of the Council of 28 January 2002 laying down the general principles and requirements of food law, establishing the European Food Safety Authority and laying down procedures in matters of food safety. Official Journal of the European Communities L 31, 1:24.

62. Council Regulation (EC) No. 104/2000 of 17 December 1999 on the common organisation of the markets in fishery and aquaculture products. Official Journal of the European Union L 17 of 21st January 2000.

63. Commission Regulation (EC) No. 2065/2001 of 22 October 2001 laying down detailed rules for the application of Council Regulation (EC) No. 104/2000 as regards informing consumers about fishery and aquaculture products. Official Journal of the European Union L 278/6 of 23 October 2001.

64. Commission Implementing Regulation (EU) No. 404/2011 of 8 April 2011 laying down detailed rules for the implementation of Council Regulation (EC) No. 1224/2009 establishing a Community control system for ensuring compliance with the rules of the Common Fisheries Policy. Official Journal of the European Communities L112, 1:147.

65. Directive 2003/89/Ec of The European Parliament and of the council of 10 November 2003 amending Directive 2000/13/EC as regards indication of the ingredients present in foodstuffs. Official Journal of the European Communities L308, 15:18.

66. Commission Directive 2006/142/EC of 22 December 2006 amending Annex IIIa of Directive 2000/13/EC of the European Parliament and of the Council listing the ingredients which must under all circumstances appear on the labelling of foodstuffs. Official Journal of the European Communities L 368. 110:111.

67. Guidance for Industry: The Seafood List – FDA's Guide to Acceptable Market Names for Seafood Sold in Interstate Commerce. 2012. Available from: URL: http://www.fda.gov/Food/GuidanceComplianceRegulatoryInformation/GuidanceDocuments/Seafood/ucm113260.htm

68. Single Laboratory Validated Method for DNA-Barcoding for the Species Identification of Fish for FDA Regulatory Compliance. 2011. Available from: URL: http://www.fda.gov/Food/ScienceResearch/LaboratoryMethods/ucm237391.htm

69. Backgrounder – Canada's Response to new EU Traceability Requirements. Available from: URL: http://www.dfo-mpo.gc.ca/fm-gp/peches-fisheries/fish-ren-peche/sff-cpd/traceability-back-tracabilite-eng.htm

70. Henson S, Mitullah W. Kenya Exports of Nile Perch: Impact of Food Safety Standards on an Export-Oriented Supply Chain. World Bank Working Paper, Washington, DC; 2004.

71. Jaffee S, Henson S. Food exports from developing countries: the challenges posed by standards In: Aksoy MA, Beghin JC, editors. Global Agricultural Trade and Developing Countries. Oxford University Press, Oxford; 2004.

72. South Africa Department of Health. Regulations Relating to the Labelling and Advertising of Foodstuffs. Foodstuffs, Cosmetics and Disinfectants Act of 1972. Pub. L. No. 54 of 1972. Government Notice No. R. 146 of 1 March 2010

73. National Regulator for Compulsory Specifications (NRCS). Compulsory Specification for Frozen fish, frozen marine molluscs and frozen products derived there from. Government Notice No. R. 979 of 4 July 2003

74. Magera A, Beaton S. Seafood Traceability in Canada. Traceability Systems, Certification, Eco-labeling and Standards for Achieving Sustainable Seafood. Ecology Action Centre; 2009. Available from: URL: http://www.ecologyaction.ca/files/images/file/Marine/Seafood_Traceability_in_Canada.pdf

75. Civera T. Species identification and safety of fish products. Veterinary Research Communications 2003;27:481–9.

76. Trutter P. National Regulator for Compulsory Specifications (NRCS). South Africa 2010. Personal communication.

77. Costa FO, Carvalho GR. The Barcode of Life Initiative: synopsis and prospective societal impacts of DNA barcoding of fish. Genomics, Society and Policy 2007;3:29–40.

78. Kochzius M, Seidel C, Antoniou A, Kumar Botla S, Campo D, Cariani A, *et al.* Identifying fishes through DNA barcodes and microarrays. PLoS ONE 2010;5:e12620.

79. Lleonart J, Taconet M, Lamboeuf M. Integrating information on marine species identification for fishery purposes. Marine Ecology Progress Series 2006;316:231–8.

80. Paquin P, Hedin M. The power and perils of 'molecular taxonomy': a case study of eyeless and endangered Cicurina (Araneae: Dictynidae) from Texas caves. Molecular Ecology 2004;13:3239–55.

81. Sites JW, Marshall JC. Operational criteria for delimiting species. Annual Review of Ecology, Evolution, and Systematics 2004;35:199–227.

82. AOAC. Official method 980.16 identification of fish Species: thin-layer polyacrylamide gel isoelectric focusing method. Journal of AOAC International 1980;63:69.

83. Rasmussen RS, Morrissey MT. DNA-based methods for the identification of commercial fish and seafood species. Food Science and Food Safety 2008;7:280–95.

84. Teletchea F. Molecular identification methods of fish species: reassessment and possible applications. Fish Biology and Fisheries 2009;19(3):265–93.

85. Anon. The label said snapper, the lab said baloney. Consumer Reports 1992;57(2):110–2.

86. Bartlett SE, Davidson WS. FINS (Forensically Informative Nucleotide Sequencing): a procedure for identifying the animal origin of biological specimens. Biotechniques 1992;12(3):408–11.

87. Greenberg P. The Catch. New York Times Magazine 23 October 2005.

88. Bossier P. Authentication of seafood products by DNA patterns. Journal of Food Science 1999;64(2):189–93.

89. Mackie IM, Prydea SE. Gonzales Sotelob C, Medinab I, Peréz Martínb R, Quinteiroc J, *et al.* Challenges in the identification of species of canned fish. Trends in Food Science and Technology 1999;10(1):9–14.

90. Dawson MN, Raskoff KA. Jacobs DK. Field preservation of marine invertebrate tissue for DNA analyses. Molecular Marine Biology and Biotechnology 1998;7(2):145–52.

91. Lockley AK, Bardsley RG. DNA-based methods for food authentication. Trends in Food Science and Technology 2000;11(2):67–77.

92. Gil LA. PCR-based methods for fish and fishery products authentication. Trends in Food Science and Technology 2007;18(11;558–66.

93. Giles RE, Blanc H, Cann HM, Wallace DC. Maternal inheritance of human mitochondrial DNA. Proceedings of the National Academy of Sciences, USA 1980;77(11): 6715–19.

94. Trotta M, Sconhuth S, Pepe T, Cortesi ML, Puyet A, Bautista JM. Multiplex PCR method for use in real-time PCR for identification of fish fillets from grouper (*Epinephelus* and *Mycteroperca* species) and common substitutes species. Journal of Agricultural and Food Chemistry 2005;53(6):2039–45.

95. Cespedes A, Garcia T, Carrera E, Gonzalez I, Fernandez A, Asensio L, *et al.* Genetic differentiation between sole (*Solea solea*) and greenland halibut (*Reinhardtius hippoglossoides*) by PCR-RFLP analysis of a 12S rRNA gene fragment. Journal of the Science of Food and Agriculture 2000;80:29–32.

96. Aranishi F. Rapid PCR-RFLP method for discrimination of imported and domestic mackerel. Marine Biotechnology 2005;7(6):571–5.

97. Armani A, Castigliego L, Tinacci L, Gianfaldoni D,Guidi A. Multiplex conventional and real-time PCR for fish species identification of Bianchetto (juvenile form of *Sardina pilchardus*), Rossetto (*Aphia minuta*), and Icefish in fresh, marinated and cooked products. Food Chemistry 2012;133(1):184–92.

98. Catanese G, Manchado M, Fernàndez-Trujillo A, Infante C. A multiplex-PCR assay for the authentification of mackerels of the genus *Scomber* in processed fish products. Food Chemistry 2010;122(1):319–26.

99. Carrera E, Garcia T, Cespedes A, Gonzalez I, Fernandez A, Asensio LM, *et al.* Identification of smoked Atlantic salmon (*Salmo salar*) and rainbow trout (*Oncorhynchus mykiss*) using PCR-restriction fragment length polymorphism of the p53gene. Journal of AOAC International 2000;83(2):341–6.

100. Hebert PD, Ratnasingham S, DeWaard JR. Barcoding animal life: cytochrome c oxidase subunit 1 divergences among closely related species. Proceedings of the Royal Society of London. Series B, Biological Sciences 2003;270 (Suppl. 1):S96–9.

101. Consortium of the Barcoding of Life. 2012. http://www.barcodeoflife.org.

102. Ward RD, Hanner R, Hebert PDN. The campaign to DNA barcode all fishes, FISH-BOL. Journal of Fish Biology 2009;74:329–56.

103. Kress WJ, Erickson DL. DNA barcodes: methods and protocols. Methods in Molecular Biology 2012;858:3–8.

104. Ward RD. FISH-BOL, A Case Study for DNA barcodes, Methods in Molecular Biology 2012;858:423–439.

105. Wolf CC, Hubner PP, Luthy JJ. Differentiation of sturgeon species by PCR-RFLP. Food Research International 1999;32(10):699–705.

106. Dawid IB, Blackler AW. Maternal and cytoplasmic inheritance of mitochondrial DNA in *Xenopus*. Developmental Biology 1972;29(2):152–6.

107. Alberts B, Bray D, Lewis J, Raff M, Roberts K, Watson JD. Molecular Biology of the Cell. Garland Publishing Inc, New York and London; 1994.

108. Kocher TD, Thomas WK, Meyer A, Edwards SV, Paabo S, Villablanca FX, *et al.* Dynamics of mitochondrial DNA evolution in animals: amplification and sequencing with conserved primers. Proceedings of the National Academy of Sciences, USA 1989;86(16):6196–200.

109. Palumbi S, Kewalo AM, Romano S, McMillan WO, Stice L, Grabowski G. The Simple Fool's Guide to PCR. Simple Fool's Guide; 2002. Available from: URL: http://palumbi.stanford.edu/SimpleFoolsMaster.pdf

110. Avise JC, Arnold J, Ball RM, Bermingham E, Lambt T, Neigel JE, *et al.* Intraspecific phylogeography: the mitochondrial DNA bridge between population genetics and systematics. Annual Review of Ecology and Systematics 1987;18:89–522.

111. Borgo R, Souty Grosset C, Bouchon D, Gomot L. PCR-RFLP analysis of mitochondrial DNA for identification of snail meat species. Journal of Food Science 1996;61:1–4.

112. Antunes A, Ramos MJ. Discovery of a large number of previously unrecognized mitochondrial pseudogenes in fish genomes. Genomics 2005;86:708–17.

113. Teletchea T, Laudet V, Hanni C. Phylogeny of the Gadidae (sensu Svetovidov, 1948) based on their morphology and two mitochondrial genes. Molecular Phylogenetics and Evolution 2006;38:189–99.

114. FishTrace. Available from: URL: http://www.fishtrace.org

115. Hanner RH, Gregory TR. Genomic diversity research and the role of biorepositories. Press Technology Tech Cell 2007;5:93–103.

116. Forster P. To err is human. Annals of Human Genetics 2003;67:2–4.

117. Nilsson RH, Ryberg M, Kristiansson E. Taxonomic reliability of DNA sequences in public sequence databases: a fungal perspective. Public Library of Science 2006;1:59.

118. Wilde LW, Norman J, Robertson J, Sarre S, Georges A. Current issues in species identification for forensic science and the validity of using the cytochrome oxidase I (COI) gene. Forensic Science, Medicine and Pathology 2010;6:233–41.

119. Hajibabaei M, Smith A, Janzen DH, Rodriguez JJ, Whitfield JB, Hebert PD. A minimalist barcode can identify a specimen whose DNA is degraded. Molecular Ecology Notes 2006;6:959–64.

120. Ward RD, Zemlak TS, Innes BH, Last PR, Hebert PDN. DNA barcoding Australia's fish species. Philosophical Transactions of the Royal Society of London. Series B, Biological Sciences 2005;360, 1847–57.

121. Folmer O, Black M, Hoeh W, Lutz R, Vrijenhoek R. DNA primers for the amplification of mitochondrial cytochrome coxidase 1 from diverse metazoan invertebrates. Molecular Marine Biology and Biotechnology 1994;3:294–9.

122. Mikkelsen PM, Bieler R, Kappner I, Rawlings TA. Phylogeny of Veneroidea (Mollusca: Bivalvia) based on morphology and molecules. Zoological Journal of the Linnean Society 2006;148(3):439–521.

123. Ivanova NV, Zemlak TS, Hanner RH, Hebert PDN. Universal primer cocktails for fish DNA barcoding. Molecular Ecology Notes 2007;7:544–8.

124. Baldwin CC, Julie HM, Smith DG, Weigt' LA. Genetic identification and color descriptions of early life-history stages of Belizean *Phaeoptyx* and *Astrapogon* (Teleostei: Apogonidae) with Comments on identification of adult Phaeoptyx. Zootaxa 2009;2008:1–22.

125. Steffens DL, Sutter SL, Roemer SC. An alternate universal forward primer for improved automated DNA sequencing of M13. Biotechniques 1993;15(4):580–82.

126. FDA Reference Standard Sequence Library for Seafood Identification. 2012. Available from URL: http://www.fda.gov/Food/FoodSafety/Product-SpecificInformation/Seafood/DNAspeciation/ucm238880.htm

127. Buhay JE. 'Coi-like' sequences are becoming problematic in molecular systematic and DNA barcoding studies. Journal of Crustacean Biology 2009;29(1):96–110.

128. Bensasson, D, Zhang D, Hartl DL, Hewitt GM. Mitochondrial pseudogenes: evolution's misplaced witnesses. Trends in Ecology and Evolution 2001;16:314–21.

129. Song, H, Buhay JE, Whiting MF, Crandall KA. Many species in one: DNA barcoding overestimates the number of species when nuclear mitochondrial pseudogenes are coamplified. Proceedings of the National Academy of Sciences, USA 2008;105:13486–91.

130. Richly, E, Leister D. NUMTs in sequenced eukaryotic genomes. Molecular Biology and Evolution 2004;21:1081–4.

131. Pamilo P, Viljakainen L, Vihavainen A. Exceptionally high density of NUMTs in the honeybee genome. Molecular Biology and Evolution 2007;24:1340–6.

132. Zhang DX, Hewitt GM Highly conserved nuclear copies of the mitochondrial control region in the desert locust *Schistocerca gregaria:* some implications for population studies. Molecular Ecology 1996;5:295–300.

133. Espiñeira M, González-Lavin N, Vieites JM, Santaclara FJ. Authentication of AnglerWsh Species (*Lophius* spp.) by means of polymerase chain reaction-restriction fragment length polymorphism (PCR–RFLP) and forensically informative nucleotide sequencing (FINS) methodologies. Journal of Agriculture and Food Chemistry 2008;56(22):10594–9.

134. Rasmussen Hellberg RS, Morrissey MT, Hanner RH. A multiplex PCR method for the identification of commercially important salmon and trout species (Oncorhynchus and Salmo) in North America. Journal of Food Science 2010;75(7):C595–606.

135. Jerome M, Lemaire C, Bautista JM, Fleurence JL, Etienne M. Molecular phylogeny and species identification of sardines. Journal of Agricultural and Food Chemistry 2003;51:43–50.

136. Sevilla RG, Diez A, Norén M, Mouchel O, Jérôme M, Verrez-Bagnis V. *et al.* Primers and polymerase chain reaction conditions for DNA barcoding teleost fish based on the mitochondrial cytochrome *b* and nuclear rhodopsin genes. Molecular Ecology Notes 2007;7(5) 730–4.

137. Michelini E, Cevenini L, Mezzanotte L, Simoni P, Baraldini M, De Laude L, *et al.* One-step triplex-polymerase chain reaction assay for the authentication of yellowfin (*Thunnus albacares*), bigeye (*Thunnus obesus*), and skipjack (*Katsuwonus pelamis*) tuna DNA from fresh, frozen, and canned tuna samples. Journal of Agricultural and Food Chemistry 2007;55(19):7638–47.

138. Lin W-F, Hwang D-F. Application of PCR-RFLP analysis on species identification of canned tuna. Food Control 2007;18:1050–7.

139. Armani A, Castigliego L, Tinacci L, Gandini G, Gianfaldoni D, Guidi A. A rapid PCR–RFLP method for the identification of *Lophius* species. European Food Research and Technology 2012c;235:253–63.

140. Abdulmawjood A, Bulte M. Snail species identification by RFLP PCR and designing of species-specific oligonucleotide primers. Journal of Food Science 2001;66:1287–93.

141. Akimoto S, Kinoshita S, Sezaki K, Mitani I. Identification of *Alfonso* and related fish species belonging to the gene Beryx with mitochondrial 16S rRNA gene and its application on their pelagic eggs. Fisheries Science 2002;68:1242–9.

142. Lopata AL, O'Hehir RE, Lehrer SB. Shellfish allergy. Clinical and Experimental Allergy 2010;40:850–8.

143. Brzezinski JL, Detection of crustacean DNA and species identification using a PCR-restriction fragment length polymorphism method. Journal of Food Protein 2005;68(9):1866–73.

144. Herrero B, Vieites JM, Espiñeira M Fast real-time PCR for the detection of crustacean allergen in foods. Journal of Agriculture And Food Chemistry 2012;60:1893–7.

Index